S0-CLH-374

Recent Advances in Acarology

Volume I

ACADEMIC PRESS RAPID MANUSCRIPT REPRODUCTION

Proceedings
of the
V International Congress of Acarology
Held August 6–12, 1978 at Michigan State University
East Lansing, Michigan

Recent Advances in Acarology

Volume I

EDITED BY

J. G. RODRIGUEZ
Department of Entomology
University of Kentucky
Lexington, Kentucky

ACADEMIC PRESS
New York San Francisco London 1979
A Subsidiary of Harcourt Brace Jovanovich, Publishers

ACADEMIC PRESS, INC.
111 Fifth Avenue, New York, New York 10003

United Kingdom Edition published by
ACADEMIC PRESS, INC. (LONDON) LTD.
24/28 Oval Road, London NW1 7DX

Library of Congress Cataloging in Publication Data

International Congress of Acarology, 5th, Michigan State University, 1978.
Recent advances in acarology.

"Includes the contents of the symposia and a selection of contributions to the workshops and submitted paper sessions" of the "V International Congress of Acarology, held at Michigan State University in August 1978."
Includes index.
1. Acarology—Congresses. I. Rodriguez, J.G., Date II. Title.
QL458.I57 1978 595′.42 79-17386
ISBN 0-12-592201-9 (v. 1)

PRINTED IN THE UNITED STATES OF AMERICA

79 80 81 82 9 8 7 6 5 4 3 2 1

Volume I

Contents

Contents of Volume II

Preface

"Recent Advances in Acarology" had its inception in the V International Congress of Acarology held at Michigan State University in August 1978. This two-volume work includes the contents of the symposia and a selection of contributions to the workshops and submitted paper sessions.

These volumes examine such timely and pertinent subjects as strategies in pest management of mites and ticks, encompassing pheromone communication, resistance of mites and ticks to acaricides, nonchemical control of ticks, new acaricides, nutritional ecology/control, and biological control. They offer new and exciting information in these important areas. Since it is international in scope, this work represents current research trends and syntheses of opinions of world authorities in the areas of subject matter covered. At the same time, it includes contributions from young researchers whose works and ideas are worthy of attention. It is not a textbook nor a methods manual, although several contributions report sophisticated research techniques applied to acarological problems. Teachers, advanced students, and researchers in acarology and entomology should find these two volumes to be valuable reference sources in fundamental acarine physiology, nutrition, virology, ecology, behavior, systematics, and structure and function.

Volume I incorporates the agricultural aspects: pest management of agricultural mites, biology of spider mites, stored food acarology, and soil mite biology. Additionally, it includes sections devoted to physiology, biochemistry, toxicology, ecology, behavior, and bionomics of acari. Volume II encompasses the medical and veterinary aspects: the management of ticks and other acari of medical and veterinary importance, biology of spotted fever ticks, disease transmission by acari, and pheromonal communication. Related topics covered in Volume II are specificity and parallel evolution of host–parasite, and systematics, morphology, and evolution of acari.

Acknowledgments

The genesis of these volumes was the V International Congress of Acarology, which was organized by a group of scientists whose active roles as architects of the congress it is a pleasure to acknowledge. These colleagues also assisted in editing the contents of the symposia and the selections from the workshops and sessions for the presentation of submitted papers: their names are listed with the corresponding sections of these volumes, and the editor is grateful for their assistance. The members of the organizing committee were E. W. Baker (president), B. A. Croft, E. A. Cross, J. E. George, J. R. Hoffman, P. E. Hunter, D. E. Johnston, R. B. Loomis, J. G. Rodriguez (secretary), D. E. Sonenshine, and R. W. Strandtmann, all representing the Acarological Society of America. The international organization was represented by R. O. Drummond, J. A. McMurtry, W. W. Moss, J. H. Oliver, Jr., and F. J. Radovsky. Special assistance was graciously rendered by W. T. Atyeo, J. E. Bath, R. M. Crowell, J. M. Magner, J. V. Osmun, and B. C. Pass, and generous support was given by staffs of the University of Kentucky and Michigan State University.

The editor wishes also to acknowledge the contributions of his wife, Lorraine, who was available as consultant and editorial assistant.

Introduction

In the past decade, pest management research on mites of agricultural crops has received increasing attention. As a result, some of the most sophisticated and effective control systems available today for any pest group have been developed. In the section on pest mangement of agricultural mites, a range of topics relating to pest management systems are examined; these topics include economic injury thresholds and tactics of control.

Among the strategies for control are host plant–mite interactions including host plant resistance, the use of selective pesticides, the development of pesticide resistance in natural predators, especially phytoseiids, and the application of simulation models to mite pest management.

Most investigations have been directed toward mites of the family Tetranychidae, which are among the most important arthropod pests of agricultural crops and ornamental plants throughout the world. They are capable of exerting tremendous stress on plants in a short-time period because of their high reproductive potential. Small in size, they can go unnoticed at low population densities; yet in warm, dry conditions, they can increase rapidly until they override their food source. Serious crop loss or even crop failure can result. Records indicate that tetranychids did much damage to cotton and hops in the mid-1800s, and chemical control began in the 1870s. The introduction of synthetic pesticides in the 1940s was quickly followed by development of resistance in the genetically plastic tetranychids. This in turn led to some excellent research in the genetics of resistance, especially to organophosphorus compounds, by workers in Switzerland and the Netherlands.

Worldwide interest in these topics is exemplified by contributions from Colombia, Denmark, Egypt, Iran, Japan, New Zealand, and the United States. *In toto,* these papers provide an excellent sampling of trends in mite pest management research and give insights into the tasks that must be accomplished in the next decade in this important area of crop protection.

Recent research efforts have generally been addressed to some aspect of reproduction, since this most awesome biological function of the tetranychid mites must be reduced or controlled if their numbers are to be maintained at a level acceptable to the grower. The section on biology of spider mites represents the state of the art, with contributions in the research areas of male reproductive behavior, silk production, pheromones, the components of reproductive success, and the effects of nutrition and temperature on tetranychid development. These studies can be considered contributions to the overall management scheme.

Beyond the production of food, feed, and fiber from agricultural crops lies the task of protecting these products against possible losses during storage as they await distribution to the peoples of the world, and the next section of this volume is devoted to stored product acarology. Meaningful solutions to the problems encountered require both basic and applied avenues of approach that reflect a broad understanding of acarine biology, nutrition, biochemistry, systematics, and ecology. The contributions to this section represent just such avenues. Control of stored product acarines is not seen as merely the selection of the best pesticides for a particular purpose, but rather even more basic strategies are examined. For example, studies involving nutrition and development have resulted not only in basic biological information but also pointed up unique possibilities of control using natural and synthetic food chemicals and hormones and their analogs to disrupt normal development. Additionally, an understanding of the overall ecology of the stored product environment may lead to the use of certain stored product mites as biological control agents of other stored product pests and fungi.

The next section is devoted to physiology, biochemistry, and toxicology. This section presents reports of innovative, effective, and even elegant techniques applied to research on acarines. Examples are the use of low-energy laser-generated x rays to measure salt concentrations in ducts of living mites, the use of labeled butanediol in metabolism studies of acarid mites, and electron microscope studies of functional morphology of ticks. Contributions on water vapor uptake and on regulation of fluid secretions in ticks form a notable component of this section. It is evident that problems in acarine physiology and biochemistry are still potentially excellent subjects for pioneering research.

The section on ecology, bionomics, and behavior truly encompasses the breadth, scope, and depth of acarology, both geographically and in terms of content. These contributions represent more than a dozen countries, and the diversity of subject matter is exemplified in such topics as the developmental cycle of sponge-associated water mites, behavior of tick larvae in relation to CO_2, and the influence of maternal age on the sex ratio of the progeny of a tetranychid. Much of the reported research deals with ecology and behavior as they relate to the biology of pest species or to development of predators for possible employment in biological control systems.

The ecological theme is continued in the final section of this volume, devoted to the reporting of recent research in soil mite biology. This research has been aimed at one or more of three general areas: basic interactions with the physical and biological soil environment; the functional position of mites in the soil system, especially as related to energy flow and nutrient cycling; and the effects of human–induced perturbations on community structure and function. Contributions deal with the effects of various physical and biological factors on orabitid mites, and a study of the marine littoral zone. The function of soil mites in energy and nutrient cycling is reviewed but not treated extensively. Other papers revolve around the ecological impact on soil mite communities of such human disturbance as radionuclide pollution, pesticide application, sewage effluent disposal, and street salting for snow and ice removal. Soil mites show promise as bioindicators of environmental disturbances.

Volume II covers topics of medical and veterinary importance; pheromonal communications, and systematics, morphology, and evolution of acari.

1.

Pest Management of Agricultural Mites

Section Editors

B. A. Croft
J. A. McMurtry
S. L. Poe
V. Dittrich

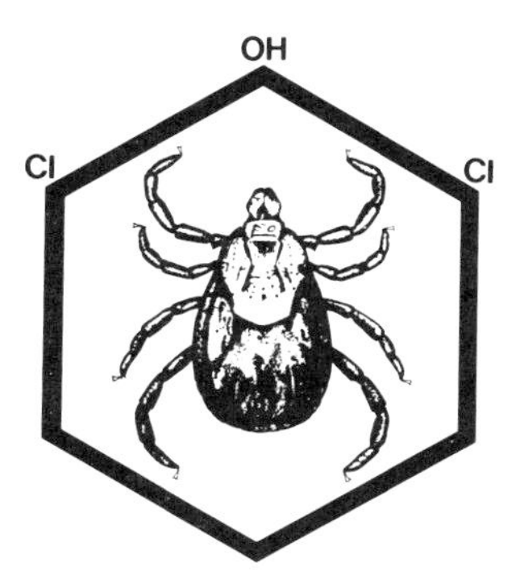

ECONOMIC INJURY LEVEL STUDIES IN RELATION TO MITES ON APPLE

S. C. Hoyt

Washington State University
Tree Fruit Research Center
Wenatchee, Washington

L. K. Tanigoshi

Science and Education Service/Federal Research, USDA
Riverside, California

R. W. Browne

Washington State University
Tree Fruit Research Center
Wenatchee, Washington

INTRODUCTION

The change from chemical control of spider mites to their management through integration of biological controls, cultural controls and use of selective chemicals has stimulated interest in the quantitative aspects of mite feeding injury on several tree crops.

A review of the literature shows many qualitative studies of mite damage designed to determine the types of injury caused by mites (van de Vrie, *et al.*, 1972). In most of these studies trees with uncontrolled mite populations were compared with trees having nearly complete control of mites. Effects of mite feeding such as reduced chlorophyll content of leaves, imbalance of growth-regulatory substances, and a reduction in CO_2 assimilation (in heavily damaged leaves) were reported, but since these physiological effects cannot be related (with present knowledge) to effects on tree growth, yield and quality of fruit they will not be discussed here.

Chapman *et al.* (1952) and Lienk *et al.* (1956) reported on a two year study of the effects of mites on apple. Cortland, Red Delicious and Rome apple trees with high mite populations had lower yield in the first year of the study and reduced bloom and yield on Red Delicious and Cortland in the second year.

Vol. I: ISBN 0-12-592201-9

The high mite population early in the first year had interfered with fruit bud formation on these varieties in the second year. These authors reported no significant difference in size of fruit, fruit drop, fruit firmness or percent soluble solids as a result of the mite populations but did find better fruit color on mite-injured trees. They suggested the lack of effect on fruit size was due to the offsetting effect of greater numbers of apples on the mite-free trees. Light and Ludlum (1972) did find a reduction in the numbers of large fruit on trees infested with *Panonychus ulmi* (Koch) in two of five years.

Evaluating the effects of mites on tree crops is complicated by the many variables which can influence yield, size and quality of fruit. Some of the more important variables are discussed below.

Crop Species

The tolerance for mites varies tremendously with the crop. Pears have a very low tolerance showing blackened leaves (transpiration injury) from just a few mites per leaf. Westigard *et al.* (1966) found reduced fruit size from relatively low populations of *Tetranychus urticae* Koch (peak population of 16.4 per leaf). In the second year of their study the number of fruit per 100 clusters, leaves per fruit, fruit size, pre-harvest drop and yield were adversely affected by mite feeding damage. The effect on fruit set was due to the mite abundance the preceding year, while other effects were influenced by current year populations. *P. ulmi* populations had no effect on yield of walnuts until the third year of high numbers (Barnes and Moffitt, 1978). A 40% loss in yield occurred the third year, but these authors found mite damage had no effect on nut quality. High mite populations on almonds (almost 100 per leaf) did not reduce yields in the year of the infestation, but the year following yields were 13 to 19% lower than on protected trees (Barnes and Andrews, 1978). Laing *et al.* (1972) found no effect on yield or sugar content of grapes from populations of *Tetranychus pacificus* McGregor up to 225 mites per leaf. These authors did not report on effects the year following the population although they recognized the possibility of adverse effects.

Variety

Differential responses of varieties to mite damage were reported by Chapman *et al.* (1952) and Lienk *et al.* (1956). We have observed a greater tendency for mite populations to develop on the Red Delicious variety but a greater effect of comparable mite populations on Golden Delicious.

Tree Vigor

Vigorous trees appear to have a much greater tolerance for mites than do those suffering from nutrient deficiencies, trees on poor soils or those suffering from freeze or mechanical injuries. No mention is made in the literature of the effects of drought or inadequate irrigation on tolerance to mite feeding injury. Adequate moisture appears to be a very important factor and will be discussed later.

Time of Attack

The time of year a mite population attacks trees is extremely important in determining the effects on production. Chapman (1959) reported a population of 67 mites per leaf on July 5 had no appreciable effect on the yield of apples. Boulanger (1958) and Lienk (1964) also pointed out it is the early-season populations which are most critical. Infestations developing after mid-July need to be controlled chiefly to protect against infestation early the following year (Light and Ludlum, 1972). Late season control may not be advisable in some areas because of adverse effects on biological control the following year.

Crop Load

Inadequately thinned trees tend to produce small apples. Since these trees have low leaf:fruit ratios, a minor loss of functional leaf surface may further reduce the poor growth rate.

Weather

Several authors have observed greater damage to foliage during periods of hot, dry weather. Huffaker and Spitzer (1950) suggested the intensified damage from high populations in hot, dry weather is due to desiccation and under less intense attack some repair of tissue may occur, retarding water loss. Boulanger (1958) reported reduced transpiration due to mite feeding. This may be the end result, but mite feeding probably reduced the resistance of the epidermis to water loss creating a water stress which then increased the stomatal resistance to water loss. High temperatures, particularly those associated with low humidities, may result in water stress on the tree retarding the growth rate of fruit and, in the short term, reduce the fruit size. This condition would be intensified by mite damge.

In studies of the effects of mites on tree performance, as many of these factors as possible should be standardized to minimize variation not associated with mite damage.

EXPERIMENTAL METHODS AND RESULTS

Mite density is another factor which must be controlled in studying the effect of mite feeding on tree crops. In many of the studies reported in the literature comparisons are made between uncontrolled mite populations and those controlled by acaricides or where controls were applied when the population reached a certain level. In the present studies a rather unique situation existed. The predatory mite *Metaseiulus occidentalis* Nesbitt had to be killed to allow populations of spider mites to develop and in some cases different numbers of *Tetranychus mcdanieli* McGregor were introduced to insure the presence of variable numbers of mites. In some cases acaricides were then applied to certain trees to maintain intermediate levels. The trees with low mite populations were generally not treated, and low populations were maintained by predation. This caused some complications because of the presence of large numbers of apple rust mites, *Aculus schlechtendali* Nalepa, on "mite-free" trees. In late studies the rust mites were reduced to low levels. Additional complications arose where both *T. mcdanieli* and *P. ulmi* were present in moderate numbers.

Since mite injury is related to the number of mites present and to the length of time they feed, the concept of mite-days is used in some of the studies. To determine the number of mite-days the number of mites present at the beginning of the sample period was added to the number present at the end of the sample period. This figure was divided by two and the result multiplied by the number of days between samples. Since most of us are used to evaluating mites in terms of population peaks, the relationship between mite-days and population peaks may prove helpful. Regressions of cumulative mite-days for the season on population peaks indicate the relationships are linear and estimates can be made as follows:

$$T.\ mcdanieli\ \text{population peak} = \frac{\text{mite-days}}{25}$$

$$P.\ ulmi\ \text{population peak} = \frac{\text{mite-days}}{27}$$

$$A.\ schlechtendali\ \text{population peak} = \frac{\text{mite-days}}{16}$$

While variations from these estimates are to be expected when applied to specific populations, errors should not be excessive. The estimates provide a simple means of relating mite densities to cumulative mite-days and are useful in that regard.

Test 1. 1972 to 1974—Smith Tract and Birchmont Orchards

Forty-eight trees at the Smith Tract and 80 trees at Birchmont of the Red Delicious variety were selected for uniformity of size and bloom in 1972. Mite populations were manipulated to vary from zero (actually 0.8 mites per leaf) to over 200 mites per leaf. Some trees had high mite populations in all three years, some moderate in all years, and some high the first year and low the second and third years. In each year 60 fruit were sampled from each tree. The fruit was weighed, measured, tested for firmness and evaluated for red color and internal chlorophyll. The sample was split into three subsamples for immediate examination and examination after three and six months of storage. These examinations included readings of soluble solids, titratable acids, pH, scald and internal breakdown. In addition, total yield was measured on each tree.

With these extensive data correlations of mites with each factor were run and a multiple regression analysis conducted. In most cases the correlation values were quite low, suggesting little or no relationship between mite populations and the fruit quality or size factors. Positive correlations occurred in one sample between mites and red color and mites and soluble solids. While these correlations indicate advanced maturity of fruit, the storage life of the fruit remained good. Chapman *et al.* (1952) also reported better color on fruit from mite-infested trees. This is probably related to a reduction in vigor on trees which are over-fertilized. While this relationship has shown up in these studies, it should be pointed out that the authors have observed poor color due to extremely high mite populations. Even those trees which had high mite populations for all three years did not show a depression in total yield.

These results suggest the tolerance for mid-summer mite populations is extremely high on vigorous apple trees. Zwick *et al.* (1976) found a high tolerance for spider mite populations on Newtown and Golden Delicious apple in a similar study. It is also possible that factors other than mites influence the size and quality of fruit to an extent that the effects of mite damage are masked. It is probable that both high tolerance for mites and effects of other factors prevented identification of mite-related effects. The results suggest anyone contemplating such studies on tree fruits limit the study to 1 or two important fruit measurements and maintain as careful control as possible of other variables.

Test 2. 1973—Tree Fruit Research Center

Four blocks of three Red Delicious trees each were established, and leaves and fruits were thinned on three limbs of each tree to give a 50:1 leaf-to-fruit ratio. This ratio was reported by Magness (1928) to be the minimum number of leaves necessary to size and mature Delicious apples. There were 10 apples per limb, but only the same five were measured each week.

Relative population levels (low, medium and high) were established by releasing *T. mcdanieli* in the medium and high level trees. After initial releases populations were not manipulated further. Mites were sampled by counting mature females on leaves on the tree.

There was a low order negative correlation between fruit growth per two-week interval and cumulative McDaniel spider mite-days at 15 (first sample), 112 and 119 days post bloom, but no significant correlation at 126 and 133 days. There was also a significant block effect at 105 and 112 days post bloom.

When cumulative fruit growth was regressed on cumulative McDaniel spider mite days the R^2 increased as the season progressed. The probability that the reduction in fruit growth was due to random effects changed from 8.3% at 105 days to 0.2% at 133 days. The slope of the regression, however, remained relatively constant at about -0.00033.

It appeared some reduction in fruit growth rate occurred prior to the first measurement, and mite populations continued to suppress fruit growth throughout the season. The effect is weak when measured over short time intervals, but it is cumulative. It should also be remembered that this effect occurred with a minimal leaf:fruit ratio.

A randomized block experimental design was used in this study. Analysis of variance, however, proved to be impractical and regression analysis was used instead. While the relative mite levels were consistent within blocks, there was a strong block effect on actual mite levels. The treatments (low, medium and high populations) were not equivalent and could not be compared to each other. The block effect due to mites was superimposed on any other block differences.

Test 3. 1973—Fischer Orchard

Mite counts were taken from brushed leaf samples over the season and fruit diameters were measured at harvest. A regression analysis showed low order negative correlations between diameter and both *T. mcdanieli* and *P. ulmi*. There was, however, a distinct east to west gradient in mite populations and fruit diameters. Reduced fruit size could have been due to mites and/or some other factor reflected by gradient.

Test 4. 1974—Tree Fruit Research Center

Mite populations failed to develop in the study trees due to excessive predation.

Test 5. 1975—Smith Tract

To facilitate year-to-year and orchard-to-orchard comparisons the time period for data analysis was standardized at 35 to 115 days post bloom. It was found that mite populations usually begin, peak and decline within this period. At 35 days post bloom fruit not affected by "June drop" can be selected and sample trees can be thinned, and 115 days post bloom is sufficiently early to be unaffected by variation in harvest dates. The difference in diameters at 35 and 115 days is defined as fruit growth, and the total mite-days accumulated during the period is defined as mite-days.

The sample units were selected limbs, four limbs per tree, and five apples per limb. Mite populations were moderate to low with a maximum of 4000 mite-days per leaf (according to earlier estimates this corresponds to a population peak of 160 per leaf) and predominately *T. mcdanieli*. Over 50% of the trees were at or near zero mite-days. The analysis was conducted on tree means. The following regression equation was used to test the data:

$$\text{Growth (in cm)} = b_o + b_1\,(\text{cm diameter at 35 days}) + b_2\,(\text{mite-days in thousands})$$

$R^2 = .379$, probability that F is significant $= .9826$

Fruit growth was found to be correlated with both mite-days and with fruit size at 35 days. Fruit that was larger at 35 days had a higher growth rate.

Estimated b	Probability the estimate is greater than 0.0
$b_o = -0.8741$	.3390
$b_1 = 0.9187$	.9430
$b_2 = -0.0675$	.9350

Fresh fruit from Washington is packed in a standard box containing approximately 45 to 50 lbs of apples based on size. A box may contain 80 large apples, 100 moderately sized apples or 125 small apples. Standard sizes are 72, 80, 88, 100, 113, 125 and 138 apples per box. A reduction in size of fruit sufficient to change it from one size to the next lower size is referred to as a loss of one box size and is a significant economic loss to the grower.

Based on the b_2 value given above, it is estimated 3000 mite-days would produce a loss of one box size and would correspond to approximately 12% loss in yield.

Based on studies conducted from 1972 to 1975, the following comments are made relative to quantitative studies of mite damage to apple. These comments may apply to other foliage feeding pests and possibly to other crops.

1. Naturally-occurring populations are unreliable. High populations usually occur on only a few trees, and in some cases appreciable populations may fail to develop on any test trees. Naturally-occurring populations tend to follow gradients which may be superimposed on other gradients in the orchard.

2. Even when populations are manipulated it is difficult to replicate given population levels. Relative population levels cannot be used, and only single tree replications are practical.

3. To reduce within-tree variation, the same fruits should be measured at each sampling date. Fruits should be selected for uniformity of size at the first measurement.

4. Some between-tree variation can be eliminated by using fruit growth rather than fruit size at harvest. Trees should be selected for uniformity of age, size and crop load.

DISCUSSION

In studies conducted to date the most significant effect of mite feeding damage on Red Delicious apple in Washington is a reduction in fruit growth rate. The amount of reduction in fruit size at harvest would depend on the magnitude of the mite damage and the time damage occurs. Our preliminary analyses suggest the longer a reduced growth rate is in effect, the greater the influence on harvest size since the effect is cumulative. As the season progresses, growth rate becomes less important, since a reduction in growth rate just prior to harvest can have little effect on fruit size.

While it is estimated that 3000 mite-days (corresponding to a peak population of about 120 *T. mcdanieli* per leaf) are required to reduce the fruit one box size, lesser numbers could reduce a percentage of the fruit in size. It is probable that significant, measurable effects begin between 80 and 100 mites per leaf (2000 to 2500 mite-days). The results of Chapman (1959) who reported no yield loss from 67 mites per leaf and Zwick *et al.* (1976) who found no reduction in fruit growth from peaks of 70 mites per leaf would support a tolerance this high. It should be remembered that the above applies to mid-season populations on vigorous trees.

While several authors have reported second year yield losses due to mite populations the preceding year on a variety of tree crops, no such effects were observed in the present study. In Washington apples are thinned extensively, both chemically and by hand, and this thinning may overcome any potential for reduced yields in the second year. Reduced yields of apples in the year following a heavy mite population were reported by Lienk *et al.* (1956), but the mite population in that case occurred sufficiently early to affect fruit bud formation for the following year. It is of interest to note that production of Red Delicious was not affected by articial debudding of 50 and 75% (Howell, 1978). On the basis of these studies it can be said that yield losses caused by mites on apple in Washington are more likely to be due to reduced fruit growth the year the population occurs rather than to reduced bloom and fruit set the following year.

Any other stresses placed on a tree will alter that tree's tolerance for mites. This effect has been particularly observed with irrigation. There is a significant reduction in fruit growth associated with inadequate soil moisture. The combined effect of inadequate irrigation and moderate mite populations will drastically reduce fruit growth. Most of the effect is probably due to irrigation, but since the mites are present they are often given as the cause. In the midwest and east where irrigation is not a widespread practice, mite problems must be aggravated during periods of low rainfall.

Adoption of even conservative estimates of economic injury levels in Washington could further influence the use of acaricides. A substantial percentage of growers do not apply acaricides. Those who do frequently make applications when populations are 10 to 30 mites per leaf. Tolerance of 50 mites per leaf in July and August would eliminate the need for acaricides in all but a few special situations. This conservative level would apply to trees with other mild stresses, but lower levels may be needed for heavily stressed trees or early season (May and June) mite populations. Where predators are eliminated or do not occur, populations would very likely exceed this level and treatment would be necessary.

In orchards untreated by acaricides, populations are usually composed of moderate numbers of *P. ulmi* (2 to 25 per leaf), relatively high numbers of *A. schlechtendali* (100 to 400 per leaf) and *M. occidentalis* (one to three per leaf). Croft and Hoying (1977) have raised the question as to whether or not these levels of apple rust mite might influence production, and the effect of apple rust mites has become of interest during the present study. Negative effects of apple rust mites would have to be substantial to offset the positive effects these mites have in the total mite-predator interaction. The presence of apple rust mites early in the season provides food for *M. occidentalis*, allowing the latter species to reach numbers capable of regulating spider mites at low densities. Additionally, Croft and Hoying (1977) and Cranham and Hoyt (unpublished data) have evidence that early feeding by moderate to high populations of apple rust mites limits the potential of *P. ulmi* to increase in numbers.

Considerable additional data on the effect of *T. mcdanieli, P.ulmi* and *A. schlechtendali* feeding on fruit growth rates of apple have been gathered and are now being analyzed. These data will be the subject of a paper which will present a more detailed analysis of the quantitative relationships between these three species and fruit growth.

SUMMARY

During the course of these investigations the term "mite-days" was used rather than population peaks, since this term takes into account both the number of mites and the length of time they are present. The most significant effect of mite feeding damage on Red Delicious apple was a reduction in fruit growth rate. This effect was cumulative with a greater reduction occurring the earlier the mite population occurred. An estimated 3000 mite-days (peak of about 120 mites per leaf) was required to produce a loss of one box size or approximately 12% loss in yield. Improved red color and higher soluble solids were positively correlated with *Panonychus ulmi* populations in one test. No other significant effects from mite feeding damage were observed on fruit quality or yield.

REFERENCES

Barnes, M. M. and Andrews, K. L. (1978). *J. Econ. Entomol.* **71**, 555-558.

Barnes, M. M. and Moffitt, H. R. (1978). *J. Econ. Entomol.* **71**, 71-74.
Boulanger, L. W. (1958). *Maine Agr. Exp. Sta. Bull.* **570**, 34 pp.
Chapman, P. J. (1959). *Proc. N. Y. State Hort. Soc.* **104**, 147-155.
Chapman, P. J., Lienk, S. E. and Curtis, O. F., Jr. (1952). *J. Econ. Entomol.* **45**, 815-821.
Croft, B. A. and Hoying, S. A. (1977). *Can. Entomol.* **109**, 1025-1034.
Howell, J. F. (1978). *J. Econ. Entomol.* **71**, 437-439.
Huffaker, C. B. and Spitzer, C. Jr. (1950). *J. Econ. Entomol.* **43**, 819-831.
Laing, J. E., Calvert, D. L. and Huffaker, C. B. (1972). *Environ. Entomol.* **1**, 658-663.
Lienk, S. E. (1964). *Proc. N. Y. State Hort. Soc.* **109**, 204-208.
Lienk, S. E., Chapman, P. J. and Curtis, O. F., Jr. (1956). *J. Econ. Entomol.* **49**, 350-353.
Light, W. I. St. G. and Ludlam, F. A. B. (1972). *Plant Pathol.* **21**, 175-181.
Magness, J. R. (1928). *Proc. Amer. Soc. Hort. Sci.* **25**, 285-288.
van de Vrie, M., McMurtry, J. A. and Huffaker, C. B.: a review. *Hilgardia* **41** (13), 343-432.
Westigard, P. H., Lombard, P. B. and Grim, J. H. (1966). *Proc. Amer. Soc. Hort. Sci.* **89**, 117-122.
Zwick, R. W., Fields, G. J. and Mellenthin, W. M. (1976). *J. Amer. Soc. Hort. Sci.* **101**, 123-125.

HOST PLANT RESISTANCE TO MITE PESTS OF CASSAVA

Anthony C. Bellotti and David Byrne

Cassava Program
Centro Internacional de Agricultura Tropical
Cali, Colombia

INTRODUCTION

A complex of spider mites, all belonging to the family *Tetranychidae* and the genera *Tetranychus, Mononychellus,* and *Oligonychus* has been identified as causing crop losses throughout the cassava growing areas of the world (Bellotti and Schoonhoven, 1978; Flechtmann, 1978). It is only in recent years that this complex has been studied in detail only at Centro Internacional de Agricultura Tropical (CIAT) and International Institute of Tropical Agriculture (IITA) and several national research institutes in Africa, Asia and the Americas. Consequently there is a paucity of information on this mite complex.

Cassava (*Manihot esculenta*) is grown in some 90 countries traditionally by small farmers, and is a major energy source for 300-500 million people throughout the tropical regions of the world. Cassava pests represent a wide range of arthropods but recent studies indicate that mites may be the most serious pest (Bellotti, 1978).

The two primary methods for controlling cassava mites under study are biological control and host plant resistance. This paper will describe the systematic evaluation of cassava germplasm at CIAT and other centers to identify resistance to several mite species.

THE CASSAVA MITE COMPLEX

Twenty four species of mites have been identified as feeding on cassava (Flechtmann, 1978; Bellotti and Schoonhoven, 1978). Undoubtedly there are several other species attacking cassava which have not been described.

Observations indicate that the most important species in South America are *Mononychellus tanajoa, Tetranychus urticae* and *Oligonychus peruvianus,* in

Vol. I: ISBN 0-12-592201-9

Africa, *M. tanajoa* and *T. urticae* and in Asia *T. urticae*. Other *Tetranychus* species are reported from Asia but information about these is scarce. *M. tanajoa*, the green cassava mite, is native to the Americas and appears limited to *Manihot* spp. but may attack other Euphorbiaceae. It has been recently introduced into Africa (Lyon, 1973; Nyiira, 1975) where it spread rapidly. *T. urticae* is a cosmopolitan species which attacks a large range of hosts. *O. peruvianus* also appears limited to *Manihot* spp. This paper will concentrate on these three species with emphasis on *M. tanajoa*.

THE HOST PLANT—MITE RELATIONSHIP

Manihot esculenta is a perennial shrub which originated in the Americas, was later taken to Africa and more recently introduced into Asia. Although the plant can be grown from seed, it is usually reproduced vegetatively for commercial purposes by planting stem cuttings. Leaves are formed at active apices and consist of an elongated petiole and a palmate leaf blade. The roots accumulate carbohydrates in the parenchyma to form swollen storage roots. Depending on ecological conditions, the plant is cultivated from 8 to 24 months. Although cassava is a tropical crop it is often grown in areas of prolonged dry periods (four months or more) and it is during these periods that severe mite attacks generally occur. Studies show that some varieties produced more leaves than needed, and there are periods when the plant can undergo high defoliation with no significant reduction in yield (Cock, 1978). However during dry periods many varieties lose foliage to compensate for the reduced availability of water. Mite attacks during this period result in additional stress to the plant. Physiological studies show that those pests that attack the plant over a prolonged period will reduce yields more than those that defoliate or damage plant parts for a brief period (Cock, 1978). The cassava plant appears better able to recover from the latter type of damage. Even after severe mite attacks, causing complete leaf loss and bud mortality, plant recovery will occur at the outset of the rains. Simulated damage studies at CIAT indicate that in a 10 mo. growing circle, insect attack after the sixth month has little effect on yield.

Most cassava mites feed on the lower surface of leaves but this varies with the species. Mites generally kill cells in the leaves causing decrease in photosynthetic ability (Cock, 1978), leaf death, decreased leaf size (Nyiira, 1972; CIAT, unpublishd data), slower development of the leaves (Nyiira, 1972), resulting in yield reduction (Nyiira, 1975; CIAT, 1975).

DAMAGE SYMPTOMS

Mononychellus tanajoa and *M. mcgregori*

These are small green mites (average body length of 350 mm.) usually found around the growing points of plants, on buds, young leaves and stems; lower leaves are less affected. Both species are similar in behavior (Bellotti and Schoonhoven, 1978; Nyiira, 1972, 1975; Bennett and Yassen, 1975). Females oviposit on the leaf undersurface, along the midrib or other veins, or in leaf concavities. Upon emerging, infested leaves are marked with yellow spots, lose their normal green color, develop a mottled, bronzed, mosaic-like appearance, and become deformed. Severely attacked leaves may develop whitish patches which eventually necrose and defoliation occurs progressively from top to bottom. High populations will cause shoots to lose their green color, stems become scarified, first turning rough and brown and eventually dieback. Apical shoots will be killed inducing lateral growth which can also be attacked and killed.

Tetranychus urticae

Attacks from this mite initially appear on the lower leaves of the plant and as the infestation progresses the population moves upward. Damage symptoms first show as yellow dots or specks along the main leaf vein, eventually spreading over the whole leaf. The leaf turns reddish, brown or rusty in color and necrosis occurs. Beginning with the basal leaves, severely infested leaves dry up, resulting in only a few leaves around the apical buds which may also be attacked. Plant mortality may occur depending upon the duration of the attack and age of the plant.

Oligonychus peruvianus

This pest also attacks the lower and intermediate leaves of the cassava plant. Mite presence is characterized by small white spots on the leaf undersurface, commonly along the central and lateral leaf veins and margins. These spots contain silk webs that the female spins under which eggs are deposited and the immatures develop. Corresponding damage symptoms on the leaf uppersurface are manifested by yellow to brown spots. Leaves may eventually necrose and defoliation of the basal leaves occur. Complete defoliation or plant mortality by this pest has not been reported nor observed.

LOSS IN PRODUCTION

Studies at CIAT (1976) with the mite complex (*T. urticae, M. tanajoa, M. mcgregori* and *O. peruvianus*) resulted in yield losses between 20-53% depending on the duration of the attack and/or the plant maturity during the attack. Nyiira (1975) demonstrated that yield losses range from 3-46% depending on the level of infestation. In general the higher the mite attack, the earlier the attack and the longer it lasts, the greater the loss in yield. Yield losses in mite susceptible versus resistant varieties have not been studied although Nyiira (1972) reported that the fresh root yield of some tolerant varieties was 27% greater than that of some susceptible varieties during a heavy mite attack.

RESISTANCE

A stable host plant resistance offers the best and most practical long term solution for controlling cassava mites because it is economical, easy to use and compatible with other control methods. To develop a resistant variety there needs to exist: 1) genetically conditioned resistance, 2) a reliable evaluation scheme and 3) breeding methods to incorporate this resistance into a commercially acceptable variety.

Genetically conditioned resistance to mite has been found in many crops (van de Vrie *et al.*, 1972). There are documented evaluations for mite resistance in cassava in South America, Trinidad, Africa, India, (Suradamma and Das, 1974), the Philippines and other areas. Most of the evaluations are unreplicated or minimally unreplicated, consisting of the only one years evaluations data and/or in one location and therefore the results are preliminary (Table I). Despite this, the results indicate that 1) mites can attack all varieties of cassava (i.e. immunity does not exist) and 2) that differences in resistance to mites do exist in cassava.

The most extensive screening for mite resistance in cassava has been done at CIAT (CIAT, 1975, 1976, 1977). The rest of this paper will discuss these studies.

In the preliminary screening for resistance to three mite species at CIAT results indicate that there is a greater level of resistance in cassava to *M. tanajoa* and *O. peruvianus* than to *T. urticae*. *M. tanajoa* and *O. peruvians* appear to have a host range limited to manihot species and other closely related plants, while *T. urticae* has a very wide host range. This higher degree of resistance might be due to a co-evolutionary process between cassava and the first two mite species.

Mononychellus tanajoa

This mite is one of the most destructive cassava pests in South America and Africa. The CIAT screening program consists of three phases.

1. Mass screening—preliminary screening
   a) Field—in germplasm bank with natural infestation
   b) Greenhouse—with potted plants and artificial infestation
2. Replicated evaluation trials of selected lines done at CIAT
3. Replicated evaluation trials done at other sites

TABLE I.
Evaluations for Resistance to Mites in Cassava.

Country	Source	Mite Species	What Evaluated[a]	No. Lines Evaluated	Remarks
Colombia	CIAT, 1976 CIAT, 1977	*T. urticae*	S—damage	2138	31 promising lines with low levels of resistance
India	Suradamma *et al*, 1974	*T. telarius* (*T. urticae*)	L—fecundity, longevity, Time for develop	20	Rated varieties R or S according to laboratory test without field testing.
Colombia	CIAT, 1975	*M. mcgregori*	G—damage	45	Found a range of resistance.
Colombia	CIAT, 1975 CIAT, 1976 CIAT, 1977	*M. tanajoa*	G—damage F—damage germplasm bank	2045 2197	Mass screening mostly unreplicated, 58 promising varieties.
Venezuela	Doreste *et al*, 1977	*M. tanajoa*	F—evaluation trial/damage	102/51	Low correlation between blocks.
Brazil	Dos Santos, *et al*, 1977	*M. tanajoa*	F—damage mite count	44	One year, one replication.
Uganda	Nyiira, 1972 Nyiira, 1975	*M. tanajoa*	F—mite counts F—damage	5 36	Replicated, one year, no primary resistance; all have recovery resistance.
Trinidad	Yaseen & Bennett, 1976	*M. tanajoa*	F—mite counts	10	Differences over a season.
Venezuela	Barrios, 1972	*Eotetranychus planki* (= *M. planki*)	F—damage	25	None immune but differences were found.
Colombia	CIAT, 1975	*O. peruvianus*	F—counts of webs per leaf germplasm bank	1884	Mass screening in germplasm bank.
	Unpublished 1978			2197	Mass screening repeated.

[a] F = Field; L = Laboratory; G = Greenhouse; S = Screenhouse.

The following damage scale is used in all evaluations:

0. No damage
1. Shoot and adjacent leaves with a few small yellow spots
2. Shoot and adjacent leaves with an intermediate number of yellow spots
2.5 Shoot and adjacent leaves with many yellow spots
3. Shoot and/or adjacent leaves with a slight yellowing, many yellow spots. Slight deformation of the apical leaves. Slight reduction of the shoot
3.5 Shoot intensely deformed and/or reduced. Apical leaf deformation. Many yellow spots
4. Shoot intensely deformed and/or reduced. A general yellowing or whitish appearance to leaves. A mosaic-like deformation of leaves
4.5 Shoot completely reduced with no leaves. Yellowing and/or defoliation in the middle of the plant.
5. Shoot dead

Mass screening in the field consists of evaluating the germplasm bank under natural infestation of this mite. This screening is characterized by escapes due to variation in mite infestation and environmental conditions and consequently shows a low correlation with replicated evaluation trials. The selections made in such a trial are preliminary. A ranking of lines with respect to resistance cannot be done.

Greenhouse screening can be useful for four reasons: (1) it requires less space, (2) it requires less time, (3) it allows for off season screening and (4) it allows for better control of screening conditions. Greenhouse screening has been used successfully to screen for resistance in cotton for *T. urticae*. It was shown that the correlation between the greenhouse results and the field results was very high (98.2%) (Schuster *et al.*, 1972). CIAT has carried out an extensive program of greenhouse screening for resistance to *M. tanajoa*. This evaluation is done according to damage ratings on vegetatively propagated cassava plants grown in small pots (10 cm dia.) that are infested with mites from the field.

Correlations between replications reveal low correlation coefficient (r^2 = 0.19, 0.25 & 0.35) which indicates the possibility of escapes. Escapes have been dealt with by a series of evaluation cycles in the greenhouse. The first selections from the unreplicated trial are evaluated in replicated trials.

The selections made in the greenhouse as a population, generally look better than the population of lines in the germplasm bank. Nevertheless only 60% of greenhouse selected resistant lines also appear to contain varying levels of resistance in field evaluations. The remaining 40% appear susceptible. In addition the majority of lines selected as containing resistance in the field do not express this resistance in greenhouse evaluations.

Preliminary data also suggest that some varieties do not react to mite attack in the greenhouse as they do in the field (Fig. 1). These differences could be

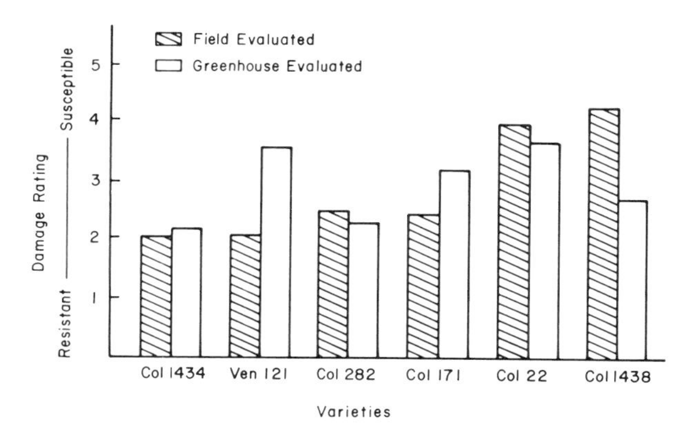

Fig. 1. Interaction between varietal mite resistance and evaluation systems.

due to the differences in growing conditions (small pots, high mite infestation, greenhouse conditions vs field grown, low mite infestation, field conditions) or in differences in plant age (small, young plant, 1-2 months old vs large plants, 4-8 months old). These interactions may be the reason for the disappointing results observed in greenhouse screening.

Lines which were selected in the preliminary screening are evaluated in replicated field trials at CIAT and other sites, using susceptible border rows to insure an even mite infestation. Results of one trial showed a good correlation between blocks ($r^2 = 0.63$) and a low coefficient of variation (C. V. = 12.11%) as compared to a similar trial in Venezuela (Doreste *et al.*, 1977) which had a low correlation between blocks ($r^2 = 0.09$) and higher coefficient of variance (C. V. = 21.04%). In the latter trial there may have been an uneven infestation of mites and therefore less confidence can be put on resistant selections.

Promising lines from preliminary screening programs are evaluated in other environmental conditions and over several years to measure resistance stability. This phase of the program is in its initial stages. In conclusion more emphasis should be put on field screening as there appears to be an interaction between plant reaction to mite attack and where the plots are evaluated (field vs greenhouse).

The mechanism of mite resistance has been studied in cotton (Schuster *et al.*, 1972b), strawberry (Rodriguez *et al.*, 1971) and other crops. Decreases in fecundity, preference, increases in mortality and time for development have been attributed to chemical factors in the leaves as well as physical factors such as pubescence, the cell size and thickness of the cell wall. The work in cassava has only progressed to the characterization stage. One study (Saradamma and Das, 1974) with *T. urticae* showed there are differences in the fecundity, longevity and time of development of the mite on different cassava varieties. Preliminary studies in CIAT with *M. tanajoa* show that resistant varieties are less preferred and mites lay less eggs on resistant varieties than on susceptible varieties. The mechanisms behind these observations are not known. It has been suggested that pubescence and HCN contents of leaves may condition resistance but no evidence has yet been exhibited.

Tetranychus urticae

Only preliminary replicated screenhouse evaluation studies have been tested for this mite. Two month old cassava plants grown in floor beds are artificially infested and subsequent damage evaluated at weekly intervals. Results of these studies show only low levels of resistance (Table I). Thirty-one promising lines have been selected for field evaluation when an adequate site can be identified.

SUMMARY

More than twenty mite species are reported as attacking cassava throughout the world. *M. tanajoa* and *T. urticae* appear to cause the most serious damage; yield losses as high as 53% have been attained in experimental plots.

Systematic studies in mite resistance in cassava is recent and preliminary results indicate that these are low to intermediate levels of resistance but no immunity. Fifty eight varieties have been identified as promising for resistance for *M. tanajoa* and 31 to *T. urticae* in the CIAT screening program.

There is a low correlation between greenhouse and field screening and only 60% of greenhouse selected resistant lines also appear to contain resistance in the field. Greenhouse screening therefore can only be used to aid in eliminating susceptible material but not for identification resistance.

It is concluded that emphasis should be placed on field screening using susceptible border rows and promising material should be evaluated in different environmental conditions and over several years to measure resistance stability.

REFERENCES

Bellotti, A. C. (1978). *Proc. Cassava Protection Workshop,* CIAT, Cali, Colombia 1977, 29-39.

Bellotti, A. and van Schoonhoven, A. (1978). *Ann. Rev. Entomol.* **23,** 39-67.

Bennett, F.D. and Yaseen, M. (1975). Trinidad, *West Indies Commonw. Inst. Biol. Control.* 12 pp.

CIAT. (1976). *Annual Report, 1975.* Cali, Colombia: Cent. Int. Agric. Trop. 57 pp.

CIAT. (1977). *Annual Report, 1976.* Cali, Colombia: Cent. Int. Agric. Trop. 76 pp.

CIAT. (1978). *Annual Report, 1977.* Cali, Colombia: Cent. Int. Agric. Trop.

Cock, J. H. (1978). *Proc. Cassava Protection Workshop,* CIAT, Cali, Colombia, 1977, 9-16.

Doreste, E., Arias, C. and Bellotti, A. (1977). *Proc. Cassava Protection Workshop,* CIAT, Cali, Colombia, 1977, 161-164.

Dos Santos, J. H. R., Almeida, F .C. G., Cavalcante, R. O. and de Pinho, J. L. N. (1977). *Fitossanidade, Fortaleza* **2**, 34-37.

Flechtmann, C. H. W. (1978). *Proc. Cassava Protection Workshop,* CIAT, Cali, Colombia, 1977, 143-153.

Lyon, W. F. (1973). *PANS Pest. Artic. News Summ.* **19**, 36-37.

Nyiira, Z. M. (1972). Mimeo. Dept. of Agric. Kawanda Research Station.

Nyiira, Z. M. (1975). pp. 27-29. Workshop on cassava improvement in Africa. IIIA, Ibadán, Nigeria.

Rodriguez, J. G., Dabrowski, Z. T., Stoltz, L. P., Chaplin, C. E. and Smith, W. O., Jr. (1971). *J. Econ. Entomol.* **64**, 383-387.

Saradamma, K. and Das, N. M. (1974). *Agric. Res. J. Kerala* **12**, 108-110.

Schuster, M. F., Maxwell, F. G., Jenkins, J. N. and Parrott, W. L. (1972a). *J. Econ. Entomol.* **65**, 1104-1108.

Schuster, M. F., Maxwell, F. G. and Jenkins, N. N. (1972b). *J. Econ. Entomol.* **65**, 1110-1111.

van de Vrie, M., McMurtry, J. A. and Huffaker, C. B. (1972). *Hilgardia* **41**, 343-432.

NITROGEN FERTILIZATION OF FRUIT TREES AND ITS CONSEQUENCES FOR THE DEVELOPMENT OF *PANONYCHUS ULMI* POPULATIONS AND THE GROWTH OF FRUIT TREES

M. van de Vrie

Research Station for Floriculture
Aalsmeer, Netherlands

P. Delver

Research Station for Fruitgrowing
Wilhelminadorp, Netherlands

INTRODUCTION

It has been repeatedly shown that the host plant condition is of great importance to the development of *Panonychus ulmi* populations. Kuenen (1946), Post (1962), Rodriguez (1964), Storms (1967), van de Vrie *et al.* (1972) and many others have associated nitrogen content of apple leaves with potentials for increase in mite populations. Host plant-tetranychid relations were reviewed by van de Vrie *et al.*; they postulated that "... if a better understanding of the food requirements of phytophagous mites is achieved, it will be possible to define more precisely the ecological significance of the altered quality of the host plant frequently observed after application of fertilizers or pesticides. We may also speculate on the possibility of influencing the population development of these mites by changing their food substrate through managing the fertilization of the host plant."

Quality of the host plant is related to many variables such as varietal differences, soil-mineral-water relations, climatic conditions and cultural practices such as pruning, chemical thinning and control of fungus diseases. Nitrogen fertilization is one of the factors which can be handled relatively easily by altering the amount of nitrogen, the period of administering and the choice of different forms of nitrogen containing fertilizers. Also soil treatments offer possibilities for manipulation of the availability of nitrogen to the fruit trees. If mite population development could be delayed by managing the

Vol. I: ISBN 0-12-592201-9

host plant quality the need for chemical control of the fruit tree red spider mite would be reduced and probably supervised control or integrated control could be achieved. The following information was collected in a study on the influence of applying equal amounts of nitrogen at different intervals in the vegetative period of young apple trees in the Netherlands. The responses of the trees and the consquences of these changes on mite development and reproduction were studied.

MATERIALS AND METHODS

Plants

Small two-year old apple trees, Beauty of Boskoop on M 2 rootstock were selected for uniform size and shape. They were planted one year prior to the experiment in a nursery under extreme low nitrogen conditions. This was achieved by the cultivation of a grass vegetation under the trees without any fertilization being added during the season. The following spring the trees were planted in pots containing approximately 23 liters of soil. The following system of fertilization was utilized:

Number	Treatment
1	No nitrogen during the experiment
2	2000 mg NO_3 per pot, mixed with the soil in March
3	2000 mg NO_3 per pot, applied 15 May
4	2000 mg NO_3 per pot, applied 19 June
5	2000 mg NO_3 per pot, applied 27 July, after the vegetative growth had stopped
6	2000 mg NO_3 per pot applied as a slow release fertilizer (Gold-N, ICI, 33.5 % N, 6 gr per pot)

Treatments 1, 2 and 4 had 14 one-tree replicates each; 3, 5 and 6 had seven one-tree replicates each, with and without mite infection, respectively. Trees without mites received identical treatment to study the response of the trees without the interference of mite populations.

Treatment 2 represents the normal situation existing in young apple orchards when N-fertilization is given at the beginning of the growing season. Fertilization at various time intervals was introduced to study the reactions of the various mite populations during the season (replicates 3, 4 and 5). In treatment 6, Gold-N was applied as a slow release N-fertilizer, to provide the trees with the same amount of nitrogen during the growing season.

The tree pots were watered weekly; the amount of water needed was determined by weighing.

Nitrogen Content

Nitrogen content of leaves was analysed frequently during the growing season. The sample size was 35 gm of fresh leaves, fully expanded and taken at random. In these leaves total N and amino acids were determined. In this report only the total N analysed is discussed in relation to mite development.

Leaf Color

Leaf color was determined frequently by a panel of at least three persons. A system with values from 1 (very light) to 10 (dark green) was employed. This determination of leaf color was used as an index of the response time of the trees to the fertilizer treatment.

Leaf Size

Leaf size was graded into 10 classes based on a standard leaf area index. This correction allowed for adjustments in mite densities/unit area, if needed.

Mite Population Development

On May 25 all trees were infested with 25 fertile females which had just started ovipositing. Mites were bred under normal nitrogen conditions and under an outdoor climate. Mites were distributed on the test trees as evenly as possible by transferring them with a fine brush. Population development was studied by counting mites (mobile and quiescent stages) and eggs separately on 50 marked leaves. This counting was performed at approximately 10-day intervals, until the end of the season when winter eggs became numerous. Winter eggs were not counted.

Since the average leaf size, and the frequency distribution of the leaf size was rather uniform within the various treatments, no correction for leaf area was considered necessary.

RESULTS

Total Nitrogen

The total N content of the leaves was analysed 14 times during the period June through October, and is represented in Fig. 1. In treatment 1, nitrogen deficiency is shown, starting at 2.25% N and decreasing gradually to 1.0% by mid-October. This gradual decrease during the season is a common phenomenon; in the spring, nitrogen is readily available to the tree from reserves in the plant tissues. Treatment 2 gave a higher N content at the beginning of the season; 2.75% N was found. Also there was a gradual

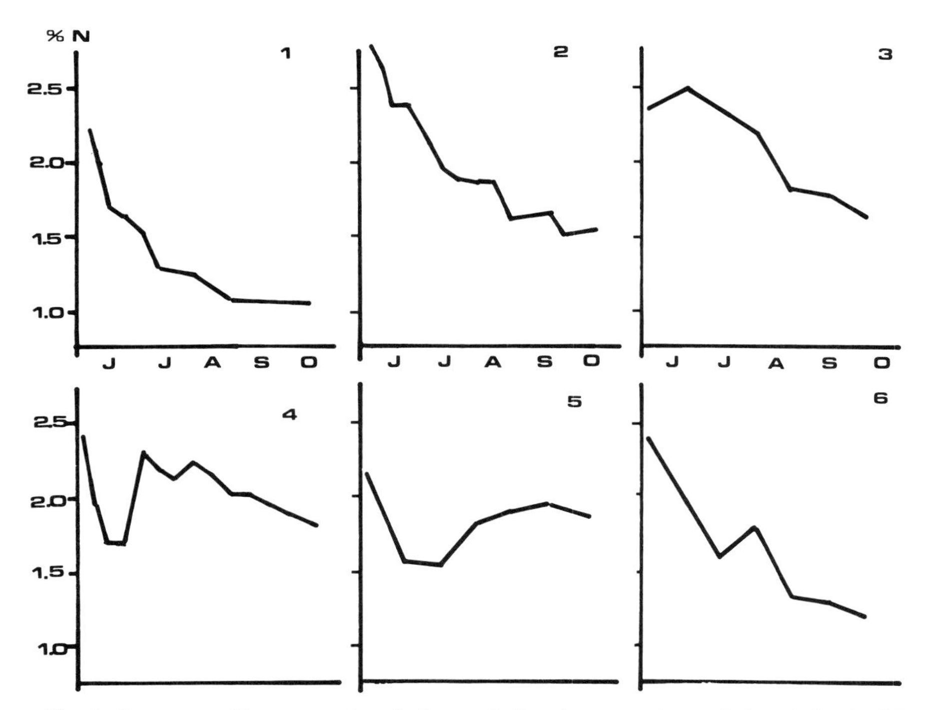

Fig. 1. Percentage N-content of apple leaves during the vegetative period as influenced by different fertilization programs.

decrease during the vegetative period. Treatment 3 when fertilized on May 15 showed a higher N level, which reached a maximum at the end of June, followed by a gradual decrease as the season progressed. Treatment 4, which was fertilized on June 9, showed a remarkable increase in N content at the beginning of July. In this treatment, a decrease in N content was also found in late season. Treatment 5, fertilized on July 24, showed a reaction to fertilization extended over a long time period, due to reduced plant growth during the latter part of the season. Almost no decrease in N content was observed thereafter. Treatment 6, having a slow release fertilizer applied at the same time as treatment 2, started with a higher N content than treatment 1 but a lower one than 2, indicating that nitrogen was available at a lower level. During the entire season this situation did not change. Also there was a gradual decrease in N content throughout the season.

Leaf Color

Leaf color was determined 11 times during the season to indicate the reaction of the trees to changes in the nitrogen content of the soil. The results are shown in Fig. 2. Treatment 1 started at a rating of 5.25 and the rating decreased continually as the season progressed. This group of trees suffered

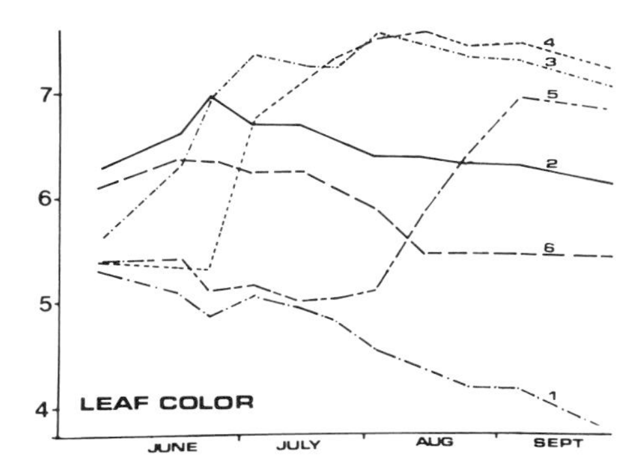

Fig. 2. Differences in leaf color as influenced by six different fertilization programs.

severe N deficiency. Treatment 2 showed a very clear reaction to fertilization at the beginning of June by having a darker green color than all the other treatments. After reaching a maximum rating at the end of June, a slow decrease over time was noted. Treatment 3 resulted in a strong darkening in the color of the leaves; this reaction extended to the end of July. Trees in treatment 4 responded at the beginning of the season very similarly to those in treatment 1, but after fertilization the color was intensified greatly. This increase continued until mid-August and remained high until the end of the season. Treatment 5 reacted in a similar way as the preceding treatment; apparently time for growth was too short at the end of the season to reach the same leaf color as treatment 4. Treatment 6 had an intermediate color reaction. At the beginning of the season the color was very similar to treatment 2, but the decrease started earlier and resulted in a generally lower level of color throughout the season.

Leaf Size

The results on leaf size are summarized in Fig. 3. This figure shows the frequency distribution of the percentages of leaves belonging to each size-class. It is clear that treatment 1 resulted in the smallest leaves, followed by treatment 5. Fertilization in treatment 5 occurred at the time that vegetative development had almost ceased. Treatments 4 and 6 had almost equal leaf class distributions; the same was found for treatments 2 and 3.

Mite Population Development

The development of the populations in the various treatments is given in Figs. 4 and 5. The average number of eggs per leaf is shown in Fig. 4. Treatment 1 had a low number of eggs throughout the summer, varying between less than 1 and 3 eggs per leaf. This was caused by the N deficiency in leaves indicated earlier. Treatment 2 allowed for a higher reproductive

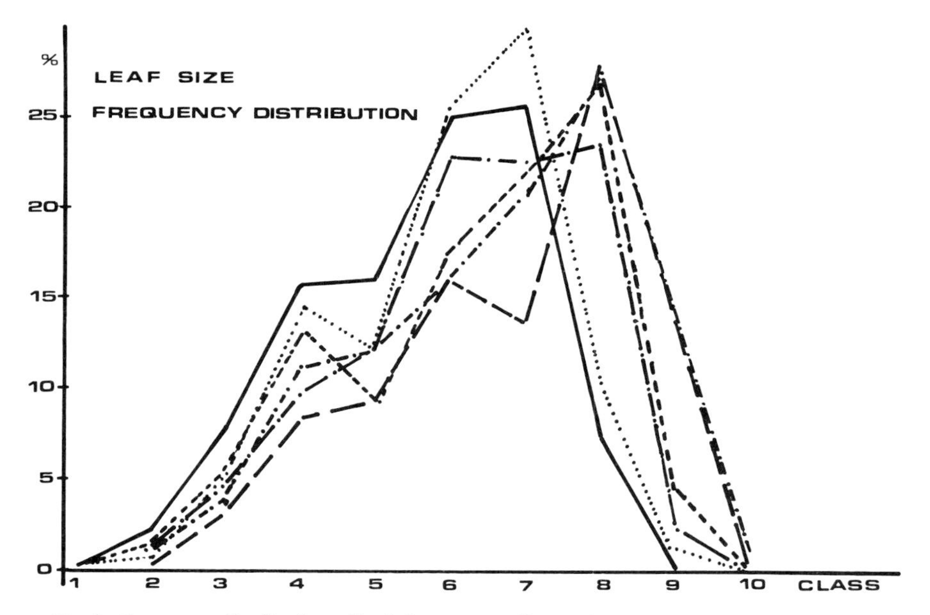

Fig. 3. Frequency distribution of leaf classes according to size.

capacity in the first generation. The number of eggs produced was 2.5 times higher than in treatment 1. During the following generations a steady population increase was found, resulting in a maximum average number of 21 eggs per leaf. Treatment 3 showed a higher level of mite population than treatment 2, resulting in a maximum average number of 28 eggs per leaf by mid-August. Treatment 4 had a mite population identical with treatment 1 at the beginning of the season. The females of this population reacted to the N-fertilization and showed a remarkable increase in oviposition. An average of 30 eggs per leaf was obtained in August by this group of mites. Treatment 5 resulted in an immediate increase in oviposition. However, as the environmental conditions for the induction of diapause prevailed by mid-August, it was too late in the season for mites to develop to a high population density. Treatment 6 showed a steady increase of *P. ulmi*, reaching a maximum average of 16 eggs per leaf in August.

The average numbers of mites (all mobile stages and quiescent stages combined) are shown in Fig. 5. Treatment 1 again showed the lowest number of mites; the average varied between one and four per leaf. Treatment 2 showed a development of three separate mite generations, each increasing in density. Treatment 3 showed almost identical timing in population peaks, although the generations showed more overlapping. Treatment 4 also showed the development of three generations and the population density was lower than in treatment 3. Treatment 5 showed an increase in mites as the summer

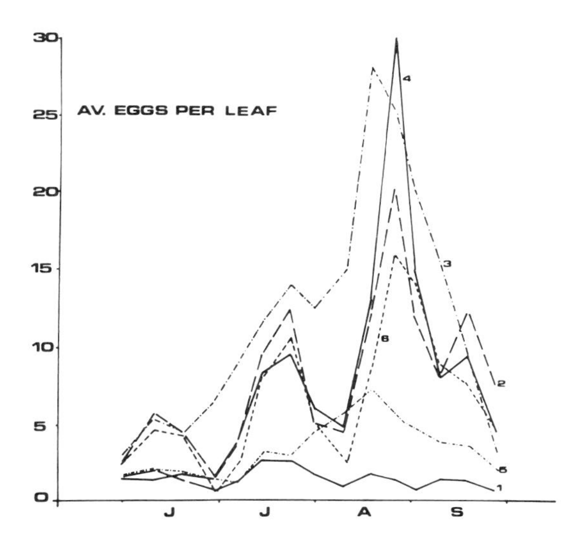

Fig. 4. Average number of summer eggs of *P. ulmi* per leaf.

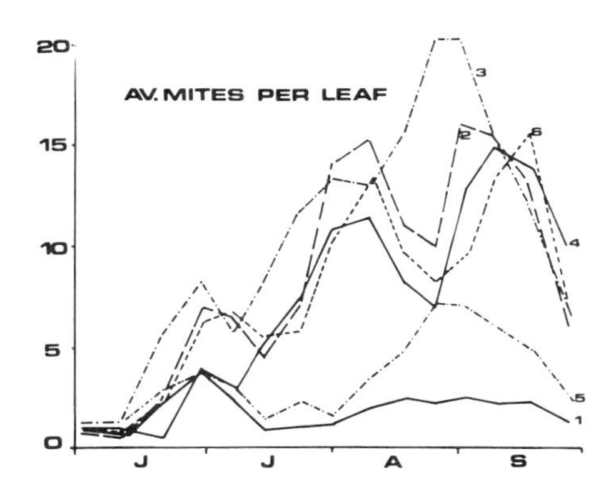

Fig. 5. Average number of mites and quiescent stages of *P. ulmi* per leaf.

eggs hatched during August, however it was too late in the season for mites to develop an extremely high density. Treatment 6 showed a steady increase in populations to a maximum average of 15 mites per leaf in August. The generations of all popuiations under the various treatments coincided in time. This is partly due to the uniform infestation with fertilized females at the beginning of the experiment. Apparently the differences in the amount and quality of the food ingested by *P. ulmi* had no influence on the rate of development of the mites.

DISCUSSION

The results presented herein demonstrate that the population density of *P. ulmi* on apple trees can be influenced by manipulation of the N-fertilization. In this experiment the N varied only in time and not amount; certainly also variation in the amount of fertilization would have consequences for mite development. Lower mite populations due to fertilization treatment would reduce the need for acaricidal applications, thus reducing the chances for development of resistant to acaricides.

The question arises whether optimal fruit production in terms of amount and quality would allow for variation in the fertilization program used. Experience, both in commercial fruit production and in experiments in the Netherlands, has shown that reducing the amounts of N has a positive influence on the quality of the fruit yield without reducing the quantity. The question is how much and for how long can lowered amounts of nitrogen fertilizer be used.

ACKNOWLEDGEMENTS

The cooperation of Dr. J. Tromp and Miss A. Ovaa in the amino acid analyses is gratefully acknowledged. Also the help of Mr. J. Oele and Mr. P. J. Bolding for the analyses of nitrogen, the determination of leaf sizes and color, and the maintenance of the whole experiment, is gratefully acknowledged.

REFERENCES

Delver, P. (1978). *Fruitteelt* **68**, 284-287, 324-327.
Kuenen, D. J. (1946). *Meded. Tuinb. Voorl. Dienst.* **44**, 68 pp.
Post, A. (1962). Dissertation, University of Leiden, 110 pp.
Rodriguez, J. G. (1964). *Acarologia, fasc. h.s. 1964,* 324-337.
Storms, J. J. H. (1967). *Fruitteelt* **57**, 862-863.
van de Vrie, M., McMurtry, J. A. and Huffaker, C. B. (1972). *Hilgardia* **41**, 343-432.

THE APPLICATION OF SIMULATION MODELS TO MITE PEST MANAGEMENT

Stephen M. Welch

Department of Entomology
Kansas State University
Manhattan, Kansas

INTRODUCTION

The current decade has seen great progress in the development of pest management programs for phytophagous mite species (Croft, 1975). An important component of this progress has been the use of computer simulation techniques (Berryman and Pienaar, 1974) to (1) predict outcomes of interactions with biocontrol agents, (2) study mite-plant interactions, (3) evaluate sampling plans, (4) complement other decision strategies, (5) aid in the training of professional pest managers and (6) improve theoretical understanding of mite population processes.

To a certain extent modeling efforts in acarine systems parallel developments in insect models (Ruesink, 1976). For example, models decompose the life cycle into discrete stage components which are treated separately. On the other hand, there are some unique features of mite models. Many models, for instance, portray the interaction of a pest with one or more predators. Economic mite species are much more likely to exhibit short continuously overlapping generations than are economic insect species. Other differences also exist.

Needless to say, it is not possible, in a publication of this length, to do justice to the whole area of mite modeling. Rather, I shall attempt to (1) convey the structure and types of processes contained in most mite models, (2) discuss briefly the procedures used to validate these models, and (3) present different methods for implementing these models in practical field programs. Examples will be drawn from mite control programs in orchard settings from the United States and Europe.

Vol. I: ISBN 0-12-592201-9

MITE MODELS AND THEIR STRUCTURE

The greatest focus of modeling attention has been on the tetranychid mites (*Tetranychus* and *Panonychus* spp.) and their phytoseiid predators (*Typhlodromus* and *Amblyseius* spp.) although other mite families (Allen, 1976) and predators (Asquith and Colburn, 1971) have been examined.

Figure 1 shows a model structure which has been widely applied by mite and insect modelers. The circles represent the life stages of the organism beginning with the egg. The arrows represent flows of individuals as they (1) develop from stage to stage, (2) as they die (predation, starvation, man-applied biocides, *etc.*) or (3) disperse from the system.

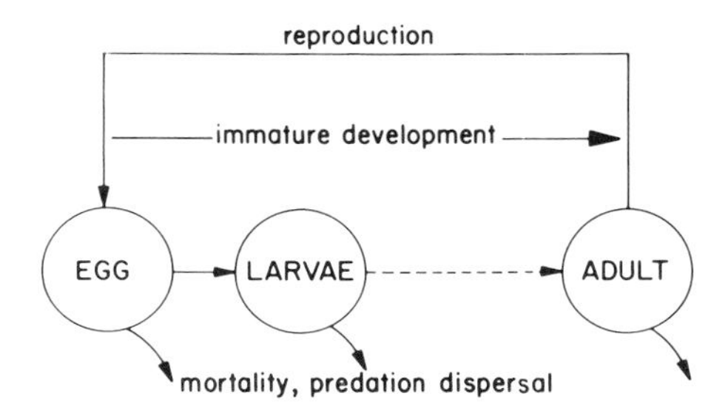

Fig. 1. Generalized model structure for a single mite species. Circles represent life stages and arrows depict flows of individuals.

The life stages in a model may be distinct morphological stages or merely units of convenience (Welch *et al.*, in press). Among mite models these divisions run from simple egg-immature-adult schemes to decompositions involving the various forms of larval and nymphal stages and quiescent periods (Logan, 1977). One model separates individuals based on their diapause characteristics (Rabbinge, 1976).

Generally, the particular division settled on depends on (1) the goal of the modeler, (2) the robustness of his existing data base, (3) resources available for further experimentation, and (4) the degree to which similar life stages are seen as combinable. Comparing model performance to structural complexity, it appears that, for existing models, any correlation is rough, at best. Examples of good and poor models exist at most levels of complexity. This indicates that accuracy in the modeling of rate phenomena is probably more important than mere structure.

DEVELOPMENTAL RATES

The phenology of life cycle events is critical to pest management models (Welch *et al.*, in press). Most mite models incorporate timing in the form of submodels describing meteorological influences on developmental rates.

In theory, the micro-weather of a fruit tree is complex; variations exist in temperatures, sunlight, wind profiles, *etc.* Most models have tended to ignore this complexity; a position which has been supported by very detailed experimental and modeling studies in the Netherlands (Rabbinge, 1976). These studies showed that (for the *Panonychus ulmi-Amblyseius potentillae* system) small oviposition rate changes have greater impacts on population dynamics than variations in micro-weather.

The conventional approach is, therefore, to fit some mathematical equation to raw developmental data assuming mean ambient temperature to be the sole driving variable. Equations used for this purpose include polynomials, spline functions, degree-day systems, and exponential forms. As an example, Logan *et al.* (1976) argued that developmental rate determining factors probably differed between low and high temperature regions. He was able to combine empirical formulas for these two regions using the mathematical technique of matched asymptotic expansions. The resultant equation proved applicable to a wide variety of temperature dependent biological rates including mite development and oviposition, sucrose oxidation by *Bacillus* spp., *etc.* It is, thus, a good example of the type of generalized submodels whose use by modelers is becoming more prevalent (Ruesink, 1976).

PREDATION

The widespread development and success in spider mite biological control programs has led to the incorporation of predation components in most models. These submodels usually include a subset of the following factors: average predation rates (Mowery *et al.*, 1975), temperature (Logan, 1977), predator functional response (Rabbinge, 1976), predator intraspecific competition, and the distribution of prey and predators among various life stages (Dover *et al.*, in press) and spatial locations (see later section).

This aspect of mite modeling benefits greatly from decades of theoretical work on prey predation relations; many standardized submodels already exist. Dover *et al.* (in press) present a fairly detailed model. Following Royama (1971), they calculated the daily feeding rate (Z) of *A. fallacis* on *P. ulmi* as

$$(1) \qquad Z = -B^{-1} \ln(\exp(-ABP) - \exp(-B(X+AP)) + \exp(BX))$$

where X is initial prey density, P is predator density, A is the maximum instantaneous feeding rate as determined by predator functional response curves, and B tells how fast this maximum rate is approached. Fitting this equation to data of Blyth and Croft (1975) showed that the constants A and B dependent strongly on temperature. By writing the coefficients of Eq. (1) as polynomials in temperature, a single equation was arrived at which included almost all of the previously listed factor. By iterating this equation over various prey stage-predator stage combinations and spatial locations estimates of the total daily prey consumption were determined.

REPRODUCTION

Explosive reproduction within mite populations is a major feature of their pestiferous nature. *Tetranychus mcdanieli,* for example, exhibits rates of increase in excess of r = 0.4/ day (Tanigoshi *et al.*, 1975; Logan *et al.*, 1976). Because r appears as an exponent in species abundance equations it is important that reproductive submodels be particularly accurate. Oviposition rates vary, of course, in response to many variables; in particular modelers have emphasized temperature, age of adult, and (for predators) consumption level.

Temperature response appears to differ qualitatively from one species (and its corresponding model) to the next. For example, *T. mcdanieli* appears to have a temperature independent oviposition period but a temperature dependent egg complement (Logan, 1977). Above 15°C, *P. ulmi* exhibits exactly the reverse behavior (Dover *et al.*, in press). *A. potentillae,* however, varies in both aspects (Rabbinge, 1976). The mathematical formalism employed in the various models reflect these differences.

Two *P. ulmi* models (Dover *et al.*, in press; Rabbinge, 1976) treat differences in adult fecundity with age (m_x in life table notation). Fecundity increases from intermediate values to a peak one quarter to one third the way through the oviposition period. Oviposition declines steadily with age thereafter.

The most complete treatment of predator consumption-oviposition interactions is due to Rabbinge (1976). As the predator *A. potentillae* ingests food, its color progressively changes from clear through deep reddish brown due to the presence of plant pigments in the prey. Quantifying this color change on a 1 to 7 scale enabled relations between consumption rates, color changes, and oviposition as a function of color to be derived for use in the model.

MITE-PLANT INTERACTIONS

The development of damage equations has proved difficult in the modeling studies presented above because of the indirect effects of leaf feeding. Progress in this area will probably have to await the appearance of detailed models of tree physiology (*e.g.,* Elfving *et al.*, in prep.).

By contrast, however, Allen (1976) has modeled the effects of *Phyllocoptruta oleivora* (Ashmead) feeding directly on Valencia oranges. Three models were examined. In the first, the percent of fruit surface damaged was assumed proportional to the time integral of mite density. The second allowed the proportionality constant to be time varying while the third incorporated the effects of fruit growth and limited fruit area. The appropriateness of both chronological and degree-day time scales was examined as was the applicability of the models to various periods during the year.

All models performed well. While no particular advantage resulted from the use of degree-days, the relation between mite density and damage did appear to be time varying. Allen (1976) related this to increasing fruit maturity. Correcting a notational error [Allen, 1976, Eq. (8)], the model finally adopted reads

$$(2) \qquad p(t) = 0.0115 \int_0^t (1 + \exp(6.92 - 0.03592))^{-1}\, m(\tau)\, d\tau$$

where p(t) is the percent surface area damage on day t. The sigmoidal factor inside the integral is the percent damage per mite per cm^2 per day. The second factor, $m(\tau)$, is the number of mites per cm^2 of fruit surface area. In a complete plant-pest model, $m(\tau)$ would simply be the instantaneous mite density as determined by the other sections of the model. Equally, this submodel could be used directly with mite population field data to infer damage indirectly.

DISPERSAL

Mite populations, particularly predaceous species, may be very mobile during the course of the season. The study and modeling of their movements are crucial to biocontrol programs which require spatial and temporal concurrence of prey and predator. For example, *A. fallacis* (unlike its prey) overwinters on the ground in orchard situations, migrates to the crown in early summer, and several generations later, returns to the ground when food or other factors become limiting.

The migration into the tree has been modeled by Harrison *et al.* (unpubl.). Mites were assumed to move randomly over the ground beneath the tree and up the trunk. A partial differential diffusion equation with empirically determined coefficients was used to portray this movement. It was found that mites more than five feet from the trunk had less than a one percent chance of reaching the crown of the tree in 45 days. Even mites beginning at the trunk-soil line had only a 20 percent chance of reaching the crown within three weeks. This was consistent with the hypothesis that predator populations in the tree develop from a very small number of individuals.

Johnson and Croft (1975) have shown that *A. fallacis* will disperse from the tree via air currents after one or more days of starvation. Incorporated into a model (Welch *et al.*, in prep.), this had the effect of more rapidly reducing predator populations in the post-peak phase of an interaction. By increasing the rate of predator decline, it enhanced the probability of some prey escaping predation resulting in low density pre-predator oscillations. Such oscillations are a common occurrence under field conditions.

SPATIAL DISTRIBUTION

The effects of spatial distribution on mite population processes is a relatively neglected area among modelers. Reasons include the increased complexity of spatial models, the feeling that mites are rather uniformly distributed, and the idea that stochastic effects will "average out" over multi-acre blocks or at moderate densities. Whether either of these latter two effects occur at or below economic densities, however, can only be determined by experiment. For the *P. ulmi-A. fallacis* system in apple, they do not (Welch *et al.*, in prep.).

Croft *et al.* (1976) present data showing (1) that moderately aggregated (k_c approximately 1.4) negative binomial distributions could be fit to within-tree poulations of both species and (2) that significant variation in mean density existed among trees in blocks up to 10 acres in size. The impact of these results on estimates of prey consumption and the duration of prey-predator interactions was studied.

To study consumption, it was assumed that Eq. (1) applied to single leaf universes. Average consumption per leaf was computed by calculating the expectation of Z when X and P have independent negative binomial distributions. Comparing this value to that obtained by substituting within-tree means for X and P, it was determined that neglecting spatial effects would result in a 45 to 65 percent overestimate of consumption at near economic densities.

By weighing the results of multiple model runs according to lognormal probabilities (Welch, 1977; Dover *et al.*, in press), it was found that the apparent duration of the prey-predator interaction was lengthened. This brought outputs more in line with field data although run times were increased by seven to 20 times. This latter effect, while costly, seems necessary for models to achieve high accuracy. The determination of (and need for) model accuracy will be explored in the next section.

MODEL VALIDATION

Validation is the process of determining whether model outputs acceptably mimic reality. The "acceptability" of a model depends, of course, on the purpose for which it is intended. At least two sets of standards have been adopted by mite modelers. The first applies to investigations whose goal is a very precise quantitative understanding of the population. An example is Rabbinge (1976). Validation experiments in this study consisted of detailed determination of predation and oviposition rates, densities of various population stages, predator color values and other factors under laboratory, greenhouse, and field conditions. All validation data were (1) completely independent of the information used to generate the model and (2) stated in terms of confidence intervals. Model outputs mimic these results to an exceptionally high degree except under field conditions where high intrinsic variation is a factor (as discussed previously).

It is, however, under exactly these highly variable conditions that most management programs must operate. Fortunately, these programs seldom require pin-point accuracy in population forecasts. Often it is necessary merely to project the general patterns of timing (Welch *et al.*, in press) of pest development in order to achieve effective management. Mowery *et al.* (1975) provides an example. Field counts were used to initiate the model which then made one week forecasts. The model was corrected by new counts each week to eliminate the build up of cumulative errors and to reset the model after miticide applications. This procedure permitted the prediction of the timing of mite outbreaks and the likelihood of control by natural enemies although predicted absolute magnitudes were only approximately correct.

This type of validation is clearly a much more liberal procedure; as biologists we might well desire more accuracy than this technique demands. There are, however, other constraints on the allowable complexity of management models. The regional scale of agriculture and geographic variation in environmental factors requires that many hundreds or thousands of computer runs be made if models are to be applied effectively. The high costs of these runs limits the degree of detail which can be built into practical mite models. Furthermore, to be useful in the agricultural extension environment, models must provide very flexible outputs. Achieving this flexibility can further complicate computer programs independently of the biology involved. At least for the foreseeable future, therefore, a degree of tolerance for management model errors will be required.

USING MODELS IN THE FIELD

Field use of models also requires striking a tradeoff between the complexity factors discussed above and the cost of communicating data and results between model and field. The more complex a model and the finer the geographic subdivisions to which it is applied, the greater the need for a centralized computer to handle the calculation load. While more realistic model outputs may be generated this way, large volumes of raw input data and output recommendations must be rapidly conveyed between growers and the computer.

A system which operates in this fashion is in place in Michigan (Brunner *et al.*, in press). This system consists of remote data terminals spread throughout agricultural areas, connected via long distance telephone lines to Michigan State University. Weather data from some 58 points are entered into the system daily by the National Oceanographic and Atmospheric Administration (NOAA). Scouting data, field reports and other data are stored on-line through the terminals. Data are geographically indexed by township, range and section coordinates. These data provide the inputs for summary programs and predictive models of various types. Outputs from these programs are directed automatically to the appropriate field staff. Data may cycle completely through the system in a few hours. Furthermore, once data are captured by the system, they are retained and can be used for other purposes (*e.g.*, research or teaching) at later times.

Communication costs for this system, however, actually exceed computation costs (Edens and Klonsky, 1977). Because of this, there are some tasks, such as local record keeping, *etc.*, which are not economical to perform on this system. Also, if the central computer becomes unavailable for any reason, all users are incapacitated.

An alternative is to use simpler models or summarize complex models in simple ways. An example is the model of Mowery *et al.* (1975). These workers executed a complex simulation model several hundred times. The results were summarized in the form of a chart which could be distributed to growers. Using this chart, growers could ascertain whether populations of the coccinellid predator *Stethorus punctum* were increasing with sufficient rapidity to control phytophagous mite populations.

Because, in this system, the "model" (*i.e.*, chart) is located in the field, communication costs are eliminated. By working in parallel, a large number of growers can generate as many recommendations per unit time as a large computer working sequentially. This parallelism also results in effective cost sharing. Furthermore, no disastrous centralized system failures are possible.

This system also has its disadvantages. For example, in order for a simplified chart to work effectively in the vast majority of cases, it must be very conservative; perhaps overly so. Also, placing a reasonable value on growers' personal time, the parallel system is probably more expensive on a regional basis than the centralized system. Finally, data, once taken and used, tend to be lost to the system.

Systems intermediate between these extremes are, however, possible and are being implemented in several parts of the United States (Croft *et al.*, 1978). These systems consist of networks of small data processors distributed at regional, county, and ultimately, firm levels. Each processor (1) assists in the collection of local data, (2) incorporates simplified models and decision aids, and (3) shares data and recommendations with other processors.

Local processors are tied in an hierarchical fashion to large central processors which carry out specialized tasks requiring great computational power. As applied to mite management such systems could function as follows. Local processors would receive inputs on local weather conditions and mite populations. Simplified models contained in these processors would yield immediate recommendations. These data would also be forwarded to the central computers where detailed models could periodically calculate adjustments so as to recalibrate the local models to the actual conditions of a particular region and season. In this way the advantages of cheap local communications, retention of input data, and more adaptive models could be combined in a single, reliable system. Systems of this type will probably constitute the arena in which the work of diverse mite modeling groups will be fused into the unitary practical management programs of the future.

SUMMARY

Computer simulation has played an important part in the development of pest management programs for phytophagous mite species in recent years. Although particular models may vary greatly, between them most aspects of mite population dynamics have been examined. Some of these include life cycle structure, development, prey-predator interactions, reproduction, mite-plant interactions, dispersal, and spatial distribution.

Techniques of model validation depend somewhat on the context in which the models is to be used; criteria applied to management models are more liberal than those applied to research models.

Finally, model characteristics (particularly complexity) have great influence on the type of implementation system necessary to use them in the field. Such systems may range from the highly automated and centralized to simple decision aids used by individual growers. In the future hierarchical systems will combine the advantages of several types of implementation structures to provide highly adaptive methods of model-based mite management.

REFERENCES

Allen, J. C. (1976). *Environ. Entomol.* **5**, 1083-1088.

Asquith, D. and Colburn, R. (1971). *Bull. Entomol. Soc. Am.* **17**, 89-91.

Berryman, A. A. and Pienaar, L. V. (1974). *Environ. Entomol.* **3**, 199-207.

Blyth, E. and Croft, B. A. (1975). *Proc. North Central Branch, Entomol. Soc. Am.* **30**, 89.

Brunner, J., Welch, S. M., Croft, B. A. and Dover, M. J. (In press). *Proc. EPPO Conference on Forecasting and Crop Protection, Paris 1977.*

Croft, B. A. (1975). C. R. 5e Symp. Lutte intégrée en vergers. OILB/SROP, pp. 109-124.

Croft, B. A., Welch, S. M. and Dover, M. J. (1976). *Environ. Entomol.* **5**, 227-34.

Croft, B. A., Welch, S. M., Miller, D. J. and Marino, M. L. (In press). *In* "Pest Management Programs for Deciduous Fruits and Nuts" (D. J. Boethel and R. D. Eikenbury, eds.). Academic Press, New York.

Dover, M. J., Croft, B. A., Welch, S. M. and Tummala, R. L. (In press). *Environ. Entomol.*

Edens, T. C. and Klonsky, K. (1977). Michigan State University Agricultural Experiment Station Report (unnumbered).

Johnson, D. T. and Croft, B. A. (1975). *Proc. North Central Branch, Entomol. Soc. Am.* **30**, 52-54.

Logan, J. A. (1977). Ph.D. dissertation. Washington State University.

Logan, J. A., Wollkind, D. J., Hoyt, S. C. and Tanigohsi, L. K. (1976). *Environ. Entomol.* **5**, 1133-1140.

Mowery, P. D., Asquith, D. and Bode, W. M. (1975). *J. Econ. Entomol.* **68**, 250-254.

Rabbinge, R. (1976). "Biological Control of Fruit Tree Red Spider Mite." Centre for Agricultural Publishing and Documentation, Wageningen, the Netherlands.

Royama, T. (1971). *Researches on Population Ecology,* Suppl. No. 1.

Ruesink, W. G. (1976). *Ann. Rev. Entomol.* **21,** 27-44.

Tanigoshi, L. K., Hoyt, S. C., Browne, R. W. and Logan, J. A. (1975). *Ann. Entomol. Soc. Am.* **6,** 972-986.

Welch, S. M. (1977). Ph.D. dissertation. Michigan State University.

Welch, S. M., Logan, J. A., Mowery, P. D., Jones, A. L. and Kroh, G. (In prep.). *in* "Integrated Pest Management of Pome and Stone Fruits" (B. A. Croft, S. C. Hoyt, Dean Asquith, and E. H. Glass, eds.). Wiley Interscience, New York.

Welch, S. M., Croft, B. A., Brunner, J. and Michels, M. F. (In press). *Environ. Entomol.*

ASPECTS OF THE FUNCTIONAL, OVIPOSITIONAL AND STARVATION RESPONSE OF *AMBLYSEIUS FALLACIS* TO PREY DENSITY

B. A. Croft and E. J. Blyth

Pesticide Research Center and Department of Entomology
Michigan State University
East Lansing, Michigan

INTRODUCTION

Monophagous or relatively obligate arthropod predators are highly adapted to sense the density states of their prey; searching and responding to varying prey levels are dominant activities in their life cycles. Scientists often break predation into discrete components for detailed study, e.g. search, prey capture, consumption, oviposition, dispersal. Predatory control or regulation, however, is a complex unified process which has its greatest relevance and application to biological control when viewed as an integrated whole.

In this regard Dover *et al.* (1979) has reported a population model for biological control of the spider mite, *Panonychus ulmi* (Koch), by the predaceous phytoseiid mite *Amblyseius fallacis* on apple. This study was based on a component analysis of predation between these two species and a resynthesis into an integrated model of the biological control process. In the course of the study, research on four major components was completed including the aspects of the functional, oviposition, starvation and dispersal response of *A. fallacis* to varying prey densities. In this paper, results relative to each of these areas are reported.

METHODS

Due to the difficulty of rearing *P. ulmi* in the laboratory, *Tetranychus urticae* Koch, a mite of comparable size and habits, was used as a surrogate prey since it can easily be reared in the greenhouse. Also, the relative value of this food source to *A. fallacis* in comparison to all active life stages of *P. ulmi* is known (Dover *et al.*, 1979). All *A. fallacis* populations were reared con-

Vol. I: ISBN 0-12-592201-9

tinuously with methods reported by McMurtry and Scriven (1965) and Scriven and McMurtry (1971).

For study of the functional and numerical response of *A. fallacis* to egg (preferred prey stage) populations of *T. urticae,* bean leaves having *ca* the spatial dimensions (surface area = 12.2 cm^2) of an apple leaf, were used as a test arena. Leaf units were placed on a water-saturated polyurethane base and the leaf was ringed with water-saturated absorbent cellucotton to form a barrier which eliminated mite migration from the leaf. Each day the appropriate density of prey eggs was added to each leaf and the corresponding consumption and oviposition recorded. At each temperature and prey level, a total of 5 replicate tests with individual adult ♀ predators of randomly selected ages were made. Data for each predator were collected for 5 consecutive days (= 25 total obs) after 2 days of preconditioning of mites to the leaf and after an equilibrium consumption and oviposition responses were attained.

For starvation tests, a common pretreatment regime was used for standardization. Mites of varying ages within a stage, but previously having unlimited exposure to food, were removed from a rearing unit with a camel's hair brush and placed single in a 1 mm dia capillary tube, 2.5 cm long or on a standard sized bean leaf (see above). The capillary tube was plugged at both ends with an absorbent cellucotton wick. In the capillary tube or on the leaf, a water source was maintained by the absorbent cellucotton wick. With predator larvae, tests were continued beyond the larval stage after molting occurred. Unfed protonymphs or deutonymphs did not molt to the next stage if no food was available. Each temperature-stage-unit-water comparison was replicated 20 times. Visual observations of mite mortality were made daily. Expiration was counted when no appendage movements were discernible. Mortality values were plotted on log-probit paper and a computer was used to obtain Lethal Time (LT) values, 95% confidence intervals and slope estimates for each treatment comparison.

In all tests, temperatures were controlled within ± 2°F and relative humidity varied from 75-95%. When experiments were conducted on leaves or in capillary tubes with water, relative humidity at the micro level approached 100%.

RESULTS

Functional and Oviposition Responses

The functional response of predators to prey density has been discussed by Holling (1965) and others. Ivlev (1955) suggested the following equation to characterize the response:

$$(1) \quad Y = Y_m(1-\exp(-B_1X)$$

where Y is the feeding rate per predator, X is initial prey density, Y_m (a constant) is the asymptoctic maximum feeding rate, and B_1 is a measure of predatory efficiency; a predator with a high B_1 reaches maximum consumption rate at lower prey densities than a predator with a low B_1. Royama (1971) proposed that Y be viewed as an instantaneous rate. Taking predator density into account and integrated from t = 0 to t = 1 day yields:

$$(2)\quad Z = -B_1{}^{-1}\ln\left[\exp(-Y_m B_1 P) - \exp\left(-B_1 (X + Y_m P\right] + (-B_1 X)\right]$$

where Z is the total amount eaten, P is predatory density, and X, Y_m and B_1 are defined as before.

Equation (2) was fitted to consumption rates of *A. fallacis* in response to varying density of eggs of *T. urticae* at 3 temperature levels (Fig. 1-A).

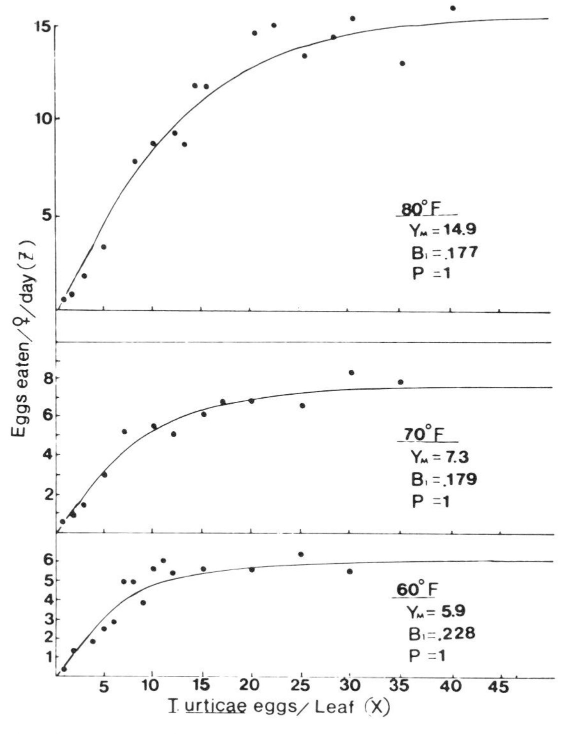

Fig. 1. A. Functional response of *A. fallacis* adult females to *T. urticae* egg densities [data fitted to consumption equation (2) in text].

Satiation of predators and plateauing of the functional response curves occurred at prey densities of *ca* 30, 15 and 10 mites/leaf at 80, 70 and 60°F, respectively. In comparison to data published by Santos (1975) on the functional response of *A. fallacis* to eggs of *T. urticae* at 77°F but on slightly larger areas, our data were remarkably similar, both in relation to the plateau level of prey density reached and the prey density at which a maximum consumption was attained.

Eggs oviposited/day/♀ when related to *T. urticae* eggs eaten/day/♀ are given in Fig. 1-B at 60, 70 and 80°F. Quadratic equations fitted to these data points also are given. As can be seen, these 2 variables were almost related in a linear fashion and maximum oviposition rates of *ca* 1, 1.5 and 3 eggs/day/♀ were obtained after predators consumed a daily average of 6, 8 and 15 prey eggs, respectively at 60, 70 and 80°F. In relation to the research of Santos (1975), oviposition rates were somewhat higher at maximum prey consumption levels by a factor of 0.5-0.75 eggs/day/♀ in our experiments.

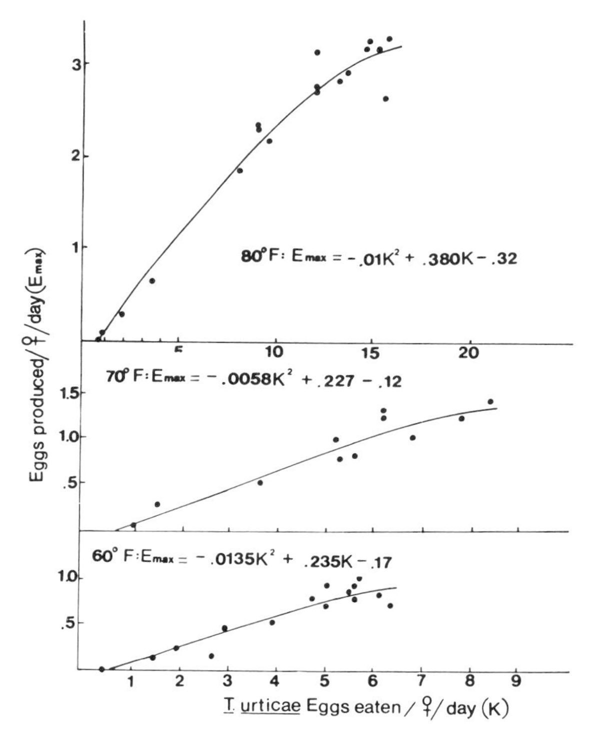

Fig. 1. B. Oviposition rate of *A. fallacis* as a function of consumption (A). Data fitted to equation

$$E_{max} = AK^2 + B_2K + C$$

where K is consumption in egg equivalents per female and A_1B_2 and C are constants. Note curve terminates at $K = K_{max}$ (see Ym in equation (2) in text).

$A = -.0000599T^2 + 00855T - .311$
$B_2 = .000118T^3 - 0238T^2 + 1.06T - 35.4$
$C = -.000109T^3 - 0126T^2 - 1.42T + 30.9$
$T = \text{temp}$ (°F)

Starvation Response

In Table 1, data for 12 treatment comparisons are given as LT_{50} and LT_{95} values, the 95% confidence interval of the LT_{50} estimate and a slope estimate in units of probit mortality per 10-fold increase in days starvation. For adult ♀ predators which are the most common stage present during a post-peak, predator/prey interaction and are the major stage involved in searching at low prey densities, LT_{50} values showed an expected inverse correlation with increasing temperatures when mites were held in capillary tubes both with or without water. Values ranged from 14.5 @ 15.5°C to 6.5 days @ 26.6°C when no water was provided and from 17.2 @ 15.5°C and 12.4 @ 26.6°C when free water was available. Water stress appeared to be most severe among mites in those tests in capillary tubes at the highest temperature tested (26.6°C).

TABLE I
Lethal Time Values (LT) for *A. fallacis* Life Stages when Starved and Exposed to Different Conditions of Temperature and Water Availability.

Treatment[a]	LT_{50}	95% C. I. LT_{50}	LT_{95}	Slope
♀A - 15.5 - C - N	14.5	13.6 - 15.7	25.8	6.6
♀A - 15.5 - C.- Y	17.2	13.2 - 21.1	30.0	6.8
♀A - 21.2 - C - N	11.8	10.9 - 13.2	22.3	6.0
♀A - 21.1 - C - Y	13.5	10.3 - 16.7	20.1	9.4
♀A - 24.0 - C - N	8.7	8.0 - 9.3	15.0	7.0
♀A - 26.6 - C - N	6.5	5.8 - 7.1	12.7	5.6
♀A - 26.6 - C - Y	12.4	11.5 - 13.5	23.8	5.8
♀A - 26.6 - L - Y	12.3	11.4 - 13.3	22.0	6.5
♀A - 26.6 - L - N	6.7	5.2 - 8.3	69.2	1.6
♀D - 21.1 - C - N	5.0	4.0 - 6.1	22.5	2.5
P - 21.1 - C - N	3.1	2.4 - 4.2	17.1	2.2
L - 21.1 - C - N	7.5	6.2 - 8.7	40.9	2.2

[a] *Treatment designation: Life stage* (♀A = adult female, ♀D = female deutonymph, P = protonymph, L = larvae); *Temp* in °C, *Method* (C = capillary tube, L = on bean leaf); *Water availability* (N = no, Y = yes).

Differences were observed when adult ♀'s were tested on capillary tubes vs. bean leaves at 26.6°C. With water, comparable values of 12.4 to 12.3 days for the LT_{50} value and slopes of 5.8 vs. 6.5 were observed. However, without water the slope of the curves, 5.6 vs. 1.6 units, was changed appreciably even though the LT_{50} were not dramatically different in the capillary tube vs. the leaf treatment, respectively (6.5 days vs. 6.7 days). Observations indicated that some predators confined to leaves without water (test A-26.6-L-N) took up fluids from the leaf. This was also confirmed by using dyes which were translocated into leaf veins and which ultimately were transferred into the bodies of the translucent predators after they had pierced the leaf tissue. Apparently, this behavior was not exhibited by all mites or at least not initially, otherwise the LT_{50} value would have been considerably higher for this test. Although mites appeared to gain some nutritive value in treatment A-26.6-L-N as compared to A-26.6-C-N when water was present, predators would not feed on the leaf for nutrients alone (compare A-26.6-L-Y with A-26.6-C-Y). The low slope value of 1.6 units and the LT_{95} value of 69.2 days in test A-26.6-L-N indicated that mites derived significant nutritional benefit from water uptake. However, the actual data points on the log-probit line only extended to 16 days when only 4 of 20 individuals were still alive. More detailed research over a longer time period would be necessary to confirm the conclusion that these mites were actually obtaining significant nutrients from their probing activities.

Comparisons between all life stages at 21.1°C, at the LT_{50} level, and without water revealed that adult female predators survived the longest, larvae were second, deutonymphs next and protonymphs were the most prone to die. The relationships of greater survival of larvae to protonymphs is to be expected when one considers their functional biology of these 2 stages. The larvae of *A. fallacis* seldom feed and even when abundant prey are present, predators molt to the protonymphal stage before taking a meal. This stage obviously is adapted to withstand conditions of food and water stress. A better way of looking at it would be to consider the larvae another "inert" stage like the egg and consider the larval-protonymphal period as one stage; as the mite approaches proto-deutonymphal molt its propensity to die from starvation stress is increased and it will not continue in development unless food is obtained. A comparison of starvation with available water between the smaller life stages was not possible in these studies because there mites easily are entrapped in a free water source and control mortality was too high in preliminary tests for comparative evaluation.

One additional comparison of interest was among the slope values for all the female adult tests (except the leaf without water comparisons which were uniformly high) versus those for the immature stages. In all cases there appeared to be greater variability as indicated by the lower slope values among the immature stages in comparison to the adult females even though the latter lived longer in all treatments. These data may reflect a compensatory mechanism to insure the survival of a few, short-lived immature mites under conditions of extreme prey scarcity and of free water availability.

SUMMARY

Features of the prey capture and consumption (functional response), oviposition (part of the numerical response) and starvation response of *A. fallacis* relative to the prey *T. urticae* demonstrate this predator to be highly sensitive to changes in the numerical density of its prey. In this respect this predator appears to be highly *r*-adapted and capable of rapidly responding to outbreaks of spider mites in the field. This conclusion has also been demonstrated in modeling (Dover *et al.*, 1979) and field sampling studies carried out over a 5-year period in Michigan (Croft and McGroarty, 1977).

Some adult female *A. fallacis* take up water from bean leaves when deprived of free water; in doing so they appear to gain some nutritional value from the leaf. There is no evidence that predators feed on the leaf under conditions of adequate water and limited food. These results are similar to those reported by Porres *et al.* (1975) for the species *Amblyseius hibisci* (Chant) when reared without food or water on avocado.

REFERENCES

Croft, B. A. and McGroarty, D. L. (1977). *Michigan State Univ. Res. Rep.* 333, 26 pp.
Dover, M. J., Croft, B. A., Welch, S. M. and Tummala, R. L. (1979). *Environ. Entomol.* (In press).
Holling, C. S. (1965). *Mem. Ent. Soc. Can. No. 45,* 60 pp.
Ivlev, V. S. (1961). "Experimental Ecology of the Feeding of Fishes." (D. Scott, Translator), Yale University Press, New Haven, CT.
McMurtry, J. A. and Scriven, G. T. (1965). *J. Econ. Entomol.* **58,** 282-284.
Porres, M.A., McMurtry, J. A. and March, R. B. (1975). *Ann. Entomol. Soc. Amer.* **68,** 871-872.
Royama, R. (1971). *Res. Pop. Ecol. Suppl.,* 91 pp.
Santos, M. A. (1975). *Environ. Entomol.* **4,** 989-992.
Scriven, G. T. and McMurtry, J. A. (1971). *J. Econ. Entomol.* **64,** 1255-1257.

MANAGEMENT OF MITE PESTS OF COTTON IN EGYPT

E.A. Elbadry

Faculty of Agriculture, Ain Shams University
Shoubra Elkheima, Cairo, Egypt

INTRODUCTION

Cotton dominates the economy of Egypt, where it occupies vast acreages of the valley and delta of the river Nile. In spite of attempts at industrialization and crop diversification, the cotton crop is still the main element of the national economy of Egypt. The area planted with cotton ranged during the last 10 years from 859,320 acres in 1966 to 1,345,990 acres in 1975, representing about one-third to one-fourth of the total area devoted to agriculture (Attiah, 1977).

Increasing population growth in Egypt is a serious problem. This, and the need to produce enough food, is a challenging problem facing the country today. Pest management in Egyptian agriculture is therefore an essential component for increased productivity of the land.

Pest management is any form of pest population manipulation invoked by man in the interest of protecting his crops. A mite pest management program on cotton in Egypt was developed during the last decade. A realistic appraisal of pest control on cotton suggests that the practice of almost total reliance on pesticides is very precarious. Major components of the current pest management program include cultural methods, biological control and use of selective pesticides. The integrated mite control system also makes use of predaceous stigmaeid and phytoseiid mites resistant to pesticides which are used against the key pests of cotton.

MAJOR PESTS OF COTTON IN EGYPT

Cotton has a large pest complex including the cotton leafworm *Spodoptera littoralis* (Boisd.), the pink bollworm *Pectinophora gossypiella* (Saund.) and the spiny bollworm *Earias insulana* Boisd. as the key pests. Pests infesting

Vol. I: ISBN 0-12-592201-9

cotton seedlings include cutworms, mole crickets, thrips, aphids and spider mites. White flies, leaf hoppers and the American bollworm are of incidental importance.

In Egypt, cotton is subject to attack by spider mites throughout April to September. Figs. 1 and 2 show the population trends of the two tetranychids *Tetranychus cucurbitacearum* (Sayed) and *Tetranychus arabicus* Attiah in two localities at lower and upper Egypt. The population starts to increase gradually during April and May. In June, it decreases slightly and then a peak follows throughout July and August. During September tetranychids become less active and the population declines.

Outbreaks of spider mites often followed the use of certain pesticides against other cotton pests (Fig. 3) (Elbadry and Khalil, 1972). In many countries the same materials tend to provoke upsets of spider mites following their application. Some authors attributed this upset to destruction of natural enemies. The different views regarding the causes for upsets of mites by pesticides, i.e., pest stimulation and natural enemy destruction, have been reviewed by Croft and Brown (1975). The consequences of insecticide use on non-target organisms were discussed by Newsom (1967).

PESTICIDES USED ON COTTON

Cotton fields are the main areas where pesticides are used in Egypt. A total of 45,516 tons of pesticides was used against cotton pests during 1970-1975 (Table I). It is of interest to note that the amount of pesticides used against cotton bollworms reached about 86% of the total pesticide tonnage used against different cotton pests.

Up to the late 1960's, it was recommended that farmers spray or dust regularly every 2 or 3 weeks from the time of squaring until all green bolls were hardened and were no longer susceptible to attack by fruit-feeding bollworms. Recommendations also included early season treatments to protect the presquaring cotton. In most cases, farmers followed automatic treatments from seedling emergence to harvest, irrespective of pest numbers or crop damage.

TABLE I.
Pesticides Used Against Cotton Pests in Egypt (Tons).

	Years						
	1970	1971	1972	1973	1974	1975	Total
Seedling pests	229	160	221	204	282	467	1563
Cotton leafworm	224	726	679	633	1671	676	4609
Bollworms	5450	5974	5406	7279	7948	7287	39344
Total	5903	6860	6306	8116	9901	8430	45516

In the early 1970's, economic thresholds for key pests were established, however, they were very conservative and encouraged the early use of insecticides to prevent losses by the cotton leafworm. Little thought was given to the unleashing of potential or secondary pests with these treatments.

Chemicals are now used in cotton fields when infestations of any pest warrant spraying. For example, to overcome the potential loss in cotton productivity caused by the bollworms, a regular program of spraying 3 times, 15 days apart, begins when 10% of the cotton bolls are infested. Methods of applying these insecticides are by ground equipment and aircraft.

Chlorinated compounds represented by DDT and BHC were used against cotton pests in the early 1950's. Toxaphene was used afterwards until it failed to control the leafworms. In 1961 it was replaced by carbaryl (Sevin®). Later on, organophosphates were applied, such as trichlorfon, methyl parathion, leptophos and EPN, as well as endrin, a chlorinated hydrocarbon. At present the pesticides curacron, chlorpyrifos, azinphos-methyl, phospholan, mephospholan, and several new pyrethroid compounds are recommended.

PESTICIDE MANAGEMENT

Safe pesticide management is an essential component and possibly a prerequisite of integrated pest management. Davies *et al.* (1976) and Davies (1977) reviewed the problems of pesticide management.

In Egypt, the number of insecticide applications per acre has been reduced from those used prior to pest management programs; however, increased applications have occurred in a few instances. The primary objective in cotton pest management is not decreased or increased insecticide use, but rather efficient use. At present, insecticides are the only quick means of reducing cotton key pest populations which have reached the economic level.

The improvement of pesticide management requires developments in research and training. Research needs include establishment of economic injury levels, evaluation and utilization of natural enemies, selective pesticides and selective use of conventional ones, and incorporation of cultural controls.

BIOLOGICAL CONTROL

Biological control is still grossly under-emphasized as a significant factor in integrated control. In the past decade, there has been an increasing appreciation of the importance of natural enemies in suppressing pest mites.

The most important natural enemies of tetranychids on cotton are the predatory mites *Amblyseius gossipi* Elbadry (Elbadry *et al.*, 1968) and *Agistemus exsertus* Gonzales (Elbadry *et al.*, 1969). Although few predators are recorded during June, these mites begin to appear in great numbers

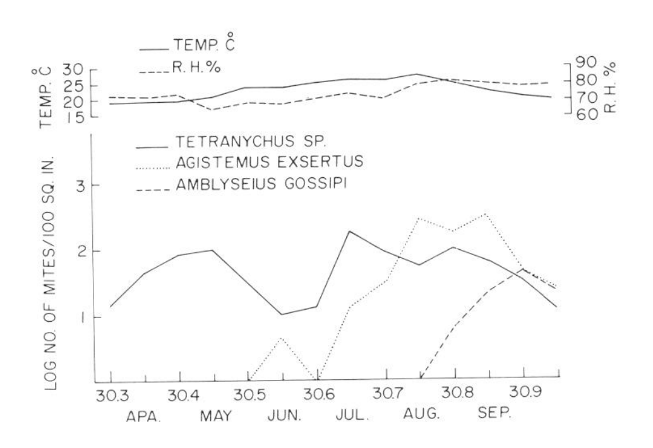

Fig. 1. Population changes of cotton mites at Kafr Elsheikh (lower Egypt) during 1970.

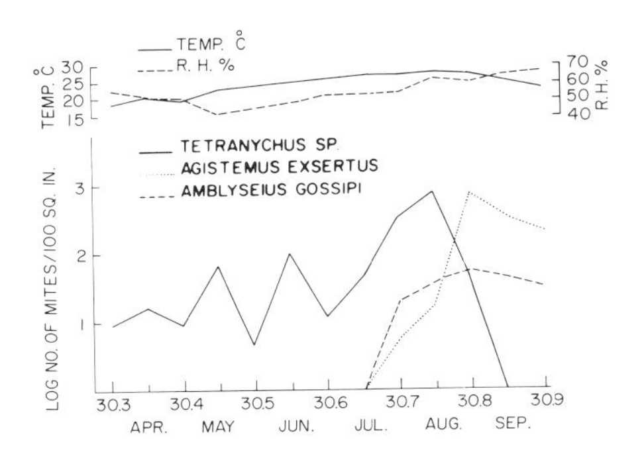

Fig. 2. Population changes of cotton mites at Beni Seuif (upper Egypt) during 1970.

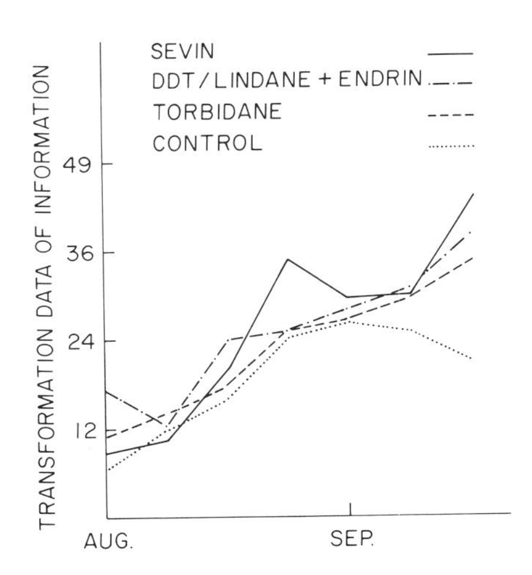

Fig. 3. Changes of phytophagous mite populations after using insecticides against the bollworms in 1970.

throughout July. The maximum population is reached about the end of August and during September. By that time, the predatory fauna are able to overcome the spider mite infestation, which declines rapidly during September.

There are a few examples of recently attempted efforts to augment and conserve indigenous natural enemies of tetranychid mites on cotton. Osman and Zohdi (1976) showed that tetranychids can be effectively suppressed on cotton in Shebin Elkom (lower Egypt) by releases of *A. gossipi* starting on May 20, before the spider mite population reaches the first peak, at the rate of 15 predators/cotton plant.

SELECTIVE PESTICIDES AND SELECTIVE USE OF CONVENTIONAL ONES

Since pesticides remain our most reliable and practical tool for immediate solution to many key pests on cotton, we cannot afford to neglect the search for suitable selective materials and ways of selectivity in using conventional ones.

The high tolerance of *A. exsertus* to many pesticides has been reported by several workers. Abo Elghar *et al.* (1971) showed experimentally that tetradifon, trichlorfon and carbaryl permitted complete survival of adult *A. exsertus* when tested at concentrations between 0.06 and 4.5%. Hassan *et al.* (1970) demonstrated also that *A. exsertus* tolerated the effect of dicofol and amidithion when applied as a direct spray with concentrations of formulations of 0.0092 and 0.09%, respectively. Several examples of physiological pesticide selectivity for use in pest management programs were reported by Croft *et al.* (1976), Kennedy *et al.* (1976) and Kennedy and Oatman (1976).

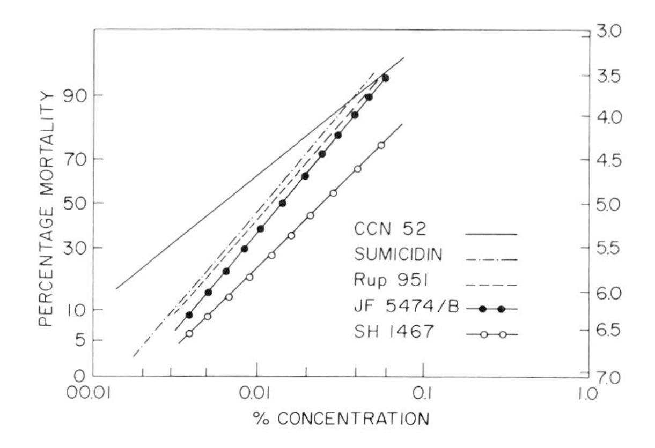

Fig. 4. Dosage mortality lines for the toxicity of CCN 52, Sumicidin, Rup 951, J. F. 5474/B and SH 1467 to adult female *Amblyseius gossipi*.

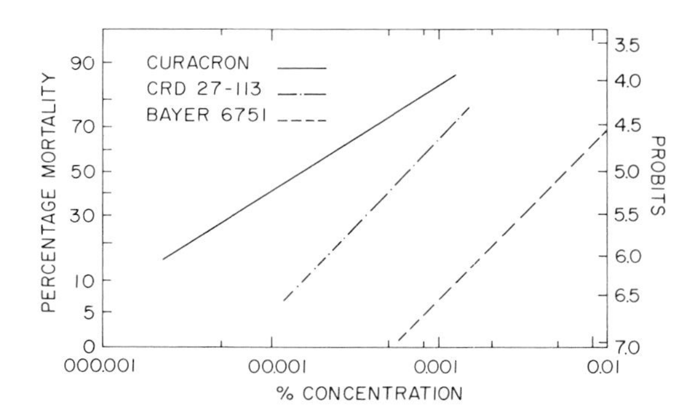

Fig. 5. Dosage mortality lines for the toxicity of Curacron, CRD 27-113 and Bayer 6751 to adult female *Amblyseius gossipi.*

Pesticides having the broadest spectrum of activity may be used in an ecologically selective manner. As far as mite pest management on cotton is concerned, it appears that the use of ecologically selective pesticides against *A. gossipi* will continue to be far more important than the use of physiologically selective insecticides. Selective action of nonselective chemicals can be obtained by manipulating doses and timing of applications.

A. gossipi was found to be very susceptible to most pesticides applied on cotton. The effect of 5 pyrethroid toxicants on *A. gossipi* showed that CCN 52 and sumicidin were the most toxic materials (LC_{50}'s 0.0061 and 0.011%), followed by Rup 951, J. F. 5474/B and SH 1467 (LC_{50}'s 0.012, 0.014 and 0.024%, resp.) (Fig. 4).

Similarly, three organophosphorus compounds were found to be very toxic to the predator *A. gossipi*. Curacron was the most toxic (LC_{50} 0.00015%), followed by CRD 27-113 and Bayer 6751 (LC_{50}'s 0.00066 and 0.0057%, resp.) (Fig. 5). Galecron is also toxic to adult *A. gossipi* (LC_{50} 0.0025%), whereas dicofol proved to be the least toxic among the tested materials (LC_{50} 0.098%).

The selective use of these toxicants is obtained by using the lowest dosages essential to produce satisfactory control of the target pests. Also, one must adjust the timing of these insecticidal treatments in cotton to minimize damage to phytoseiid populations.

CULTURAL METHODS

The cultural practices, including modifications of the mosaic of crop or variety plantings, affect the pest species, their natural enemies, or alternate prey or hosts. Adding habitat diversity may be helpful, e.g., small corn patches planted next to cotton fields have encouraged the build-up of phytoseiid and stigmaeid predatory mite populations and the achievement of better biological control of tetranychids on cotton (Elbadry and Elbenhawy, 1968).

Strip cropping of onions and cotton has been used for many years in Egypt to decrease mite infestation on cotton, since onions are not infested with cotton tetranychids. When the system is properly managed, the need for pesticide treatment of cotton to control mites early in the season can be virtually eliminated since the mites are kept at levels where they cause no damage. The main principles involved in cultural control of arthropod pests have been reviewed by Stern *et al.* (1977) and Adkisson (1977).

Field observations have shown that several cultural practices will enhance the susceptibility of cotton to spider mite attack. These include regular biweekly irrigation which favors mite infestation. Tillage increases the incidence of mite attack as compared to no-tillage. Finally, more dense cotton populations affect the micro-climate and thus alter the development of mite attacks.

IMPLEMENTATION PROGRAMS

For more than 10 years, a great variety of efforts have been made in Egypt to take integrated pest control research to the practical level in farmers' fields. This has involved a number of scouting and supervised control programs. Later, integrated control programs involving predators and selective use of chemicals were derived and implemented. Recent research has emphasized reduction of insecticide input through better timing of applications. By scouting each field and treating as needed, insecticide use against key pests is minimized. In several areas of Egypt about 8 applications were commonly applied. The average now is about 4 insecticide applications annually, with scouting.

Several pest management programs have been adopted for cotton in developed countries (Lincoln *et al.*, 1975; Frisbie *et al.*, 1976; Woodham *et al.*, 1977). Emphasis has been placed on systems and modeling approaches as tools in interrelating the complex interdisciplinary components of the desirable pest management programs (Rudd, 1975; Baker, 1976; Croft *et al.*, 1976; Valentine *et al.*, 1976). In nearly all cases, pest management programs have come about as the result of a gradual evolution in which new technology has been introduced as a step-by-step process rather than through the introduction of a complete fully-formed system.

SUMMARY

Cotton dominates the economy of Egypt; it occupies *ca* 25-30% of the total land devoted to agriculture. Cotton is usually subject to attack by several key pests. Outbreaks of these pests regularly require chemical treatment. The use of wide spectrum pesticides, however, leads to outbreaks of the tetranychids *T. arabicus* Attiah and *T. cucurbitacearum* (Sayed), when predators which keep those mites in check were destroyed. The most effective and prevalent natural enemies of phytophagous mites are *A. exsertus* Gonzales and *A. gossipi* Elbadry.

A pest management program on cotton in Egypt was developed which includes integrated use of predaceous stigmaeid and phytoseiid mites and suitable selective pesticides and selective use of conventional ones. *A. exsertus* is highly tolerant to several pesticides, whereas *A. gossipi* is very susceptible to most pesticides recommended for cotton. These nonselective chemicals are used by manipulating doses and timing of application.

Cultural practices such as intercropping of onions among cotton has noticeably decreased spider mite infestation during April to June. Small corn patches planted next to cotton fields have shown the potential of encouraging the build up of predatory mites and of achieving better biological control of tetranychids.

REFERENCES

Abo Elghar, M. R., Elbadry, E. A., Hassan, S. M. and Kilany, S.M. (1971). *J. Econ. Entomol.* **64**, 26-27.

Adkisson, P. L. (1977). *Proceedings of the UC/AID—University of Alexandria, A. R. E. 1977,* 30-44.

Attiah, H. H. (177). *Proceedings of the UC/AID—University of Alexandria, A. R. E. 1977,* 61-74.

Baker, E. A. (1976). *Bull. Oepp.* **6**, 47-55.

Croft, B. A. and Brown, A. W. A. (1975). *Ann. Rev. Entomol.* **20**, 285-335.

Croft, B. A., Brown, A. W. A. and Hoying, S. A. (1976). *J. Econ. Entomol.* **69**, 64-68.

Croft, B. A., Howes, J. I. and Welch, S. M. (1976). *Environ. Entomol.* **5**, 20-34.

Davies, J. E. (1977). *Proceedings of the UC/AID—University of Alexandria, A. R. E. 1977,* 50-60.

Davies, J. E., Freed, V. H. and Smith, R. F. (1976). *UC/AID pest management and related environmental protection Project, the Philippines,* 1975, 1-8.

Elbadry, E. A. and Elbenhawy, E. M. (1968). *Ent. exp. & appl.* **11**, 273-276.

Elbadry, E. A., Afifi, A. M., Issa, G. I. and Elbenhawy, E. M. (1968). *Z. Angew. Entomol.* **62**, 189-194.

Elbadry, E. A., Abo Elghar, M. R., Hassan, S. M. and Kilany, S. M. (1969). *Ann. Entomol. Soc. Amer.* **62**, 660-661.

Elbadry, E. A. and Khalil, F. A. (1972). *Z. Angew. Entomol.* **72**, 319-323.

Frisbie, R. E., Sprott, J. M., Lacewell, R. D., Parker, R. D., Buxkemper, W. E., Bacley, W. E. and Norman, J. W. (1976). *J. Econ. Entomol.* **69**, 211-214.

Hassan, S. M., Zohdi, G. I., Elbadry, E. A. and Abo Elghar, M. R. (1970). *Bull. Soc. Entomol. Egypte. Econ. Ser.* **4**, 213-217.

Kennedy, G. G. and Oatman, E. R. (1976). *J. Econ. Entomol.* **69**, 767-772.

Kennedy, G. G., Oatman, E. R. and Voth, V. (1976). *J. Econ. Entomol.* **69**, 269-272.

Lincoln, C., Boyer, W. P. and Miner, F. D. (1975). *Environ. Entomol.* **4**, 1-7.

Newsom, L. D. (1967). *Ann. Rev. Entomol.* **12**, 257-286.

Osman, A. A. and Zohdi, G. (1976). *Z. Angew. Entomol.* **81**, 245-248.

Rudd, W. G. (1975). *Math. Biosci.* **26**, 283-302.

Stern, V. M., Adkisson, P. L. Beingolea, O. G. and Viktorov, G. A. (1977). In "Theory and practice of biological control" (C. B. Huffaker and P. S. Messenger, eds.), Academic Press. New York, NY.

Valentine, H. T., Carlton, T., Newton, M. and Talerico, R. L. (1976). *Environ. Entomol.* **5**, 891-900.

Woodham, D. W., Rubinson, H. F., Reeves, R. G., Bond, C. A. and Richardson, H. (1977). *Pestic. Monit. J.* **10**, 159-167.

THE ROLE OF INSECTICIDE-RESISTANT PHYTOSEIIDS IN INTEGRATED MITE CONTROL IN NEW ZEALAND

D. R. Penman

Department of Entomology, Lincoln College
University College of Agriculture
Canterbury, New Zealand

C. H. Wearing, Elsie Collyer, and W. P. Thomas

Entomology Division
Department of Scientific and Industrial Research
Auckland; Nelson; Christchurch; New Zealand

INTRODUCTION

Control of phytophagous mites in New Zealand is a continuing problem and until recently has relied on the use of chemicals. The development of resistance to commonly used miticides is well documented, and the increasing cost of new chemicals has led to the search for alternative controls, particularly in apple orchards. Any reduction in the use of miticides through adoption of integrated methods of control should reduce direct costs to the grower and delay the development of resistance.

The relative importance of the phytophagous mite species and their population dynamics varies considerably between the main orchard areas which experience widely differing climatic conditions. European red mite (ERM), *Panonychus ulmi* (Koch), is the main mite pest in pip and stonefruit orchards throughout New Zealand with twospotted spider mite (TSM) sometimes reaching damaging levels particularly in late season. However, TSM is the most important species in berryfruit crops. Other mites such as *Bryobia rubrioculus* Scheut. and rust mites are only of minor importance.

A wide range of natural enemies of phytophagous mites occur in apple orchards in New Zealand (Collyer, 1964a). Most natural enemies, including *Stethorus bifidus* Kapur and the phytoseiid mites are only abundant in unsprayed situations. In common with the situation in apple orchards in North

Vol. 1: ISBN 0-12-592201-9

America (Hoyt, 1969; Croft, 1975), the use of broad spectrum insecticides has eliminated or reduced the controlling influence of natural enemies. It is only with the development of resistance in some phytoseiids to some commonly used insecticides that control systems utilizing predator mites have become feasible.

Research in New Zealand has been concentrated on utilizing an existing phytoseiid, *Typhlodromus pyri* Scheut., and evaluating imported phytoseiids, *Amblyseius fallacis* Garman and *T. occidentalis* Nesbitt. This paper reports New Zealand experiences with these insecticide-resistant strains of predator mites.

PHYTOSEIID MITES

Typhlodromus pyri Scheut.

T. pyri is the most abundant phytoseiid mite found in Nelson apple orchards and the interaction of the predator with ERM populations has been studied there since 1962. Integrated control experiments using ryania as a selective insecticide for control of codling moth and leafrollers have shown that *T. pyri* was capable of controlling ERM in most seasons (Collyer and van Geldermalsen, 1975). However, in the 1970-71 season ERM caused severe damage probably due to the unusual hot dry early-season conditions favouring *P. ulmi* development while inhibiting the growth of *T. pyri* which is known to prefer humid conditions (Collyer, 1976).

Insecticide-resistant *T. pyri* were first detected at the Appleby Research Orchard in Nelson in 1968 in a block of Dunn's Favourite which had been regularly sprayed with azinphosmethyl for several years (Collyer, 1976). An initial resistance level of 9.7 x in this strain was insufficient to permit adequate survival of *T. pyri* under repeated applications of azinphosmethyl (Hoyt, 1972) and practical utilization of *T. pyri* for field control was not possible at that time (Hoyt, 1973). However, *T. pyri* numbers in the Dunn's Favourite proved sufficient to check ERM populations in the absence of acaricides in the 1973-74 season (Collyer, 1978). Later field surveys also showed that resistant *T. pyri* were widespread in Nelson commercial orchards (Wearing and Penman, 1975) and were present in Hawke's Bay (Wearing and Gunson, 1976). Subsequent toxicity tests gave LC_{50} values up to 0.3% a.i. and a resistance level considered sufficient to permit the predator to form the basis of an integrated mite control program. The apparent logarithmic increase in resistance with increasing exposure to azinphosmethyl prompted a recommendation that integrated mite control using *T. pyri* should not be attempted unless the apple block had had at least 8-10 years exposure to regular azinphosmethyl applications (Penman *et al.*, 1976).

With the evidence of *T. pyri* giving control of ERM in the Appleby Research Orchard, Nelson (Collyer, 1978), the commercial utilization of *T. pyri* in an integrated mite control program became a possibility. In the 1975-76, 1976-77, and 1977-78 seasons, 5 commercial export orchards in Nelson were monitored by Department of Scientific and Industrial Research (D. S. I. R.) personnel for levels and ratios of *T. pyri* and *P. ulmi* (Wearing *et al.*, 1978a, b). Miticides were applied only when specified spray thresholds were exceeded. Use of the selective miticide, cyhexatin, improved predator:prey ratios and savings of $75-80/ha in materials were estimated for all three seasons compared with standard commercial practice. There was no down-grading of fruit from export.

Full commercial utilization of integrated mite control was first implemented in the 1976-77 season (Wearing *et al.* 1978b). Growers using mite counting services and paying close attention to spray thresholds and recommendations by extension personnel saved an average of more than two miticide applications over that season (Table I). In the 1977-78 season a saving of only one miticide application was estimated for those orchards. This contrasted with the greater reduction of miticides in the D. S. I. R.-monitored blocks which was probably due to closer monitoring and more accurate timing of sprays. Generally, the 1977-78 season was difficult for integrated mite control in that the cool early season did not favour build up of *T. pyri* and the advent of dry conditions in January-March allowed *P. ulmi* to develop rapidly. However, initial results with integrated mite control show sufficient promise to warrant its inclusion in standard orchard management practice.

Many variables such as climate, aspect and cultivars influence integrated mite control but the results show that azinphosmethyl-resistant *T. pyri* can have a substantial effect in reducing ERM populations. Whether *T. pyri* can be regarded as an "efficient" predator is still unresolved (Chant, 1959; Collyer,

TABLE I.
Number of Miticide Applications in Commercial Apple Orchards Practising and Not Practising Integrated Mite Control, Nelson, New Zealand

Season	Integrated Mite Control		No Integrated Mite Control (i.e. without *T. pyri*)
	Using Counting Service	Monitored by D. S. I. R.	
1975-76	N. A.	1.2 (range 0-4)	3.8 (range 1-14)[a]
1976-77	1.4 (range 0-4)	1.5 (range 0-4)	4.4 (range 3-7)
1977-78	3 (range 1-5)	1.8 (range 0-4)	>5 (range 4-7)

[a]Some of these orchards contained *T. pyri*, others not. (From Penman and Ferro (1977) and Wearing *et al.* [1978a, b]).

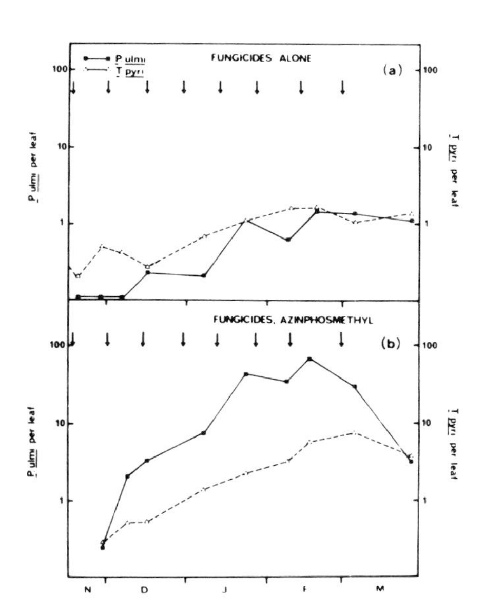

Fig. 1. Interaction of *P. ulmi* and *T. pyri* populations (a) where fungicides alone are applied at regular intervals, and (b) where azinphosmethyl is added to the spray program (cv. Sturmer Pippin) (adapted from Walker and Penman (1978)).

1964b, c). Recent studies on the efficiency of predation by *T. pyri* (Walker, 1978) suggest that the within-tree, between leaf dispersion of *T. pyri* and *P. ulmi* are not closely associated early in the season. Data from total leaf samples may indicate favourable predator:prey ratios but actual control of ERM populations on individual leaves may not be as great as expected, particularly early in the season.

Where only the fungicides, dodine and colloidal sulphur, were regularly applied to apple trees, *T. pyri* gave season-long control without the use of miticides and with minimal assistance from insect predators (Fig. 1a) (Walker and Penman, 1978). The addition of regular applications of azinphosmethyl appeared to disrupt the efficiency of predation by a resistant strain of *T. pyri* and ERM numbers built up to a maximum of 68 per leaf in mid-February (Fig. 1b). In spite of high numbers of *T. pyri* (up to almost eight per leaf in early March), the azinphosmethyl treated trees showed irregular leaf bronzing from mid-February. This suggests that ERM populations were largely free of predators in these pockets and that the use of azinphosmethyl may have disrupted the distribution of the predator.

T. pyri is also difficult to rear in large numbers, particularly in contrast to other insecticide-resistant phytoseiids (Croft, 1976b), so that large-scale field releases may be difficult. However, *T pyri* has been successfully reared in Australia on apple seedlings inoculated with ERM (J. L. Readshaw, pers. comm.). Information is also being gathered on the chemical susceptibility of

T. pyri from field experiments (Collyer, 1978; Wearing, unpubl. data) and from laboratory tests using field collected specimens (Wearing, unpubl. data). The development of mass rearing would aid chemical testing considerably.

Although not readily distributed, *T. pyri* is of considerable use in integrated mite control in Nelson, an area where ERM is the dominant phytophagous mite. *T. pyri* is well adapted to that area but it may not be effective in regions where mixed populations of ERM and TSM, or populations of TSM alone, are dominant. Likewise, *T. pyri* may not be well adapted to some of the less humid climatic areas. It was to provide predators for those situations that the two organophosphate insecticide resistant phytoseiids, *A. fallacis* and *T. occidentalis* were introduced to New Zealand.

Amblyseius fallacis Garman

A. fallacis was first successfully introduced into New Zealand in 1973 and the rearing, quarantine and release procedures have been published elsewhere (Thomas and Chapman, 1978). Populations for field release were obtained from 186 surviving active stages received from Michigan State University in 1973 and a further 116 active stages received in 1975.

The results of releases over several seasons are now available and the following five points regarding the establishment of *A. fallacis* have emerged.

(1) *Within-season release.* A single release of *A. fallacis* into the trees early in the season was effective in controlling ERM in that season (Fig. 2a), (Penman, unpubl. data). Following an early application of cyhexatin approximately 300 *A. fallacis* per tree were released into 10-year-old trees (cv. Hawke's Bay Red Delicious) at Lincoln College, Canterbury, in mid-December, 1976. The build up of predator and prey was closely associated and ERM did not exceed 12 per leaf. In contrast, trees with no predators released showed a rapid build up of ERM in late season to over 300 per leaf (Fig. 2b) (Penman, 1978). *A. fallacis* migrated into the trees too late in the season to affect ERM numbers and reduce the extensive leaf bronzing and defoliation. Similar results have been obtained with releases into considerably older trees in North Canterbury (Thomas and Chapman, 1978).

(2) *Season following release.* Control of ERM by *A. fallacis* in the year following release was inadequate. In the year following successful establishment of *A. fallacis*, ERM built up unchecked in the release trees through early summer and *A. fallacis* was not recovered until mid-January (Fig. 2c). By the time predator numbers built up, ERM exceeded 300 per leaf and extensive damage had occurred.

Maintenance of some ground cover beneath the trees to provide a source of *A. fallacis* is regarded as essential if integrated mite control using this predator is to be successful (Croft and McGroarty, 1977). A white clover, *Trifolium repens* L., dominant ground cover was maintained beneath the release trees, but periodic sampling of the clover showed few predators and TSM present

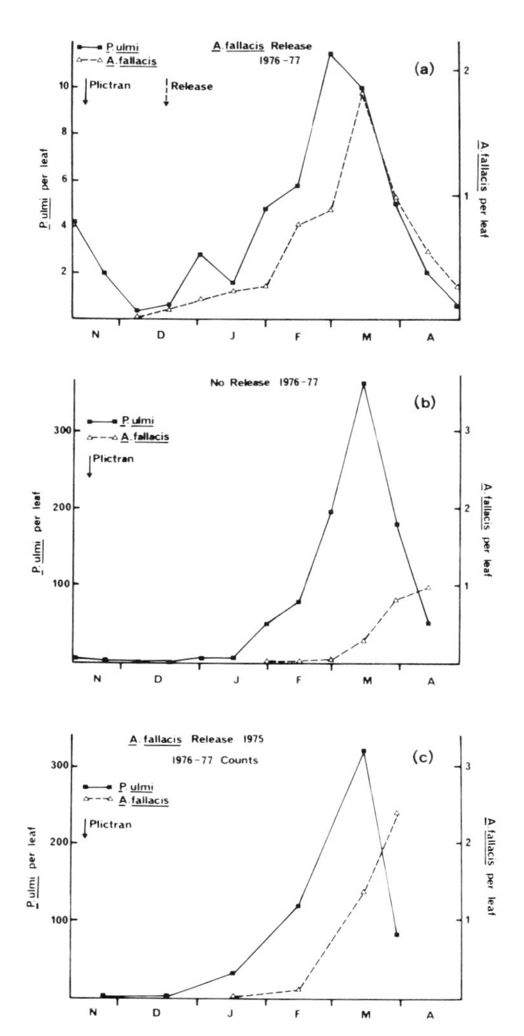

Fig. 2. Seasonal population densities of *P. ulmi* and *A. fallacis* in cv. Hawke's Bay Red Delicious apples (1976-77): (a) Release of *A. fallacis* directly into the tree. (b) No release of *A. fallacis.* (c) Recover of *A. fallacis* after release in the previous season. (From Penman, unpubl. data).

early in the season. *A. fallacis* poulations slowly built up in the ground cover but there was little apparent link between populations in the ground cover and subsequent predator populations in the trees.

(3) *Overwintering.* Overwintering mortalities appear to be crucial in effective establishment of *A. fallacis*. Winters in the release areas are relatively mild in comparison to Michigan, the source of our *A. fallacis* strain. While frosts are common, the daily maximum temperature may average 12°C for

TABLE II.
Tetranychus urticae and *Amblyseius fallacis* per Trap Band on Apple Trunks (cv. Kidd's Orange) during Winter, 1976-77.

Sample Date	*T. urticae*	*A. fallacis* Alive	*A. fallacis* Dead
13 July	147.3	356.8	0
14 September	1.3	30.3	139.3
22 September	0	0	0

(After W. P. Thomas and L. M. Chapman, unpubl. data)

July in Christchurch, Canterbury (Robertson, 1976). At such temperatures *A. fallacis* can become active and feeding and oviposition has been observed throughout the winter (Thomas, unpubl. data). At the onset of winter large groups of TSM and *A. fallacis* have been found in corrugated cardboard shelters placed around the base of the trees. Both predator and prey remained plentiful in the bands during the winter but declined rapidly with the advent of spring (Table II) (Thomas, unpubl. data). By 22 September, 1977, neither predators nor prey were detected in the bands or ground cover. Such high overwintering mortality may prevent an early season build up of *A. fallacis* populations in the ground cover. Consequently *A. fallacis* numbers in the trees may not be sufficient to control ERM in early season.

Diapause in *A. fallacis* is apparently regulated by photoperiod with temperature as a modifying influence (Rock *et al.*, 1971). The critical conditions of 11.75 to 12 hrs light per day at 15.6°C were reached in the release orchards but complete diapause was not apparent. Within the trap bands several *A. fallacis* were observed to feed and oviposit. Since our strain of *A. fallacis* was obtained from Michigan, a strain from a zone with warmer winters may be better adapted for winter diapause in New Zealand and so reduce winter mortality.

(4) *Interaction with T. pyri. A. fallacis* failed to establish on trees where *T. pyri* were already present in Nelson. This may have been due to the lack of TSM in the release orchard to provide an alternative food source but the relatively high density of *T. pyri* in the release trees may also have contributed to this result (Thomas and Chapman, 1978).

(5) *Food preference.* Where mixed populations of ERM and TSM occur, the species composition was changed following the release *A. fallacis* (Thomas and Chapman, 1978; F. A. Gunson, pers. comm.). The hot, dry late season of 1977-78 was favourable for build up of TSM. In no-release trees populations of ERM and TSM reached high levels (Fig. 3a). *A. fallacis* arrived from neighbouring trees late in the season with minimal effect. With early season releases of *A. fallacis* (Fig. 3b), ERM still far exceeded damage thresholds while TSM populations were held in check. The build up of *A. fallacis* closely followed the growth of TSM populations in the trees. A similar pattern was observed with multiple releases of *A. fallacis* (Fig. 3c). Where an early season

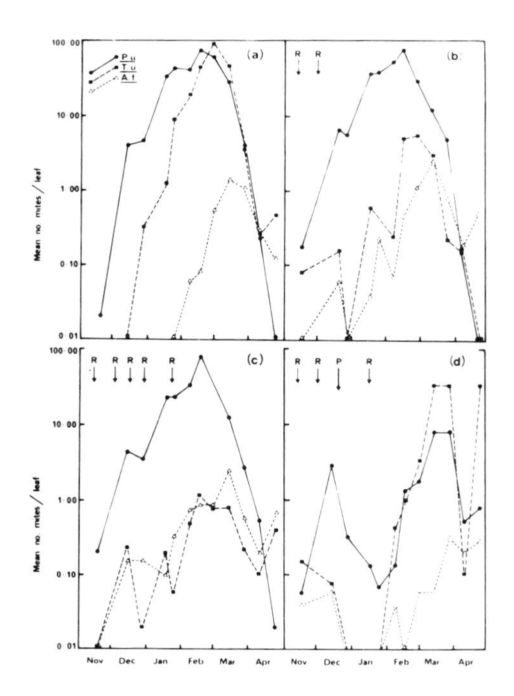

Fig. 3. *P. ulmi, T. urticae* and *A. fallacis* populations in cv. Hawke's Bay Red Delicious, Rangiora, North Canterbury, 1977-78: (a) No *A. fallacis* releases and no miticides. (b) Two *A. fallacis* releases and no miticides. (c) Five *A. fallacis* releases and no miticides. (d) Three *A. fallacis* releases and one early season miticide (P = cyhexatin). (All trees were regularly sprayed with azinphosmethyl.) (Adapted from Thomas and Chapman (1978)).

cyhexatin spray was applied in conjunction with an early predator release (Fig. 3d) the phytophagous mite populations were depleted and some predators probably died from starvation. ERM numbers did not exceed 9 per leaf for the remainder of the season but TSM built up to a maximum of 40 per leaf. An additional *A. fallacis* release in late January re-established the predator but numbers were probably too low to make a significant contribution to control of phytophagous mites.

These and other field observations suggest that *A. fallacis* prefers TSM as prey but will feed on ERM in the absence of TSM. Observations on inter-leaf distributions suggest a close association between occurrence of TSM and occurrence of *A. fallacis* (J. T. S. Walker and F. A. Gunson, unpubl. data). New Zealand orchards also lack the build up of rust mites as alternate prey which is important in maintaining high *A. fallacis* numbers in Michigan (Croft 1975).

Typhlodromus occidentalis Nesbitt

Where TSM is the main phytophagous problem, release of *T. occidentalis* first imported in 1976, may provide a solution. Releases have been made in a range of horticultural crops such as apples, strawberries, raspberries and hops.

At Lincoln College, Canterbury, *T. occidentalis* was released into a 3 yr old strawberry bed in the 1976-77 season. In the absence of predator releases (Fig. 4a), TSM reached 21.4 per leaflet (Penman, unpubl. data). Some phytoseiids were found late in the season but these were predominantly *A. cucumeris* (Oudms) with a few *A. longispinosus* (Evans). In the release blocks, *T. occidentalis* was recovered early (Fig. 4b) but an accidental application of carbaryl for leafroller control decimated the predator population. A subsequent release in January did allow *T. occidentalis* numbers to build up and hold TSM to a maximum of 9.1 per leaflet. The susceptibility of *T. occidentalis* to available aphicides limited attempts to monitor overwintering recovery. A severe aphid problem in the 1977-78 season necessitated several applications of dimethoate which were harmful to the predators.

T. occidentalis has had considerable success in controlling TSM in apples in Australia (Readshaw, 1975). In the limited situations where TSM is found as the main mite pest in apples in New Zealand, *T. occidentalis* has been established and has given effective control in the season of release. Also observations suggest that *T. occidentalis* is capable of much faster dispersal to other trees than is *A. fallacis*.

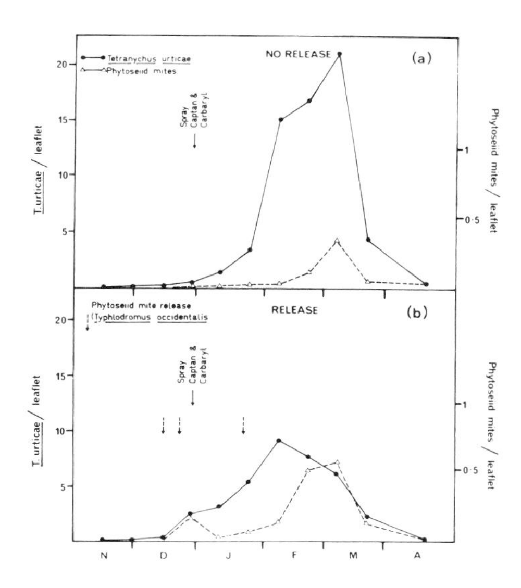

Fig. 4. Population densities of *T. urticae* and phytoseiid mites in a thre year old strawberry bed (cv. Red Gauntlet) at Lincoln College, Canterbury, 1976-77 season. (a) No *T. occidentalis* release, no miticides. (b) *T. occidentalis* releases before and after carbaryl application, no miticides. (From Penman, unpubl. data).

CONCLUSIONS

The use of insecticide-resistant *T. pyri* in integrated mite control in Nelson is now a commercial reality giving significant savings in miticide applications. Growers have readily accepted the program and in view of the considerable orchard to orchard variation in mite populations and management practices, the ultimate aim must be grower operated integrated mite control (Yuill, 1975). The performance of *T. pyri* as a predator of ERM means that sole reliance cannot be placed on this species. Regular monitoring will be essential to correctly time the remedial chemical applications which may be needed. *T. pyri* and the other phytoseiids are likely to be only an interim solution. Under a heavy spray schedule of azinphosmethyl for control of leafrollers and codling moth (a mean of 9.4 applications in Nelson in 1975-76 [Penman and Ferro, 1977]), it is likely that these pests could develop resistance to organophosphate insecticides. Most of the alternative chemicals are toxic to *T. pyri*. Research must, therefore, continue to develop selective controls for all pests in the tree-fruit ecosystem.

T. pyri is only effective in controlling ERM and is being spread to those areas where ERM is the dominant phytophagous species. Where ERM and TSM occur together, *T. pyri* and *T. occidentalis* may offer compatible and complementary predation of the two pest species but problems may arise in developing non-disruptive spray programs. *A. fallacis* does appear to offer a single predator solution to both ERM and TSM problems but the apparent high overwintering mortality and prey preference, and the slow movement of predators from the ground cover into the trees are limiting factors. Further investigation into these influences are necessary before the full potential of *A. fallacis* can be assessed. *T. occidentalis* appears to offer a solution to TSM problems on a number of crops where this species is dominant provided a spray program can be designed which is compatible with the predator-prey interaction.

Insecticide-resistant phytoseiids adapted to a range of climatic conditions and prey preferences are now present in New Zealand. The future for *T. pyri* and *T. occidentalis* looks assured provided resistance to azinphosmethyl in leafroller species does not develop in the near future. Any proposed expansion of *A. fallacis* releases will require careful evaluation of the factors presently limiting the effectiveness of this predator.

SUMMARY

Integrated control of European red mite (ERM) using an insecticide-resistant phytoseiid, *Typhlodromus pyri* Scheut. is now a commercial reality in commercial apple orchards in Nelson, New Zealand. After 8-10 years exposure to azinphosmethyl, resistance levels in *T. pyri* are considered sufficient for use of this predator in mite control. Where sufficient resistant *T. pyri* are present and where growers have paid close attention to spray thresholds, considerable savings in miticide applications have been demonstrated.

An insecticide-resistant strain of *Amblyseius fallacis* Garman has been introduced for evaluation for control of ERM and twospotted spider mite (TSM). Control of ERM in the season of release has been demonstrated but problems with overwintering mortality in the ground cover and the delay in movement of the predators into the trees in summer may limit the effectiveness of the predator. Insecticide-resistant *Typhlodromus occidentalis* Nesbitt have effectively controlled TSM populations in the season of release in strawberries and apple trees.

REFERENCES

Chant, D. A. (1959). *Can. Entomol.* **91,** (Supplement 12).

Collyer, E. (1964a). *J. Agric. Res* **7,** 551-568.

Collyer, E. (1964b). *Proc. 1st Intern. Congr. Acarology, 1963* pp. 363-371.

Collyer, E. (1964c). *Entomol. Exp. Appl.* **7,** 120-124.

Collyer, E. (1976). *N. Z. J. Zool.* **3,** 39-50.

Collyer, E. (1978). *N. Z. J. Zool.* **5,** (in press).

Collyer, E. and van Geldermalsen, M. (1975). *N. Z. J. Zool.* **2,** 101-134.

Croft, B. A. (1975). *In* "Introduction to Insect Pest Management" (R. L. Metcalf and W. H. Luckmann, eds), pp 471-507. Wiley, New York.

Croft, B. A. (1976). *Entomophaga* **21,** 383-399.

Croft, B. A. and McGroarty, D. L. (1977). *Mich. State Univ. Agric. Expt. Sta. Res. Rept.* 333.

Hoyt, S. C. (1969). *Proc. 2nd Intern. Congr. Acarology, 1967* pp 117-133.

Hoyt, S. C. (1972). *N. Z. J. Sci.* **15,** 16-21.

Hoyt, S. C. (1973). *N. Z. J. Exp. Agric.* **1,** 77-80.

Penman, D. R. (1978). *N. Z. J. Zool.* **5,** (in press).

Penman, D. R. and Ferro, D. N. (1977). "Orchard Pest Control: A survey of Nelson Pip-Fruit Orchardists." Lincoln University College of Agriculture Press.

Penman, D. .R., Ferro, D. N. and Wearing, C. H. (1976). *N. Z. J. Exp. Agric.* **4,** 377-380.

Readshaw, J. L. (1975). *J. Aust. Inst. Agric. Sci.* **41,** 213-214.

Robertson, N. G. (1976). *N. Z. Official Yearbook 1975,* 13-21.

Rock, G. C., Yeargan, D. R. and Rabb, R. L. (1971). *J. Insect Physiol.* **17,** 1651-1659.

Thomas, W. P. and Chapman, L. M. (1978). *Proc. N. Z. Weed and Pest Control Conference* **31,** (in press).

Walker, J. T. S. (1978). Unpubl. M. Hort. Sc. thesis, Lincoln University College of Agriculture, Canterbury, New Zealand.

Walker, J. T. S. and Penman, D. R. (1978). *Proc. N. Z. Weed and Pest Control Conference* **31,** (in press).

Wearing, C. H. and Gunson, F. A. (1976). *Orchardist N. Z.* **49,** 157-158.

Wearing, C. H. and Penman, D. R. (1975). *Orchardist N. Z.* **48,** 122.

Wearing, C. H., Walker, J. T. S., Collyer, E. and Thomas, W. P. (1978a). *N. Z. J. Zool.* **5,** (in press).

Wearing, C. H., Collyer, E., Thomas, W. P. and Cook, C. (1978b). *Proc. N. Z. Weed and Pest Control Conference* **31,** (in press).

Yuill, H. (1975). *Orchardist N. Z.* **48,** 384.

MITE POPULATIONS ON APPLE FOLIAGE IN WESTERN OREGON

M. T. AliNiazee

Department of Entomology
Oregon State University
Corvallis, Oregon

INTRODUCTION

The European red mite, *Panonychus ulmi* (Koch), and the yellow mite, *Eotetranychus carpini borealis* (Ewing) are the two most important mite pests of apples in the Willamette Valley of Oregon. Although a number of other phytophagous mite species, such as the twospotted mite, *Tetranychus urticae* Koch, apple rust mite, *Aculus schlectendali* (Nalepa) and the McDaniel mite, *Tetranychus mcdanieli* McGregor, are commonly encountered in apple orchards, none are serious pests. The predatory mite, *Typhlodromus occidentalis* Nesbitt, is rarely seen in the apple orchards of the Willamette Valley, but another phytoseiid mite, *Typhlodromus arboreus* Chant is very common and is an important factor in reducing populations of phytophagous mite species (AliNiazee, 1978).

The studies reported here were initiated in 1973 to determine the long term effects of elimination of all insecticide and fungicide sprays from a block of apple trees on phytophagous and predatory mites.

METHODS

A portion of our experimental orchard located at Corvallis, OR was selected for this study. Mite counts in the fall of 1972 indicated a very low population of predatory mites, and moderate populations of European red mite and yellow mite. The orchard had been under a chemical control program for over 10 years.

We selected 6 trees for sampling. The samples were collected at approximately weekly intervals from all sides of the tree. Twenty-five mature leaves

Vol. I: ISBN 0-12-592201-9

were collected from each tree and mite counts made using a mite brushing machine (Henderson and McBurnie, 1943). Counts were initiated in June and terminated in September.

RESULTS AND DISCUSSION

Although weekly counts were made of all mobile stages and eggs of the 4 species of mites, the weekly count data are not presented here. Rather, the data are given as averages of early, mid, and late season counts (Table I).

Population Trend of *A. schlechtendali*

Apple rust mite overwinters as an adult in cracks and crevices on the old wood, lateral buds on terminal shoots. It measures 160-175 μm long in the protogyne state (Jeppson *et al.*, 1975). The adults emerge in early spring and move to opening buds. The eggs, laid on the undersides of the leaves, hatch in 10-12 days depending upon the temperature. A number of generations occur per year.

Data (Table I) indicate that elimination of the spray program resulted in a large increase of the population of *A. schlechtendali* during 1973. The early spring counts reached as high as 286 mites/leaf. There was a rather sudden drop of the apple rust mite population by the end of July, apparently in response to increasing numbers of *T. arboreus*. No other factor seems to be responsible for this change. In later years, the population of *A. schlechtendali* stabilized at relatively low level (0.5-15.8 mites/leaf) and no population flare ups were noticed during the rest of the study period. Normally, two population peaks were observed per year.

TABLE I.
A Seasonal Comparison of Mite Populations of a 6 Year Study of an Unsprayed Apple Orchard in Covallis, OR. The Population Averages Pertain to Seasonal Dates, i.e., "Early" is the Season Prior to July 15, "Mid" is July 16-Aug. 15, and "Late" is Aug. 16-Sept. 15.

	Average Number of Mobile Mites and Eggs Per Leaf Derived from Weekly Collections											
	Apple Rust			*European Red*			*Yellow*			*T. arboreus*		
	Early	Mid	Late	Early	Mid	Late	Early	Mid	Late	Early	Mid	Late
1973	172.8	23.8	7.6	0.1	0.4	0.6	2.6	4.8	8.7	2.1	3.1	3.8
1974	3.5	3.7	0.9	0.3	2.5	1.3	0.7	2.5	2.3	3.1	3.3	3.3
1975	5.3	6.5	5.2	4.4	10.4	9.9	0.1	0.5	1.9	0.9	2.7	2.5
1976	0.5	5.6	2.0	1.5	11.8	3.8	0.5	5.6	2.0	0.9	1.6	1.3
1977	1.8	7.2	5.5	12.3	6.8	4.8	0.5	2.7	1.1	1.3	2.0	0.8
1978	5.4	1.5	—	1.2	1.0	—	1.6	3.0	—	0.5	0.6	—

Relationship of *A. schlechtendali* to Other Phytophagous Mites and *T. arboreus*

Studies conducted over a number of years in different commercial orchards of western Oregon, failed to establish any direct relationship between the apple rust mite and the other two major phytophagous mite species, the yellow mite and the two spotted spider mite. However, it seems that early season feeding of apple rust mites might make the foliage slightly less susceptible to European red mites. Data (Table I) indicate a direct relationship between the populations of *A. schlechtendali* and *T. arboreus*. As the population of the latter species increased, the number of the former species decreased. For instance, in early 1973, the population of *T. arboreus* was small, while the population of *A. schlectendali* was large. As the season progressed, the population of *T. arboreus* increased and *A. schlechtendali* decreased. Under a stable condition, however, the apple rust mite population is never eliminated by *T. arboreus*, but are reduced to low numbers.

A number of earlier studies (Anderson and Morgan, 1958; Hoyt, 1969; Lord *et al.*, 1978) noticed that a high population of another predatory mite, *T. occidentalis,* will develop by feeding on *A., schlechtendali.* It seems that a similar relationship exists between *A. schlechtendali* and *T. arboreus* in apple orchards of the Willamette Valley. Therefore, the presence of apple rust mite could be beneficial, by providing essential alternate food supply for *T. arboreus* when other food is scarce, as reported by Hoyt and Caltagirone (1971) for *T. occidentalis.*

Population Trend of *Panonychus ulmi*

P. ulmi is one of the major mite pests of apples in the Willamette Valley. It overwinters as an egg, laid mostly on spurs, small branches, and cracks and crevices of the bark. The eggs hatch in early spring, and young nymphs are visible by the end of April, mostly on basal leaves. They complete one generation in 2-4 weeks depending upon temperatures and may complete 5-7 generations/year. Hoyt (1969) observed that in June, about 81% of the European red mites were on spur foliage, but by August this declined to 57%, and the remaining 33% were on the terminal growth. Anderson and Morgan (1958) report one half of the eggs and active forms on the upper surface of the foliage.

In 1973 after the elimination of insecticides, the population density of *P. ulmi* stayed at fairly low numbers (Table I), ranging from 0.1 to 0.5, probably because of low number of overwintering eggs due to constant use of chemical pesticides in the previous years. Toward the end of 1974, however, the population started to build up, and by September 2, 1.6 mites/leaf were noticed. A gradual increase occurred in 1975, and the trend continued in 1976. A peak of 22.3 mites/leaf was recorded on 8/6/1976. Then population slowly declined, started to stabilize, and further declined in September. In 1978, the population of European red mite was fairly low. It is not known whether it was the result of a natural decline and stabilization of the populations or due to a temporary climatic effect. In either case, the 1978 population of 1-2 mite/leaf is much below any damage threshold. Sudden elimination of insecticides caused an increase in population of *P. ulmi* over a period of *ca* 3 years, then the population declined and stabilized.

Relationship of *P. ulmi* to Other Phytophagous Mites and *T. arboreus*

A competitive relationship could exist between *P. ulmi* and *E. carpini borealis,* and *P. ulmi* and *A. schlechtendali,* particularly on spur foliage. Such a competition will favor *E. carpini borealis* because of its higher reproductive potential. However, due to low productions in early season and availability of ample foliage, there probably would be relatively little competition. Hoyt (1969) reported a competitive relationship between *P. ulmi* and *Tetranychus mcdanieli,* particularly in the high central areas of the tree. Since *E. carpini* does not form any webbing, it is incapable of trapping other mite species and reducing the population. Coexisting populations of *P. alumi* and *E. carpini* are mostly seen on the fully grown leaves.

Population Trend of *Eotetranychus carpini borealis*

Data indicate (Table I) that immediately after the elimination of pesticides, the population of yellow mite showed a tendency to increase. However, the presence of a large number of *T. arboreus* apparently kept them under control. The population of *E. carpini* increased to about 9.4 mites/leaf late in 1973, but it declined the next year. Throughout 1974 and 1975, the yellow mite population was low. There was some increase in the population towards the later part of 1976, concurrent with a slight decline in the population of *T. arboreus.* Yellow mite populations stayed fairly low throughout 1977 and 1978. These data indicate that in the absence of pesticide treatments, *E. carpini* can be kept under control by *T. arboreus.*

Population Trend of *T. arboreus*

T. arboreus quickly responded to the elimination of pesticides and availability of prey. The population increased with time and large numbers were found by the end of 1973. A fairly high survival was noted in the spring of 1974, resulting in a large population build-up in mid summer. A low survival was apparent in early 1975, probably due to starvation caused by the elimination of prey in the fall of 1974. This caused an easing of pressures on both apple rust mites and European red in 1975. The population of *T. arboreus* increased by the end of 1975; however, slightly lower numbers were seen in 1976, 1977, and 1978. This seems to have been caused by the lowering of available food supplies and populations of different mite species.

Relationship of *T. arboreus* to Phytophagous Mites

Considering the overall summary of mite populations over the 6 year period the data indicate that the elimination of pesticide sprays caused an initial increase in the population of *T. arboreus,* then a slight decrease because of food limitation (Table I). Field observations indicated that *T. arboreus* not only feeds on the 3 mite species discussed here, but also feeds on a number of other mites including tydeids. It is difficult to say which one of these mite species is the preferred food of *T. arboreus,* although apple rust mites seem to be highly favored. The slow increase of European red mites over a period of the first 3 years in spite of presence of *T. arboreus* might indicate that European red mite is least favored by the predator mite.

Long term elimination of pesticides had an overall positive effect on the biological control of orchard spider mites. All major species were reduced to levels much below economic thresholds, and a rather impressive build up of the *T. arboreus* population was noticed.

SUMMARY

A six year (1973-1978) study conducted in the Willamette Valley of Oregon indicated that the elimination of all pesticide sprays had a positive effect on overall biological control of spider mites. Long term population trends showed an abrupt increase in apple rust mite population during the first year, then a slow decline and stabilization of the population at low numbers. The yellow mite populations were low throughout the study and never became a problem. The population of European red mite increased slowly, peaked during the third year, and declined to low numbers during the fifth and sixth years. The predatory mite, *Typhlodromus arboreus,* responded to the elimination of insecticides and ample availability of food by increasing its numbers. A build-up of *T. arboreus* was noticed during the first four years of the study. During the latter years *T. arboreus* populations slowly declined, probably because of lack of prey.

Results of this study indicated that in the absence of pesticides, *T. arboreus* was capable of increasing to large numbers and kept phytophagous mites below economic level.

REFERENCES

AliNiazee, M. T. (1978). *Proc. IV Intern. Congr. Acarol.* Saalfelden, Austria, 1974. In Press.

Anderson, N. H. and Morgan, C. V. G.. (1958). *Proc. 10th Intern. Congr. Entomol.* Montreal 1956. **4**, 659-665.

Henderson, C. F. and McBurnie, H. V. (1943). *U.S.D.A. Circ. 671*, 1-11.

Hoyt, S. C. (1969). *Proc. 2nd Intern. Congr. Acarol. 1967.*

Hoyt, S. C. and Caltagirone, L. E. (1971). *In* "Biological Control" (C. B. Huffaker, Ed.), pp. 395-420. Plenum Press. New York.

Jeppson, L. R., Keifer, H. H. and Baker, E. W. (1975). "Mites Injurious to Economic Plants." Univ. of Calif. Press. Berkeley.

Lord, F. T., Herbert, H. J. and MacPhee, A. W., (1958), *Proc. 10th Intern. Congr. Entomol.,* Montreal (1956). **4**, 617-622.

RESISTANCE OF DIFFERENT VARIETIES OF SUNFLOWER AND SAFFLOWER TO *TETRANYCHUS TURKESTANI* IN SOUTHERN IRAN

H. A. Navvab-Gojrati and N. Zare

Department of Plant Protection,
College of Agriculture
Pahlavi University, Shiraz, Iran

INTRODUCTION

Sunflower, *Helianthus annuus* L., and safflower, *Carthamus tinctorius* L. have been under commercial production in Iran since 1966 and 1967 respectively (Ghorashy and Kheradman, 1972). Since then, research projects have been underway to find varieties most suitable for Iran. Naturally, one of the desired qualities of the selected varieties would be their resistance to plant pests. On an international scale, a great deal of work has been done regarding analysis of various sunflower and safflower varieties having resistance to diseases and pests (Anon., 1969; Ashri, 1973; Banihashemi, 1975; Orellana and Bear, 1969). However, very little information is available regarding varietal resistance to pest mites. Mites are known to occur on safflower in many regions of the world, with varying degrees of infestation and damage (Weiss, 1971). Thus, *Tetranychus urticae* (Koch), the twospotted spider mite, and *T. pacificus* (McGreg.) the Pacific mite, occasionally occur in very large numbers on safflower in California (Knowles and Miller, 1965). The carmine spider mite, *T. cinnabarinus* (L.), has also been recorded on safflower in many countries of the Middle East and Africa, including Iran. However, reportedly it does not damage safflower (Weiss, 1971).

Recently, *T. turkestani* Ugarov and Nikolski, the strawberry spider mite, has been found to infest both safflower and sunflower in southern Iran. However, no information is available in the literature regarding the extent of damage done by this pest and the resistance of different varieties of safflower and sunflower to it.

This paper reports results of work done to determine relative degree of

Vol. I: ISBN 0-12-592201-9

tolerance by safflower and sunflower varieties to *T. turkestani*. infestation in southern Iran.

MATERIALS AND METHODS

Fifteen varieties of sunflower and 16 varieties of safflower were screened in this study for their relative tolerance to *T. turkestani*. The sunflower varieties are shown in Table I. Each variety was cultivated in a 5.5m^2 plot in a completely randomized block design with 4 replications for each variety. Sampling for mites was done twice a week. For sampling, 12 leaves were collected at random from each replicated plot. Thus, a total of 48 leaves was picked from each variety of sunflower and safflower. Both sides of a leaf were examined for *T. turkestani* and the number of mites was recorded. Initially infestation was determined in the field using a magnifying glass, and later when the population of mites increased, the samples were brought to the laboratory for counting the mites under a microscope. Mite populations were monitored for a period of 32 days when the plants became ready for harvest.

TABLE I
Varieties of Sunflower, *Helianthus annuus* L. Screened for their Tolerance to *T. turkestani*. Counts Reflect Average Number of Mites/Leaf for a Sampling Period of 32 Days on Sunflower.

Sunflower	Mites/Leaf	Sunflower	Mites/Leaf
Hybrid 53	90	Chernianka 66	120
Record 1. Gen.	96	NS-P-317	120
Record Cert.	100	Armarisky	123
Louck	104	Record Elite	125
Orizont	105	Vniimk 8931	126
Hybrid 52	106	NS-Peredovik	135
Record	110	Vniimk 6540	142
Mhjak	115		

RESULTS AND DISCUSSION

Up to 105 days after planting, no mites were observed on any of the sunflower varieties. The first *T. turkestani* infestation was noted on Vniimk 6540 variety 105 days following planting. However, other varieties were infested soon thereafter, and a week after the appearance of the first infestation, all varieties developed mite infestations. In general, the mite populations increased initially, reaching a peak density of 19 mites/sample on 124-day old plants. The population then declined a little and then stabilized at about 16 mites per sample until the last sampling, 137 days following planting.

Although none of the sunflower varieties included in this test was totally immune to the mite attack, some differences in the degree of infestations were recorded. Total number of mites recorded on each variety during the entire 32-day period of sampling is shown in Table I. Thus, the total mite population counted and recorded on Vniimk 6540 variety was the highest and that on Hybrid 53 variety was the lowest of all varieties. The former variety had a mite population almost 2x that of the latter variety. This indicates relative susceptibility of Vniimk 6540 and relative resistance of Hybrid 53 variety. In addition to Vniimk 6540, a few other varieties, namely NS-Peredovik, Vniimk 8931 and Record Elite also showed a relatively higher degree of susceptibility to mite attack than the other varieties. The rest of the varieties had a somewhat similar tolerance, and there were no great differences in the mite populations counted on them.

The infestation of safflower by *T. turkestani* also started quite late and no mites were observed for a period of 97 days following planting. Later when the mite infestation did occur, *T. turkestani* could be found on all 16 varieties, but the mite populations did not reach a damaging level. The mite caused only a partial folding of the leaves of some varieties. All safflower varieties showed a more or less uniform light infestation. The varieties/selections under study were the following: Arak 2811, Dart 45, Esfahan 2819, Ferio 3176, Gila 2983, Lida 41, Local 3151, Marand 3184, Mashhad 3150, Nebraska 3178-10, Nebraska 8125-3123, Pacific 1-3164, Rezaieh 3174, U.S. 1, U.S. 10-3163 and Ute 3174.

According to previous observations, when the mite infestation starts early and the mid-season temperatures are high, *T. turkestani* occasionally can develop high populations. In light of available information, it seems advisable to plant sunflower and safflower as early in the season as possible to minimize *T. turkestani* damage. This early planting would allow the plants to become large enough to tolerate a late season build-up of the mite population, and would be able to set and mature a full set of buds and seed heads without acaricidal treatments.

SUMMARY

Varieties of sunflower and safflower were screened for their relative tolerance to *Tetranychus turkestani* Ugarov and Nikolski. There were differences in the degree of infestations and none of the sunflower varieties was immune to the mite attack. The Vniimk 6540 variety of sunflower was found to be the most susceptible and Hybrid 53 variety the most tolerant of all varieties. NS-Peredovik, Vniimk 8931, and Record Elite also showed a relatively higher degree of susceptibility to mite attack than the other sunflower varieties. The rest of the varieties showed a somewhat similar level of tolerance.

All varieties of safflower were lightly infested with mites, but populations did not reach damaging levels during the growing season.

ACKNOWLEDGEMENTS

This research has been financed in part by a grant from the Ministry of Science and Higher Education, Iran, Project No. 201-50.

The authors would like to express their appreciation to Drs. Mir S. Mulla and J. A. McMurtry, Department of Entomology, University of California, Riverside, for their critical review of the manuscript.

REFERENCES

Anonymous. (1969). "Sunflower: A Literature Survey, Jan. 1960-June 1967." U.S.D.A. Natl. Agric. Lib., Beltsville, Maryland 133 pp.

Ashri, A. (1973). Final research report. The Hebrew University of Jersusalem, Rehovot, Israel, pp. 78-106.

Banihashemi, Z. (1975). *Plant Dis. Rep.* **59**, 721-4.

Ghorashy, S. R. and Kheradnam, M. (1972). *World Crops* **24**, 156-7.

Knowles, P.F. and Miller, M. D. (1965). "Safflower." Calif. Agric. Exp. Stn. Ext., Circ. **532**, 51 pp.

Orellana, R. G. and Bear, J. E. (1969). "Sunflower Diseases and Pests: A Selected Bibliographical List." U.S.D.A., A.R.S., C.R.D., Beltsville, Maryland 91 pp.

Weiss, E. A. (1971). "Castor, Sesame and Safflower." Intertext Publishers, Leonard Hill, London. 704 pp.

GEOGRAPHICAL DISTRIBUTION OF THE CITRUS RED MITE, *PANONYCHUS CITRI* AND EUROPEAN RED MITE, *P. ULMI* IN JAPAN

Norizumi Shinkaji

Faculty of Horticulture, Chiba University
Matsudo, Chiba, Japan

INTRODUCTION

The citrus red mite, *Panonychus citri* (McGregor) is found on hosts such as citrus, rose, almond, and pear in Japan, China, North and South America, USSR, and South Africa. The European red mite, *P. ulmi* (Koch) occurs in most deciduous fruit orchards in Europe, India, Bermuda, China, Japan, Georgia in the USSR, North and South America, Tasmania, and New Zealand (Pritchard and Baker, 1955; Jeppson *et al.,* 1975). In Japan, *P. citri* occurs throughout the country, *P. ulmi* in the colder areas (Ehara, 1964).

In the present paper, an attempt is made to complete in detail the geographical distribution of the citrus red mite and the European red mite in Japan based on the studies reported in previous papers (Shinkaji, 1961a and 1961b) in Japanese with English summary and on studies conducted recently.

MATERIALS AND METHODS

The *Panonychus* mites were collected from major fruit trees such as citrus, pear, peach, cherry and apple in various districts in Japan during the period from August to September, were later mounted on slides by Hoyer's modification of Berlese's medium. According to Pritchard and Baker (1955), the collected *Panonychus* mites were classified by the relationship between the ratio of the length of the clunal to the inner sacral and the ratio of the length of the outer sacral to the inner sacral in hysterosoma.

In order to make clear the distribution of the diapause type of the citrus red mite, the overwintering eggs were collected from the twigs or bark of pear trees

Vol. I: ISBN 0-12-592201-9

during the period from November to December. The overwintering eggs were measured for the diameter, were tested for hatchability at 25°C, and were tested with the Ehrlich's diazo reaction reported by Shinkaji (1961b).

RESULTS AND DISCUSSION

Figure 1 shows the relationship between the ratio of the length of outer sacral to the inner sacral and the ratio of the length of the clunal to the inner sacral. From Fig. 1, it is inferred that the individual population having 0.30-0.50 as the ratio of the length of the outer sacral to the inner sacral corresponded to *P. citri* and the populations having 0.55-0.70 *P. ulmi.* The citrus red mites were collected from fruit trees: citrus, pear, peach and apple. The European red mites were collected from apple, cherry, pear and peach but not from citrus trees. McGregor and Newcomer (1928) state that the death of the deciduous-fruit *Paratetranychus* (= *P. ulmi*) colonies should be attributed to inability of the deciduous-fruit mites to adjust to the orange foliage.

It was suggested by Shinkaji (1961b) for the first time that there are two strains of the citrus red mite. One is a non-diapause strain, the other is a diapause strain. The diapause strain overwintered on the twigs or bark of

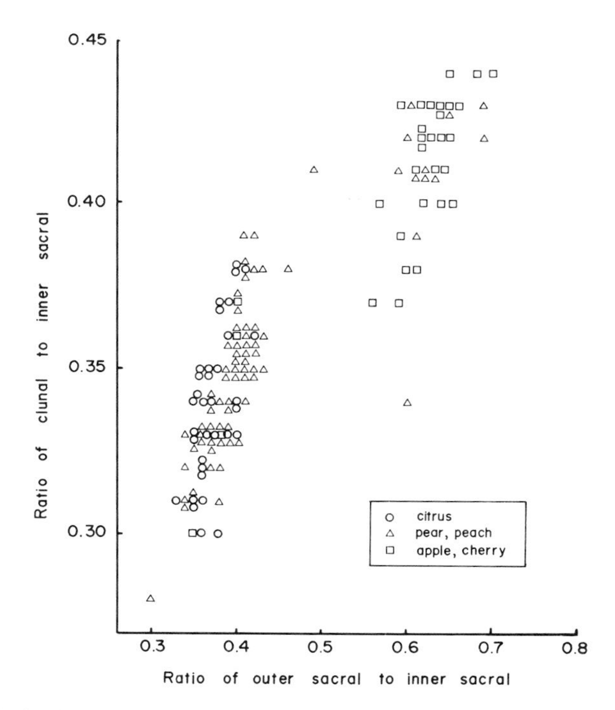

Fig. 1. Relationship between the ratio of the length of the outer sacral to the inner sacral and the ratio of the length of the clunal to the inner sacral in *Panonychus* mites.

deciduous fruit trees as the egg stage and did not hatch until the middle of April, whereas the eggs of the non-diapause strain hatched continuously throughout the winter. It is possible to distinguish the eggs of the diapause strain from non-diapause eggs in two ways. Firstly, (Table I) the diapause strain eggs were slightly larger than the eggs of the non-diapause strain. Such relationship was observed between the winter (diapause) eggs and the summer eggs of *P. ulmi* by Blair and Groves (1952). Secondly, as Table II shows, both the overwintering eggs and the females which deposited these eggs in the diapause strain gave a positive Ehrlich's diazo reaction. This reaction became weaker as the time passed after egg deposition. On the other hand, the eggs of the non-diapause strain showed a negative Ehrlich's diazo reaction throughout the winter season. Further, in recent tests (unpublished), the critical photoperiod for diapause incidence of the diapause strain at 20°C was found to be about 13 hrs on the mite in Tottori district located near 35°30′ in the north latitude.

TABLE I.
Diameter of Eggs in *Panonychus* Mites

	Diamter of Egg (μm)					
	Twig			Leaf		
	Date	Mean	± s.d.	Date	Mean	± s.d.
Panonychus citri						
Diapause Strain	Feb. 12	164.74	± 6.13	Aug. 7	154.26	± 4.84
Non-Diapause Strain	Dec. 2	142.56	± 8.92	Sept. 18	138.54	± 4.29
Panonychus ulmi	Feb. 19	157.93	± 5.31	Aug. 20	142.73	± 4.81

TABLE II.
Ehrlich's Diazo Reaction on the Females and the Overwintering Eggs of the Citrus Red Mite

			Female		Egg							
			Oct. 10	Nov. 18	Dec. 4	Dec. 25	Jan. 12	Jan. 28	Feb. 9	Feb. 23	Mar. 8	Mar. 23
Ehrlich's Diazo Reaction	Diapause Strain	+	9	16	15	18	16	3	3	6	4	5
		±	1	4	4	2	3	10	7	6	7	7
		−	0	0	1	0	1	7	10	14	9	8
	Non-Diapause Strain	+	0	0	0	0	0	0	0	0	0	0
		±	0	2	1	1	5	1	3	1	1	1
		−	3	18	19	19	15	19	17	19	19	19

Figure 2 shows the geographical distribution of the citrus red mite and the European red mite on major fruit trees drawn on a map. The non-diapause citrus red mite was distributed all over the citrus growing area. On pear trees, the non-diapause citrus red mite was found in the lower latitude zone while the diapause strain was found in the higher latitude zone. The line of demarcation

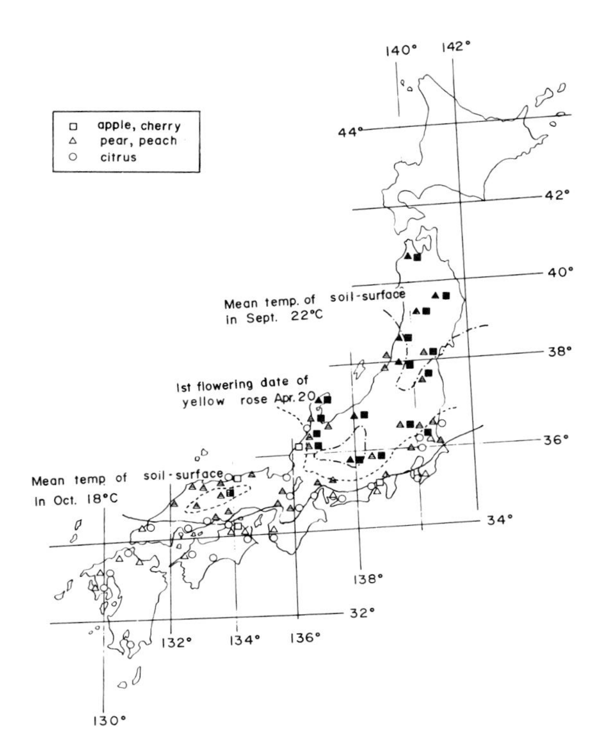

Fig. 2. Geographical distribution of the citrus red mite (open and semi-shaded symbol) and the European red mite (shaded symbol).

between these strains was fitted in closely with a line where the mean temperature of the soil-surface in October is 18°C in the atlas of agricultural meteorology presented by Daigo (1947). The European red mites on pear trees were collected from a zone of higher latitude than the diapause citrus red mite distribution. The distribution of the citrus red mite on peach trees was the same as that on pear trees. This mite was also found on apple trees when they were grown near to other fruit trees which were infested by the citrus red mite. Ehara (1956) collected the citrus red mite from mulberry, *Morus bombycis* Koidz., in Hokaido, Honshu and Kyushu. The citrus red mites were found abundant on *Picrasma quassioides* Benn. in Aomori district by Yamada (1967). It was observed by each reporter that the mite overwintered in egg stage.

The southern limit of the European red mite distribution on apple trees fitted closely with the zone where the first flowering-date of the yellow rose, *Kerria japonica* Dc., is April 20 in Daigo's atlas (1947). This corresponds to the zone of 37° in the north latitude. The distribution of the European red mite in Japan was in conformity with the American distribution of this mite, which did not occur on deciduous plants south of latitude 37°N, presented by Newcomer and Yothers (1929). In the case of pear trees, the European red mite was not collected from areas south of the zone where the mean temperature of the soil-surface in September is 22°C.

The *Panonychus* mites, which had some characteristics of both the citrus red mite and the European red mite, were also collected. The color of dorsal tubercles is the same as the body color of the citrus red mite, but mounted specimens have clunals distinctly shorter than the outer sacrals as for the European red mite and have the longer inner sacrals than the European red mite (Fig. 3). They were collected from apple, pear and cherry trees near the zone where the mean temperature of the soil-surface in September is 22°C (Fig. 4).

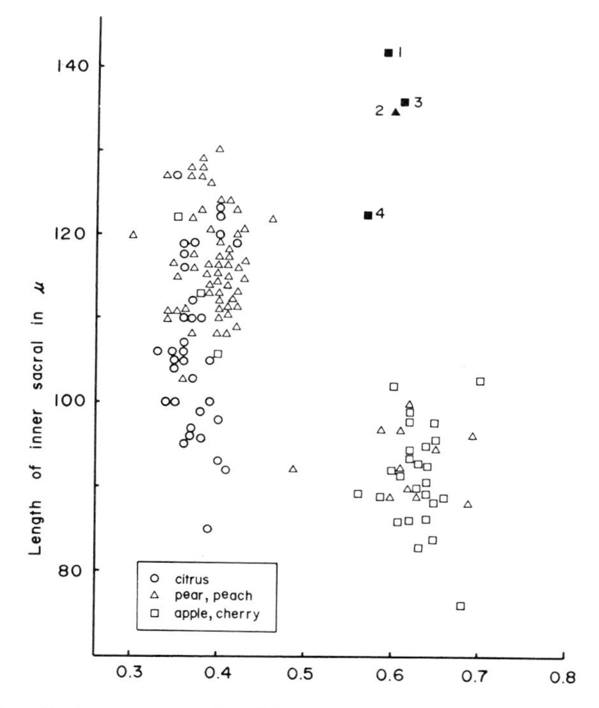

Fig. 3. Relationship between the ratio of the length of the outer sacral to the inner sacral and the length of the inner sacral in *Panonychus* mites. The shaded symbols show the *Panonychus* mites having some characteristics of both *P. citri* and *P. ulmi*.

SUMMARY

The *Panonychus* mites were collected from citrus, pear and apple trees and were classified according to Pritchard and Baker (1955). There were two strains of the citrus red mite, *P. citri* (McGregor), in Japan. One was the diapause type, which overwintered as a diapause egg stage. The other was non-diapause type, which passed the winter in developmental stages. The diapause type was distributed in higher latitude zones than the non-diapause

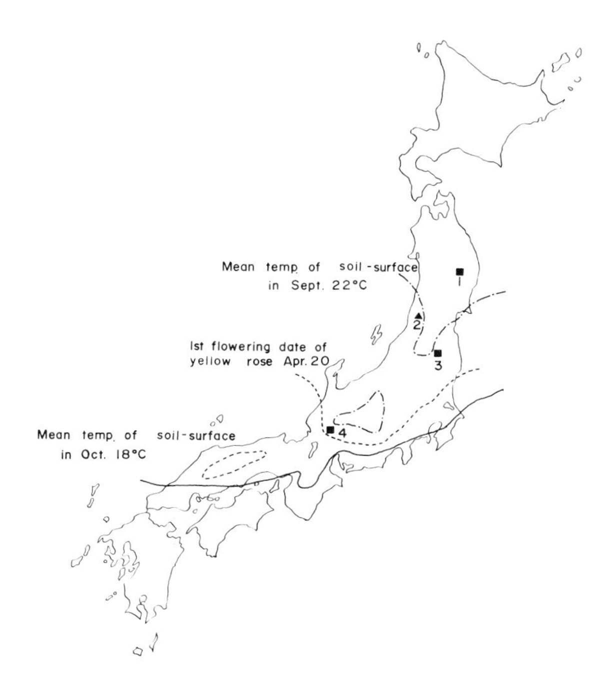

Fig. 4. Geographical distribution of the *Panonychus* mites, which had some characteristics of both the citrus red mite and the European red mite. The shaded symbols and the figures beside them show the same as those shown in Fig. 3.

type. The line of demarcation of the citrus red mite distribution was represented by the line where mean temperature of the soil-surface in October is 18°C. The European red mite, *P. ulmi* (Koch), was not found on apple trees which were grown in the south of the zone where the first flowering-date of the yellow rose (*Kerria japonica* Dc.) is April 20, corresponding to the zone of 37° in the north latitude. The European red mite on pear trees was collected in the zone of higher latitude than the diapause type citrus red mite distribution. The *Panonychus* mites, which had some characteristics of both the citrus red mite and the European red mite, were also collected from apple, pear and cherry trees near the zone of demarcation between the diapause type citrus red mite and the European red mite.

REFERENCES

Blair, C. A. and Groves, J. R. (1952). *Hort. Sci.* **27**, 14-43

Daigo, Y. (1947). "Handy Atlas of Agricultural Meteorology in Japan." Kyoritsu Pub. Co. Ltd. (Tokyo), 234pp.

Ehara, S. (1956). *Fac. Sci. Hokkaido Univ. Ser. 6 Zool.* **13**, 15-23.

Ehara, S. (1964). *Acarologia* **6**, 409-414.

Jeppson, L. R., Keifer, H. H., Baker, E. W. (1975). ''Mites Injurious to Economic Plants.'' Univ. Calif. Press (Berkeley), 614 pp + 74plts.
McGregor, E. A. and Newcomer, E. J. (1928). *J. Agric. Res.* **36**, 157-181.
Newcomer, E. J. and Yothers, M. A. (1929). *Tech. Bull. U. S. Dep. Agric.* **89**, 1-69.
Pritchard, A. E. and Baker, E. W. (1955). *Pacif. Coast Ent. Soc. Mem. Ser.* **2**, 1-472.
Shinkaji, N. (1961a). *Bull. Hort. Stn.,* Nat. Tokai-Kinki Agric. Exp. Sta. **6**, 49-63.
Shinkaji, N. (1961b). *Bull. Hort. Stn.,* Nat. Tokai-Kinki Agric. Exp. Sta. **6**, 49-63.
Yamada, M. (1967). *Aomori-ken Seibutsu Gakukaishi* **9**, 1-4.

STUDIES ON PESTICIDE RESISTANCE IN THE PHYTOSEIID *METASEIULUS OCCIDENTALIS* IN CALIFORNIA

Marjorie A. Hoy and Nancy F. Knop

Department of Entomological Sciences
University of California
Berkeley, California

INTRODUCTION

Spider mites have become increasingly important agricultural pests during the past 25 years. Major reasons for this increase in importance appear to be the ability of spider mites to tolerate or develop resistance to pesticides applied for their control and the destruction of their predators by pesticides (Huffaker *et al.*, 1970; McMurtry *et al.*, 1970). Thus, determining the effects of pesticides on spider mites and on predators of spider mites is an important component in the development of pest management programs. Selective use or reduced rates of pesticide applications may allow spider mite predators to survive to control spider mites and still give effective control of the target pest(s) (Hoyt, 1969).

The development of resistance to pesticides by spider mite predators is important to pest management programs. Such genetic resistance has been rarely observed in parasites or predators (Croft and Brown, 1975), but several species of phytoseiids have developed resistance in the field, particularly to organophosphate (OP) insecticides (Croft, 1976). Resistance to carbaryl, a carbamate insecticide, has been found for *Amblyseius fallacis* Garman (Meyer, 1975). Possible carbaryl resistance for *Metaseiulus (Typhlodromus) occidentalis* (Nesbitt) has been suggested based on survival in the field but has not been confirmed yet by laboratory tests (S. C. Hoyt, pers. com.).

During the past year we have been engaged in a genetic selection program designed to increase the usefulness of the spider mite predator, *M. occidentalis.* This predator is important in orchards and vineyards in the western U.S.A. and if its pesticide resistance can be defined and extended pest management programs can be developed or improved. This project consists of several steps. One step evaluated the impact of inbreeding on this species (Hoy, 1977), since a genetic selection program would be hampered by abnormal sex ratios

Vol. I: ISBN 0-12-592201-9

such as those exhibited by honey bees upon inbreeding. Another step involved determining the current status of resistance in field populations, and this paper reports part of those efforts. This paper also presents preliminary results of a selection program in which resistance to the synthetic pyrethroid, permethrin, is sought.

MATERIALS AND METHODS

Resistance levels to OP insecticides were determined for *M. occidentalis* collected during 1977 from wild blackberries and pears from diverse areas of California. Usually 20-50 adult females were used to initiate a colony, which was maintained on the twospotted spider mite (*Tetranychus urticae* Koch) reared on pinto beans. Predators were reared at 22-25°C under a long day on paraffin-coated paper discs on moist cotton. As soon as sufficient gravid adult females were available (500-600), LC_{50} values were obtained using a slide dip analysis (Anon., 1968). Formulated pesticides evaluated included azinphosmethyl (50 WP) and diazinon (50 WP) since these OPs have been used commonly in pears. Two colonies were tested with the OP dimethoate as well. Twenty females were placed on their backs on a sticky tape mounted on a microslide, dipped, and held at 27°C and 18 hrs light for 48 hours. Mortality was determined by lack of movement of the females' appendages when prodded with a fine sable brush. Eighty to 100 females were tested at each dose, and LC_{50} values were determined using logit analysis (Berkson, 1953).

Roush and Hoy (1978) obtained slide dip-determined LC_{50} values for the synthetic pyrethroid, permethrin (2E), for 3 colonies of *M. occidentalis*. These colonies were very susceptible to permethrin, showing LC_{50} values of 0.72, 1.32 and 1.48 gm active ingredient (AI) per 100 liters of H_20, respectively. These predators were 20-40 fold more sensitive than were 2 colonies of, *T. urticae* tested at the same time.

Colonies of *M. occidentalis* were established from California pears, grapes and blackberries in the summer of 1977. In the winter 1977-78 a colony was established from mites collected in the bark of apple trees by Dr. S. C. Hoyt. These colonies were screened for susceptibility to permethrin beginning in the fall 1977. As resistant predators must survive on treated leaves in the field we decided to conduct the screening/selection program using dipped leaves. Pinto bean leaf squares 3 cm^2 infested with abundant *T. urticae* as prey were dipped for 5 seconds into freshly prepared formulated permethrin doses of 0.8, 1, 2, 3, 6 and 12 g AI/100 liters of H_20 placed bottom up on wet cotton and allowed to air dry for 3 hours. Fifty gravid females were transferred from colonies to each leaf square and 500-1000 females were usually selected on one date. The leaves were held for 48 hrs at 24-27°C room temperature and 18 hrs light. Mortality was assessed as above. Survivors were transferred individually to small untreated leaf discs, provided with prey, and held to determine if oviposition occurred. Females surviving to oviposit were transferred to paper disc cultures. Progeny of these survivors were reared and selection was repeated.

Since we were aware that slide dip LC_{50} values are often different from LC_{50} values obtained by other methods, an LC_{50} for *M. occidentalis* was developed using 20♀♀/2 cm. diameter dipped bean leaf disc using one colony collected from wild blackberries in Berkeley, CA.

RESULTS AND DISCUSSION

Most predator colonies evaluated were resistant to the OP insecticides tested (Table I). Higher levels of resistance to diazinon than to azinphosmethyl were observed from colonies collected from pears in California even though diazinon has been used rarely in recent years. Azinphosmethyl is applied at reduced rates for control of the codling moth, *Laspeyresia pomonella* (L.), in the California Pear Integrated Pest Management (IPM) program.

The colony collected from wild blackberries showed low tolerance to both azinphosmethyl and diazinon (Table I), which was expected because exposure to pesticides is rare in this habitat.

Two colonies of *M. occidentalis* collected from pears were tested for resistance to dimethoate, an OP which is not applied to pear orchards. LC_{50} values were 74.3 and 75.5 g AI/100 liters water, with confidence intervals of 59.9-89.9 and 56.3-100.7, respectively. These LC_{50} values lie in the middle of the range of LC_{50} values obtained for *M. occidentalis* collected from California vineyards, where dimethoate is used commonly. This moderate level of

TABLE I.

LC_{50}s of *M. occidentalis* Collected During 1977 from California Pear Orchards and from Wild Blackberries Using Slide Dip Analysis

Predator Source		LC_{50}s (95% confidence interval)[a]	
County	Orchard	Azinphosmethyl	Diazinon
El Dorado	Marchini	203.7 (131.8-287.6)	1246.2 (802.8-1941.2)
Lake	Ivicvich	191.7 (143.8-263.6)	455.3 (323.5- 635.1)
	Jones, Home	443.4 (275.6-695.0)	707.0 (311.5- 719.0)
Sacramento	Randall Island	215.7 (167.8-287.6)	599.1 (455.3- 802.8)
	Sacramento River	515.3 (359.5-742.9)	—
Solano	Ames & Allbright	215.7 (179.7-275.6)	766.9 (575.2-1030.5)
	Neitzel	95.9 (24.0-359.5)	359.5 (203.7- 635.1)
Yuba	Campbell	239.7 (203.7-299.6)	1030.5 (778.9-1390.0)
	DiGiorgio, N. E.	131.8 (83.9-203.7)	1210.2 (802.8-1821.4)
	Levake	431.4 (323.5-599.1)	982.6 (659.0-1497.8)
Alameda	Wild blackberries	107.8 (71.9-179.7)	335.5 (203.7- 539.3)

[a] Expressed as g AI/100 liters water.

TABLE II.
Response of *M. occidentalis* Collected from Pears, Grapes, and Blackberries in California to Selection with Permethrin on Bean Leaf Discs

Dose g AI/100 Liter	Total ♀♀ Treated	% Females Surviving to	
		48 Hrs	Oviposition
12	1000	0.70	0
6	3855	0.91	0.05
3	1705	2.17	0.18
2	155	7.74	0
1	2655	14.20	3.24
0.08	970	33.20	—

resistance to dimethoate without a previous history of exposure supports the hypothesis that *M. occidentalis* is generally resistant to OP insecticides in agricultural habitats.

The leaf disc LC_{50} obtained for permethrin for the wild blackberry predator colony was 0.22 g AI/100 liters (0.10-0.49). This is lower than that obtained for the same colony (0.72 g) using a slide dip analysis (Roush and Hoy, 1978), as expected, and may be a more realistic predictor of their survival under field conditions.

The predator colonies collected from pears and grapes during the summer of 1977 gave disappointing results when selected with permethrin. Results of screening (= first round of selection) of predator colonies with permethrin are given in Table II. For example, 2655 females were tested using 1 g AI/100 liters water. After 48 hours, an average of 14.2% survived but only 3.24% of those tested survived and oviposited. Also, the progeny were reared and reselected one or more times, but no consistent improvement in survival is evident in any of these lines to date.

A mixed colony started with females from all available pear, grape and blackberry colonies was selected 4 times at .08 g AI/100 liters and twice at 1 g AI/100 liters without giving a consistent or stable increase in survival. The sixth selection (at 1 g) gave only 10% survival of 250 females, a rate below the average for the initial selections at 1 g (Table II).

The Washington apple colony is giving consistently better results (Table III). Twenty-one percent of 750 females survived a dose of 1 g AI permethrin/100 liters. The progeny of these females were reared and in the second selection round, 47.8% of 680 females survived the same dose. Fourteen percent of the females tested laid eggs and females have been tested in a third selection round. Survival of these females ranged from 32 to 57%. These results are the most consistent and promising we have had.

To eliminate the impact of experimental errors such as differences in solution strength and day to day variation, same day comparisons were made on two different dates using selected and the original (unselected) Washington apple colonies. These results (Table IV) show differences in survival between the

TABLE III.
Response of *M. occidentalis* Collected from Apples in Washington to Selection with 1 g AI Permethrin/100 Liter Water on Dipped Bean Leaf Discs

Selection Date		No. Tested	% Females Surviving to	
			48 Hrs	Oviposition
I	28 Mar.	500	20.4	6.0
	6 Apr.	250	22.8	4.0
II	11 May	180	51.1	21.0
	15 June	500	39.0	11.0
III	29 June	150	56.7	25.3
	27 July	500	31.8	—

TABLE IV.
Same Day Comparisons of the Effect of Permethrin on Selected and Unselected Washington Apple Colonies of *M. occidentalis*

Date Tested	No. Times Colony Selected at 1 g	Dose g AI/100 Liters Water	Total ♀♀ Treated	% Alive at 48 Hrs
10 July	0	2	50	10.0
	2	2	100	36.0
	2	4	100	20.0
26 July	0	1	20	5.0
	2	1	30	23.3
	3	1	50	54.0

unselected and selected colonies, and support the hypothesis that this colony exhibits a genetic resistance to permethrin. However, future responses to selection and analysis of genetic crosses will confirm the hypothesis, or show that the response is only vigor tolerance, and show how high this resistance, if genuine, may go.

Currently recommended field application rates for permethrin range from 1.2-12 g AI/100 liters. If a colony can be developed that can survive moderate to low rates of permethrin, viable IPM programs can be developed. For example, the pear psylla (*Psylla pyricola* Foerster) is resistant to currently registered pesticides in the western U.S.A. and the synthetic pyrethroids show promise as a control agent. However, field trials have repeatedly shown that spider mite outbreaks occur after application of synthetic pyrethroids. These outbreaks may be due to predator mortality, to physiological stimulation of spider mite reproduction, or to a combination of both factors. A pyrethroid resistant predator could be very important to pear pest management programs. The same may be true for grapes where experimental use of permethrin has been associated with spider mite outbreaks (Hoy *et al.*, unpubl.).

SUMMARY

Colonies of the phytoseiid, *Metaseiulus occidentalis,* collected from pears in California were evaluated for their resistance to organophosphate insecticides. Most colonies tested were resistant to azinphosmethyl and diazinon. Tolerance of these colonies to the synthetic pyrethroid, permethrin, was low. A survey of these colonies, and of colonies from California vineyards, gave only 14.2% survival to one gram AI permethrin/100 liters of H_2O. Survival of a colony from apples in Washington State was higher (21.2%) and after two rounds of selection with permethrin, survival increased further. Selection of this line is continuing and a future genetic analysis may support our hypothesis that genetic resistance to permethrin is present. The potential for multiply-resistant *M. occidentalis* in IPM programs is discussed.

ACKNOWLEDGEMENTS

These studies were supported by the Pear and Almond Advisory Boards and by the Agricultural Experiment Station. We thank R. Roush, M. Wilson and C. Nawalinski for their assistance, and S. C. Hoyt for providing a colony of *M. occidentalis* from a Washington apple orchard. We also thank I. C. I. for the permethrin.

REFERENCES

Anonymous (1968). *Bull. Entomol. Soc. Amer.* **66**, 1109-1113.
Berkson, J. (1953). *J. Am. Stat. Assoc.* **48**, 565-599.
Croft, B. A. (1976). *Entomophaga* **21**, 383-399.
Croft, B. A. and Brown, A. W. A. (1975). *Ann. Rev. Entomol.* **20**, 285-335.
Hoy, M. A. (1977). *Intern. J. Acarol.* **3**, 117-121.
Hoy, M. A. *In* "The Use of Genetics in Insect Control," (M. A. Hoy and J. J. McKelvey, eds.) Rockefeller Press, New York. (In press).
Hoyt, S. C. (1969). *J. Econ. Entomol.* **62**, 74-86.
Huffaker, C. B., van de Vrie, M., and McMurtry, J. A. (1970). *Hilgardia* **40**, 391-458.
McMurtry, J. A., Huffaker, C. B., and van de Vrie, M. (1970). *Hilgardia* **40**, 331-390.
Meyer, R. H. (1975). *Environ. Entomol.* **4**, 49-51.
Roush, R. T., and Hoy, M. A. (1978). *Environ. Entomol.* **7**, 287-288.

NEW ACARICIDES TO CONTROL RESISTANT MITES

D. H. C. Herne

Research Station
Agriculture Canada
Vineland Station
Ontario, Canada

J. E. Cranham and M. A. Easterbrook

East Malling Research Station
East Malling, Maidstone
Kent, England

INTRODUCTION

"Spider mites are a continuous potential danger in many crops because of their ability to develop resistance to chemicals which initially give effective control." These words by Helle and van de Vrie in a 1975 publication state very well the situation that has plagued researchers, chemical companies, growers and extension personnel since the early 1950's; the need for a continual search for new compounds, monitoring and testing for resistance, and evaluation of new compounds.

The resistance situation with mites has been amply reviewed by Helle (1965), Helle and van de Vrie (1974), and Dittrich (1975). In brief, resistance or cross-resistance has been detected in many tetranychid mites to most of the a) organophosphorus compounds (OPs), b) organochlorines, e.g. chlorbenside, ovex, fenson, tetradifon, c) nitrophenols, e.g. binapacryl (Morocide, Acricid), d) the quinoxalines, e.g. oxythioquinox (Morestan), e) the formamidines, e.g. chlordimeform (Fundal, Galecron). The situation exists in all parts of the world where these chemicals have been used extensively. However, growers usually have a choice of acaricides, and thus mites will often differ in their response to acaricides within regions.

Most evidence for resistance has come from common and repeated control failures in the field and not from standard laboratory tests such as the E.S.A. or F.A.O. standard test methods.

The most common method used to slow resistance development to

Vol. I: ISBN 0-12-592201-9

acaricides and increase their useful life is to alternate their usage. Combining acaricides with different chemical structures has also been tried, especially in Japan. Activity is often increased but resistance ultimately develops to both acaricides in a mixture.

The earlier hopes for compounds with negative cross-resistance have not been realized, nor have known synergists proved of value.

It is encouraging that chemical companies are discovering many new potentially good acaricides with novel chemical structures compared to other pesticides. However, the rejection rate of experimental chemicals after a few years of evaluation, or even after registration, is alarming. For example, chemicals discontinued or withdrawn by companies in recent years have included the following: U-27415 (Banomite, Upjohn); thioquinox (Eradex, Chemagro); AC 44858 (American Cyanamid); Cela S 2957 (Celamerk), benzoximate, benzomate (Citrazon, Nippon Soda), and chlordimeform (Galecron, Ciba-Geigy). The processes necessary for developing and registering new compounds on different crops and mite species are increasingly more complex and costly. The rapid development of IPM (Integrated Pest Management) programs, with an increased emphasis on acaricides and insecticides with low toxicity to beneficial species is narrowing the range of suitable chemicals, and causing concern in the agricultural chemical industry.

The many researchers from around the world who were queried for this review were more optimistic about future mite control for mainly two reasons. One, the general emphasis today on integrated pest management of crops has reduced the need for several acaricide sprays each season on many crops. Thus the selection pressure on mites with new acaricides is greatly reduced. Many workers are confident that by the reduced usage, and the alternation of available acaricides, the resistance problem is much reduced. Secondly, tetranychid mites have not developed resistance to some of the newer chemicals, e.g. organotin compounds, and propargite (Omite), after many consecutive applications over five or more years. However, there are indications that on some crops, e.g. apple, we may be relying too heavily on too few acaricides, and many IPM programs could be jeopardized if resistance did develop. The need for new and even more selective acaricides is still very real.

NEW ACARICIDES AGAINST TETRANYCHIDS

The major mite pests of deciduous fruit in North America are 1) the European red mite, *Panonychus ulmi,* (ERM) e.g. in Ontario, Nova Scotia, British Columbia and New York), 2) the two-spotted spider mite *Tetranychus urticae,* (2-spot) e.g. in Oregon and 3) the McDaniel mite *Tetranychus mcdanieli,* e.g. in British Columbia and Washington. Other minor species, which are sporadic pests, include the four-spotted spider mite *Tetranychus canadensis* in Ontario, and the Yellow mite *Eotetranychus carpini borealis* in Oregon.

Resistance is general to most of the older sulfur-containing bridged-diphenyl acaricides which include ovex, tetradifon (Tedion), Mitox, and Genite. The exceptions are propargite (Omite), and possibly tetrasul (Animert) both of which are still effective in most areas. Resistance to organophosphorus compounds is also widespread though mite suppression is still obtained in most regions with demeton, phosalone, demeton-methyl, dimethoate, and ethion. Dicofol (Kelthane) is still widely recommended but resistance or control failures to it have been reported in Ontario, Nova Scotia, Quebec, and British Columbia, and in Washington. Resistance to *T. mcdanieli* and *P. ulmi* has also been reported in Canada and the U.S.A. to Morestan (oxythioquinox, quinomethionate, chinomethionat), binapacryl (Morocide), and the formamidine acaricides. No resistance to the oils has been reported. In Michigan and British Columbia oils are used extensively in early season but in Ontario and Nova Scotia are recommended only every 3 to 4 years.

In the IPM programs for apple, pear, and peach in most regions of North America acaricides are used only as necessary. The acaricides usually recommended are cyhexatin (Plictran), fenbutatin-oxide (Vendex), and propargite (Omite), which favour predacious phytoseiids. Other compounds with some acaricidal activity, e.g. dinocap, Dikar, OPs, help suppress mite populations when used for other pests or diseases. For growers not in IPM programs other acaricides which are still recommended in different regions include oxythioquinox, dicofol, formetanate (Carzol), tetradifon, tetrasul (Animert), and chlorobenzilate (Acaraben). The choice of acaricides is less critical in IPM programs relying on the mite predator *Stethorus* as in Pennsylvania.

New acaricides considered to be promising by the majority of researchers include the following: fentrifanil, (Fusilade, PP199, ICI Plant Prot. Div.; Chipman Chem. Ltd. Canada). Fusilade is a polysubstituted diphenylamine compound with excellent activity against all stages of the different tetranychid pests, including those resistant to dicofol and OP compounds. Most reports indicate it is compatible in IPM programs at dosage rates still effective for pest mites. Although it proved toxic to *T. occidentalis* strains in British Columbia, in laboratory tests at Vineland, Ontario and Geneva, New York, it was relatively non-toxic to *Amblyseius fallacis*. Some phytotoxicity occurred to citrus in field trials in Texas but weather conditions may have been partly responsible.

Another novel compound XE333 (Chevron Chem. Co.), a diphenyl ether, is also effective against all life stages of pest species and is reported by the company to be safe to beneficial species. It has slow knockdown activity but long residual action. Initial tests at Vineland Station, Ontario, indicated some cross-resistance with dicofol R mites, but this must be confirmed. A DuPont material, DPX3792, has been extensively tested against different tetranychid species with excellent results. It is reported toxic to *A. fallacis* at high rates, e.g. 16 fl. oz. form/100 US gal. at Geneva, New York, but is effective agasint *P. ulmi* at lower rates. Malonoben (GCP5126), a Gulf Chem. Co. product, was also listed as promising by many workers in North America. It is a malodinitrile compound which kills active stages but not eggs. It has low toxicity to *A. fallacis*. Zardex or cycloprate (Zoecon Corp.) is considered a good candidate for IPM programs because of its long residual activity. It is mainly ovicidal but exposed female mites lay non-viable eggs. It has slow activity and is most suitable against initially low mite populations. Amitraz (Mitac, BAAM, Upjohn) with a structure derived from chlordimeform is active against all mite stages and like chlordimeform it controls *Psylla pyricola*. Low toxicity to mirids and anothocorids was confirmed at East Malling, England. Hostathion (HOE2960, triazophos, Hoechst), an insecticide-acaricide, is toxic to most tetranychid spp. including those resistant to OPs.

Many other new compounds have shown promise in initial trials. For example at Geneva, New York, XE567, XE626, and RH6564, were highly toxic to *P. ulmi*, and some safety to *A. fallacis* was indicated. Further testing may prove the value of these and other experimental compounds.

In Europe *P. ulmi* is the major pest of fruit trees and vines. It is also a pest in New Zealand, Tasmania and is spreading in Australia. It was recently introduced to South Africa. As in North America *P. ulmi* is generally resistant to all OPs but control is still obtained with some of them where the resistance factor is low. For example, omethoate (Folimat), methidathion (Supracide), phosalone, triazophos, dioxathion, chlorpyrifos are effective for 2-3 years before failures occur. Resistance to dicofol, tetradifon, and most organochlorines and quinomethionate have been reported in some regions of the U.K., Holland, Switzerland and Bulgaria, but these compounds and tetrasul continue to have a limited use in Europe. Also dinocap, and binapacryl still give worthwhile mite suppression.

The tin compounds cyhexatin, fenbutatin-oxide and azocylotin are widely used against *P. ulmi* in Europe and South Africa.

Bromopropylate (Neoron, Acarol), and oils are currently used in Europe (except in the U.K.), Australasia, and in South Africa. Amitraz is a promising new compound for *P. ulmi* control.

The main pest on hops and small fruits is *T. urticae*. On hops grown mainly in West Germany, Czechoslovakia, U.K. and Yugoslavia, mites are resistant to all OPs, but less generally to dicofol. Current control is obtained with cyhexatin, dicofol-tetradifon mixture and aldicarb.

T. urticae is the dominant species in Australia, where it is generally resistant to most of the older acaricides, e.g. OPs, organochlorines. There is a report of resistance in *T. urticae* to chlordimeform in Australia with a probable cross-resistance to amitraz but no resistance to cyhexatin and propargite. *Tetranychus cinnabarinus* is an important pest of tree fruit in South Africa and Israel where cyhexatin and propargite are also the main acaricides used in recent years.

In Europe, South Africa, and Australasia, many of the new acaricides considered promising are similar to those in North America, e.g. amitraz, fentrifanil (PP199), Zardex, nitrilocarb, and malonoben.

In Japan, the mite pests of deciduous fruit are *P. ulmi* and *T. urticae.* As in the other parts of the world resistance has occurred to OPs and most of the acaricides used extensively from the 1950's to the present. For example, there is resistance in some regions to OPs including phenkapton, vamidothion (Kilval), and to tetradifon, dicofol, chlorfenethol (Dimite), oxythioquinox, binapacryl, and chlorfensulphide. Some of these acaricides and other organochlorines alone or in combination are still effective in some areas. Currently recommended as alternatives, where effective, are dicofol, acarol (Bromopropylate, Neoron), chlordimeform, cyhexatin, fenbutatin-oxide, benzomate (Citrazon, developed in Japan), chlorfenson (ovex), and binapacryl.

Combinations or mixtures of acaricides as formulations have been used extensively in Japan and may have led to much of the cross-resistance that exists. Examples of mixtures include Milbex (CPAS + BCPE), NA66 (chlorfenethol + an OP), chlorfenethol + chlordimeform, Mitecidin C (polynactin + chlorfenson).

Resistance in the citrus red mite, *P. citri,* has occurred in different regions of Japan to tetradifon, dicofol, Phenkapton (and other OPs), oxythioquinox, chlorobenzilate, vamidothion and chlorfenson. An isolated case of resistance to benzomate (Citrazon) has been reported. However, most of the above acaricides, plus chlordimeform, chloropropylate, binapacryl and propargite were recommended up to 1975. Formulations of mixtures recommended were azoxybenzene + Omite, azoxybenzene + Smite (PPPS), and aramite + azoxybenzene. Twenty different kinds of acaricides are currently available to growers in Japan and alternating acaricides is recommended. Many new compounds are tested annually against all tetranychid species and in 1978 Zardex, Osadan (chemical unknown), and Acrex (dinobuton) are being subjected to related test programs. Ando (1975) has published a history of acaricide development in Japan from the time ovex, the first acaricide, was marketed in 1958.

The most important tetranychid pest of citrus in North America is the citrus red mite, *Panonychus citri.* Resistance in *P. citri* has been confirmed in California, to tetradifon, ovex, oxythioquinox and most OPs, and in Arizona to dicofol. In California standard test methods for resistance give poor correlations because control of *P. citri* depends on long exposure to acaricide residues. A lesser pest, the Texas citrus mite, *Eutetranychus banksi* is tetradifon-resistant in Texas. The false spider mite, *Brevipalpus,* is an occasional problem in Texas. Mites in Texas, California, and Arizona, including the important citrus rust mite (covered under eriophyids), are difficult to control or "knock out" with most acaricides. Poor coverage due to adverse weather conditions is a factor in poor control.

Current recommendations include oil, propargite, Plictran, dicofol, ethion and chlorobenzilate (Acaraben). Plictran must be applied in California before young fruit appear because of phytotoxicity problems.

New, promising compounds being tested include most of those described under deciduous fruit: Zardex with its long residual action and safety to beneficials is being considered for inclusion in California IPM, but results in Arizona have varied from excellent to fair; Fusilade (PP199), malonoben (GCP5126), amitraz (BAAM), and DPX3792. Aldicarb soil treatments have given excellent long-term control of citrus mites in three years of testing in Arizona. Application of methi-dathion (Supracide), and phosalone for other pests have been followed by mite infestations, in Arizona.

The mite pests of cotton in the U.S.A. are the strawberry spider mite, *Tetranychus turkestani*, the carmine spider mite, *T. cinnabarinus*, the Pacific spider mite, *T. pacificus,* and the two-spotted spider mite, *T. urticae.* Another six tetranychid species are of lesser importance. These mites are resistant to some degree to most OPs, with the exception of phorate (Thimet) seed or soil treatment. There is localized resistance, in Arkansas, to monocrotophos (Azodrin), and dicofol resistance in *T. pacificus* in California. However, chlorpyrifos (Lorsban), and methidathion are still recommended in most regions for mites on cotton. Propargite is also used and is reported safe for beneficials. Chlordimeform is still available in some areas, but its use is limited and closely supervised.

Of new materials DPX3792 was less effective than dicofol against the carmine and strawberry mites in Arizona. Di-Syston seed treatments were promising for control of the strawberry mite.

With the large complex of tetranychids on cotton, and difficult control conditions, there is a real need for new effective acaricides with low toxicity to beneficials.

NEW ACARICIDES AGAINST MITES UNDER GLASS

The major mite pest under glass in Europe is *T. urticae.* In Holland, Belgium and U.K. resistance of this mite is widespread to OPs, tetradifon, dicofol, formetanate, binapacryl and oxythioquinox (quinomethionate, Morestan). The mite is resistant to aldicarb on year-round chrysanthemums in the U.K.

The problem with *T. urticae* mites on tomatoes and cucurbits in greenhouses has been alleviated by the introduction of predacious *Phytoseiulus persimilis* in integrated programs in the U.K. and Canada. The systemic carbamates aldicarb and oxamyl (Vydate) are registered for vegetables in the U.K. In floriculture, dienochlor has only recently become available but has been used longer on the Continent. In Canada few acaricides are officially registered for the use of growers not following the integrated program. Available compounds do not provide adequate control in British Columbia and Ontario greenhouses. Companies are reluctant to gather data necessary for registration of low-acreage crops. The organotin compounds Plictran and Vendex have proven effective, as well as Temik (aldicarb), for some crops. Compounds relatively more toxic to *T. urticae* than *P. persimilis* are urgently required for the integrated program. The tin-containing acaricide XE567 appears to have these characteristics in Ontario tests. In greenhouse trials in Quebec, the new compound UC 44858, now discontinued, gave excellent control of *T. urticae* and surprisingly there was no resistance to Kelthane. There is little evidence for resistance to dienochlor (Pentac) against *T. urticae.*

The withdrawal of greenhouse registration for oxythioquinox (Morestan) in North America was a setback for the integrated program for greenhouses, involving *T. urticae* and whitefly.

ERIOPHYIDS

Several probable cases of resistance developing in eirophyid mites have been reported. There is strong evidence for resistance in the apple rust mite, *Aculus schlechtendali*, to parathion (British Columbia), and other OPs, to endosulfan (Thiodan) and chlordimeform in Washington State. Also, failure to control the pear rust mite *Epitrimerus pyri* with OPs and endosulfan were common and widespread in Washington State. In Israel zineb, used for many years, failed to control the citrus rust mite, *Phyllocoptruta oleivora*, in 1967 trials. In Texas, chlorobenzilate has been less effective on the citrus rust mite for the last three years and the need for increased application rates indicate resistance.

In Japan one instance of resistance to chlorobenzilate was confirmed in the Japanese citrus rust mite, *Aculus pelekassi.*

It seems likely that other cases of resistance in eriophyids exist, but the difficulty of testing for resistance in these mites has prevented these being recognized.

Sulphur, in various forms, particularly lime sulphur, has proven very effective against eriophyids. However, problems such as phytotoxicity have resulted in a great reduction in its use.

Of the organochlorines, dicofol (Kelthane) has been used against a wide range of eriophyid species, e.g. *E. pyri, P. oleivora,* and *A. pelekassi.* The related compounds chlorobenzilate (Akar), chloropropylate (Acaralate, Rospin), bromopropylate (Acarol, Neoron), and propargite (Omite) are also extremely effective.

The organotin compound cyhexatin (Plictran) controls eriophyids including *A. schlechtendali,* the pear rust mite, *E. pyri* and *P. oleivora.*

Formetanate (Carzol) has been recommended for control of eriophyids on apple and citrus in North America.

Against the Japanese citrus rust mite, dialifos (Torak), Sipdan (chlordimeform + OP), cyhexatin, and oil have all proven effective.

Organophosphorus compounds have given rather conflicting results against eriophyids. However, examples of good control achieved are as follows: vamidothion (Kilval) and ethion against *E. pyri, A. schlechtendali* and against

Phyllocoptes gracilis; azinphos-methyl (Guthion) against *Eriophyes pyri*; dimethoate (Rogor) against *Aculus fockeui* and *Epitrimerus pyri;* and Diazinon against *Epitrimerus pyri.*

Oxythioquinox (Morestan) has proved successful against many eriophyids including *P. oleivora, Epitrimerus pyri,* and *A. schlechtendali.*

Fungicides which have had some success against eriophyids are mancozeb (Dithane) and especially zineb against *P. oleivora,* maneb and zineb against *Epitrimerus pyri,* and benomyl (Benlate) against *Aceria sheldoni.* However, zineb is reported to kill the fungus active against *P. oleivora* in Texas. The dinitrophenol 'derivatives' dinocap (Karathane) and binapacryl (Morocide) are known to have activity against some eriphyids, including *Epitrimerus pyri* and *A. schlechtendali.*

Endosulfan (Thiodan) is certainly one of the most widely used and successful compounds for eriophyid control. Examples of its use are against *Acalitrus essigi* and *Epitrimerus pyri.* It is sometimes used in combination with oil.

The so-called 'bud-mites', i.e. those eriophyids that spend most of their life cycle within the buds of the host, are particularly difficult to control. Often control can only be achieved by spraying during the short dispersal phase of the mite. For example in tests in England only 9 compounds out of 77 gave reasonable control of the black currant gall mite, *Cecidophyopsis ribis.* Compounds which have given success against bud mites are endosulfan, e.g. against *C. ribis* and *A. sheldoni;* endrin against *C. ribis* and the hazel bud mite *Phytoptus avellanae*; chlorobenzilate and chloropropylate, e.g. against *A. sheldoni;* oxythioquinox against *A. sheldoni* and carbaryl (Sevin) against the bud form of *Eriophyes pyri.*

The blueberry bud mite, Acalitus vaccinii, an occasional problem in New Jersey, is tolerant of specific acaricides such as tetradifon, dicofol, propargite and chlorobenzilate, but is easily controlled by oil, ethion, and especially endosulfan.

There are more recent materials which have been reported to control eriophyids: fenbutatin-oxide (Vendex, Torque) proved effective against *A. schlechtendali, Epitrimerus pyri,* and especially *P. oleivora;* further Comite (a propargite formulation), and oxamyl (Vydate) which controlled *P. oleivora* and *A. schlechtendali.* Fentrifanil (Fusilade, PP199) gave good control of *P. oleivora* in Arizona and Texas, but some injury to immature lemon leaves occurred during one season's tests in Arizona. No injury occurred in Texas.

Diflubenzuron (Dimilin; Philips-Duphar, Thom.-Hayward), a new molt inhibitor, is reported to be very effective against the citrus rust mite, and also the apple rust mite. Zardex is considered a promising acaricide for citrus pest management in California. Aldicarb (Temik) as a granular application gave control of the citrus rust mite for 3 months in Texas and is also being considered for IPM programs of citrus. Amitraz (Mitac, BAAM) is reported effective in the control of *A. schlechtendali,* and *A. essigi.* Its near relative

Galecron, now withdrawn, is also very toxic to eriophyids. Hostathion (Hoe 2960) shows promise against the Japanese citrus rust mite *A. pelekassi.*

There are occasions when it may be useful to preserve populations of free-living eriophyids to provide food for predacious mites, whilst controlling spider mites. A few chemicals can give this selective effect.

Oil sprays applied at ½″ green bud stage, tetrasul (Animert), and tetradifon (Tedion) applied at prebloom period are reported to kill tetranychid spp. but spare *A. schlechtendali,* an alternate host of major phytoseiid predators in British Columbia, Nova Scotia, and Washington.

Conversely, edosulfan (Thiodan) is relatively more toxic to *A. schlechtendali* than to *P. ulmi.* Chlorobenzilate, very toxic to citrus bud and rust mites is realtively non-toxic to tetranychids on citrus in California.

THE FUTURE FOR NEW ACARICIDES

'Bud mites' are likely to remain the most important eriophyid pests and new materials of increased persistence and/or systemic activity will be necessary to improve control.

Rust mites should become increasingly important. Populations are likely to increase with the change to non-acaricidal fungicides. This is happening on apple and pear in the U.K. where binapacryl and dinocap, which exert some control on eriophyids, are being replaced by chemicals such as bupirimate (Nimrod) which are not acaricidal. The importance of rust mites as a component of integrated pest management programs is also increasing. If populations of rust mites are to be allowed to survive to provide food for predacious mites, more knowledge will be needed of the selectivity of chemicals to eriophyids compared to tetranychids and phytoseiids and also of the economic damage levels of rust mites, on which little research seems to have been done.

SUMMARY

Of the many acaricides or insecticide-acaricides developed since the 1950's, few are still effective, or available to growers, because of resistance development, toxicological factors or cost factors. To the groups heavily used now for many years such as the tin compounds, propargite and dienochlor, others have recently been added, e.g. fentrifanil, amitraz and benzoximate and yet more new groups are under development. Integrated pest management offers a means of conserving selective acaricides, but the very need for selectivity is narrowing the range of suitable chemicals. There is more need than ever for cooperation with chemical companies in the evaluation of new compounds if our growers are to have effective 'back-up' acaricides for the future.

ACKNOWLEDGEMENTS

The authors greatly appreciate the many researchers in North America, Europe, South Africa, the Middle East, and Australasia who kindly contributed information, and regret that space limitations did not permit personal references to them.

REFERENCES

Dittrich, V. (1975). *Z. angewandte Entomologie,* **78**, 28-45.
Helle, W. (1965). *Advances in Acarol.* **2**, 71-93.
Helle, W. and Vrie, van de M. (1974). *Outlook on Agriculture* **8**, 119-125.
Ando, Meiki (1978). Japanese pesticide Information, *Japan Plant Prot. Assoc.* **34**, pp. 14-21.

NEW PESTICIDES: CHANCES OF DISCOVERY, DEVELOPMENT AND COMMERCIALIZATION

Michihiko Sakai

Pesticide Research Laboratories
Takeda Chemical Industries, Ltd.
Sakyo-ku, Kyoto, Japan

INTRODUCTION

It is expected that pesticides will continue to play an essential role in pest control, though an alternative technology offers advantages over total reliance on pesticides. Also, emphasis on the safety of pesticides has been markedly increased concomitantly with the growing public consciousness of the hazards of chemical substances. Thus, greater effort has been expended to develop safer, more selective and more efficient pesticides. It is apparent now that pesticide research and development (R&D) in industry is subjected to social and economic trends.

In this paper the present situation and the prospects of pesticide R&D are discussed.

APPROACHES TO THE DISCOVERY OF NEW PESTICIDES

The change in the number of chemicals screened to obtain a commercial pesticide in the course of the last 20 years is shown in Table I. We need now to screen 6.5 times more chemicals than in 1956. The higher number implies an increasing difficulty to satisfy the requirements for the chemicals to be more effective to targets but safer to mammals and the environment.

Inevitably the methods to prepare the chemicals and the biological screening procedure should be improved to increase the chance of discovery of candidates which will finally reach the market.

The approaches to discover a new pesticide can be categorized schematically into three classes: 1) isochemical synthesis in which the analogues of known effective chemicals are screened; 2) random or shotgun approach where various kinds of chemicals are screened indiscriminately; and 3) rational approach by which chemicals are screened which were prepared according to

Vol. I: ISBN 0-12-592201-9

completely novel concepts derived from basic biological research.

The method which is most widely used at present and is likely to continue to be used in the future is the isochemical approach. As is obvious in the history of pesticides, this approach was highly successful in the development of better chemicals. However, this appraoch frequently implies the frustrating problem that the chemicals thus prepared are liable not to surpass their model substances in biological activity. The shotgun approach may have been utilized in expectation of finding an unconventional pesticide, but the usefulness of this approach may be questionable because of its low probability for discovery of a new product. There is no concrete precedent for the rational approach excepting the juvenoids, but this approach is increasingly important as I will mention later.

As for chemical sources of recently discovered insecticides and acaricides (Table II), the organophosphate-type is found the most frequently of modern insecticides and will remain so in the future. Frequent substitution of acaricides is mainly due to the development of resistance of mites. However, in regard to the fact that acaricides have been discovered from various chemical groups, further discoveries of novel acaricides may be expected in the future.

TABLE I.

Number of Chemicals Screened to Obtain One Commercial Pesticide (Gilbert, 1978).

Year	No. of Chemicals
1956	1800
1964	3600
1967	5500
1970	7400
1973	10200
1977	12000

TABLE II.

Transition of Registration Status of Insecticides and Acaricides in Japan (Asakawa, 1978)

Chemical Category	No. of Chemicals in Indicated Status			
	Registered as of 1966	1966-1977		Registered As of 1977
		Newly Registered	Void	
Arsenicals	2	0	1	1
Natural substances	7	0	2	5
Chlorinated hydrocarbons	9	0	8	1
Organophosphates	30	20	8	42
Carbamates	6	9	7	8
Fluorine compounds	3	1	4	0
Acaricides	21	13	16	18

THE RESEARCH AND DEVELOPMENT PROJECT

The flow of the pesticide development commences with the discovery of a new active chemical. The organization and flow of the R&D project is planned not only to be effective regarding discovery and evaluation of marketability of a candidate, but also to accommodate to the complex registration requirements (Fig. 1). The components of the organization involve different fields of biology, toxicology, chemistry and economics, and they are integrated into a critical path schedule. The patent strategy is also important to ensure a return on the investment in R&D.

In each part of the path, the assessment is made rigorously. The rigorousness is indicated not only by the desire for higher efficacy but also by stringent requirements for registration. A vast amount of the complex data, as exemplified in Japanese registration requirements (Table III), has to be furnished from the R&D project for registration purposes. Along with these, the profitability, i.e. the market potential, manufacturing cost and investment are analyzed to make a decision in whether to continue or terminate the R&D (see Wellman, 1967; Gilbert, 1978).

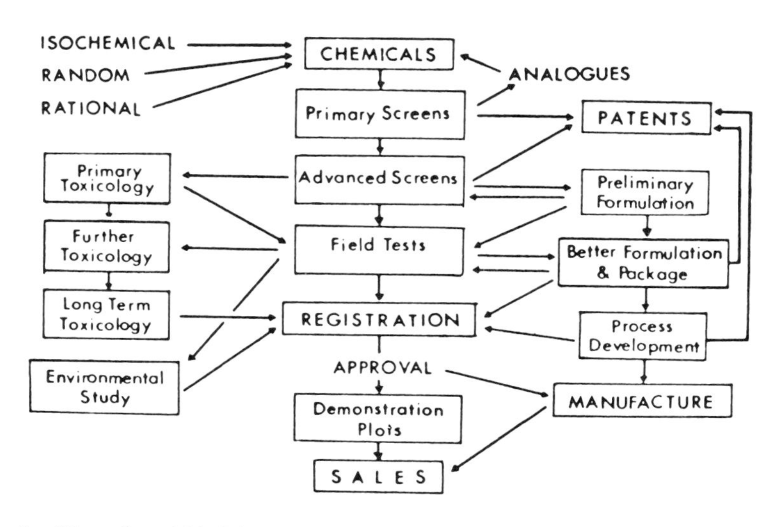

Fig. 1. Flow of pesticide R&D project.

PROBLEMS FACED BY INDUSTRY

According to recent evidence it has not been unusual in Japan for industrial firms to spend more than 8 years to complete the R&D project, because complex and long-term studies are needed to verify the safety of the candidate

TABLE III.

Data Required for Pesticide Registration in Japan.

I **Chemical Data**
1. Active Ingredient
   a. Identity
   b. Physical and chemical properties
   c. Method of analysis
2. Impurities
   a., b. and c. as for 1.
3. Formulation
   a. Composition
   b. Properties
   c. Method of analysis
   d. Package

II. **Efficacy Data**
1. Spectrum of activity
2. Dose for efficient control
3. Application technique
   a. Timing
   b. Frequency
   c. Spray method and/or site
4. Crop Tolerance

III. **Toxicological Data**
1. Acute toxicity
   a. Oral (2 species)
   b. Subcutaneous
   c. Intravenous
   d. Inhalation
   e. Skin irritation
   f. Mucous membrane irritation
2. Subacute toxicity
   (1 month feeding as preliminary test of chronic toxicity test)
3. Chronic toxicity
   a. 2 year feeding study
   b. 3 generation reproduction study
   c. Hematological and biochemical examinations
   d. Pathological and histological examinations
   e. Carcinogenic study
   f. Mutagenic study
   g. Teratologic study
4. General Pharmacology
5. Fate in animal
   a. Distribution and accumulation
   b. Metabolism

IV. **Environmental Science Data**
1. Fate in plant
   a. Residue analytical method
   b. Residues
   c. Metabolism and degradation
   d. Effect on flavor (tea and tobacco)
2. Fate in soil
   a. Residue analytical method
   b. Residues in field
   c. Half-life in laboratory
3. Wildlife studies
   a. Acute fish toxicity
   b. Acute bird toxicity
   c. Toxicity to natural enemies
   d. Air, water and soil contamination

V. **Special Studies**
   a. Residual toxicity to silkworm on mulberry leaves
   b. Honeybee toxicity
   c. Effect on livestock

pesticide to mammals and the environment. Moreover, the development may occasionally be arrested or slowed down, if a certain part of the critical path is faced with difficulty. The same trend exists in many other countries.

As a result of the expansion of the R&D program in its area and term, its cost to develop a product is dramatically rising year by year, e.g. 1.2, 4.1 and 14 million US dollars in 1956, 1969 and 1977, respectively (Waitt, 1975; Gilbert, 1978). Therefore, the R&D strategy is directed to commercialize a pesticide marketable in a profitable market to ensure a positive return on investment. This is the reason why industries attach importance to major crops, but preclude minor uses for development of a new compound.

Substitution of a marketed pesticide by a new one increases the risks of the pesticide business. Since the appearance of a competitor from another company is hardly predictable, the solution to decrease the risk is again to develop more effective pesticides in major crops.

It is likely, unfortunately, that acaricides cannot be a main subject of the R&D, because of their usable life-time is likely to be shortened by rapid development of resistance.

Establishment of a way, which leads to safer chemicals and safer use, cannot be found for the successful exploration of new concepts, their transformation into pesticide development, and the solution of highly complex chemical structure-activity relationships, except by more interdisciplinary exchange of knowledge among academic, government, and industry researchers in recognition of their joint responsibility.

SUMMARY

The present situation and the prospect of pesticide research and development (R&D) in industry is discussed with respect to the chance of discovery of new products and the organization of R&D. More chemicals have to be screened to satisfy the demands for more efficient and safer chemicals, and to fulfill the increasingly stringent registration requirements. Inevitably the area and time of R&D on pesticides have expanded. To call in the increasing costs and to invest in discovery of a new product, industry is liable to favor candidates which fit profitable markets, whereas the same economics preclude pesticide development for minor uses. However, industry feels that they have a fair chance of success, because pesticides are essential in modern agriculture and a vast amount of chemicals is demanded all over the world. To develop more effective and safer pesticides and application techniques in future, it is urgently needed to promote basic research by coordination among academic resources and that in governmental and industrial agencies with a sense of responsibility from all concerned.

REFERENCES

Anonymous (1977). *Farm Chem.*, Sept. 38-43.
Asakawa, S. (1978). (in Japanese) *Kongetsu no Noyaku. Jan.* 14-19.
Braunholtz, J. T. (1976). *Proc. XV Intern. Cong. of Entomol., Washington,* 747-755.
Eichers, T. R., and Andrilenas, P. A. (1978). *USDA Agr. Econ. Rpt. No. 399.* 20 pp.
Gilbert, C. H. (1978). *Farm Chem.* Apr. 20-27.
Waitt, A. W. (1975). *Pestic. Sci.* **6**, 199-208.
Wellman, R. H. (1967). *In* Fungicides, Vol. I, 126-151. (D.C. Toreson, ed.) Academic Press, New York and London.

BIONOMICS OF *STENEOTARSONEMUS SPINKI* ATTACKING RICE PLANTS IN TAIWAN

C. N. Chen, C. C. Cheng and K. C. Hsiao

Entomology Division, Plant Protection Center, Taiwan
Wufeng, Taichung Hsien, Taiwan
Republic of China

INTRODUCTION

The tarsonemid mite, *Steneotarsonemus spinki* Smiley was recently found to attack rice plants in Taiwan, especially second crop rice growing in Southern parts of the island. The outbreak in 1976 caused severe damage to the rice crop. The area of infestation in the second rice crop increased from 17,100 ha. in 1976 to 19,146 ha. (*ca* 4.5% of the total cropping area) in 1977. The percentage of empty grains (including partially filled grains) ranged from 20 to *ca* 60%, with a grain loss equivalent to 20,000 metric tons, valued at *ca* US $9,200,000.

In addition to its direct damage, the mite usually carries spores of rice sheath rot fungus (*Acrocylindrium oryzae* Sawada), which causes brownish spots on rice sheath and grains (Hsieh *et al.*, 1977), damage termed "sterile grain syndrome." The syndrome is manifested by 1) a loose and brownish flag leaf sheath; 2) a twisted panicle neck; and 3) impaired grain development resulting in empty or partially filled grains with diseased brown spots and the panicles standing erect.

The appearance of each stage of the mite was briefly described by Ou *et al.* (1976), although adults had been previously described by Smiley (1967) and Gutierrez (1967). Detailed external and internal morphology was studied by Chow *et al.* (1976) and Lee (1977) by a scanning electron microscope. The mite with its puncturing-sucking mouthparts can suck sap from inner parts of a rice leaf sheath (Lee and Chow, 1977). Lo and Hor (1977) made a preliminary field investigation and suggested that the mite was the main causal agent in this "sterile grain syndrome."

The present paper reports bionomic data in Taiwan, including effects of temperature and humidity on mite development and mortality, its role in the "sterile grain syndrome," the screening tests of pesticides in the greenhouse, and some preliminary observations made in the field.

Vol. I: ISBN 0-12-592201-9

MATERIALS AND METHODS

A rice leaf sheath free from any pest was cut into sections of 15-20 mm. Five sections were put inside a 9 cm dia. petri dish containing a piece of wet filter paper. A quiescent female was transferred onto each section by a fine camel hair brush employing a binocular microscope. Five to 10 petri dishes, each containing 10 females were put inside a growth chamber, with a light intensity of 10 lux and photoperiod of 12 L : 12 D. Experiments were conducted at 25°C, 28°C and 30°C. Each day at 8:30 a.m., 12:30 p.m. and 4:30 p.m., the petri dishes were examined under a binocular microscope. Once the eggs were laid, the females were removed to another petri dish until no more eggs were laid. The leaf sheath sections were renewed at 3-day intervals. The development of each stage, longevity of adults and its reproduction were recorded.

An experiment designed to study interactions of temperature and humidity on the mortality of the mites utilized temperatures of 25, 28, 30 and 32°C in 4 growth chambers. Four plastic containers, 13.5 cm dia. x 27 cm height, each containing 2 plastic jars 7.5 cm dia. x 10.5 cm height, were filled with *ca* 200 ml of water or one of 3 saturated salt solutions, viz. KCL, $NaNO_2$ and CrO_3, to regulate the R.H. in the container at 100%, 85%, 65 ± 2% and 40%, respectively. Two petri dishes each with 20-70 females were then apportioned to each container. Mortality was checked once every hour for 6 consecutive hours and then at 24 hours.

An experiment to relate the mite to rice yield loss and quality was conducted in an isolated area where no rice plants had previously been planted. There were 6 experimental plots as well as a check plot and each plot contained 40 pots with one hill of rice plants per pot (16 cm dia. x 9.5 cm height). The plots were randomly assigned to treatment. The check plot was sprayed with a combination of tricyclohexyltin hydroxide diluted 2000x plus 5% benomyl diluted 500x at weekly intervals to suppress mite infestation and disease infection. The rest of the plots were infested by inserting 5 mite-infested sheath sections of 10 cm each into 5 hills (*ca* 1000 mites). Thus, T-I and T-II plots were infested at the tillering stage. T-II was sprayed with pesticides in a treatment identical to that of the check plot while T-I received no pesticide treatment. B and H plots were infested at the booting and heading stages, respectively. Neither of these plots were treated with any pesticides. The F plot was infested at the flowering stage by introducing mites directly on the panicles. It also received no pesticidal treatment. The mite density was estimated by randomly selecting 45 flag leaves from 15 hills in each treatment. Only half of the flag leaf sheath was examined to estimate the density of the mite just before harvest. The following data were also obtained from these 15 hills: the percent of twisted panicle necks, diseased leaf sheaths, empty grains, and the weight of 100-grain, weight of a panicle and the percent yield reduction.

RESULTS AND DISCUSSION

The development of the mite was greatly influenced by the ambient temperature. On the average, the mite required 17, 4 and 2.75 days to complete its development from egg to adult stage at 25°C, 28°C and 30°C, respectively (Table I). At 25-28°C, the adult female and male respectively lived for 15 ± 1.0 days and 7.6 ± 0.4 days; while both sexes lived for only 5 days at 30°C. The preoviposition period was about 2 to 5 days and egg-laying period varied from 2 days at 30°C and 7 days at 28°C to 13 days at 25°C. Each female could lay from 0 to 78 eggs during its life span, with an average of 30.8 ± 3.4 eggs (Table I).

TABLE I.
Influence of Temperature on Development, Longevity of Adults and Reproduction of the Rice Tarsonemid Mite.

Temp. (°C)	Developmental Stage (Days)			Adult Longevity (Days)	
	Egg	Larva	Quiescent	♂	♀
25	10 ± 0.4 (8-14)	6 ± 0.3 (5-9)	1.0	7 ± 0.3 (5-9)	18 ± 1.8 (5-31)
28	2 ± 0.2 (1-5)	1.5 ± 0.2 (1-4)	0.5	8 ± 0.7 (4-14)	13 ± 0.5 (7-15)
30	0.5 ± 0.06 (0.25-1.5)	2 ± 0.3 (1-4)	0.25	5 ± 0.3 (2-7)	5 ± 0.4 (3-8)

Reproduction

Temp. (°C)	Preoviposition Period (Days)	Oviposition Period (Days)	Eggs/♀
25	5 ± 0.4 (2-7)	13 ± 1.5 (3-20)	23 ± 4.7 (0-53)
28	2 ± 0.1 (1.5-4)	7 ± 0.4 (5.5-11)	48 ± 5.2 (2-78)
30	3 ± 0.2 (1.5-3.5)	2 ± 0.25 (1.5-4.5)	18 ± 3.4 (0-39)

The mite was found to be highly sensitive to humidity. The mortality rate of the mite increased within the range of 25°C to 32°C, as the temperature increased and the relative humidity decreased (Fig. 1). Such a relationship can be expressed in a multiple linear regression equation $\hat{Y} = 60.89 + 1.20 X_1 - 0.94 X_2 + 1.92 X_3$ ($R^2 = 0.6352^{**}$), where $\hat{Y}$ = % mortality; X_1 = temp. (°C); X_2 = % RH; and X_3 = duration (hrs). In general, when the RH is less than 40% all of the mites will die within 4 hrs., if the temperature is above 30°C; and within 6 hrs when the temperature is from 25°C to 28°C. Hence, temperatures between 28°C-30°C and a RH above 80% is optimal for the mite.

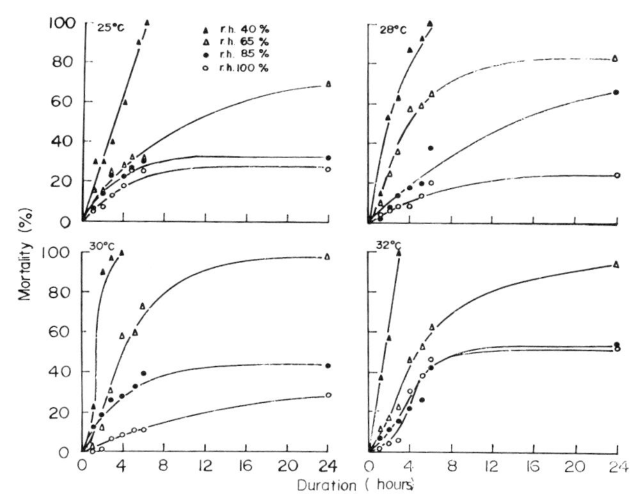

Fig. 1. Influence of temperature and humidity interactions on the mortality of *Steneotarsonemus spinki.*

The correlation coefficients between each component of the "sterile grain syndrome" are shown in Table II. The density of mites (X_1) and the percent empty grains (X_2) are correlated with yield reduction (Y) and the X_1 and X_2 are also highly correlated. In fact, the relationship between the % yield reduction (Y) and the mite density (X_1) can be expressed as $Y = -82.09 + 49.24 \log (X + 1)$. ($r^2 = 0.9698^{**}$) (Fig. 2). Furthermore, the relationship between % diseased leaf sheaths (X_4) and % twisted panicle necks can be expressed as $1/Y = 0.0661 + 0.1868/X$ ($r^2 = 0.9967$) (Fig. 3). The inference from these analyses is made that the tarsonemid mite plays a major role in causing the "sterile grain syndrome". Its sucking damage resulted in empty grains while the fungus spores carried by it caused the hardening of the flag leaf sheath which with its tightness hindered the heading of the panicle. As a result the fast growing and extending panicle neck is forced to become twisted inside the leaf sheath and once the heading is completed the flag leaf sheath becomes loose.

TABLE II.

Correlation Coefficient Between Each Component of the "Sterile Grain Syndrome." X_1 = Density of Mites (No./Flag Leaf Right Before Harvest), X_2 = % Empty Grain (Including Partially Filled Grains), X_3 = % Panicle Neck Twisted, X_4 = % Diseased Leaf Sheaths, Y = % Yield Reduction.

	X_1	X_2	X_3	X_4
Y	0.9730**	0.8307**	0.6295	0.3277
X_1	1.0000	0.8169*	0.4624	0.1652
X_2		1.0000	0.4984	0.0445
X_3			1.0000	0.6942

*, **Significant at the 5% and 1% level.

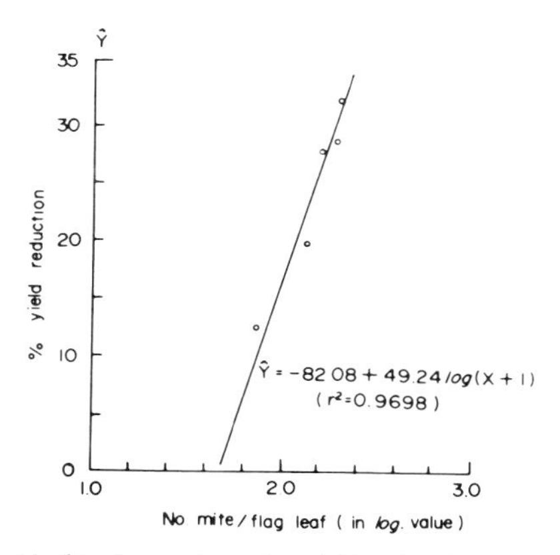

Fig. 2. Mite density just before harvest vs. rice yield reduction.

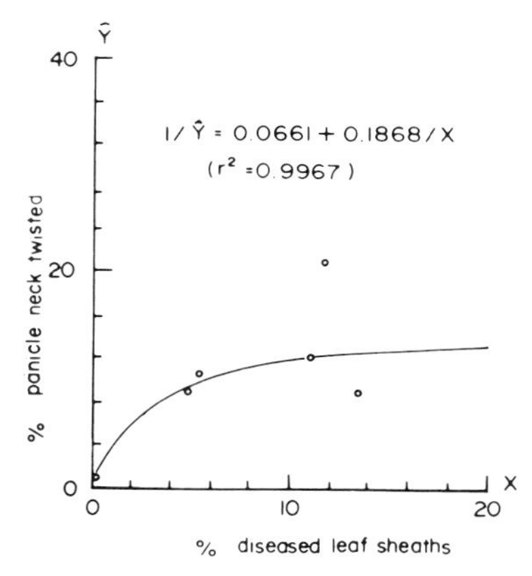

Fig. 3. Percentage of diseased rice leaf sheaths vs. percentage of twisted panicle neck.

The peak population of the mite in the fields is from heading to milking stage which occurs during the translocation of synthesized nutrients. The damage reduces the percent of filled grains as well as the weight of 1000-grain and consequently as great a yield reduction as evidenced by the damage of the brown planthopper (Chen and Cheng, 1978).

A comparison of the rice quality of mite-damaged grains with those of normal grains showed that the percentage of empty and partially filled grains in mite-damaged grains was significantly increased. The percentage of marketable rice (including white, green and rusty rice) from mite-damaged grains ranged from 30% to 60%, while that of normal rice amounted to more than 80% (Fig. 4).

SUMMARY

The mite *Stenotarsonemus spinki* required 17, 4 and 2.75 days to complete its development from egg to adult stage at 25°C, 28°C and 30°C, respectively.

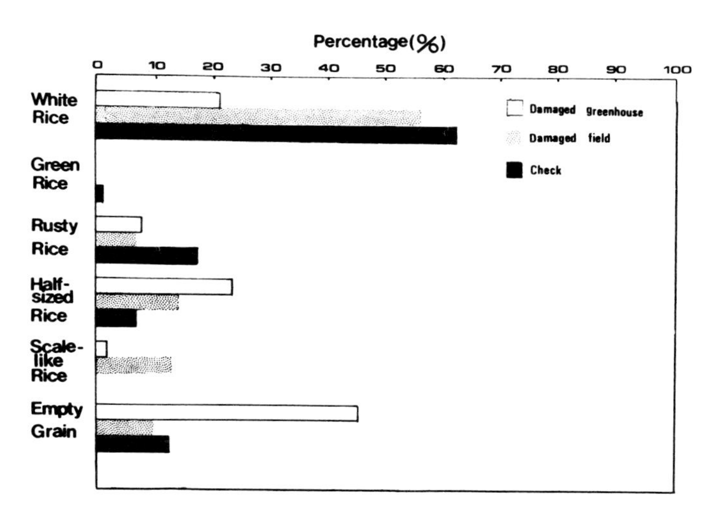

Fig. 4. Comparison of rice quality of mite damaged grains to those of normal ones.

At 25-28°C, the adult female and male respectively lived for 15 ± 1.0 days and 7.6 ± 0.4 days; while both sexes lived for only 5 days at 30°C. The preoviposition period was about 2 to 5 days and egg laying period varied from 2 days at 30°C and 7 days at 28°C to 13 days at 25°C. Each female could lay from 0 to 78 eggs during its life span, with an average of 30.8 ± 3.4 eggs.

The % mortality ($\hat{Y}$) influenced by both temperature (X_1, °C) and R.H. (X_2, %) and by duration (X_3, hrs) could be expressed as $\hat{Y} = 60.89 + 1.20 X_1 - 0.94 X_2 + 1.92 X_3$ ($R^2 = 0.6352^{**}$). Temperature between 28-30°C and a R.H. above 80% is optimal for the mite.

The mite's role in causing the so-called "sterile grain syndrome" and its relation to yield loss and quality of rice were also discussed.

The mite was most abundant during rice heading to milking stage in the field. After harvest, mites were found on the leaf sheath of rice stubbles and then on ratoons. No flora other than rice, and possible wild rice, so far surveyed were found to be host plants.

In the field the mites were most abundant during heading to milking stage, and as the rice approached its maturity the mite density declined rapidly. After harvest they were found on the leaf sheaths of rice stubbles and then on ratoons.

Extensive surveys of the weed flora (serving as its potential host plant) around mite-infested fields were made in 1977. Of 16 species belonging to five families so far examined, no rice tarsonemid mite has been found on any, although Lee and Chow (1977) suggested that the mite might survive on wild rice.

ACKNOWLEDGEMENTS

We thank Mr. Wen-ying Su, Mr. Michael Flanagan, Mrs. Y. R. Kao, for assistance in conducting this study. This project was supported by the Sino-American Joint Committee on Rural Reconstruction.

REFERENCES

Chen, C. N. and Cheng, C. C. (1978). *I. Plant Protec. Bull.* (R.O.C.) (In Press)

Chow, Y. S., Lee, T. L., Wu, H. P., Hsieh, K. C., and Wang, C. H. (1976). *Natl. Sci. Council Monthly* (R.O.C.) **4**, 38-43.

Gutierrez, J. (1967). *Bull. Entomol. Soc.* France **72**, 323-330.

Hsieh, S. P. Y.., Liang, W. J., and Chang, S. Y. (1977). *Plant Protec. Bull.* (R.O.C.) **19**, 30-36.

Lee, T. L. (1977). Master's thesis, Institute of Biological Sciences, National Taiwan Normal Univ., Taipei, R.O.C. 59 pp.

Lee, T. L. and Chow, Y. S. (1977). *Natl. Sci. Council Monthly* (R.O.C.) **5**, 960-963.

Lo, K. C. and Hor, C. C. (1977). *Natl. Sci. Council Monthly* (R.O.C.) **5**, 274-284.

Ou, Y.T., Fang, H.C. and Tseng, Y. H. (1977). *Plant Protec. Bull.* (R.O.C.) **19**, 21-29.

Smiley, R. L. (1967). *Proc. Entomol. Soc. Wash.* **69**, 127-146.

ACQUISITION AND RETENTION OF *PSEUDOMONAS MARGINATA* BY *ANOETUS FERONIARUM* AND *RHIZOGLYPHUS ROBINI*

S. L. Poe

Department of Entomology & Nematology
University of Florida
Gainesville, Florida

W. E. Noble

Department of Ornamental Horticulture
California Polytechnic State University
San Luis Obispo, California

R. E. Stall

Department of Plant Pathology
University of Florida
Gainesville, Florida

INTRODUCTION

Several species of bulb and soil inhabiting astigmatid mites have been named as associates of decaying, pathogen infected gladiolus corms from the sandy soils of Florida. Engelhard (1969) listed mites of the *Rhizoglyphus solani* group, *Tyrophagus palmarum* (Oudemans), *T. putresentiae* (Schrank) and *Histiostoma* (= *Anoetus*) *feroniarum* (Dufour). Subsequent collections (Poe, 1971) revealed *R. robini* Claparede and *A. feroniarum* as the two principal species. These mites have been implicated as potential vectors of phytopathogens from field to field, within fields and from plant to plant in field and storage.

Noble and Poe (1973) reported a high degree of attractancy of *R. robini* and *A. feroniarum* to *Fusarium oxysporum, F. gladioli* and the scab organism *Pseudomonas marginata* McCullough, respectively. All active stages of the mites readily invaded petri dish cultures of the pathogens, however, the most frequent migrant was the uniquely adapted deutonymph (hypopus). *Anoetus*

Vol. I: ISBN 0-12-592201-9

mites fed and reproduced well on bacterial cultures. However, *Rhizoglyphus* preferred *Fusarium* cultures. These data suggest that each mite species has a preferred microfauna on the host corms. Both species are associated with and seem to prefer soft, moist, decaying tissue.

Following these discoveries, experiments were designed to determine 1) if viable bacteria were ingested, 2) if any bacteria that passed through the alimentary canal would remain viable, and 3) the length of time after feeding that a mite would disseminate viable bacteria. All tests were done using the scab-causing organism *Pseudomonas marginata.*

MATERIALS AND METHODS

Two isolates of *Pseudomonas marginata* McCullough, Br-1SR (Streptomycin resistant mutant) and F-1 (an isolate which formed a white precipitate (WP) on Nutrient Agar (NA) were used as hosts for *A. feroniarum* and *R. robini.* Four tests were done. In the first 2 tests separate cultures of the two mite species were fed on each bacterial isolate for 48h. Three mites observed probing the bacterial lawn were removed and surface sterilized by dipping in solutions of 95% ethanol or 0.78% sodium hypochlorite for 5 min. Mites were then removed, air dried for 10 min. and placed on separate NA plates. Two individuals were crushed with sterile probes and separately streaked over the medium. The third individual was rolled or dragged over the medium. Plates were incubated at 30°C for up to 72h with daily observation for bacterial growth.

In the second test separate surface sterilized mites fed as above were aseptically transferred to dilution tubes containing 0.45 ml of beef serum, crushed, and a dilution series run. The contents were placed on NA + 100 ppm Streptomycin and 100 ppm Penicillin (NASP) and incubated at 30°C for up to 72 hr.

In test 3 colonies of Br-1SR on NASP and F-1 on NA were infested separately with 20 *A. feroniarum* and 20 *R. robini.* Petri dishes were taped closed and mites allowed to feed 24 hr. Mites observed in contact with lawns were individually removed and placed on separate NA or NASP plates. At 24 hr intervals, each mite was aseptically transferred to another plate of similar medium. This procedure was repeated daily until the death or escape of the mite. The agar plates on which mites had resided were incubated at 30°C for 24 hr then examined for colonies of bacteria forming a white precipitate. Data from all tests were pooled and the mean number of days which a mite would transmit a bacterium forming a precipitate was determined for each species and isolate.

In test 4, "Beverly Ann" gladiolus plants were inoculated with an injection of 10^8cells/ml of *P. marginata* F-1, placed in a moist chamber and allowed to develop neck-rot symptoms. As the disease progressed, *A. feroniarum* mites were observed on the deteriorating tissue. Twenty-five mites were removed and placed on individual NA plates, incubated at 30°C for 24h, then removed. The plates were further incubated for 7 days and examined for the presence of precipitate forming colonies. Similarly, *Rhizoglyphus* mites were removed from symptomatic tissue of previously inoculated plants and individually plated on NA. Mites remained on the plates for 2 successive 48h periods at 30°C. Plates were observed daily for 10 days for the presence of WP colonies.

RESULTS AND DISCUSSION

Of the 17 mites (13 *A. feroniarum* and 4 *R. robini)* crushed on NA plates in test 1, white precipitate forming colonies appeared from 13 mites (Table I). From the 9 intact mites that had been surface sterilized, only one was positive for a precipitate and only a single colony was observed. In most of the trials of test 1, crushed mites of both species were positive for precipitate forming bacteria while the single mite left intact was negative. These data indicate that the bacterium was present within the body of the mites and that the sterilization treatment effectively eliminated the pathogens from the external body of the mites.

TABLE I.
Assay for Presence of *Pseudomonas marginata* McCullough Within the Body of *Anoetus feroniarum* and *Rhizoglyphus robini* Mites

Species	Bacterial Isolate	Colonies Producing a White Precipitate in Nutrient Agar			
		Crushed		Intact	
		No. Mites[a]	Rating[b]	No. Mites[a]	Rating[b]
Anoetus feroniarum	F-1	2	PP	1	P
		2	PP	1	A
		2	AA	1	A
		3	PPA	2	AA
	Br-1SR	2	PP	1	O
		2	PP	1	A
		2	PA	1	A
Rhizoglyphus robini	F-1	2	PP	1	A
Total:		17	13 P	9	1 P

[a] Mites surface sterilized by dipping 5 min. in 0.78% sodium hypochlorite or 95% ethanol.
[b] Rating P = Presence of bacteria with white precipitate
A = Presence of bacteria without white precipitate
O = No bacterial growth

In the dilution assay, test 2, colonies from *A. feroniarum* fed on 7 day old Br-1SR cultures were observed to form 2 and 4 white precipitate colonies in the first dilution (Table II). Fifty percent of *R. robini* mites contained the precipitate forming bacteria in their body but these were interspersed among many non precipitate forming colonies. The two "positive" mites were taken from 7 day old cultures, the two "negative" mites from 8 day old cultures. In

TABLE II.
Dilution Plate Assay of Bacteria from Mites after Exposure to *Psuedomonas marginata* Br-1SR

Mite Species	Test	Age of Br-1SR Bacterial Culture in Days	Number of Bacterial Colonies From Crushed Mites	
			Dil No. 1	Dil No. 2
Anoetus feroniarum	1	7	4 A[a]	0
	2	7	2 A	0
Rhizoglyphus robini	1	7	TNTC P	TNTC P
	2	7	TNTC P	TNTC P
	3	8	TNTC A	TNTC A
	4	8	TNTC A	TNTC A

[a] TNTC = Too numerous to count
P = Presence of bacterial colonies forming WP in NASP media
A = Presence of bacterial colonies not forming WP in NASP media
O = No bacterial growth

the former case colonies of bacteria were too numerous to count (TNTC, Table II).

The retention time of bacteria in the bodies of *A. feroniarum* and *R. robini* mites is shown in Table III. After exposure to F-1 bacterial colonies in test 3, *A. feroniarum* disseminated bacteria for an average of 2.3 days, minimum 1 day and maximum of 5 days. With the Br-1SR isolate the range was 1-3 days with a mean of 1.8 days. *R. robini* disseminated the F-1 bacteria from 1 to 3 days with an average time of 1.3 days; however, with the Br-1SR isolate the average was 2.4 days, with a minimum of 1 and a maximum of 5 days (Table III).

TABLE III.
Bacterial Retention Time of Mites after Exposure to *Pseudomonas marginata* Cultures

Number and Species of Mites Tested	Bacterial Isolate Exposed to:	Mean Number of Days Bacterial Producing WP in NA[a]
15 *Anoetus feroniarum*	F-1	2.3
6 *Anoetus feroniarum*	Br-1SR	1.8
15 *Rhizoglyphus robini*	F-1	1.3
7 *Rhizoglyphus robini*	Br-1SR	2.4

[a] WP = White precipitate
NA = Nutrient Agar

In a similar test but with the bacterium *Erwinia carotovora, A. feroniarum* disseminated the organism for 1.5 days and *R. robini* for 2.8 days. *E. carotovora* colonies were distinguished by formation of pits in a pectate medium.

Bacterial colonies resulted on all 25 plates of *A. feroniarum* removed from diseased gladiolus (test 4), however, none of the colonies formed a WP on NA, typical of the F-1 isolate. Further, bacterial colonies selected from these plates and tested for pathogenecity on gladiolus caused no symptoms. Three of the

TABLE IV.
Assay for Presence and Dissemination of *Pseudomonas marginata* by *Rhizoglyphus robini* Obtained from Diseased Gladiolus

Bacterial Isolate	No. Mites Tested	1st 48 Hr Exposure Negative	1st 48 Hr Exposure Positive	2nd 48 Hr Exposure Negative	2nd 48 Hr Exposure Positive	No Growth
S-D	10	9	1[a]	8	0	2
Br-1	5	5	0	4	0	1
F-1	10	8	2	—	—	—
Total	25	22	3	12	0	3

[a] 2 white precipitate colonies present on the plate with the mite.

25 *R. robini* isolated from symptomatic plants disseminated WP forming colonies (Table IV). Two WP colonies were isolated from a NA plate and tested on the host plant. Bacteria from one colony resulted in pathogenic symptoms. Two WP colonies when recovered and tested gave similar pathogenecity, physiological and biochemical test results.

The low incidence (12%) of *R. robini* mites positive for dissemination of *P. marginata* from diseased gladiolus to NA suggests that this mode of dissemination is not efficient. The completely negative results obtained with *A. feroniarum* diminish the likelihood of these mites being capable of any long-term dissemination of *P. marginata.* Of the two species, *R. robini* appears better suited for retaining viable ingested bacteria, probably because of the different food preference and digestive physiology (Noble and Poe, 1973). However, as a suspect vector of *P. marginata,* the low recovery frequency and limited retention time make only short term dissemination of any consequence, and that under a most restricted set of conditions.

The age of the bacterial culture that mites fed on appeared to be important for recovery of *P. marginata.* The gut flora of mites is suspected of competitively inhibiting *P. marginata* growth since mites fed on colonies 8 days or older appeared negative for the bacterium.

The positive results from surface sterilized and crushed mites indicated that *P. marginata* had been ingested and had survived for at least 12 hr within the mite gut. The less consistent detection and retrieval might result because *A. feroniarum* is a bacterial feeder and undoubtedly possesses enzyme systems to digest consumed cells. This would appear to be in general agreement with the dilution plate assay where *R. robini* contained a much greater concentration of viable bacteria than did *A. feroniarum.* When considered in light of feeding preferences demonstrated by Noble and Poe (1973), *R. robini* for fusarium and *A. feroniarum* for bacteria, there appers to be good agreement.

Of more importance, however, is that these experiments demonstrate that both mite species are capable of ingesting *P. marginata,* when exposed to bacterial colonies. Further, the bacteria are retained or disseminated differentially by the two species for as long as 5 days, but averaging no more than 3 days. Most often by the fourth transfer day, continuous trails of non *P.*

marginata bacterial colonies were present, suggesting a transition of flora within the mite intestine.

The results of these experiments cast serious doubt on the probability of active mites or hypopi serving as a mobile reservoir of pathogenic innoculum. Because of the higher bacteria density observed and being a fungus feeder, *R. robini* appears more suspect as a possible vector, either mechanical or via the alimentary canal.

SUMMARY

Experiments were performed to determine if two species of mites, *Anoetus feroniarum* (Dufour) and *Rhizoglyphus robini* Claparede commonly associated with diseased gladiolus corms in Florida could ingest and pass through the ailmentary tract the pathogenic scab organism *Pseudomonas marginata* McCullough.

Both species of mites ingested bacteria from laboratory petri dish cultures and from infected corms. Furthermore, after ingestion, viable organisms were recovered from the gut for periods of one to five days. *Pseudomonas marginata* was recovered more frequently, in higher densities and for slightly longer periods of time from *R. robini* than from *A. feroniarum*. Dissemination was short term averaging no more than three days for either species. Consequently, the potential of these mites as vectors of *P. marginata* over long periods of time in other than local situations appears slight.

REFERENCES

Engelhard, A. W. (1969). *Phytopathology* **59**, 1025 (abst).
Poe, S. L. (1971). *The Fla. Entomol.* **54**, 127-133.
Noble, W. E., and Poe, S. L. (1973). *Proc. Fla. State Hort. Soc.* **85**, 401-404.

MITES FROM PLOTS SUPPLIED WITH DIFFERENT QUANTITIES OF MANURES AND FERTILIZERS

Niels Haarløv

Zoological Institute
Royal Veterinary and Agricultural University
Copenhagen, Denmark

INTRODUCTION

Studies on ecology of soil inhabiting animals have mostly been based on investigations in habitats influenced as little as possible by human activities. In the present case, however, the opposite approach has been taken, as the research has been limited to arable soil belonging to an experimental farm in Denmark called Askov Forsøgsstation.

The approach of these studies has been an attempt to demonstrate possible correlations between application of manures and fertilizers to the experimental plots of the farm, and abundance plus species-composition of the soil inhabiting animals found there. Only micro-arthropods, with special reference to mites and Collembola, will be mentioned. Besides, from the same experimental plots, the ecology of microorganisms and earthworms has been studied.

METHODS AND MATERIALS

The soil of the experimental plots is of glacial origin and a rather sandy loam. It is grown with cereals and root crops and plowed each year.

The plots are fertilized with NPK and manured with farmyard manure and slurry in quantities corresponding to 0 kg, 80 kg, 100 kg, 150 kg, and 200 kg N per ha each year, and every second year with 500 kg and 1000 kg N per ha, plus every fourth year with 2000 kg N per ha.

Until now three series of samples have been taken, i.e. in all 141 samples. Yet the material is still too small to allow a more thorough statistical treatment. Sampling dates were: 31/10 1976, 3/5 1977, and 13/10 1977.

The dimensions of samples for extraction of animals were 1×10^{-3} sq.m.

Vol. I: ISBN 0-12-592201-9

and a height of *ca* 8 cm. Each sample was divided and put into two Tullgren glass funnels which had a maximum of 8 cm. The heat source was a 60 w. electric bulb placed 40 cm above the sample, which was gradually heated to about 35°C during 2-3 days.

RESULTS

From a general survey of the results, no differences could be seen between abundance or species composition of mites and Collembola from plots which were either fertilized or manured. Therefore, all plots with the same calculated contents of N per ha were averaged together. Fig. 1 shows average values of mites plus collembolans per 1×10^{-3} sq.m. With the lowest abundance in plots with no dressing, values are almost the same up to 250 kg N per ha. Then there seems to be a more positive correlation between higher doses and number of mites and collembolans.

Excluding the more irregularly occuring species, the material was divided into four groups: 1) Collembola, 2) Eupodidae, 3) Gamasidae and 4) Oribatidae. In Fig. 2, it is clearly demonstrated that collembolans were the dominant group and Oribatidae the least dominant, while Eupodidae and Gamasidae were intermediate.

DISCUSSION

It is well known (Dunger, 1964; Tischler, 1965; Wallwork, 1976; Edwards and Lofty, 1975) that application of manures and fertilizers may have a positive effect on the abundance of soil animals. In the present case, this is verified, but a pronounced increase was evident only at dosages of above 250 kg N per ha per year. In practice, a Danish farmer will use only about 100 kg maximum. The finding that collembolans are the dominant group of microarthropods in relation to mites in fields with annuals, contrary to conditions in crops of perennials as, for instance, grassland, is also in accordance with previous results (Persson and Lohm, 1977).

Concerning the absolute number of species and individuals of mites and collembolans, the results are in agreement with earlier studies, namely, that the level is very low compared with, for instance, habitats with well developed litter- and fornalayers as in spruce forests (Tischler, 1965; Hammer, 1949). However, numbers as high as 68 collembolans and 10 mites have been found in single samples of less than 1×10^{-3} sq.m. from certain microhabitats. These wide variations are of course a result of the mosaic structure of the soil.

In South Africa, Loots and Ryke (1967) found, as a general rule, that in soils with little organic material the abundance of Trombidiformes will be relatively higher than that of the oribatid mites, contrary to what is found in more organic soils. In the present study only small numbers of individuals

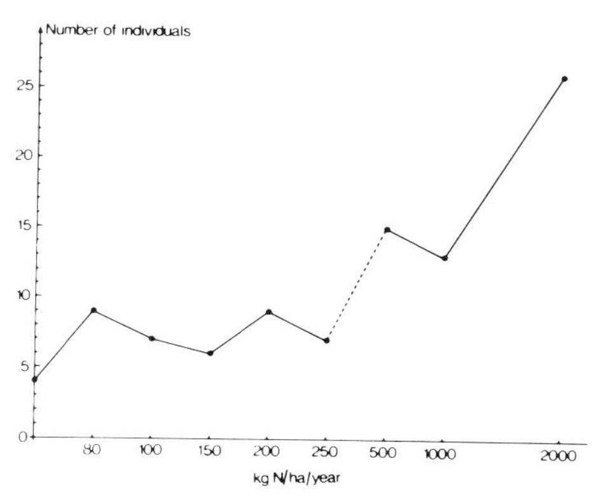

Fig. 1. Average number of mites plus collemboles per 1×10^{-3} m^2.

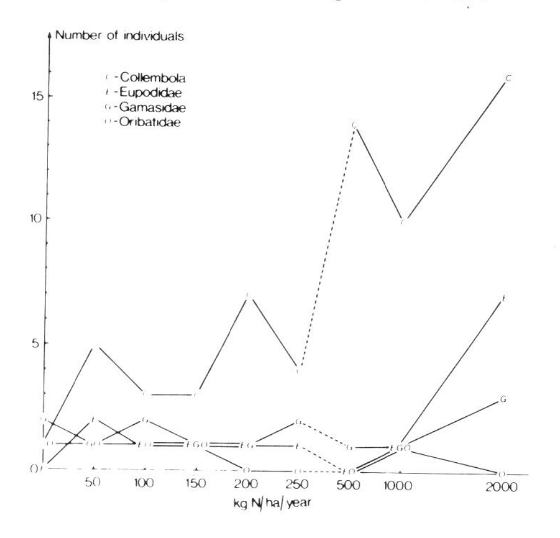

Fig. 2. Average number of collemboles and different groups of mites per 1×10^{-3} m^2.

from each of these groups were found, but, nevertheless, the results are in accordance with the above-mentioned statement, with Eupodidae as representative for Trombidiformes. Both Eupodidae and Gamasidae are, as a general rule, predators on other microarthropods, nematodes, enchytraeidae, etc. So it seems natural that they have been found in almost the same numbers. Collembola and oribatid mites are both saprovores/microvores, and yet they are found in quite different numbers.

As an attempt to elucidate this problem, it is natural to consider the very convincing experiment by Karg (1967) from which he concludes that the fauna of mites in fields with annuals corresponds principally to what is found in a woodland soil excluding those species living in the upper, highly organic litter- and fornalayers down to A_1. Thus, the species of mites in arable soil are represented to the very surface by members of the eudaphon (Dunger, 1964). In the present study, this corresponds well with the presence of the small *Oppia*

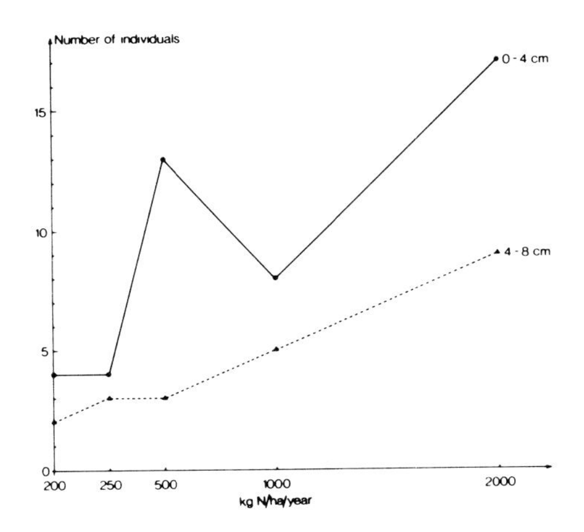

Fig. 3. Average number of mite plus collemboles at depths of 0-4 cm and 4-8 cm per 1×10^{-3} m^2.

species as almost the only oribatid mites in the material, and besides, that the Gamasidae regularly are represented by Rhodacaridae.

That the viewpoint of Karg also can be extended to the collembolans is well documented by the high number of dominant *Tullbergia krausbaueri* in the material, supplemented with the more irregular dominance of the small, more or less pigmented, and mostly blind species of *Folsomia* and Isotomidae. The fact that the mites and collembolans are found rather regularly down to layers below 4 cm points in the same direction (Fig. 3). Although it is not known why the collembolans are dominant in relation to the oribatid mites, this difference can perhaps be correlated with different food preferences: The mites are probably more dependent on mycelia, which they bite off with their relatively strong chelicerae, while the collembolans may be more adapted to scrape off films of bacteria, algae, etc. on the soil particles.

REFERENCES

Dunger, W. (1964). "Tiere im Boden. Die neue Brehm-Bucherei." A. Zimsen Verlag. pp. 1-265.
Edwards, C. A. and Lofty, J. R. (1975). Rep. Rothamsted Exp. Stn. for 1974. pp. 133-154.
Hammer, M. (1949). *Ann. Ent. Fenn. suppl.* **14**, 75-80.
Karg, W. (1967). *Pedobiologia* **7**, 198-214.
Loots, G. C. and Ryke, P. A. J. (1967). *Pedobiologia* **7**, 121-124.
Persson, T. and Lohm, U. (1977). *Ecological Bull./NFR* **23**, 1-211.
Tischler, W. (1965). "Agrarokologie." Gustav Fischer Verlag. pp. 1-499.
Wallwork, John A. (1976). "The Distribution and Diversity of Soil Fauna," pp. 355. Academic Press, New York.

EFFECTS OF THE INSECTICIDE DIFLUBENZURON ON SOIL MITES OF A DRY FOREST ZONE IN BRITISH COLUMBIA

Valin G. Marshall

Department of Fisheries and the Environment
Canadian Forestry Service
Pacific Forest Service Centre
Victoria, B. C.

INTRODUCTION

Experimental spraying with the insecticide diflubenzuron during an outbreak of the Douglas-fir tussock moth, *Orgyia pseudotsugata* (McDonnough), at Kamloops, British Columbia, provided an opportunity to study the effects of this insecticide on soil mites. The objective was to study qualitative and quantitative changes in the mite fauna in the top six cm of the soil, where numbers of individuals and species were unsually high. Results for three and 12 months following two application rates are reported here.

Diflubenzuron, a stomach insecticide which is effective against many insects, inhibits chitin synthesis and kills insects by disrupting the moulting process (Post and Vincent, 1973), but little information is available on its effects on the soil biota (Marx, 1977). Lipa and Chmielewski (1976), in a laboratory study, found that only 60-70% of larval *Tyrophagus putrescentiae* (Schrank), whose food contained over 1000 ppm diflubenzuron, reached the adult stage and they attributed this mortality to the mechanical effect of diflubenzuron dust, which covered the mites' bodies and also lowered the humidity of the food. They suggested that lower dosages of diflubenzuron were not inimical to *T. putrescentiae,* because some mites of the family Acaridae lack chitinous hypostracum in the cuticle; therefore, the effect of this insecticide on these mites will be different from that on insects.

Since chitin occurs in bacteria, lichens, blue green algae, fungi and arthropods (Allison, 1957; Tracey, 1968; Muzzarelli, 1977), large-scale use of diflubenzuron may reduce soil mites directly through its effect on cuticle synthesis or indirectly through changes in the food chain, both of which could adversely affect soil fertility.

Vol. I: ISBN 0-12-592201-9

MATERIALS AND METHODS

The work was done in a Douglas-fir, *Pseudotsuga menziesii* (Mirb.) Franco, Forest Zone, a few km north of Kamloops (50.30° N, 120.24° W), where the mean annual precipitation is generally less than 300 mm and mean monthly temperatures are −6°C for January and 21°C for July. Two sites were selected: one at Palmer Forsyth Road and the other at Tod Mountain Road, 11 km away. Both sites were about 600 m above sea level and supported a 100-year-old forest of mainly Douglas-fir, with small components of ponderosa pine, *Pinus ponderosa* Laws., and lodgepole pine, *Pinus contorta* Douglas.

The Palmer Forsyth Road stand averaged 462 stems per hectare, with dominant trees 19.5 m. tall. The soil was an Orthic Grey Luvisol with an organic layer (L, F, H) averaging 2 cm. The insecticide contained 140 g (a.i.) of diflubenzuron per hectare and was applied as a water-base emulsion at the rate of 11.2 litres per hectare. The emulsion contained 10% ethylene glycol and 3.8 g. of rhodamine B-Extra-S dye, and was applied by a fixed-wing plane with conventional boom and T8010 nozzles, on 18 June 1976.

The Tod Mountain Road stand averaged 495 stems per hectare, with dominant trees 30 m tall. The soil was a degraded Utric Brunisol with combined organic layers averaging 1.9 cm. The diflubenzuron insecticide was applied as in the previous site, but at the rate of 280 g. (a.i.) per hectare, on 15 June 1975.

In a check and a diflubenzuron-sprayed area, 320 m apart, a circular 60-m-diameter plot was selected and divided into three sectors. On each sampling date, six randomly-selected cores were removed from each sector, giving 18 cores per treatment. Cores were taken by a 5-cm-diameter sampler similar to that described by Vannier and Vidal (1965). At the Palmer Forsyth Road site, samples were taken on three occasions: one pre-spray sampling (76/5/19), one month before spraying, and two post-spray samplings, 2 weeks (76/6/29) and 3 months (76/9/22) after diflubenzuron application. At the Tod Mountain Road site, samples were taken only once, one year after diflubenzuron application.

Cores were placed in polyethylene bags and transported 400 km by car to the laboratory. The cores were then divided into two sections: 0-3 cm (mainly L, F, H layers) and 3-6 cm of the upper mineral horizons, and each section was placed intact in an inverted position into a canister-type apparatus (Lussenhop, 1971). Extraction was carried out for 7 days. Specimens were counted in petri dishes under binocular microscopes at 50x magnification. Identification of species was normally made under binocular microscopes, but samples of representative species were permanently mounted for detailed examination with a compound microscoope.

Population counts were transformed to square roots, to compensate for widely varying counts including zeros, and analyzed as a completely randomized design. Significance ($P \leq 0.05$ and 0.01) of each mean difference was determined by the Student-Newman-Keuls' multiple range test (Steel and Torrie, 1960).

RESULTS AND DISCUSSION

Fifty-six taxa were recovered from the Palmer Forsyth Road site and 64 from the Tod Mountain Road site, but only 30 of these showed significant changes from diflubenzuron application (Table I). The values for total mites per core, which also includes species not significantly affected, are given to show the relative abundance of significantly affected species. The net decrease or increase in taxa, relative to the check, for the individual soil layers is shown under the appropriate columns; those for the 0-6 cm soil depth are between columns.

In the Palmer Forsyth Road site, the direct effect of diflubenzuron was evident 2 weeks after application from the reduction of 6 species when both soil layers are considered. These values omit *Belba* sp. and *Liochthonius* n. sp. 1, which were significantly higher ($P \leq 0.05$) in the 0-6 cm soil depth of the prespray calibration of the check (2:0 and 13:3, respectively), and *Propelops pinicus* and *Trhypochthonius tectorum,* which were significantly affected in the 3-6 cm layer, but not for the total 0-6 cm soil depth. The diminution did not increase substantially 3 months after application when 7 species were reduced. During this 3-month period, only 3 species, *Scutacarus quandrangularis, Stigmalychus* n. sp. and *Schwiebia terrana,* increased in numbers.

The initial reduction of 7 species in the Palmer Forsyth Road site 3 months after spraying was also matched by 7 species in the Tod Mountain Road site 1 year later. These values omit *Coccorhagidia pittardi* and *Eremaeus foveolatus,* which showed a signficant change in one layer, but not in the 0-6 cm soil depth. In the Tod Mountain Road site, however, 6 species had increased over the check, almost equalling the 7 species that were reduced. These included the prostigmatids *Cocceupodes breweri, Eupodes* sp., *Stigmalychus* n. sp. and a Bdellidae sp. and the cryptostigmatids *Belba* sp. and *Tectocepheus velatus. Liochthonius* n. sp. 2 increased in the top soil layer, but not in total 0-6 cm soil depth. The reduction of individual species did not affect the total mite population, which showed no significant changes, during the 3-month period after spraying, nor one year later. Increases in certain other species apparently compensated for those that were reduced.

Except for *Nanorchestes* n. sp., Erythraeidae sp., *S. terrana, Liochthonius* n. sp. 3 and *Suctobelbella* sp., all other significantly affected mites inhabited both sides, but only a few species showed consistent trends in population changes following the spray application. Increases in *S. quadrangularis* and *Tarsonemus* n. sp. were temporary, while that in *S. terrana* appeared to be sustained. Unfortunately, *S. terrana* was not found in the Tod Mountain Road site to allow further interpretation of this early response. *Stigmalychus* n. sp. increased greatly on most sampling dates, while *Tectocepheus velatus* was reduced initially, but showed higher numbers in the sprayed plot at Tod Mountain Road one year after diflubenzuron application. *Eremaeus foveolatus* showed reductions from spraying in both sites for 2 sampling periods,

TABLE I.
Population Density of Mite Taxa Showing Significant Changes to Diflubenzuron Spraying in Two Douglas-Fir Sites in British Columbia

Taxa	Mean Number of Mites Per Core (Check Plot : Sprayed Plot)[a]					
	Palmer Forsyth Road				Tod Mountain Road	
	76.6.29[b]		76.9.22[c]		76.5.19[d]	
	0-3 cm	3-6 cm	0-3 cm	3-6 cm	0-3 cm	3-6 cm
MESOSTIGMATA						
Amblyseius nicola Chant	0:0	0:0	0:1	0:0	1:0**	0:0
Zerson spp.	2:2	1:0	5:1**	1:1	4:3	0:0
PROSTIGMATA						
Alicorhagia usitata Th., Me & Ry.	26:27	18:19	47:33	40:12**	5:9	2:4
Cocceupodes breweri Strandtmann	1:1	0:0	4:4	2:1	1:4	0:1*
Coccorhagidia pittardi Strandtmann	0:0	0:0	0:1	0:0	0:0	1:0*†
Eupodes sp.	6:6	7:3	18:13	9:7	3:6*	1:1
Nanorchestes n. sp.	—	—	—	—	1:0*	0:0
Scutacarus quadrangularis (Poali)	3:10*	1:9*	14:11	5:7	30:9	10:1
Stigmalychus n. sp.	73:49	39:165**	52:55	83:175	1:14*	1:33*
Tarsonemus n. sp.	4:5	2:3	11:21	2:5	5:1**	0:0
Bdellidae	7:8	3:1	11:11	4:4	1:5**	0:1*
Cunaxidae	2:0**	0:0	1:0	0:0	1:2	0:0
Erythraeidae	3:0**	1:0*	0:0	0:0	—	—
ASTIGMATA						
Schwiebia terrana Jacot	4:10**	1:2	7:12*	1:3	—	—
CRYPTOSTIGMATA						
Belba sp.	3:0**	1:0**	9:0**	2:0*	0:1*	0:0
Brachychthonius bimaculatus Willmann	13:16	9:4*	20:10	9:1*	4:4	3:8
Eremaeus foveolatus Hammer	13:8	1:0**	24:21	4:3	12:2*†	0:0
Joshuella sp.	1:0	0:0	2:0**	0:0	0:0	0:0
Liochthonius n. sp. 1	4:9	6:3	8:6	10:4*	0:0	0:0
Liochthonius n. sp. 2	0:0	0:0	0:0	0:0	4:5	0:4*†
Liochthonius n. sp. 3	—	—	—	—	2:0*	1:0*
Oppiella nova (Oudemans)	2:1*	3:0**	7:2**	6:1*	2:1	0:0
Opiella simplissimus (Jacot)	0:0	1:0	0:0	1:0	1:0**	4:0*
Parachipteria nivalis (Hammer)	2:0*	1:0**	2:1	0:0	0:0	0:0
Propelops pinicus Jacot	1:1	1:0*†	1:0	0:0	2:4	0:0
Quadroppia ferrumequina Jacot	0:0	0:0	0:0	1:0	3:1*	0:0
Scheloribates sp.	0:0	1:0*	0:0	0:0	0:0	0:0
Suctobelbella sp.	—	—	—	—	1:0*	0:0
Tectocepheus velatus (Michael)	5:7	1:0	25:9**	2:1	6:17*	0:0
Trhypochthonius tectorum (Berlese)	6:7	3:1*†	14:12	5:2	5:2	0:0
No. of species (decrease/increase)	4/2	9/2	4/0	4/0	7/5	3/4
No. of species for 0-6 cm soil depth (decrease/increase)	6/3		7/1		7/6	
Total mites per core	277:262	138:259	393:314	252:291	130:128	30:57

[a] Mean of 18 cores, each 19-6 cm² sampled to a depth of 6 cm.
[b, c, d] Two weeks, 3 months and 1 year after spraying, respectively
*, ** Significance at P < 0.05 and 0.01
† Significant change in one layer not reflected in the combined 0-6 cm soil depth

although a reduction in the Tod Mountain Road site one year after spraying was evident only in the 0-3 cm soil layer.

Response within acarine orders was varied. The Mesostigmata were generally adversely affected, the significantly changed taxa showing a reduction in numbers. Among the Prostigmata, an equal number of species increased as decreased. All Cryptostigmata showed a reduction in numbers immediately following diflubenzuron application, but 2 species had increased compared to the check one year after application in the Tod Mountain Road site. The only abundant representative of the Astigmata, *S. terrana,* increased after spraying, supporting Lipa and Chmielewski's (1976) thesis that low dosages of diflubenzuron would not be inimical to some Acaridae.

In the 0-3 cm soil layer, 4 species were reduced in the Palmer Forsyth Road site, whereas 7 species were reduced in the Tod Mountain Road site. Although the 3-month post spray data for the 0-6 cm soil depth for the Palmer Forsyth Road site and the Tod Mountain Road site showed a reduction of 8 and 10 species, respectively, the greater species dimunition in the 0-3 cm layer of the latter site possibly resulted from the higher application rate. Data on the precise vertical distribution on each species would be difficult to obtain, because large numbers of samples would be required to overcome the patchy horizontal distribution of many mite species, seasonal movement in the soil profile, variation in soil organic matter, and random penetration of the spray in the soil. However, the total reduction of most of these species for the 0-6 cm layer does not indicate a downward movement from the 0-3 to the 3-6 cm layer, but rather a sustained effect of diflubenzuron on mites in the top soil layer at Tod Mountain.

Affected species were not restricted to a particular trophic level. Significant increases occurred among predators such as *C. breweri* and Bdellidae sp., and saprophages such as *Stigmalychus* n. sp. and *Liochthonius* n. sp. 2. Similarly, decreases occurred among predators such as Cunaxidae sp. and Erythraeidae sp., and saprophages such as *Alicorhagia usitata, E. foveolatus* and *Oppiella nova.* Only 4 species recovered in the check plot were not found in the sprayed areas and these might have been eliminated by the spray. Again, these included predators (*Amblyseius nicola* and *Nanorchestes* n. sp.) and saprophages (*Joshuella* sp. and *Suctobelbella* sp.).

The study demonstrates the importance of working at the species level in determining the effects of agricultural and forestry practices on soil mites. Although the total mite population was unchanged up to one year after spraying, some species were reduced, while others seem to have disappeared from the sprayed plots. Changes effected in the check because of defoliation, for example, more litter fall and possibly higher ground temperatures, might have caused quantitative faunal changes, but these would be unlikely to increase the species composition of the site.

It is concluded that a single application of diflubenzuron up to a rate of 280 g (a.i.) per hectare had no drastic effect on any of the trophic levels of the mites. However, there is still need for vigilance, because a few species seem to have been eliminated from the sprayed area, while others had not returned to the "normal" level one year after diflubenzuron application. It is not clear how much changes in the soil mite population will affect soil fertility.

SUMMARY

Effects of 2 rates of diflubenzuron, 140 and 280 g (a. i.)/ha, applied by aircraft, were examined on soil mites in a Douglas-fir *(Pseudotsuga menziesii)* forest. Total mite population was not significantly affected two weeks, three months and one year after application, but a few species found in the check were not found in the sprayed plot. Other species, which were significantly reduced initially, had not recovered to the "normal" level of the check. Generally, members of the Mesostigmata and Cryptostigmata were adversely affected. Among the Prostigmata, the same number of species increased as decreased; whereas an increase was noted in the Astigmata. At least two trophic levels were altered, with increases and decreases occurring among predators and saprophages.

ACKNOWLEDGEMENTS

I thank Miss Andrea Hall and Mr. L. A. Bjerstedt for technical assistance and Dr. C. Simmons for the statistical analysis.

REFERENCES

Allison, F. E. (1957). pp. 85-94, *In* "Soil, the Year Book of Agriculture 1957." USDA, Washington.

Lipa, J. J. and Chmielewski, W. (1976). *Bull. Acad. Pol. Sci., Sie. Sci. Biol.* **24**, 381-384.

Lussenhop, J. (1971). *Pedobiologia* **11**, 40-45.

Marx, J. L. (1977). *Science* **197**, 1170-1172.

Muzzarelli, R. A. A. (1977). "Chitin." Pergamon Press, Oxford.

Post, L. C. and Vincent, W. R. (1973). *Naturwissenschaften* **60**, 431-432.

Steel, R. G. D. and Torrie, J. H. (1960). "Principles and Procedures of Statistics with Special Reference to the Biological Sciences." McGraw-Hill Book Co. Inc., New York.

Tracey, M. V. (1968). *In* "Advances in Comparative Physiology and Biochemistry." (O. Lowenstein, ed.). Academic Press, New York.

Vannier, G., and Vidal, P. (1965). *Rev. Ecol. Biol. Sol.* **2**, 333-337.

2. Biology of Spider Mites

Section Editor

W. W. Cone

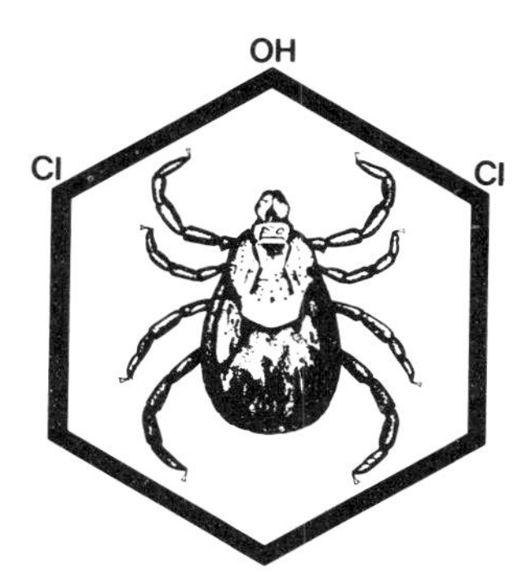

REPRODUCTIVE BEHAVIOR AND SEXUAL SELECTION IN TETRANYCHINE MITES

Daniel A. Potter

Acarology Laboratory
The Ohio State University
Columbus, Ohio

INTRODUCTION

In polygynous mating systems, the ability to conquer or intimidate other males is an important component of male fitness (Williams, 1966). Among the arthropods, male aggression is a common phenomenon, having been studied extensively in the crustaceans and the insects. Often, males in these groups possess specialized morphological adaptations which function in intrasexual competition, or display elaborate patterns of precopulatory behavior (Richards, 1927; Wilson, 1975; Parker, 1970).

In contrast, behavioral studies of the Acari have been hindered by the small size and cryptic habits of most mites, so that the reproductive biology of many groups is still poorly known. As a consequence, aggression between conspecific males has been observed in only a few scattered instances (e.g., Costa, 1967; Woodring, 1968).

In the spider mites, however, agonistic behavior has been known for over 80 years. As long ago as 1897, Perkins reported that males battle for possession of quiescent female deutonymphs. Despite brief mention by a number of other authors (Ewing, 1914; Smith *et al.*, 1967; Cone *et al.*, 1971b) the adaptive significance of this intriguing behavior has remained unstudied.

As part of a comprehensive analysis of precopulatory behavior in tetranychine mites, I have studied the role of male aggression in the spider mite mating system. This paper provides a detailed description of these aggressive encounters, and demonstrates their importance for male reproductive success. In addition, I examine a number of previously unrecognized factors which greatly increase intra-male competition, and discuss how certain components of spider mite precopulatory behavior contribute to the over-all success of these mites as colonizing organisms.

Vol. I: ISBN 0-12-592201-9

METHODS

Initial observations were based on a large colony of *Tetranychus cinnabarinus* (OSU strain) maintained on kidney bean plants in the greenhouse. Mites from this colony have been the object of previous investigations (Mitchell, 1973; Wrensch and Young, 1975) and their fecundity and growth characteristics are known. In the present studies, *T. cinnabarinus* was used in those experiments requiring an intact population on host plants.

Later, when it became necessary to distinguish between individual males and determine which were successful in mating, colonies of sexually compatible wild-type (red-eyed) and albino (white-eyed) *T. urticae* (Sambucus strain) were obtained from the Netherlands. These were maintained on detached leaf cultures. Preliminary experiments confirmed that mating between these mutant and wild-type *T. urticae* stocks is non-assortative, and that eggs from albino x wild-type crosses developed normally.

T. cinnabarinus and *T. urticae* are sibling species with virtually identical reproductive behaviors. Generalizations based on experiments with one species are therefore applicable to the other.

Most laboratory studies were conducted on 12 or 17mm bean leaf discs, pressed onto water-soaked cotton and incubated at 27 ± 1°C under 24 h fluorescent illumination. To ensure that male response would be as normal as possible, in those experiments involving male competition or guarding behavior, the females were always transferred as active deutonymphs and allowed to become quiescent on the test discs.

RESULTS

Guarding and Aggression

Female spider mites spin strands of silk, settle on the leaf surface and become quiescent during their final molt. Guided by a sex attractant (Cone *et al.*, 1971a) and by the female's webbing (Penman and Cone, 1972), males search for and guard quiescent female deutonymphs, awaiting their emergence as sexually receptive adults. When a male encounters a quiescent female, he usually initiates characteristic guarding behavior, remaining motionless beside the female or climbing on top and assuming a resting position with his palps and front legs in contact with her dorsum. At ecdysis, the guarding male often helps to pull the old cuticle from the emerging female. Mating nearly always occurs as soon as the female is free from the exuviae.

This prolonged precopulatory guarding enables a male to retain access to a potential mate until she becomes receptive. In this respect, it is similar to the sequestering of females by certain male insects (Parker, 1970), amphipods (Hynes, 1955) and aquatic crabs (Hartnell, 1969). Spider mites differ from most insects however, in that generally only the first mating is effective (Boudreaux, 1963; Helle, 1967). For a male to mate successfully, therefore, he must locate and gain control of virgin females prior to ecdysis.

Intense fighting may occur when two or more males attempt to guard a single female (Potter *et al.*, 1976a, b). When a newcomer is attracted to an attended deutonymph, he may, upon detecting the resident guarder, leave without incident. More often he attempts to assume a guarding position or tries to dislodge the guarding male. Should the original male fail to respond, the two competitors may become coguarders, vying for a favorable position on the female. Commonly, however, the intruding male is threatened or attacked.

Fights generally begin when the guarding male backs down from the quiescent female and confronts the intruder with outspread front tarsi and extruded cheliceral stylets. The threatened male either retreats or responds by adopting a similar stance. The mites then lunge at each other, grappling violently with their forelegs and jousting with the extruded mouthparts. While most fights are brief, sometimes this sparring continues as long as several minutes.

Another commonly employed tactic is to use the palps to apply strands of silk to the mouthparts and legs of the opponent. A male so entangled may have his movements impeded and be forced to retreat to clean himself.

Despite the rugged nature of these encounters, injuries are rarely inflicted. On several occasions, however, I have observed a male puncture the integument of his opponent with the extruded stylets. Mites so injured appeared crippled and usually died within a short time.

More often, fights terminate with the retreat of one combatant and the return of the other to the quiescent female. Should an encounter end in stalemate, the two antagonists may guard the female jointly. Additional fighting between these males or with other intruders may break out at a later time.

Factors Increasing Male Competition

The evolution of prolonged precopulatory guarding and male aggression is generally associated with conditions of high male density and intense competition for matings (Parker, 1970, 1974). Since in most spider mite populations, females outnumber males by about 3:1 (Laing, 1969; Overmeer, 1972) it would appear that potential mates are in abundance and that frequent and intense male aggression need not occur.

In spider mites, however, it is not the tertiary sex ratio which determines the degree of male competition, but rather the ratio of sexually capable males to available females. While males are sexually long-lived and may mate repeatedly, at any given time the number of available females is limited to quiescent deutonymphs and teneral virgins. Since males ordinarily do not migrate they accumulate during a colonizing episode until they greatly outnumber the available quiescent females, so that the functional sex ratio (no.

sexual capable males:no. of available females) becomes increasingly male-skewed (Potter *et al.*, 1976a; Potter, 1978).

In an earlier study (Potter, 1978) I monitored the functional sex ratio of a population of *T. cinnabarinus* as the mites colonized and eventually destroyed a large pot of beans. By classifying the mites into behavioral categories as the samples were counted, a record was obtained of how changes in functional sex ratio affected male guarding behavior. From this study, there emerged two important factors which greatly increase male competition for matings. First, as had been hypothesized, adult males outnumbered available females throughout most of the colonizing episode (Fig. 1). Despite this excess of males, however, throughout the infestation a large proportion of the available females was left unguarded at any one time. It was found that even during the period of highest density (days 28-34 from the initial infestation) males directed their guarding predominantly at those deutonymphs nearest to ecdysis and left the younger deutonymphs unguarded.

A significant positive correlation was found between the functional sex ratio and the proportion of females guarded jointly by two or more males. Most multiple guarding occurred late in the colonizing episode (days 28-36) at a time when, despite very high male density, from 39 to 52% of the available females were unguarded. In many instances the multiple-guarded females had taken on the wrinkled, silvery appearance characteristic of the period just prior to ecdysis. This suggests that males will take positions as co-guarders

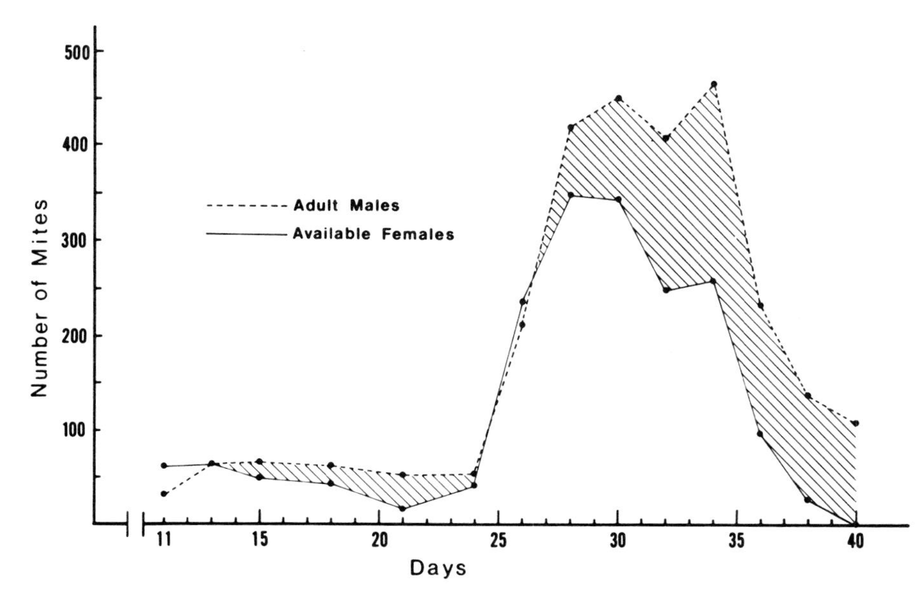

Fig. 1. Numbers of adult males and quiescent female deutonymphs during a colonizing episode on bean plants. The periods when sexually capable males outnumber available females are indicated by shading.

around the females closest to ecdysis or continue to search in preference to guarding the remaining less mature deutonymphs.

Other evidence that attractiveness to males increases as a quiescent female deutonymph matures was presented by Cone *et al.* (1971b) and Potter *et al.* (1976a). That males should discriminate between females that have only recently become quiescent and those on the verge of ecdysis is not surprising, since such behavior would enable a male to avoid waiting may hours before each opportunity to mate. As a consquence, however, males are competing for only a small subset of the available females at any one time. It is this intense competition for females close to ecdysis, compounded by the male-skewed functional sex ratio, which accounts for the frequent and intense struggles between males.

Guarding, Functional Sex Ratio, and the Fertilization of Dispersants

Because the great majority of females mature, mate and abandon the host plant late in a colonizing episode, their development coincides with a period when the functional sex ratio has become strongly male-skewed. This has important consequences for the fertilization of dispersants.

As the leaves of an infested host plant become heavily webbed and severely damaged, great numbers of adult females congregate at the leaf tips in jostling masses, from which they drop off or descend on strands of silk. Concurrent with the monitoring of functional sex ratio, I collected a large sample of female dispersants as they abandoned a second similarly infested pot of beans. An additional sample of teneral (newly emerged) adults was taken from the leaf surfaces. Of the several hundred females collected, virtually 100% of both groups were found to have been inseminated.

Since a single inseminated female can quickly give rise to a new colony, it is important for females to mate before migrating. As has been shown, the accumulation of adult males toward the end of a colonizing episode and the focussing of male courtship behavior on the more mature deutonymphs lead to intense competition for matings. As an important consequence, virtually all females are inseminated at ecdysis and can quickly become potential colonizers.

Consequences of Solitary vs. Joint Guarding

Guarding males may respond to intrusions either by attacking the trespassing male or by clinging tightly to the quiescent deutonymph. Typically, when two or more males remain at ecdysis, there is a scramble as each tries to maneuver into mating position. To determine the consequences of solitary vs. joint guarding, experiments were conducted in which two males were allowed to compete for a single quiescent deutonymph. It was found that guarders that establish dominance and retain sole possession during the final hour before ecdysis nearly always mated (93% success). Guarders who occupied the preferred position on top of the female but who tolerated or were unable to repel the other male had a reduced frequency of mating sucess (63%) (Potter *et al.*, 1976a, b).

For guarders, another disadvantage of tolerating intruders is that mating itself may be disrupted. Once copulation begins, the remaining, unsuccessful males will often climb over or pull at the coupled mites. Sometimes this harassment forces the mating male to disengage prematurely, permitting a competitor to replace him.

Late in a colonizing episode, when the density on the upper leaves of an infested plant may exceed six males/cm^2 and males may outnumber available females by more than 5:1 (Potter, 1978), the likelihood of disrupted matings increases. Under such conditions, it was found that about one female in seven (*ca* 14%) will receive two effective inseminations (Potter and Wrensch, 1978).

These observations underscore the disadvantages to guarding males of tolerating co-guarders. Given the chances of losing the female as she emerges and the possibility of a disrupted mating, it is apparent that if a male is to maximize his chances of inseminating the female he is guarding, he must establish his dominance over other interested males before the female ecdyses.

Determinants of Male Mating Success

In polygynous mating systems a male's fitness is a function of the number of females he can inseminate in his lifetime. Since prolonged precopulatory guarding reduces the rate at which other females are encountered, for this behavior to persist it would be expected to result in a greater probability of mating than would withdrawal for further searching. To determine if becoming the first guarder ensures or significantly increases a male's mating chances, experiments were conducted in which males (alternately red-eyed or albino) were introduced to discs holding a single quiescent female. Later, after guarding had been established, a second male of the opposite genotype was added. Of the 70 replicates in the study, 71% of the females were ultimately inseminated by the first male to establish guarding (Potter *et al.*, 1976a, b).

In many mating systems, the outcome of aggressive encounters is dependent

TABLE I.
Size and Success of Male *Tetranychus cinnabarinus* in Aggressive Encounters.

	No. of Fights in Which Winner Was:			
Character	Larger	Equal Size	Smaller	χ^2 (1 d. f.)[a]
Length tarsus I	13	2	3	6.25 (P = .012)
Length tarsus IV	16	1	1	13.24 (P = .0003)
Distance between prodorsals II	11	2	3	4.57 (P = .033)

[a] Cases of equal size not considered.

TABLE II.
Mating Success of Old, Experienced vs. Young, Virgin Male *Tetranychus urticae.*

	No. of Wins	χ^2
Old, experienced vs. Young, virgin	55 : 37	3.52 (P = .061)
Larger vs. Smaller[a]	56 : 36	4.35 (P = .037)
Old, experienced & larger vs. Young, virgin & smaller	23 : 22	8.02 (P = .005)
Old, experienced & smaller vs. Young, virgin & smaller	23 : 22	.03 ns

[a]The length of tarsus I was used as a size criterion.

upon the relative size of the combatants. To determine the relationship between male size and fighting success in spider mites, leaves heavily infested with *T. cinnabarinus* were examined for fighting males. The winners and losers from 20 decisive fights were compared with regard to the size of three length characters (Table I). Based on these criteria, winning males were consistently larger (Potter *et al.*, 1976a, b).

Finally, experiments were conducted to determine if old, experienced males are able to outcompete younger, virgin opponents. Large numbers of 1-2 day old virgin albino and wild-type *T. urticae* were obtained by rearing eggs from unfertilized females. These were paired with older (6-7 days) experienced males taken from crowded, mixed-sex cultures, and placed together on 12mm leaf discs containing a single quiescent female (albino). The study consisted of 110 replications, 55 in which the older male was wild-type and 55 in which he was albino. After ecdysis and mating, the males were collected and measured and the progeny reared to determine which male had been successful.

The data, summarized in Table II, indicate that older, experienced males had a greater overall rate of success than did their younger, virgin adversaries. Size was again found to be important, with the larger of the two males winning in the majority of cases. Males who possessed both of these attributes, that is, older experienced males who were also the larger of the pair, were overwhelmingly more successful. Smaller males who are older and more experienced, however, can apparently compete equally with larger opponents.

The question of male mating competitiveness is of more than academic interest. Particularly in the Netherlands, much attention has been paid in recent years to the possibility of controlling greenhouse populations through the release of sterilized or genetically incompatible males (Helle, 1969; Overmeer and van Zon, 1973, 1977; Overmeer, 1974; Smith, 1975; Feldmann, 1977). The present findings suggest that young mites from all-male cultures may be unable to compete equally with older males from crowded populations. This would represent a potential deterrent to the success of genetic control measures.

SUMMARY

Male spider mites search for and guard quiescent female deutonymphs, awaiting their emergence as sexually receptive adults. Intruding males may be threatened or attacked. Fights involve vigorous grappling with the forelegs, jousting with the extruded cheliceral stylets and the use of silk to entangle and repel the opponent.

Experiments revealed the reproductive advantages of first discovery, and of establishing dominance over other males prior to the female's ecdysis. Large males were found to be superior in aggressive encounters, and older, experienced males were found to outcompete younger, virgin opponents.

Because males are sexually long-lived and individual females available only briefly, males accumulate and are in excess during most of a colonizing episode. In addition, guarding is concentrated on those females closest to ecdysis. Together, these factors greatly increase the competition for matings, and ensure that virtually all females are inseminated at or shortly after emergence.

ACKNOWLEDGEMENTS

I am much indebted to Dr. D. E. Johnston, Dr. D. L. Wrensch and Mr. Michael P. Murtaugh for their thoughtful comments and criticism of the manuscript. Dr. W. Helle, Laboratory of Applied Entomology, Amsterdam, provided colonies of the Sambucus strain. Thanks are due to Ms. Peggy H. Bennett, Acarology Laboratory, for aid in preparation of the manuscript.

REFERENCES

Boudreaux, H. B. (1963). *Ann. Rev. Entomol.* **8**, 137-154.

Cone, W., McDonough, L. M., Maitlen, J. C. and Burdajewicz, S. (1971a). *J. Econ. Entomol.* **64**, 355-358.

Cone, W., Predki, S. and Klostermeyer, E. C. (1971b). *J. Econ. Entomol.* **64**, 379-382.

Costa, M. (1967). *Acarologia* **9**, 304-329.

Ewing, H. E. (1914). *Oregon Agric. Exp. Sta. Bull.* **121.**

Feldmann, A. M. (1977). *Entomol. Exp. Appl.* **21**, 182-191.

Hartnell, R. G. (1969). *Crustaceana* **16**, 161-181.

Helle, W. (1967). *Entomol. Exp. Appl.* **10**, 103-110.

Helle, W. (1969). *Publ. Organiz. European Plant Protect. Paris Ser. A.* **52**, 7-15.

Hynes, H. B. M. (1955). *J. Anim. Ecol.* **24,** 352-387.
Laing, J. E. (1969). *Acarologia* **9,** 32-42.
Mitchell, R. (1973). *Ecology* **54,** 1349-1355.
Overmeer, W. P. J. (1972). *Entomol. Ber.* **32,** 240-244.
Overmeer, W. P. J. (1974). *In* "The Use of Genetics in Insect Control." (R. Pal and M. J. Whitten, eds.), pp. 45-56. Elsevier Publ. Co., Amsterdam.
Overmeer, W. P. J. (1977). *Z. Ang. Entomol.* **84,** 31-37.
Overmeer, W. P. J. and van Zon, A. Q. (1973). *Entomol. Exp. Appl.* **16,** 389-394.
Parker, G. A. (1970). *Biol. Rev.* **45,** 525-567.
Parker, G. A. (1974). *Behavior* **48,** 157-184.
Penman, D. R. and Cone, W. W. (1972). *Ann. Entomol. Soc. Am.* **65,** 1289-93.
Perkins, C. H. (1897). *In* "Report of the Entomologist," 10th Ann. Rep. *Vt. Agric. Exp. Sta.* pp. 75-86.
Potter, D. A. (1978). *Ann. Entomol. Soc. Am.* **71,** 218-222.
Potter, D. A. and Wrensch, D. L. (1978). *Ann. Entomol. Soc. Am.* (In press).
Potter, D. A., Wrensch, D. L. and Johnston, D. E. (1976a). *Ann. Entomol. Soc. Am.* **69,** 707-711.
Potter, D. A., Wrensch, D. L. and Johnston, D. E. (1976b). *Science* **193,** 160-161.
Richards, O. W. (1927). *Biol. Rev.* **2,** 298-364.
Smith, F. F., Boswell, A. L. and Webb, R. E. (1967). *Proc. 2nd Intern. Cong. Acarol.* pp. 155-159.
Smith, J. W. (1975). *Environ. Entomol.* **4,** 588-590.
Williams, G. C. (1966). "Adaptation and Natural Selection: A Critique of Some Current Ecological Thought." Princeton Univ. Press, Princeton, N. J.
Wilson, E. O. (1975). "Sociobiology." Belnap Press, Harvard Univ., Cambridge, Mass.
Woodring, J. P. (1969). *Ann. Entomol. Soc. Am.* **62,** 102-108.
Wrensch, D. L. and Young, S. S. Y. (1975). *Oecologia* **18,** 259-267.

SOME POSSIBLE ROLES OF FARNESOL AND NEROLIDOL IN THE BIOLOGY OF TWO TETRANYCHID MITES

Samuel Regev

Washington State University
Irrigated Agriculture Research and Extension Center
Prosser, Washington

INTRODUCTION

The two spotted spider mite *Tetranychus urticae* Koch and the carmine spider mite *Tetranychus cinnabarinus* (Boisd.) are serious pests of many crop and ornamental plants. Three physiological factors are discussed here which the author believes contribute to the success of these mites. These are: 1) Male mating behavior and the role of farnesol and nerolidol as sex attractants, 2) gonadotropic effect of farnesol, and 3) utilization of farnesol as a precursor for the synthesis of other chemicals.

MALE MATING BEHAVIOR AND THE ROLE OF FARNESOL AND NEROLIDOL AS SEX ATTRACTANTS

Arrhenotoky is the major sex determining mechanism in the Tetranychidae. Unfertilized females produce only male progeny whereas fertilized females produce both sexes, with the sex ratio being in favor of female progeny.

The first few eggs laid by a fertilized female are male since they are already mature and covered with an egg shell at the deutonymphal stage. Subsequent eggs may be fertilized. Helle (1967) indicated the first insemination supplies the female with enough sperm to produce diploid offspring (i.e., females) until she dies. The males produced by a fertilized female reach the adult stage before their sisters, some of whom are at the last quiescent molting stage (pharate females). The males are strongly attracted to these pharate females (Fig. 1a). The male remains with the pharate female until emergence when mating occurs.

Cone *et al.* (1971a, b) found that the males were attracted to small clumps

Vol. I: ISBN 0-12-592201-9

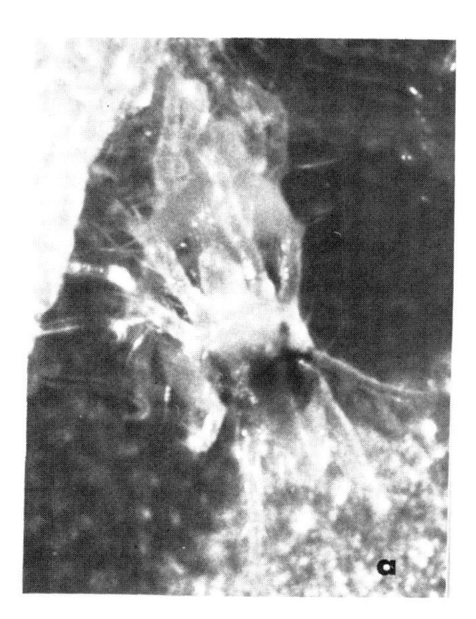

Fig. 1a. *T. cinnabarinus* male attending quiescent female deutonymph.
b. *T. urticae* male attending PVP clump soaked with 100 ppm synthetic farnesol.

of polyvinylpyrrolidone (PVP) soaked with crude ether extracts of pharate females (Fig. 1b). Male bioassay of a dilution series of crude ether extract of pharate females soaked on PVP clumps revealed several peaks of attractiveness (Cone *et al.*, 1971a), suggesting that more than one chemical with sex attractant effect is present in the crude ether extract.

Preliminary simple chemical tests provided some clues to the chemical identification of the sex attractant:

a. *Bromination*—rendered the extract unattractive, indicating the presence and the importance of double bond.
b. *Acetylation*—destroyed its attractiveness which was then restored by saponification with NaOH.

The results of these tests indicated that the sex attractant could be a polyene alcohol. Later the terpenoid alcohols farnesol, nerolidol and geraniol were isolated and identified from crude ether extracts of *T. urticae* pharate females (Regev and Cone, 1975a, 1976a). Thin layer chromatography (TLC), gas-liquid chromatography (GLC), and mass spectrometry analyses were used to identify those chemicals. The same alcohols were isolated and identified from crude ether extracts of *T. cinnabarinus* pharate females (Regev, unpublished).

Bioassays with synthetic standards of these chemicals soaked on PVP clumps revealed that both farnesol and nerolidol were significantly attractive to *T. urticae* (Regev and Cone, 1975a, 1976a) and *T. cinnabarinus* within a concentration range of 10-100 ppm. Geraniol did not attract males of either species. Since farnesol and nerolidol are found in the crude ether extract of *T. urticae* and *T. cinnabarinus* pharate females, and since both strongly attract the males of these species, they could contribute to population growth by increasing the frequency of mating and the production of fertilized eggs (female progeny).

GONADOTROPIC EFFECT OF FARNESOL

Although farnesol is important as a sex attractant it seems to have an even greater role as a gonadotropic agent.

During the chemical identification work of *T. urticae* sex attractant, preliminary field and greenhouse observation indicated differences in the population build-up of *T. urticae* on different varieties of hops. Later a study was designed to investigate if the differences observed in mite populations could be associated with varietal differences in foliar farnesol content.

Quantitative analysis of leaf farnesol, using TLC and GLC techniques, demonstrated a strong correlation between varietal farnesol content and population development of *T. urticae* on hops (Regev and Cone, 1975b) and *T. cinnabarinus* on strawberry (Regev, 1978). The most resistant varieties had the lowest foliar farnesol content and the most susceptible varieties had the highest (Fig. 2a, b). The same relationship was demonstrated for varieties of roses.

The influence of the high foliar farnesol content on mites could be expected due to the gonadotropic effect of farnesol. Juvenile hormone (JH) is known to have a gonadotropic effect on adult females of many insects. The first chemical discovered with JH activity was farnesol (Schmialek, 1961). Wigglesworth (1961) showed that farnesol induced yolk formation in decapitated adult *Rhodnius* females. Gonadotropic activities of farnesol derivatives were reported by Wigglesworth (1961, 1963), Bowers *et al.* (1965) and Bracken and Nair (1967). It is conceivable that the high oviposition rate found on farnesol-rich plant varieties was brought about by an increase in yolk deposition in the ovaries of the egg-laying females. The net result would be increased egg production.

Experiments were conducted to determine the gonadotropic effect of farnesol on *T. urticae* (Regev and Cone, 1976b). Preliminary testing using 200 ppm farnesol in vegetable oil (usual with insects) shortened the female life span. Therefore a 40% aqueous solution containing 200 ppm farnesol was applied to virgin adult females soon after emergence (*ca* 0.036 μg/female). Females treated topically with 40% aqueous ethanol only, served as the control. The number of eggs laid by the farnesol-treated females was significantly greater compared with the control ($\bar{x} = 59.8$ in treated females, $\bar{x} = 41.16$ control for the 1974 experiment, and $\bar{x} = 76.4$ in treated females against $\bar{x} = 53.9$ in the control for the 1975 experiment).

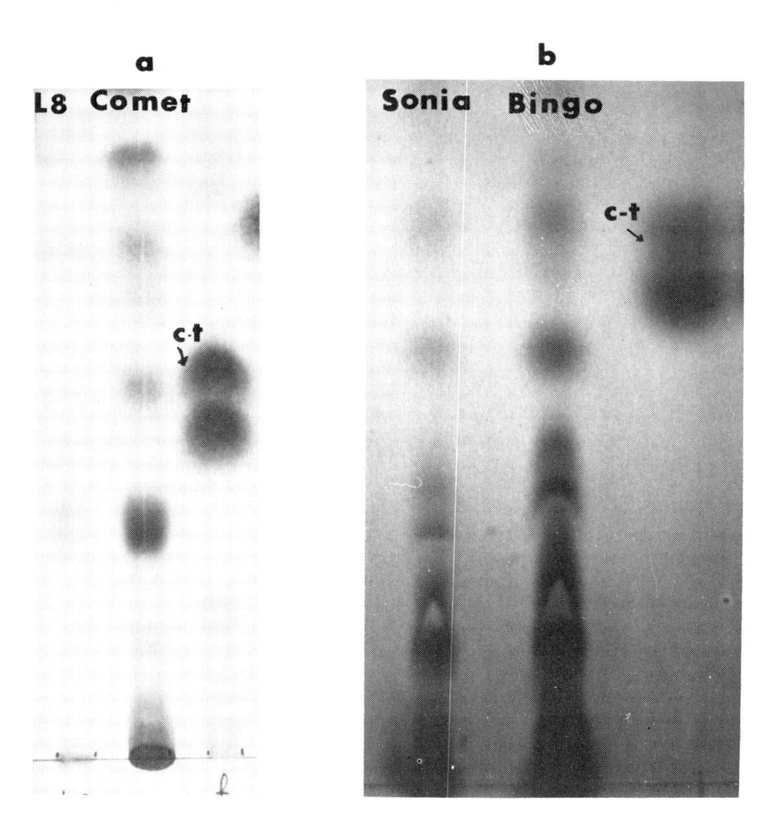

Fig. 2. Silica gel thin layer chromatogram demonstrating a varietal difference in foliar farnesol content of: a) two varieties of hops, L8 (resistant) and Comet (susceptible); b) two varieties of roses, Bingo (B) (susceptible) and Sonia (S) (resistant). The susceptible varieties show a more intense spot for the *cis-trans* isomer of farnesol. The developing solvent was 20% ethyl acetate in benzene followed by charring with 5% H_2SO_4.

An attempt to introduce exogenous farnesol into leaf-discs by placing them on cotton pads soaked with a water solution containing 50 ppm farnesol, 1% sucrose, and 0.1% Tween 80® was unsuccessful. The experiments revealed no significant differences in the number of eggs produced by the females grown on the farnesol-treated or normal leaf-discs.

UTILIZATION OF FARNESOL AS A PRECURSOR FOR THE SYNTHESIS OF OTHER CHEMICALS

Steroids

Arthropods require exogenous sterols in their diet. Farnesol is an acyclic terpene alcohol composed of three isoprene units. The derivative farnesylpyrophosphate is a precursor of squalene, itself a precursor of cholesterol and other polyisoprenoids. The obligatory requirement of all arthropods for an exogenous supply of sterols stems from the fact that they lack an enzymatic system that cycles the open chain hydrocarbon squalene into the molecuar structure of sterols (e.g. lanosterol). The pathway for the synthesis of squalene from two farnesylpyrophosphate molecules is shown in Fig. 3. Farnesylpyrophosphate is synthesized from acetyl Co-A via mevalonic acid. Detailed studies showed that acetate could be replaced by mevalonic acid in bacteria that require acetate for their growth. These bacteria convert mevalonic acid to squalene via farnesylpyrophosphate, and then synthesize cholesterol from squalene. Although such a symbiotic relationship has not been found between the spider mites and their intestinal bacterial flora it is conceivable that the mite is capable of utilizing farnesol for the synthesis of sterol in that manner. Brand *et al.* (1975) demonstrated a symbiotic relationship between several bark beetle species and the bacterium *Bacillus cereus* which lives in the hind gut of the insect. These authors hypothesized that the bacteria synthesized the pheromone verbanol from α-pinene which was acquired from the host plant.

Juvenile Hormone

Reibstein and Law (1973) were able to demonstrate an enzymatic synthesis, *de novo,* of JH from small precursors such as acetate, propionate and methionine using a homogenate of corpora allata of the tobacco hornworm. The same authors showed, by using several JH precursors that farnesyl pyrophosphate is presumably cleaved to farnesol and then oxidized to farnesenic acid, which is subsequently epoxidized enzymatically to JH III acid in the presence of NADPH and oxygen (Reibstein *et al.*, 1976) (Fig. 4).

The conversion of epoxy farnesenic acid into the hormone is done by alkylation of the carboxyl group by enzymes from the corpora allata (homogenates) and extracts from certain bacteria. It is conceivable that the alkylation of the epoxy farnesenic acid into JH could be achieved in the mite by an organ equivalent to the corpora allata (not found yet) or by the intestinal flora.

Farnesenic acid can be esterified into methyl-farnesenate by corpora allata. However, there is no hormone synthesis from methyl-farnesenate. The accumulation of methyl-farnesenate occurs only when NADPH and probably

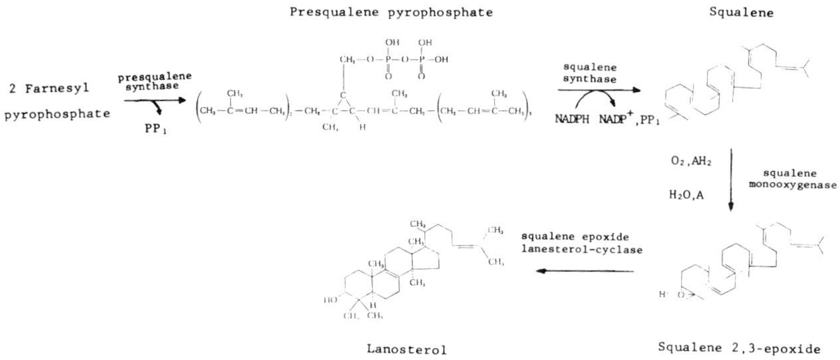

Presqualene pyrophosphate

Squalene

2 Farnesyl pyrophosphate

presqualene synthase

PP_i

squalene synthase

NADPH $NADP^+, PP_i$

O_2, AH_2

H_2O, A

squalene monooxygenase

squalene epoxide lanesterol-cyclase

Lanosterol

Squalene 2,3-epoxide

Fig. 3. Formation of lanosterol from farnesol phyrophosphate via squalene.

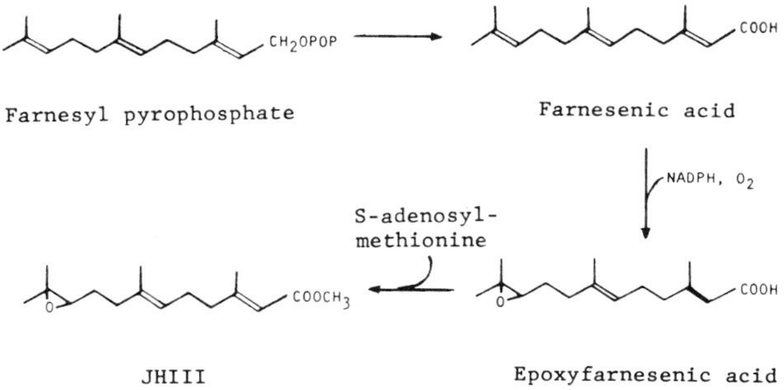

Farnesyl pyrophosphate

Farnesenic acid

NADPH, O_2

S-adenosyl-methionine

JHIII

Epoxyfarnesenic acid

Fig. 4. Enzymatic conversion of farnesol phyrophosphate into JH III (after Reibstein *et al.* 1976).

oxygen are in short supply. Apparently, methyl-farnesenate is used as a storage form for farnesenic acid during a shortage of NADPH (Reibstein *et al.*, 1976).

The same authors reasoned that the cleavage of farnesylpyrophosphate to farnesol prior to the oxidation of the latter to farnesenic acid, would be followed by an alcohol and aldehyde dehydrogenase reaction most likely employing NAD as a cofactor. In order to utilize farnesol, the mite must possess enzymatic systems capable of using farnesol as substrate. At least two such enzymatic systems were found in *T. urticae* by using acrylamide gel electrophoresis. The first enzymatic system, octanol dehydrogenase (Fig. 5a), used farnesol as a substrate turning it into its aldehyde farnesal. According to Madhavan *et al.* (1973) this enzyme catalyzes the oxidation of primary alcohols and was found to oscillate in phase with ecdysis in Drosophila. The second enzymatic system, aldehyde oxidase, readily oxidizes farnesal and other aldehydes. Yamamoto and Jacobson (1962) confirmed the JH activity of farnesol on *Tenebrio molitor*.

Although JH and ecdysone have not been isolated and identified yet in mites, it is conceivable that they do exist in this group of arthropods. Both

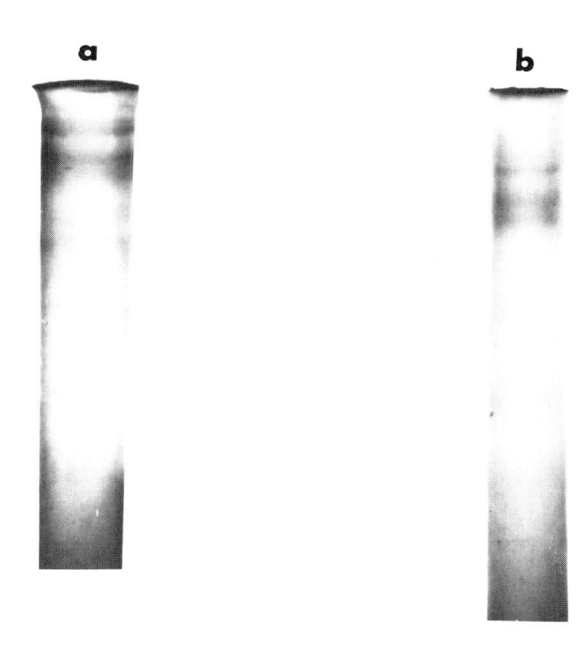

Fig. 5. Polyacrylamide gel zymograms stained with farnesol for octanol dehydrogenase (a) and with farnesol for aldehyde oxidase (B).

arthropod classes, Insecta and Acarina progress through developmental stages ending with the molting of their integument. The shedding of the integument does not occur naturally beyond the adult stage. Mango *et al.* (1976) were able to produce "super ticks" by application of exogenous ecdysone, i.e., the normal females underwent an extra molt after they received an injection of ecdysone. The "super tick" females produced more eggs than the normal ones. This study demonstrated that the tissue of the ticks (the closest group to mites) responded to ecdysone. Apparently, JH and ecdysone are present in mites and ticks and function in a way similar to insects. The very small weight of the mites (*ca* 3 μg) could be the main reason why both hormones have not yet been isolated and identified from this group. This problem can be appreciated even more by recalling that it took 500 kg of pupae and developing adult silkworms *Bombyx mori* to obtain 25 mg of molting hormone (Butenandt and Karson, 1954). Hopefully, such an analytical task will be easier with the future advancement of analytical tools and techniques.

SUMMARY

Some possible roles of farnesol and nerolidol in the biology of two tetranychid mites are discussed. These are: 1) male sex attractant functions, 2) gonadotropic effect, and 3) utilization of a precursor for synthesis of sterols and JH.

REFERENCES

Bowers, W. S., Thompson, M. J. and Uebel, E. C. (1965). *Life Sci.* **4**, 2323-2331.
Bracken, G. K. and Nair, K. K. (1967). *Nature* **216**, 483-484.
Brand, J. M., Bracke, J. W., Markovetz, A. J., Wood, D. L. and Browne, L. F. (1975). *Nature*, **254**, 136-137.
Butenandt, A. and Karlson, P (1954). *Z. Naturforsch.* **9b**, 389-391.
Cone, W. W., McDonough, L. M., Maitlen, J. C. and Burdajewicz, S. (1971a). *J. Econ. Entomol.* **64**, 358-361.
Cone, W. W., Predki, S. and Klostermeyer, E. C. (1971b). *J. Econ. Entomol.* **64**, 379-382.
Helle, W. (1967). *Ent. Exp. Appl.* **10**, 103-110.
Madhavan, K., Conscience-Egli, M., Sieber, F. and Ursprung, H. (1973). *J. Insect Physiol.* **19**, 235-241.
Mango, C., Odhiambo, T. .R and Galun, R. (1976). *Nature* **260**, 318-319.
Regev, S. (1978). *Ent. Exp. Appl.* **24**, 22-26.
Regev, S. and Cone, W. W. (1975a). *Environ. Entomol.* **4**, 307-311.
Regev, S. and Cone, W. W. (1975b). *Environ. Entomol.* **4**, 697-700.
Regev, S. and Cone, W. W. (1976a). *Environ. Entomol.* **5**, 133-138.
Regev, S. and Cone, W. W. (1976b). *Environ. Entomol.* **5**, 517-519.
Reibstein, D. and Law, J. H. (1973). *Biochem. Biophys. Res. Communs.* **55**, 266-272.
Reibstein, D., Law, J. H., Bowlus, S. B. and Katzenellbogen, J. A. (1976). *in* "The Juvenile Hormones," (L. I. Gilbert, ed.) Plenum Press, NY. 572 pp.
Schmialek, P. (1961). *Z. Naturforsch.* **16**, 461-464.
Yamamoto, R. T. and Jacobson, M. (1962). *Nature* **196**, 908-909.
Wigglesworth, V. B. (1961). *Stal. J. Insec. Physiol.* **7**, 73-78.

COMPONENTS OF REPRODUCTIVE SUCCESS IN SPIDER MITES

Dana L. Wrensch

Acarology Laboratory
The Ohio State University
Columbus, Ohio

INTRODUCTION

Spider mites of the *Tetranychus urticae* complex, notable the two-spotted spider mite, *T. urticae* Koch and the carmine spider mite, *T. cinnabarinus* (Boisduval) are polyphytophagous colonizing species of tremendous economic importance. These species and close relatives cause millions of dollars of damage annually and contribute greatly to the nearly 39% pre-harvest loss in world agricultural food and fiber production (W. W. Cone, pers. comm.).

As successful colonizing species, spider mites are adapted for opportunistic exploitation of ephemeral and patchily distributed resources. Such success leads to outbreaks characterized by extremely rapid population growth and devastating impact on the host.

A high intrinsic rate of increase, r, is a pervasive feature of spider mite populations during outbreaks. This report considers 1) the biological traits that influence r as a function of environmental conditions and 2) how these traits ultimately affect the reproductive success of the species. The traits considered are fecundity, hatchability, duration of oviposition, longevity, rate of development, survivorship, and sex ratio.

On the whole, distinctions will not be emphasized among species and strains of mites in the *T. urticae* complex because the qualities discussed, and conclusions drawn, are expected to be general features of their biology.

FECUNDITY

The number of offspring produced is crucial to a population's ability to increase. Many of the factors influencing variation in fecundity have been identified. Daily fecundity is in general positively correlated with temperature

Vol. I: ISBN 0-12-592201-9

(Hazan *et al.*, 1973). Lowered humidity also increases fecundity (Hazan *et al.*, 1973; Boudreaux, 1958). Davis (1952) showed that the average number of eggs produced per female decreases during the latter part of colonization due to crowding. Attiah and Boudreaux (1964) found that individuals in dense colonies with adequate supplies of food produced fewer eggs. This observation was confirmed by Wrensch and Young (1978). Wrensch and Young (1975) showed that females from good resources produced 11% more eggs than females from poor resources and that fertilized females produce 12% more eggs than unfertilized females. They also showed that fecundity was significantly correlated with the duration of oviposition and longevity. Dittrich *et al.* (1974) showed convincingly that insecticide residues can stimulate fecundity. Host plant conditions influencing fecundity have been studied by Henneberry (1962), Poe (1972), and Watson (1964). This list is not complete but serves to illustrate how sensitive fecundity is to environmental conditions.

HATCHABILITY

Hatchability, the survivorship of eggs, is an important parameter influencing net fitness (number of adult offspring). It has been shown to be unaffected by humidity (Boudreaux, 1958) but reduced by crowding (Davis, 1952). Wrensch and Young (1975) found that it is uniformly high (*ca* 95%) regardless of host quality during parental development or the fertilization status of females. Inbreeding tends to reduce the hatchability of haploid (male) eggs (Helle, 1965). Interstrain incompatibilities are reflected in the differential hatchability of the F_2 eggs (Helle and Pieterse, 1965; Boudreaux, 1963). Interspecific hybrids, when not sterile, tend to have reduced hatchabilities (Wrensch and Murtaugh, 1978; Murtaugh and Wrensch, 1978). Lines carrying an allele for eye color introduced from *T. cinnabarinus* into *T. urticae* (albino) initially had very low hatchability but after 26 generations of stabilization had hatch equal to that of the *T. urticae* strain (Murtaugh and Wrensch, in prep).

DURATION OF OVIPOSITION

Although the duration of oviposition is likely to be affected by many environmental variables, it has not been as extensively studied as has fecundity. Laing (1969) found an average of 15.7 days, while Lehr and Smith (1957) found the duration to be 14.7 days. Wrensch and Young (1975) showed that resource quality during parental development did not affect length of oviposition but that a female's fertilization status did. Unfertilized females oviposited on the average three days longer than fertilized females. But even when the influence of the duration of oviposition on fecundity is removed by covariance analysis (they were highly correlated), differences in fecundity were highly significant (see section on Fecundity). Thus the duration of oviposition is unimportant in determining variation in fitness.

Egg production curves in spider mites are essentially triangular (Boudreaux, 1969; Hazan *et al.*, 1973; Wrensch and Young, 1975). By fitting orthogonal polynomials to daily fecundity data and taking the second derivative to obtain the maximum, turnover rates can be found. Wrensch and Young (1975) found that unfertilized females have earlier turnovers than fertilized females and that females from good resources have turnovers about a half-day earlier than females from poor resources. The importance of relative turnover rates will be shown in the final section.

At 22°C it takes about 18 days of oviposition to reach a correlation of 0.9 between partial and total egg production, but correlations at the time of peak egg production of approximately 0.5 to 0.6 are reached within four to six days from onset (Wrensch and Young, 1975).

LONGEVITY

Longevity is highly correlated with the duration of oviposition (Wrensch and Young, 1975). Cagle (1949) reported a maximum longevity of 68 days for females and 94 for males. It has been found that both density (Attiah and Boudreaux, 1964) and high humidity (Hazan *et al.*, 1973) tend to reduce longevity. Wrensch and Young (1975) found that although fertilized females from good resources produce 21% more eggs than unfertilized females on poor resources, their longevity is reduced by 17%. This shorter life span was interpreted as due to reproductive exhaustion. By conserving energy (lower fecundity rates and longer life spans) unfertilized females may be able to exploit mating opportunities with a son, leading to a pulse of daughters late in life. The intrinsic rate of increase under such conditions was found to be *ca* 50% that of initially fertilized females (Wrensch and Young, 1975).

RATE OF DEVELOPMENT

Lewontin (1965) showed that rate of development was the most important single variable in influencing the intrinsic rate of increase. In terms of changes in r, a small change in developmental rate is approximately equal to a 10-fold change in total fertility.

Rate of development, from egg to adulthood, is a linear function of, and inversely correlated with, temperature (Boudreaux, 1963) and is retarded by high humidities (Hazan *et al.*, 1973). Wrensch and Young (1978) found that on the average at 22°C, development takes 16 days but that the rate was extremely sensitive to their environmental treatments. Although males consistently developed faster than females, the differences were more pronounced when parents developed on good resources and oviposited on good resources at low density (conditions similar to those early in a colonizing episode). At this time, functional sex ratio (number of adult males to number of quiescent female deutonymphs (Potter, 1978)) is very low and inbreeding would be expected to be high. A relatively larger difference in male and female developmental rates would be important to ensure successful insemination of all sisters. Conversely, the conditions near the end of a colonizing episode are poor with high mite density. Male developmental rate is close to that of females so that the likelihood of outbreeding is enhanced, especially with the abundance of males to available females (Potter, 1978). Outbreeding at this time would enable a wide spectrum of genetic variability to be available to the dispersing females, i.e., the potential founders.

Both sexes develop more rapidly if host quality is poor during parental development, good during oviposition and development of offspring, and density is low (conditions encountered by the progeny of the founding female (Wrensch and Young, 1978)). The difference in developmental rate between mites at low and high density is almost one day. It was shown that developmental rate is density dependent and the effects of density on develomental rate are direct results of net offspring density or crowding, first from differences in parental fecundity as functions of the resource quality under which they developed, and second, from the numbers of offspring from contrasting numbers of ovipositing females.

SURVIVORSHIP

Hazan *et al.* (1973) found that survivorship decreases with higher humidities and extremely low or high temperatures. They found that as temperature drops, it becomes relatively more important to survivorship than relative humidity. Boudreaux (1958) also demonstrated the deleterious effects of high humidity. Davis (1952) found that the survivorship of nymphs was decreased by crowding. Wrensch and Young (1978) found that survivorship was profoundly influenced by the quality of the host upon which offspring developed and by their density. Survivorship, measured as no. of adults/no. of eggs x 100, ranged from 20 to 70%. Good resources for offspring and low density resulted in higher proportions of survivors.

Survivorship can be studied with finer resolution by examining the relative success of each developmental stage. Wrensch and Young (in prep.) found that all developmental stages, whether mobile or quiescent, survive better on good quality resources, but that when resource quality is poor, quiescent stages survive better. In addition they found that mobile and quiescent protonymphs and mobile deutonymphs have higher survivorships if their parents developed on good resources. This suggests that the resource quality during maternal development influences the quality of offspring as well as the quantity.

SEX RATIO

The factors influencing sex ratio are important because sex ratio determines the number of dispersing females. In addition, because feeding ovipositing females cause the greatest damage, sex ratio reflects potential impact of a population on the host.

Boudreaux (1963) stated that there is no "normal" sex ratio. Studies since then have helped to identify the causes of variability. Overmeer and Harrison (1969) in reciprocal crosses found that more females were produced in an A x B cross than B x A. This indicated the existence of a maternal effect. Mitchell (1972) studied the inheritance of sex ratio and found that two loci were involved. He too found a maternal effect.

Length of copulation (D. A. Potter, pers. comm.; Overmeer, 1972) also influences sex ratio, for if a mating is incomplete, insufficient sperm are available for fertilization. In addition, the number of daughters has been shown to increase with maternal age (Hazan *et al.*, 1973). Sex ratio differ between inter- and intraspecific matings between *T. cinnabarinus* and *T. urticae* (Murtaugh and Wrensch, 1978) with more females produced intraspecifically.

Sex ratio is also extremely sensitive to mite density and conditions under which parental mites develop (Wrensch and Young, 1978). Over a four-day period of oviposition, more female than male offspring were always produced but sex ratio (females/males) ranged from 1.20 to 3.89, e.g., when offspring density was low, three or four times as many daughters were produced. It was shown that females ovipositing in crowded conditions produced fewer fertilized eggs and that differences in the competitive ability of immature males and females led to differential survivorship which lowered the sex ratio. Females from good resources produced relatively fewer females suggesting female control of sex ratio. Although there is no known mechanism for control of sex ratio by the female, the lack of mitochondria (and hence motility) in the sperm (Pijnacker and Drenth-Diephius, 1973) would suggest that such a mechanism must exist.

To investigate further the influences of host quality on sex ratio, albino *T. urticae* females mated to wild-type males (enabling differentiation of sex in dead immatures) were isolated on leaf discs of either good (undamaged) or poor (damaged by previous feeding) quality. Females were left for the duration of their oviposition. Those females ovipositing on poor leaves had consistently higher sex ratios through time than did females ovipositing on good leaves (Fig. 1). This response to environmental conditions appears to be adaptive. Females on a poor resource are responding to their environment by producing as many daughters, or potential dispersants, as possible, and thus optimizing their fitness. It is because of the high functional sex ratio late in colonization, when resource quality is poor (Potter, 1978) that the ovipositing females are able to reduce the proportion of sons, assuring that their

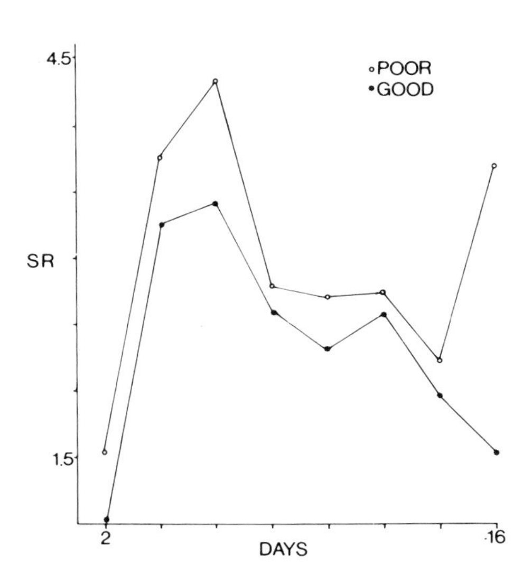

Fig. 1. Changes in sex ratio (SR) through time for females that developed on good and poor resources.

dispersing daughters, as potential founders, are inseminated before they disperse.

If one compares the sex ratio of the progeny from females that have developed on contrasting good vs. poor resources, when then placed on good resources, an interesting feature emerges (Fig. 2). Females from a good resource after 17 days and females from a poor resource after 12 days start producing more sons than daughters. The implication here is that the females either ran out of sperm or that the remaining sperm denatured. Nonetheless, the females were incapable of accepting a second, late insemination (Potter and Wrensch, 1978) even though virgin females of similar age were capable of being inseminated (Wrensch, unpubl. data).

A very high stable sex ratio of approximately nine females to one male was found in a Brazilian spider mite similar to *T. cinnabarinus* (Wrensch and Flechtmann, 1978). Given that the accumulation of males relatively early in a colonizing episode and the specificity of guarding ensure the insemination of virtually all females (Potter, 1978), there is no reason why the sex ratio should be as low as it generally is (*ca* 3:1 in most populations). However, the amount of heterozygosity available is dependent to a large extent on sex ratio insofar as sex ratio affects effective population size (Fig. 3). The loss of heterozygosity increases as sex ratio decreases, thus decreasing the amount of genetic variability available in the population. Too little variability deprives the population of the opportunities for selection. Conversely, as sex ratio increases above 3:1, the change in heterozygosity becomes drastically slowed, precluding the rapid fixation of beneficial mutants or the extinction of

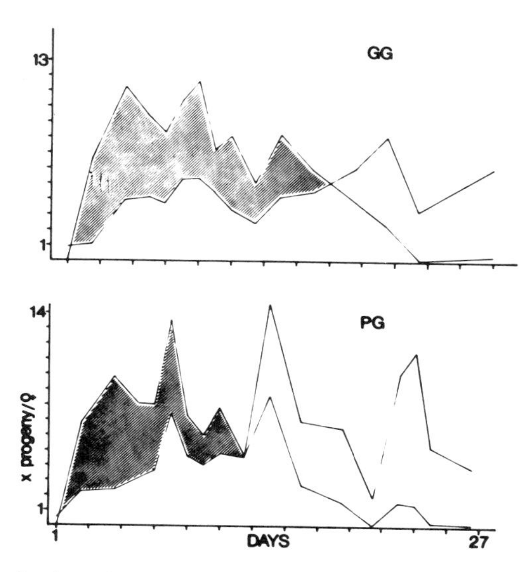

Fig. 2. Life time production of sons and daughters by females from poor (PG) and good (GG) resources, ovipositing on good resources. Shaded areas indicate when females are in excess.

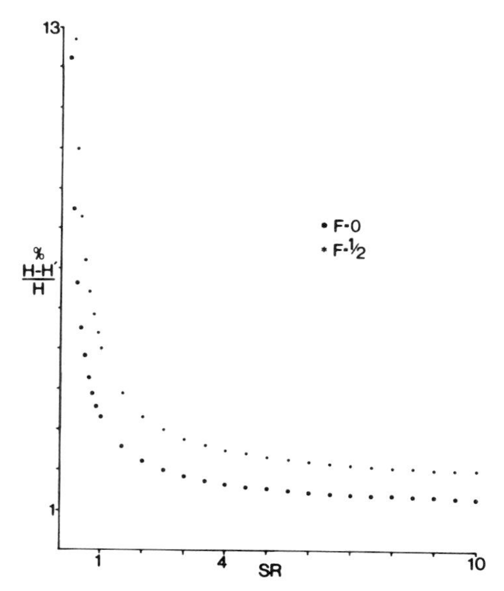

Fig. 3. Changes in the rate of loss of heterozygosity ($\frac{H-H'}{H}$) as a function of sex ratio (♂ / ♀). "F" is the coefficient of inbreeding.

deleterious ones. Thus a sex ratio near 3:1 appears to be the optimum and, as most published reports confirm, is the general case. Note that when inbreeding is included, the rate of loss of heterozygosity is accelerated (Fig. 3).

COLONIZING ABILITY AND THE INTRINSIC RATE OF INCREASE

Colonizing ability is measured by a set of biological traits that influence *r*. Spider mites are successful colonizing species in part because they have a characteristically high *r*.. In McArthur-Wilson terminology (1967) they are classic examples of the *r* strategist.

The Lotka equation is conventionally used to measure *r*. As an effective estimate for comparisons of subtle population differences, however, the method of Lewontin (1965) is appropriate. Lewontin derived from the Lotka equation a new formulation using rate of development, turnover rate, duration of oviposition, and fecundity as the biological parameters from which a numerical analysis produced the *r* value. The two assumptions are that the fecundity schedule is triangular and that the sex ratio is 1:1. The former assumption is valid for spider mites but fecundity must be adjusted for the unequal proportions of females and males normally found.

The value of *r* has been shown to vary with temperature and humidity (Hazan *et al.,* 1973). They found values ranging from .106 to .355. Shih *et al.* (1976) found an *r* of .336 at 27°C and Laing (1969) found an *r* of .143 at 20.3°C. Watson (1964) showed the influence of improved host nutrition on increasing *r* and found values from .177 to .256. Wrensch and Young (1975) found *r* values of .259 for females from good resources and .242 for females from poor resources using Lewontin's method. They varied the parameters one at a time to determine the changes necessary to produce an increase in *r* of the females from the poor environment equal to that of females from the good environment. Demonstrating the relative importance of the individual parameters, it was found that a 1.15 day decrease in developmental rate, a 3.30 day increase in turnover rate, a 6.46 day decrease in duration of oviposition, or a 26% increase in female offspring would so increase *r* in the females from the poor resource.

Five of the eight treatment combinations from Wrensch and Young (1978) represent discrete "slices" of a colonizing episode. These can be ordered to reflect the sequential stages of host destruction and population density as they alter during colonization. Using data on fecundity, turnover rate, and duration of oviposition (Wrensch and Young, 1975) and developmental rate and sex ratio (Wrensch and Young, 1978), *r* was determined for the contrasting environmental combinations (Table I).

The capacity to increase is highest at the outset of a colonizing episode and declines steadily thereafter. Founding females (from a poor resource, ovipositing on a good resource at low density), although having reduced fecundity, produce progeny with the fastest developmental rate and highest proportion of daughters. The greater relative importance of rate of development over fecundity, as noted earlier, results in the highest intrinsic rate of increase. This optimization is an essential quality of founders, making

TABLE I.
The Intrinsic Rate of Increase (r) for Five Stages of a Colonizing Episode.

	Stage of Colonization				
Treatment	Upon Founding	Early	Middle	Late	Dispersing
Leaf quality during parental development	poor	good	good	good	poor
Leaf quality during oviposition and development of offspring	good	good	good	poor	poor
Density	low	low	high	high	high
r	.217	.216	.195	.181	.175

possible the rapid establishment of the population and enhancing chances of success of the colonizing attempt.

Variations in r are a basic feature of spider mite populations. Clearly r is limited by environmental conditions, causing spider mites to maintain stable populations during non-outbreak conditions, but permitting rapid exploitation of hosts when circumstances become optimal.

ACKNOWLEDGEMENTS

I appreciate the constructive criticisms provided by Professor Donald E. Johnston and Daniel A. Potter.

REFERENCES

Attiah, H. H. and Boudreaux, H. B. (1964). *J. Econ. Entomol.* **57**, 53-57.
Boudreaux, H. B. (1958). *J. Insect Physiol.* **2**, 65-72.
Boudreaux, H. B. (1963). *Ann. Rev. Entomol.* **8**, 137-154.
Boudreaux, H. B. (1969). *Proc. 2nd Intern. Cong. Acarol.*, 485-490.
Cagle, L. R. (1949). *Virginia Agric. Expt. Sta. Tech. Bull.* **113**, 31 pp.
Davis, D. W. (1952). *J. Econ. Entomol.* **45**, 652-654.
Dittrich, V., Streibert, P. and Berthe, P. A. *Environ. Entomol.* **3**, 534-540.
Hazan, A., Gerson, U. and Tahori, A. S. *Acarologia* **15**, 414-440.
Helle, W. (1965). *Entomol. Exp. App.* **8**, 299-304.
Helle, W. and Pieterse, A. H. (1965). *Entomol. Exp. Appl.* **8**, 305-308.
Henneberry, T. J. (1962). *J. Econ. Entomol.* **55**, 134-137.
Laing, J. E. (1969). *Acarologia* **11**, 32-42.
Lehr, R. and Smith, R. F. (1957). *J. Econ. Entomol.* **50**, 634-636.
Lewontin, R. C. (1965). *In* "The Genetics of Colonizing Species." (H. G. Baker and G. L. Stebbins, eds.) Academic Press, New York. pp. 77-91.
McArthur, R. H. and Wilson, E. O. (1967). "The Theory of Island Biogeography." Princeton Univ. Press, Princeton, N. J., 203 pp.
Mitchell, R. (1972). *Entomol. Exp. Appl.* **15**, 299-304.
Murtaugh, M. P. and Wrensch, D. L. (1978). *Ann. Entomol. Soc. Am.* (in press).
Overmeer, W. P. J. (1972). *Entomol. Berichten* **32**, 240-244.

Overmeer, W. P. J. and Harrison, R. A. (1969). *N. Z. J. Sci.* **12,** 920-928.
Pijnacker, L. P. and Drenth-Diephius, L. J. (1973). *Netherlands J. Zool.* **23,** 446-464.
Poe, S. L. (1972). *Fla. Entomol.* **54,** 183-186.
Potter, D. A. (1978). *Ann. Entomol. Soc. Am.* **71,** 218-222.
Potter, D. A. and Wrensch, D. L. (1978). *Ann. Entomol. Soc. Am.* (in press).
Shih, C. T., Poe, S. L. and Cromroy, H. L. *Ann. Entomol. Soc. Am.* **69,** 362-364.
Watson, T. F. (1964). *Hilgardia* **35,** 273-322.
Wrensch, D. L. and Flechtmann, C. H. W. (1978). *Ciencia y Cultura* (in press).
Wrensch, D. L. and Murtaugh, M. P. (1978). *J. Heredity* **68,** 329-330.
Wrensch, D. L. and Young, S. S. Y. (1975). *Oecologia* **18,** 259-267.
Wrensch, D. L. and Young, S. S. Y. (1978). *Environ. Entomol.* **7,** 499-501.

TETRANYCHID DEVELOPMENT UNDER VARIABLE TEMPERATURE REGIMES

L. K. Tanigoshi

Boyden Entomology Laboratory
Federal Research, SEA, USDA
Riverside, California

J. A. Logan

Department of Zoology and Entomology
Colorado State University
Fort Collins, Colorado

INTRODUCTION

The critical importance of temperature to development of tetranychid populations has long been recognized, and considerable research effort has been directed toward measuring the effects of static temperature levels on the life stage development of spider mites. In this paper, we have used data sets derived from studies of *Tetranychus mcdanieli* McGregor (Tanigoshi *et al.*, 1975a; 1976) in our discussion of the general applicability of various mathematical functions to tetranychid temperature-dependent life history parameters. The choice reflects the fact that the McDaniel spider mite is potentially the most destructive mite occurring on apples in Washington State. However, since 1966 it has been controlled there continuously by a widespread organophosphorus-resistant phytoseiid predator, *Metaseiulus occidentalis* (Nesbitt).

The curvilinear temperature-dependent relationships for embryonic and postembryonic development of *T. mcdanieli* concur with the empirical relationships reported by English and Turnipseed (1941), Cagle (1949), Shinkaji (1959) and Parent (1965) for several tetranychid species. Application of the day-degree concept to the forementioned empirical methods has indicated that the concept's inherent linearity between rate of development and temperature is not valid for spider mite populations (Logan *et al.*, 1976). The empirical methods most commonly used to describe the functional relationship between developmental rate and temperature for tetranychid species have

Vol. I: ISBN 0-12-592201-9

either been the exponential equation forms and/or least squares polynomials. In every case, the behavior of these curves was questionable at temperatures near or above the optimum temperature for development. Deductive equations approximating sigmoid curves were derived and proposed by Pradhan (1945) and Stinner *et al.* (1974) to functionally simulate insect developmental time estimations. However, both equations are unsatisfactory because of their symmetry about the thermal optimum, a growth rate situation that is obviously not correct for the *Tetranychidae.*

EFFECTS OF TEMPERATURE ON POPULATION INCREASE OF *TETRANYCHUS MCDANIELI*

Constant Temperature Regimes

The developmental curves for the egg and immature life stages are quartic in form, with the latter curves being asymmetrical about the minimum developmental point. Curves for the quiescent stages and pre-oviposition period are cubic in form and differ from the quartic curves in that developmental period decreases rather than increases beyond 33°C (Fig. 1).

Variable Temperatures Regimes

A generalized variable temperature developmental model (SIMDEV) using polynomial regression equations as reported by Tanigoshi *et al.* (1976) for *T. mcdanieli* developmental rates was used to obtain estimates of developmental periods at both the alternating and mean of the daily alternating temperatures. The alternating temperature regimes selected were 5-20°, 10-25°, 15-30°, and 15-35°C. SIMDEV (Fig. 2) estimates the percentage of egg development that would occur in the first hour after deposition at the temperature given for that time interval. SIMDEV repeats the operation at each successive hour until 100% development is attained and then continues through each subsequent life stage until deposition of the first egg.

A mathematical generalization of variable temperature developmental models such as SIMDEV can be made by assuming existence of some developmental relationship (r) that describes the specific developmental rate of a life stage as a function of temperature (T). If temperature is then described as a function of time (t), by definition of developmental rate, the proportion of the life stage completed during a time interval ($Dt = t_2 - t_1$) is given by

$$PC = \int_{t_1}^{t_2} r\,[T(t)]\,dt. \quad (1)$$

Conversely, the amount of time required to complete a life stage may be found by setting PC = 1 and solving for Dt.

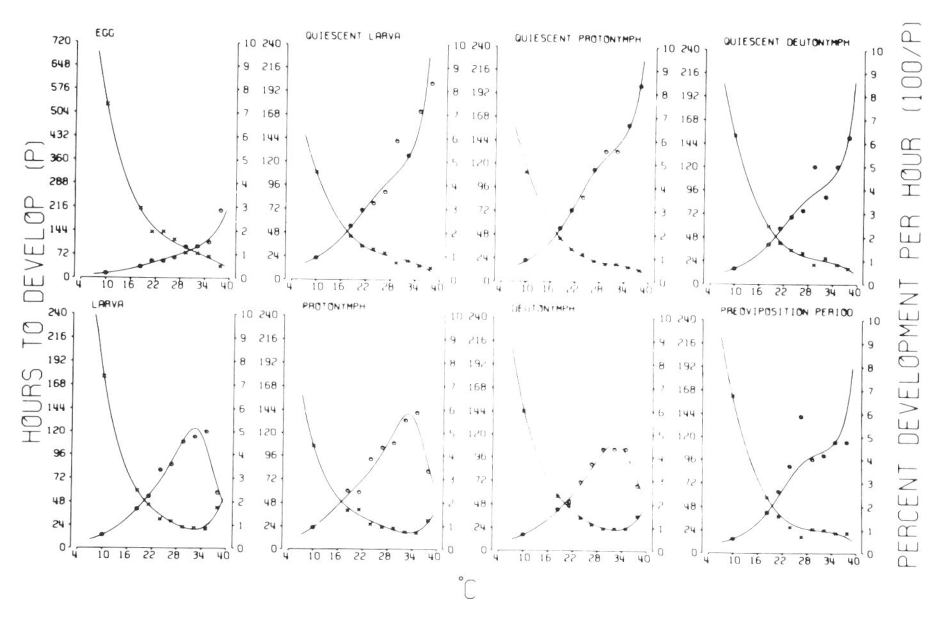

Fig. 1. Developmental curves of immature female *T. mcdanieli.* (Tanigoshi *et al.*, 1975a).

Data required for determination of the functional form of r are usually obtained by measuring the rate of development (1 / time required to complete a life stage) for a range of fixed, constant temperatures. Two implicit assumptions must be met for valid utilization of such data in Eq. (1). The first of these is that under variable temperatures, the organism adjusts "instantaneously" to variation in ambient conditions. Under natural temperature regimes, where temperature changes tend to be gradual and continuous, this is generally a valid assumption. Equation (1) also assumes that the temperature of the microhabitat of the organism can be accurately measured, at least in a correlative fashion, and furthermore that the adaptive behavior of the organism under test conditions mimics that in the natural environment. The second of these assumptions is more tenuous, and validity should be judged on an individual basis. For example, validation of SIMDEV indicates that both these assumptions are at least approximately satisfied for *T. mcdanieli* (Tanigoshi *et al.*, 1976). Once these assumptions are deemed reasonable, all that remains is determination of the functional forms of r and T.

Perhaps the most common approach to description of variable temperatures in arthropod developmental models has been to descretize the continuous temperature function (for example, the familiar day- or hour-degree models). SIMDEV assumes a constant temperature for a small time interval (one hour) and by so doing reduces the integration of Eq. (1) to a summation. Another approach is to actually fit a continuous function to

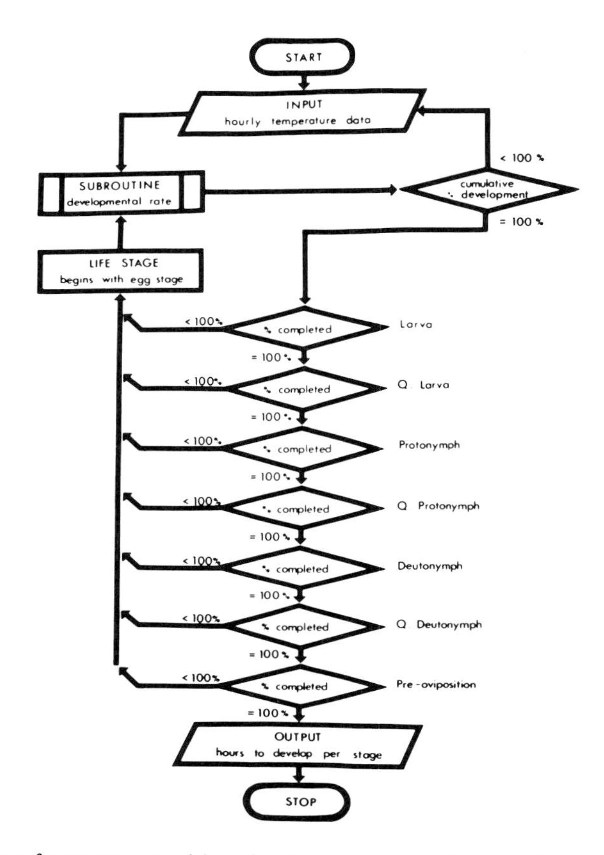

Fig. 2. Flow chart for temperature driven developmental model (Tanigoshi *et al.*, 1976).

observed data. For example, the diurnal temperature cycle may be adequately represented by the Fourier series,

$$T(t) = a_o + \sum_{i=1}^{n} [a_i \cos(kit) + b_i \sin(kit)] \qquad (2)$$

where temperature (T) is a function of time (t) and k is a Fourier constant describing the period of the function. Although Eq. (2) is purely empirical, it does provide an analytic expression for T of Eq. (1) and additionally, it provides a temperature profile that is easily modified. For example, if the series of Eq. (2) has temperature range R_1, then

$$T(t) = a_o + (R_2/R_1) \sum_{i=1}^{n} [a_1 \cos(kit) + b_1 \sin(kit)] \qquad (3)$$

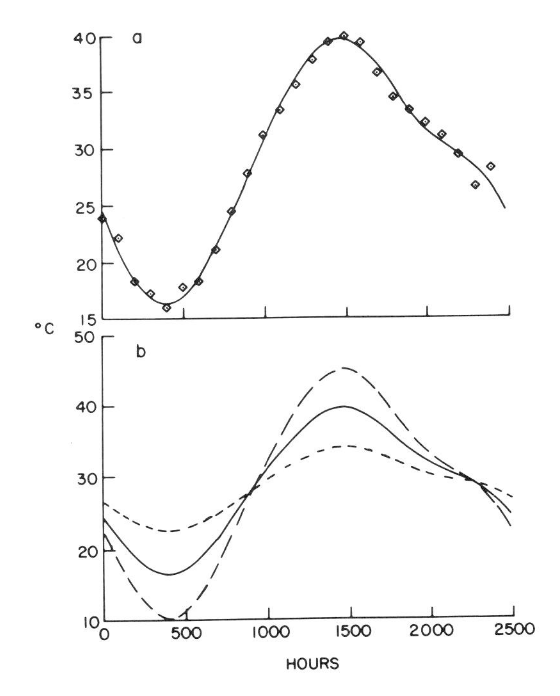

Fig. 3. (a) Third order Fourier Series fitted to hourly temperatures for Aug. 18, 1969, Wenatchee, Washington. (b) Effect of modifying the function shown in Fig. 8a (———) by "R" factors (Eq. 3) of 0.75 (- - -) and 1.25 (— — —).

is of the same form as Eq. (2) with mean equal to a_0 but with range R_2. Fig. 3a shows a third order fit of Eq. (2) to a temperature profile recorded at Wenatchee, Washington on 18 August 1969; the modified series given by Eq. (3) is plotted in Fig. 3b.

Determination of a suitable functional form for r has been a recurrent theme in entomological/ecological literature for most of the present century. The earliest useful relationship was the day-degree concept which assumes that rate of development is proportional to temperature, i.e., a linear relationship between r and T. It has been found that many poikilothermic organisms do exhibit a linear response over much of their normally encountered temperature range. In fact, in 1976, Gilbert *et al.* stated, "Over the intermediate range of temperatures which the insect normally experiences in the field, the rate of development increases linearly with temperature . . . at unusually high temperatures, the rate of development declines, but this rarely happens in the field: the temperature requirements of different species and populations are adjusted to suit the local temperature regimes. Thus the rate of development is directly proportional to the ambient temperature, measured above the threshold temperature."

This statement may be true for some arthropod species, but should not be accepted as paradigm. In fact, adaptation of an organism to its environment may depend on a nonlinear development response to temperature. This is true for periods of environmental stress (i.e., diapause, estivation, etc.), but it may be of equal importance for temperatures that are considered "intermediate" and are commonly encountered during active phases of a particular organism's life system. For example, with respect to the three tetranychid species commonly occurring on apple foliage in Washington State (*Tetranychus urticae* Koch, *T. mcdanieli* and *Panonychus ulmi* (Koch)), there is evidence that competition for a finite resource is reduced, at least in part, by partitioning along a thermal gradient. *T. urticae* is adapted to cool periods (spring), *P. ulmi* to intermediate temperatures (early summer), and *T. mcdanieli* to hot periods (late summer).

Nonlinearity in response to temperatures can also be of direct anthropocentric importance. Integrated mite control in Washington State is facilitated by encouraging spring populations of apple rust mites, *Aculus schlechtendali* (Nalepa), which serve as an early season food resource for *M. occidentalis* (Hoyt, 1969). This strategy of encouraging build up of a phytophagous mite is accepted by growers, in part, because of the typical collapse of apple rust mite populations which occurs with hot summer temperatures, even in the absence of predator populations. Further integrated pest management implications of the nonlinear effect of temperature on *T. mcdanieli* and *M. occidentalis* populations will be discussed in a later section of this paper.

Tanigoshi *et al.* (1975a) and Logan *et al.* (1976) clearly demonstrate the statistical significance of nonlinearity in developmental rate with respect to temperature for *T. mcdanieli.* To demonstrate the implicit biological impact of the explicit statistical significance, we simulated development of *T. mcdanieli* protonymph comparing an appropriate nonlinear model (Logan *et al.*, 1976) with a linear fit of the same data (Fig. 4). The linear model accounts for almost 80% (correlation coefficient of 0.89) of the total variation in the data, and might be judged an adequate fit without additional information. However, with commonly encountered field temperatures (temperature profile used in simulation is that shown in Fig. 3a and is typical of a late summer "hot spell" for the Wenatchee, Washington area), solution of Eq. (1) with linear r provides specious results.

If a linear form is assumed for r when in fact it is nonlinear, two sources of error may be introduced in solution of Eq. (1). The first of these is the absolute difference between models in predicted developmental rate at a given constant temperature, and the second involves the effect of confounding r with T. Figure 5 shows a plot of the difference between the two developmental models (i.e., $f(T) = r_1 - r_2$, where r_1 is Eq. (6) from Logan *et al.* (1976) and r_2 is the linear fit). Over a range of 10 to 40°C, the difference function has roots at 24.4 and 36.0°C. In other words, at temperatures near these two points, the linear

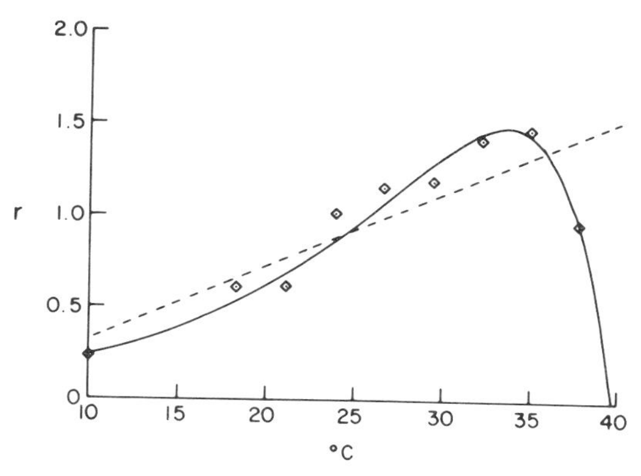

Fig. 4. Observed developmental rates for *T. mcdanieli* fitted by the nonlinear model of Logan *et al.* (1976) and by a least squares determined linear function.

model should provide good results. Also, the absolute value of f(T) between these two points is not large. Since mean daily summer temperatures seldom, if ever, fall outside this range it could be argued that the linear model should provide good results. This may not at all be the case since the difference between the two descriptions of r in Eq. (1) is confounded by the nature of variation in the temperature cycle T(t).

Figure 6a makes it possible to further explore the effect of variation in temperature profile on model Eq. (1), since it shows the difference between the total model integrated over a 24 hour Dt plotted as a function of the variation in the temperature cycle. For example,

$$g(R) \int_{t=0}^{t=24} = r_1\,[T(t;R)]dt - \int_{t=0}^{t=24} r_2\,[T(t;R)]dt \qquad (4)$$

where $R = R_2/R_1$ of Eq. (3) and T $(t;R = 1)$ as shown in Fig. 3a. Several points of interest should be noted in Fig. 6a. First, the value of g at $R = 0$ is the same as that of Fig. 5 for temperature equal to the mean of the profile shown in Fig. 3 (i.e., $T = 29.55°C$). Secondly, the difference between the linear and nonlinear based model for $R = 1$ is 0.16. Therefore, if the nonlinear model is correct, the linear representation would result in an error amounting to 16% of the total life stage during the 24 hour time interval. However, a temperature profile with an identical mean but a range = 0.84 would theoretically result in zero error (again assuming that the nonlinear model is correct).

The effect of changing the mean temperature to 24.4°C (the lower root) is shown in Fig. 6b. It can be seen from Fig. 6 that assumption of a linear model could result in the false conclusion that development under variable temperatures could take three different courses. The development could advance

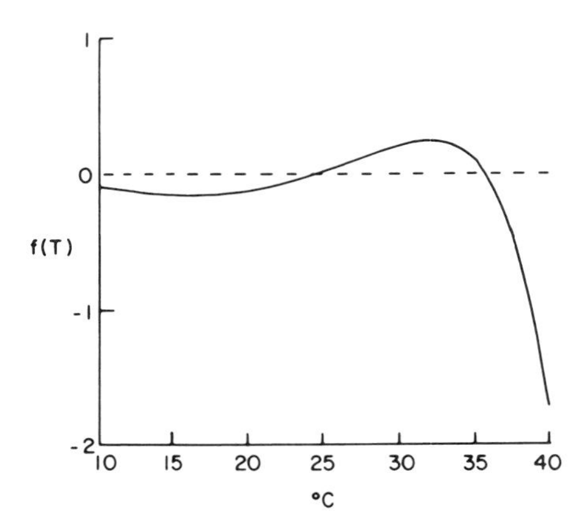

Fig. 5. Plotted difference between the nonlinear and linear developmental models as a function of temperature.

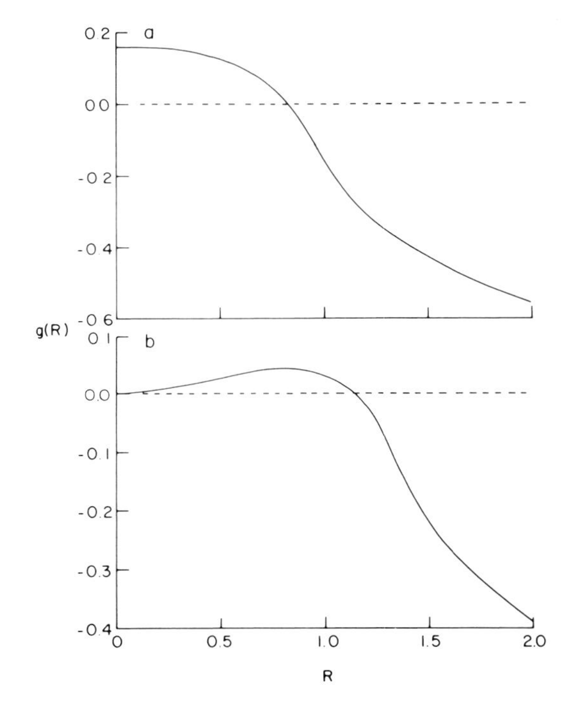

Fig. 6. (a) Plot of the difference between solution of model Eq. 1 assuming linear and nonlinear rate functions shown in Fig. 5 and varying R over the range (0-2). (b) Similar to function shown in Fig. 6a with mean temperature (a_0 of Eq. (3)) changed to the lower root shown in Fig. 5.

slightly compared with fixed temperature, have no effect, or be retarded. This would depend solely on characteristics of the temperature regime. Such disparate results have been reported in the literature for arthropod development under variable temperatures.

Thus for active stages of *T. mcdanieli,* developmental rates are not only nonlinear, but they are also asymmetric with respect to the optimum temperature. This asymmetry emphasizes the importance of accurately representing the circadian temperature cycle. Even for a linear r, asymmetry of the temperature curve will result in a PC of Eq. (1) different from the curve that would be obtained by using a constant temperature equal to the mean temperature. For developmental curves that are obviously asymmetric about the optimum, temperatures even slightly above optimum may have a significant effect on computation of PC. The relative importance of temperatures over the given range in Fig. 4 is indicated by the relative arc length of the curve. Although temperatures above optimum (T = 33.24°C) account for a relatively small segment of the total independent variable range, a large amount of the arc length of the curve is accounted for over this range.

Effect of Temperature on Population Development

By using life stage developmental studies Tanigoshi *et al.*, (1975b) and Tanigoshi and Browne (1978) demonstrated the applicability of SIMDEV to *M. occidentalis.* Within the constraints imposed by growth chamber environments and the assumptions inherent to the original data upon which the model is based, SIMDEV: (1) simulates population growth where all Malthusian conditions except temperature are held constant; (2) does not include the density dependent processes of intra- and interspecific competition; (3) assumes that generations do not overlap; and, (4) contains no function describing the predator's maximum consumption response. Moreover, by assuming that both prey and predator coexist on both leaf surfaces and are subject to the same ambient temperatures, the computer simulation for estimating r_m may serve to identify orchard temperature conditions that may "favor" one species over the other. Comparisons of the SIMDEV generated r_m values for both *T. mcdanieli* and *M. occidentalis* (Fig. 7) indicate that at about 34°C, further temperature increase has a negative effect on *M. occidentalis* population growth. Thus during extended periods when daily maximum temperatures are 35°C or higher, the comprehensive multidisciplinary apple pest management system of the Pacific Northwest (Hoyt, 1969) may favor the natural increase of *T. mcdanieli* over that of *M. occidentalis.* Such a potential for temperature-induced "prey escape" is well-illustrated in a mathematical simulation model developed by Logan *et al.* (pers. comm.) based primarily upon parameters estimated in Logan (1977). Fig. 8a shows a phase space plot of two simulations that were initiated with identical predator-to-prey ratios but were exposed to different temperature

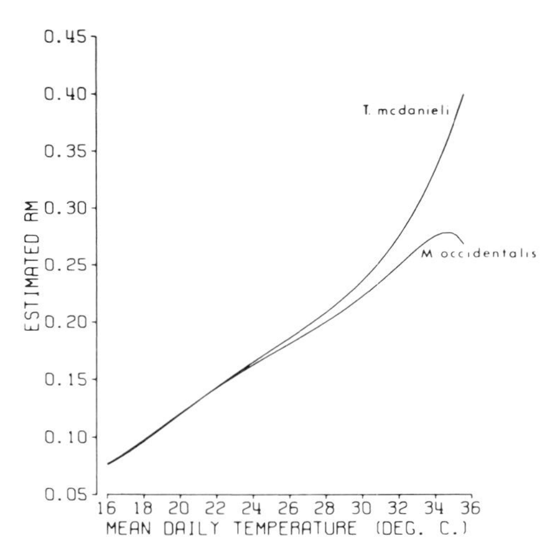

Fig. 7. Estimated r_m values for *M. occidentalis* and *T. mcdanieli* at simulated "standard" temperature cycles (Tanigoshi and Browne, 1978).

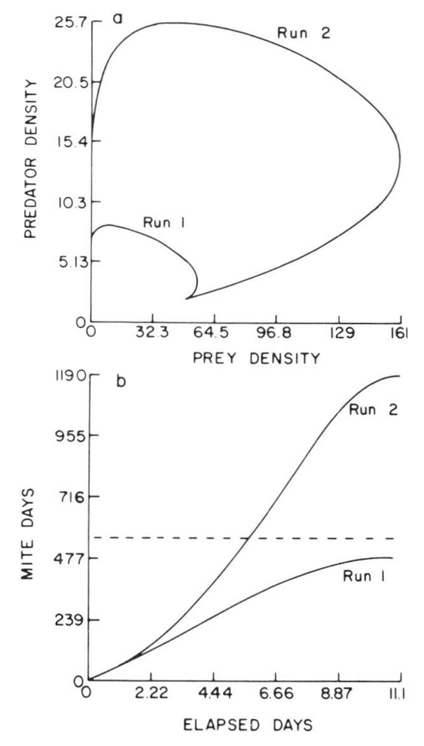

Fig. 8. (a) Phase plane plot of simulated interaction between *T. mcdanieli* and *M. occidentalis.* Run 1 for a constant temperature of 21.1°C and run 2 for a constant temperature of 33.0°C. (b) Accumulated mite days for simulations shown in Fig. 8a.

conditions (run 1 at a constant 21.1°C and run 2 at a constant 33.0°C). Under the high temperature simulation, the population of adult prey reached almost a three-fold greater density than under the low temperature simulation. The economic impact is perhaps better illustrated in Fig. 8b in which accumulated mite-days are plotted for both simulations. If a level of 500 adult mite-days is used as an approximate economic threshold, it is seen that the low temperature simulation never reaches the threshold, but the high temperature simulation exceeds the economic threshold in 5.5 days. Nevertheless, by encouraging the application of advances in biological and environmental monitoring schemes for the maintenance of favorable predator:prey ratios and by minimizing disruptive pesticide practices, the integrated mite-control program will help to offset the effect of occasional "hot spells" during the fruit growing season.

SUMMARY

When rate of development of *T. mcdanieli* was plotted as a function of temperature, the curves proved to be nonlinear and asymmetric over the temperature range commonly encountered in the field. Such a response is probably not uncommon for *Tetranychidae,* particularly in highly variable environments. The results emphasize both the importance of determining the rate of development for the entire temperature range likely to be encountered and the importance of adequate representation of the actual curve.

REFERENCES

Cagle, L. R. (1949). *Tech. Bull. Virginia Agr. Expt. Sta.* **113,** 1-31.

English, L. L. and Turnipseed, G. F. (1941). *J. Agr. Res.* **63,** 75-77.

Gilbert, N., Gutierez, A. P., Frazer, B. D. and Jones, R. E. (1976). "Ecological relationships." W. H. Freeman and Co., San Francisco.

Hoyt, S. C. (1969). *J. Econ. Entomol.* **62,** 74-86.

Logan, J. A. (1977). Ph.D. Thesis. Washington State University, Pullman. 137 p.

Logan, J. A., Wollkind, D. J., Hoyt, S. C. and Tanigoshi, L. K. (1976). *Environ. Entomol.* **5,** 1133-40.

Parent, B. (1965). *Ann. Soc. Entomol. Quebec* **10,** 3-10.

Pradhan, S. (1945). *Proc. Nat. Inst. Sci. India* **11,** 74-80.

Shinkaji, N. (1959). *Bull. Hort. Res. Sta.* **5,** 129-142.

Stinner, R. E., Gutierrez, A. P. and Butler, G. D. (1974). *Can. Entomol.* **106,** 519-524.

Tanigoshi, L. K and Browne, R. W. (1978). *Ann. Entomol. Soc. Amer.* **71,** 313-316.

Tanigoshi, L. K., Hoyt, S.C., Browne, R. W. and Logan, J. A. (1975a). *Ann. Entomol. Soc. Amer.* **68,** 972-978.

Tanigoshi, L. K., Hoyt, S. C., Browne, R. W. and Logan, J. A. (1975b). *Ann. Entomol. Soc. Amer.* **68,** 979-986.

Tanigoshi, L. K., Browne, R. W., Hoyt, S. C. and Lagier, R. F. (1976). *Ann. Entomol. Soc. Amer.* **69,** 712-716.

SILK PRODUCTION IN *TETRANYCHUS* (ACARI: TETRANYCHIDAE)

Uri Gerson

Faculty of Agriculture
Hebrew University of Jerusalem
Rehovot, Israel

INTRODUCTION

Webbing (or spinning) by spider mites was recognized by Linnaeus who incorporated the ability to spin into the description of *Acarus telarius*. The act of webbing and the form of the webs were used in describing some species during the last century. However, no special studies were devoted to this subject until quite recently. The web confers substantial advantages upon the spinning mites (dispersal, colony establishment, mate finding and guarding, protection against natural enemies and other adverse environmental factors). It affects the host plants and is affected by them. In order to explore the web-affected life style of spider mites, an experimental model was required. This was provided by studies on webbing by the carmine spider mite, *Tetranychus cinnabarinus* (Boisduval), and more recently by Saito's work on *Tetranychus urticae* Koch.

METHODS

Possibly the greatest drawback to studies on spider mite webbing is the difficulty of obtaining quantitative data. The threads are 0.03-0.06 μm in diameter, transparent, and are not produced in any consistent regularity. An indirect method of estimation was therefore developed, utilizing the mites' fecal pellets (Hazan *et al.*, 1974). *Tetranychus cinnabarinus* produces two kinds of anal excretions: yellow, roundish, viscid pellets ("white pellets") and black, berrylike ones ("black pellets") (Fig. 1). Similar pellets were observed and figured by Wiesmann (1968) in *T. urticae*. Both kinds of pellets preserve their respective shapes when held by the webbing, where they also serve to

Vol. I: ISBN 0-12-592201-9

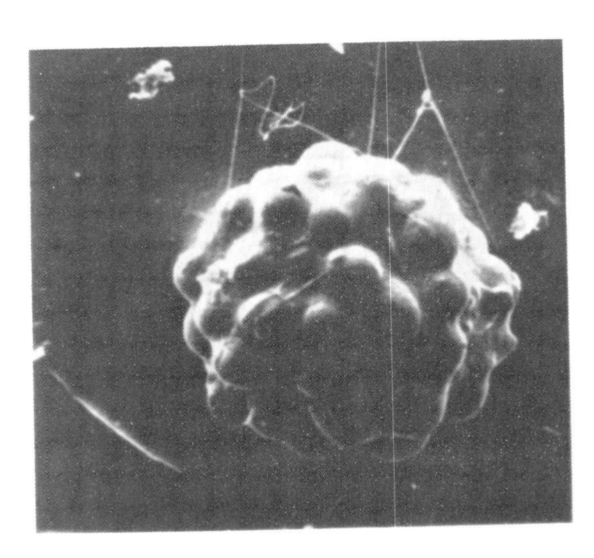

Fig. 1. A "black pellet" produced by *Tetranychus cinnabarinus* (scanning electron microscope micrograph, 580 X).

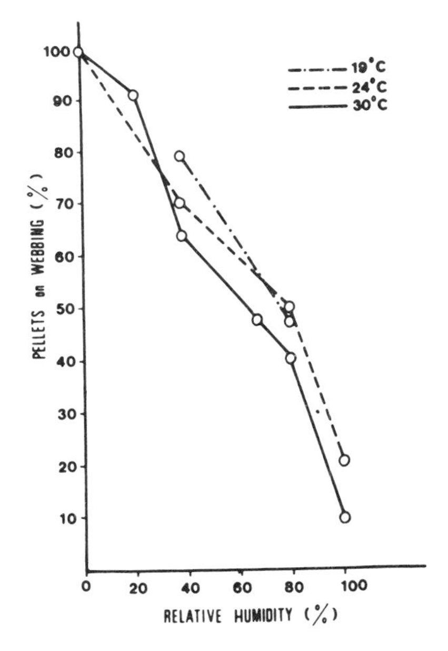

Fig. 2. Percent of pellets deposited by *Tetranychus cinnabarinus* on the webbing, from the total pellet production, at various temperatures and relative humidities.

anchor silk threads. Upon falling onto the leaf surface, both pellets lose their shapes but remain recognizable. They may then be counted.

Production of silk as well as pellets was observed to depend upon the prevailing humidities: more silk was secreted and more pellets expelled under dry as compared to humid conditions (Fig. 2).

A similar method based on the ability of the viscid webs to retain particles was developed by Fransz (1974). Webbed bean leaf discs (5 cm^2) were sprayed with a known amount of a fine-grained yellow powder normally used as a fluorescent. The grains stuck to the silk thread and their proportion (from the total amount sprayed) was used as an indirect measure of web density. Fransz (1974) gave this proportion values between zero and one. Saito (1977a) has developed a different approach to the problem. A grid of 1 mm^2 cells was made from thin (50 μm) nylon threads and placed on the leaves. Females of *T. urticae* and *Panonychus citri* (McGregor) were then placed on these leaves and kept there for certain periods under constant conditions of temperature, humidity and illumination. After the mites were removed the threads transversing each cell were counted, using dark-field microscopy. The obtained value was employed as a quantitative estimation of silk production by these mites.

WEB PRODUCTION

Temperature and Humidity

T. cinnabarinus were reared on bean leaves at four different temperatures and six different relative humidities (Hazan *et al.*, 1973). Using our pellet-based method we found that all active stages of *T. cinnabarinus* spin, although webbing by males was not actually investigated (Hazan *et al.*, 1974). The amount produced increased from stage to stage, being always larger at lower as compared to higher relative humidities (Fig. 3). The same pattern was observed in regard to females whose webbing peaked during the first week of adult life. Although more webs were spun at 0% R.H. as compared to 38% R.H., females survived much longer at the latter humidity, and thus produced more silk. We then incorporated life table data (Hazan *et al.*, 1973) into web production data and computed the total amount of silk ("S") a single female may produce during her life. The result (Fig. 4) shows that 24°C and 38% R.H. are optimal. Saito (1977b) reported that *T. urticae* produced its largest amounts of silk at 33% R.H.

Light

The effects of light on web spinning by *T. cinnabarinus* was studied by totally depriving some mites of light (except during the daily leaf change done

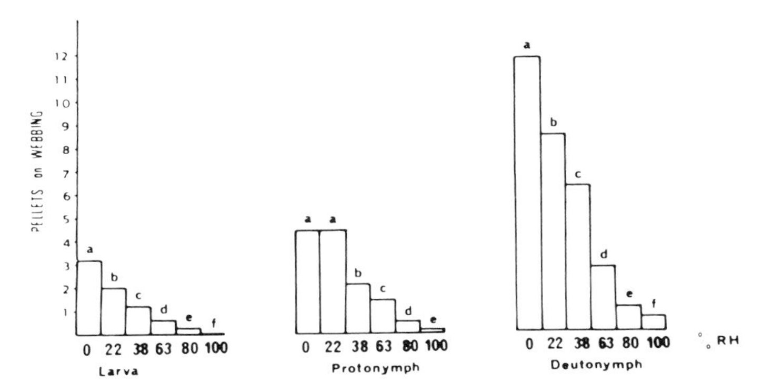

Fig. 3. Webbing by preadults of *Tetranychus cinnabarinus* as affected by various relative humidities at 30°C. (Identical letters on columns indicate means which do not differ at the 0.05 level of significance.)

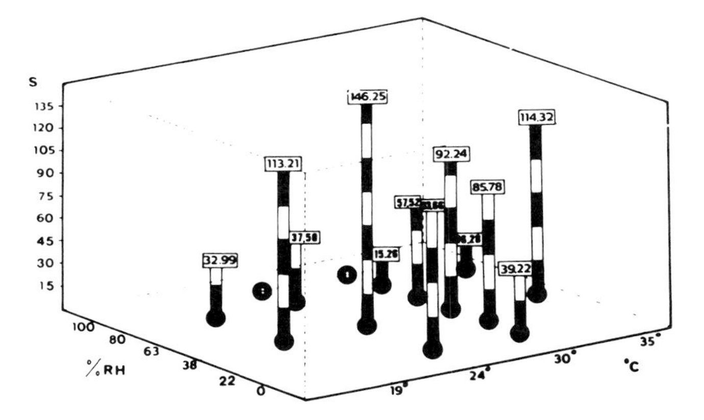

Fig. 4. Amounts of webbing ("S") produced by female progeny of *Tetranychus cinnabarinus* throughout their lives at various combinations of temperatures and relative humidities.

under 15 w illumination). Silk production by these mites was compared to others held under conditions of continuous fluorescent light, and to mites maintained under an alternating light regime (14L:10D). The largest amount of silk was spun under continuous light conditions, less under the alternating light regime, and least in total darkness. The differences were statistically significant (P greater than or equal to .05). There was no difference between webbing produced on upper or lower surfaces of bean leaves.

Oviposition

Throughout these experiments there was a significant correlation between estimated amount of web and number of eggs deposited by the same females (Fig. 5). This close relationship between spinning and ovipositing, together with the fact that eggs are always deposited so as to be in touch with webs, indicates that some benefit accrues to the eggs from the silk. To explore this possibility we carefully removed newly-deposited eggs from the web and placed them on bean leaves. Untouched, web-covered eggs on similar leaves served as controls. The experiment was run at two temperatures and six relative humidities, with a minimum of 100 eggs per treatment. Hatch of de-webbed eggs was significantly reduced at the lower humidities as well as under saturation conditions at 30°C (Fig. 6). The same relationship was observed at 24°C. There was no difference in hatchability under the intermediate humidity conditions. The web thus appears to serve as a humidity-regulating factor in the immediate vicinity of the eggs. Hazan *et al.* (1975a) postulated that such activity may be due to the hygroscopic activity of the mites' "white pellets," as well as related to moisture-absorbing characteristics of the silk's protein.

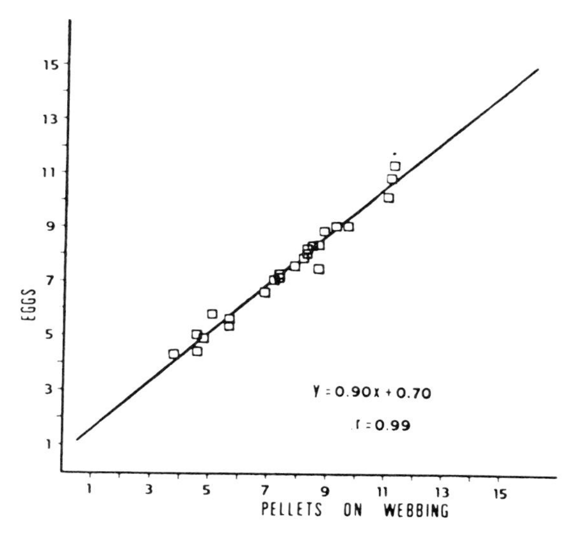

Fig. 5. The relationship between webbing and ovipositing by *Tetranychus cinnabarinus* under conditions of 38% relative humidity at 24°C.

Locomotion and Feeding

Saito (1977a) noted a very close correlation between the walking activities of *T. urticae* females and their spinning. Similar observations relating to the males of this species, were made by Fransz (1974). Egg deposition, walking and spinning thus appear to be closely interconnected. Another major activity, feeding, is at its optimum at 24°C and 38% (Hazan *et al.*, 1975b), conditions which, as noted, best promote webbing. Thus all major biological functions of these mites appear to be geared to each other. Starved mites however, spin a copious amount of silk before they die (Hazan *et al.*, 1974), implying that feeding is not a prerequisite for spinning, and that proteins required for this silk are carried over from the preadult stages.

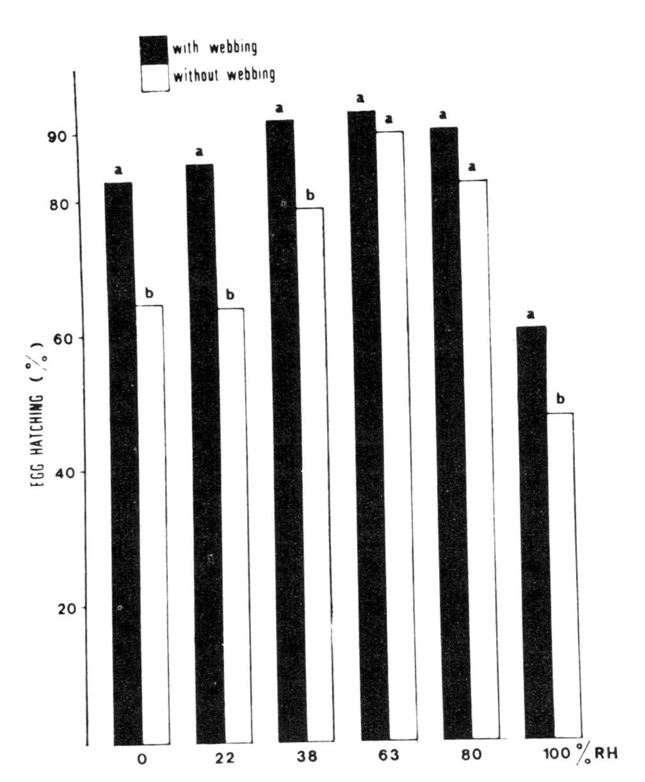

Fig. 6. The effect of *Tetranychus cinnabarinus* webbing on egg hatchability at 30°C and under various relative humidities. (Identical letters on columns indicate means which do not differ at the 0.05 level of significance.)

Protection of Eggs and Immatures

Web production by the Tetranychinae apparently first served to cover and thus protect the eggs; this is the only known spinning activity of the Tenuipalpoidini, regarded as the most primitive tribe in the Tetranychinae (Gutierrez *et al.*, 1970). Singer (1966) described how females of *Tenuipalpoides dorychaeta* cover their eggs with a fine, loose network of silk. More advanced Tetranychinae also cover single or groups of eggs with webbing. The summer eggs of *Panonychus ulmi* (Koch) are webbed down by "guy ropes," the function of these silken ropes being to maintain the egg in an upright position for a brief period. The embryo meanwhile waterproofs the eggs from within, closing a small hole which the "ropes" had held appressed to the substrate (Beament, 1951). McGregor and Newcomer (1928) noted very similar egg roping by females of *Panonychus citri* (McGregor). Species of the more advanced *Oligonychus, Eotetranychus* and *Tetranychus* regularly cover their eggs with webbing (Gutierrez *et al.*, 1970).

Larvae of *T. cinnabarinus* begin to spin soon after emergence (Hazan *et al.*, 1974). In *Tetranychus* webbing continues throughout life. The larvae and nymphs of some species attach their anterior claws to the substrate with some silk prior to molting. This helps the emerging mites to free themselves from their old skins.

Mating Behavior

As they mature, female deutonymphs and males utilize silk for several purposes. Preparing for its final molt, the deutonymph of *T. urticae* spins a silken cover upon which it rests. At that time it secretes a pheromone (Cone *et al.*, 1971). Male behavior is modified by female web, its movements towards the molting deutonymph becoming more linear (Penman and Cone, 1972). Once its potential mate has actually been located the male spins an extensive mat across the deutonymph, the webbing being always within two male body lengths (Penman and Cone, 1974). The web thus serves as a physical contact between the sexes, should a male become separated from a female deutonymph. Cone *et al.* (1971) noted that these deutonymphs attracted more males under dry conditions than under humid ones, such effects being enhanced by the greater amounts of silk produced at low relative humidities. Silk may well serve as a pheromone substrate or carrier (Penman and Cone, 1974). Thus, the drier it becomes, the more silk would be produced, and more pheromone would be available to attract males.

The webs spun by males around molting deutonymphs may be interpreted as a form of territoriality (Penman and Cone, 1974). Support for this interpretation (which implies competition) comes from Lee (1969) and Potter *et al.* (1976), who reported on intraspecific fights between males of *T. urticae* over molting deutonymphs. Such males may at times apply strands of silk to the mouthparts and legs of their opponents, forcing them to withdraw from the battle (Potter *et al.*, 1976). Thus the male which spins more silk faster wins—a factor favoring selection for more and faster spinning.

Colonization

Emergent females web, oviposit and spin some more. As the larvae and nymphs also web, a canopy of silk arises on the substrate. This canopy, increasing as the family increases, serves as a nest, a place to live in, for the colony. Once established, mites tend to aggregate and remain within their web (Foott, 1962), thereby enhancing its protective effect for them. The silk defines the limits of the colony, members gradually spreading out and thus enlarging its diameter. The mites web throughout their lives (Hazan *et al.*, 1974). Even diapausing females are covered by silk (Nielsen, 1958). Amounts of silk produced depend on air humidity and temperature (Hazan *et al.*, 1974), on host plant (Dosse, 1964) and on the smoothness of the substrate. On the whole, spinning proceeds more slowly on a smooth than on a rough surface (van de Vrie *et al.*, 1972).

Protection from Climatic Factors

Webs protect individual mites from being blown off their host plants by winds, and colonies from being wetted by rain (Davis, 1952; Linke, 1953). The ability of individuals or small colonies to maintain themselves under the web, despite harsh climatic conditions, contributes to their subsequent rapid population increase (Davis, 1952). The cast skins and dirt particles which accumulate in the webs of *Eotetranychus populi* (Koch) were believed by Reeves (1963) to increase the protectiveness of the overlying silk cover.

Webs in Interspecific Relationships

Webs play a role in competition between *T. urticae* and *P. ulmi* (Foott, 1962, 1963). The latter species, which spins very little, has larger populations on peach foliage than on apple leaves, apparently because *T. urticae* webs less on peach. All active stages of *P. ulmi* were hindered by the threads, which also reduced their areas of occupation and sometimes caused their death. Many bodies of *P. ulmi* were in fact found in the webs, usually suspended above leaf surface. Hoyt (1969) noted that silk produced by *T. mcdanieli* (McGregor) not only trapped *P. ulmi,* but also reduced its feeding and oviposition sites. Aggressive behavior of *T. urticae* males against *P. ulmi* was reported by Lee (1969). She noted that these males webbed down *P. ulmi* females and then used their chelicerae to probe the captives' dorsum. When *Tetranychus* invaded fruit trees already infested by *Bryobia,* populations of the latter always declined, becoming restricted to silk-free leaves (Georgala, 1955). It appeared that *Bryobia* was unable to traverse leaf surfaces covered by webbing. Hoyt (1969) reported that webs of *T. mcdanieli* do not interfere with the activities of *Aculus schlechtendali* Nalepa on apple. Sterlicht (1969), on the other hand, stated that silk produced by *T. cinnabarinus* reduced populations of *Aceria sheldoni* (Ewing) on lemon plants in the laboratory.

Natural Enemies and Webbing

Webbing protects *Tetranychus* colonies from various natural enemies (Ewing, 1914). Such protection is more evident in regard to general feeders, like lacewing larvae, the anystid mite *Anystis agilis* (Banks) and *Balaustium putmani* Smiley (Erythraeidae). The latter ate all stages of *T. urticae*, but eventually became entangled in the webbing. The same happens with many phytoseiid mites, which also become disturbed and tend to avoid the silk (McMurtry *et al.*, 1970). On the other hand, specific predators seem to be attracted to and undisturbed by the silk. The beetle *Stethorus punctillum*

Weise readily penetrates *Tetranychus* webbing and remains underneath to feed (Putnam, 1955).

Several predator phytoseiid mites are attracted to and thrive within webbed *Tetranychus* colonies. These include *Amblyseius fallacis* (Garman), *Phytoseiulus persimilis* Athias-Henriot and *Typhlodromus occidentalis* Nesbitt (Putman and Herne, 1966; Huffaker *et al.*, 1969).

The reactions of two phytoseiids to webbing spun by a common prey, *Tetranychus pacificus* McGregor, were explored by Takafuji and Chant (1976). They reported that *Iphiseius degenerans* Berlese consistently avoided leaf surfaces covered by silken threads, and never placed any of its eggs into these webs. *P. persimilis,* on the other hand, always distributed itself on silk-covered leaves, where they invariably deposited their eggs. Evidence of further interaction between *P. persimilis* and *Tetranychus* silk was presented by Schmidt (1976), who found that webs attracted the predator more than prey eggs or exuviae. The silken threads apparently led *P. persimilis* to *Tetranychus,* possibly by a tactile sense. Such observations suggest that certain predators have adapted to this defense mechanism and are, in fact, using it for their own purposes. Oviposition of predators in the silk may enable their eggs to benefit from the webs' hygroscopic properties as well as to be next to an abundant food supply.

Webbing and Acaricides

The silken mats produced by *Tetranychus* afford it some protection from pesticides (Davis, 1952; Linke, 1953). The latter author even advocated the destruction of webs during routine spider mite control procedures. However, in view of the rapid development of pesticide resistance by *Tetranychus* it might be of interest to compare web production between susceptible and resistant spider mite races.

Some acaricides, like formamidines, induce spider mite spindown at high concentrations. Gemrich *et al.* (1976) reported that several chemicals of this group (so far only known as U-42564 and U-42558) all with a thioaryl (thiophenyl) moiety, caused mites to lower themselves on silken threads from treated leaves, one hour post treatment. We obtained considerable and persistent spindown with these formamidines as well as with Plictran® , all applied at a concentration of 150 ppm (unpublished). The physiological nature of this spinning reaction is not understood.

Dispersal

Colony growth leads to crowding and then dissemination. Although many spider mites disperse by ballooning, *T. urticae* and *T. cinnabarinus* do not use this method (Fleschner *et al.*, 1956; van de Vrie *et al.*, 1972). Wene (1956) reported migration of *Tetranychus marianae* McG. (= evansi Baker and

Pritchard) by thick (2-4 inch) units of silk, each enclosing many live mites. Spider mites in greenhouses often form silk ropes while leaving heavily infested plants (Hussey and Parr, 1963). Mites aggregate to form a ball at the upper apices of infested plants. As individual mites drop off, they spin silken threads which become progressively thicker as more and more mites move along them. Most active stages partake in this form of migration; Hussey and Parr (1963) counted 1350 protonymphs, 156 deutonymphs and 14 females of *Tetranychus* in one silken rope. They also reported that more roping occurred at lower (70%) relative humidity than under saturation conditions.

Web-plant Interactions

The interactions between webs and plants include injury caused by the former, and silk production affected by the latter. Web damage to plants was incorporated into the description of *T. telarius* by Linnaeus. Oatman *et al.* (1967) and Qureshi *et al.* (1969) provided descriptions and photographs of tomatoes and other plants actually strangled by webs of *T. evansi.* Economic injury to commercial flowers in greenhouses, due to dense *Tetranychus* webbing, was described and figured by Hussey *et al.* (1969). Reeves (1963) thought that web placement might help in identifying the various spinning spider mites in the field. This was done by Leigh (1963), who distinguished between several species of *Tetranychus* on cotton by the amount and form of silk their colonies produced. Such differences imply that various species spin differently on the same host plant, differences that may affect their specific level of injury. On the other hand, *Tetranychus* is known to produce different amounts of silk when feeding on various host plants (Dosse, 1964; Foott, 1962). Quantities of webbing spun may depend on nutritional as well as topographic factors, the mites spinning more silk on rough than on smooth surfaces (van de Vrie *et al.*, 1972, citing several sources).

SUMMARY

Webbing by the carmine spider mite, *Tetranychus cinnabarinus* (Boisduvall), was studied as a model for silk production by spider mites. All active stages spin, the amount produced increasing from stage to stage and peaking during the first week of adult female life. More silk was produced at lower relative humidities. Females held in continuous light secreted more silk than those kept in continuous darkness or under an alternating light regime. Webbing was not affected by leaf side. Separating eggs from web reduced hatchability only at very low and at very high humidities. These observations, supplemented by data from the literature, are used to discuss the role of webbing in the biology of *Tetranychus*. The webbing covers and protects deposited eggs. Emerging mites molt under its canopy and use it to free themselves from the old skin. Web serves as a sex pheromone carrier and is used by males to protect their claim on females from other males. Web enables mite colonies to be set up on plants, protects them from adverse climatic conditions and maximizes use of leaf surface. Dispersal is facilitated by threads or silk balls. The web protects *Tetranychus* from many natural enemies, but serves to attract its more specific predators. Webs confer an advantage upon *Tetranychus* while competing with non-spinning phytophagous mites. The webbing causes some economic injury to cultivated plants, and it may protect mite colonies from pesticides.

In conclusion, *Tetranychus* has gained a distinctive selective advantage by evolving a life style based on webbing.

ACKNOWLEDGEMENTS

I would like to thank Dr. E. G. Gemrich II, The Upjohn Company, Kalamazoo, Michigan, for supplying the formamidines, Mr. H. H. Keifer, of Sacramento, California, for information on webbing eriophyids, and Dr. Y. Saito of Hokkaido University, Sapporo, for an English translation of his 1977a paper.

REFERENCES

Beament, J. W. L. (1951). *Ann. Appl. Biol.* **38**, 1-24.

Cone, W. W., Predki, S. and Klostermeyer, E. C. (1971). *J. Econ. Entomol.* **64**, 379-382.

Davis, D. W. (1952). *J. Econ. Entomol.* **45**, 652-654.

Dosse, G. (1964). *Z. Agnew. Entomol.* **53**, 455-461.

Ewing, H. E. (1914). *Oregon State Agric. Exp. Sta. Bull.* **121.**

Fleschner, C. A., Badgley, M. E., Ricker, D. W. and Hall, J. C. (1956). *J. Econ. Entomol.* **49**, 624-627.

Foott, W. H. (1962). *Can. Entomol.* **94**, 365-375.

Foott, W. H. (1963). *Can. Entomol.* **95**, 45-57.

Fransz, H. G. (1974). "The Functional Response to Prey Density in an Acarine System." *Centre for Agricultural Publishing and Documentation,* Wageningen.

Gemrich, E. G. II, Lamar, L. B., Tripp, T. L. and Bande Streek, E. (1975). *J. Econ. Entomol.* **69**, 301-306.

Georgala, M. B. (1955). *Union South Africa Dept. Agri. Science Bulletin,* No. 360.

Gutierrez, J., Helle, W. and Bolland, H. R. (1970). *Acarologia* **12**, 732-751.

Hazan, A., Gerson, U. and Tahori, A. S. (1973). *Acarologia* **15**, 414-440.

Hazan, A. Gerson, U. and Tahori, A. S. (1974). *Acarologia* **16**, 68-84.

Hazan, A., Gerson, U. and Tahori, A. S. (1975a). *Acarologia* **17**, 270-273.

Hazan, A., Gerson, U. and Tahori, A. S. (1975b). *Bull. Entomol. Res.* **65**, 515-521.

Hoyt, S. C. (1969). *Proc. 2nd Intern. Cong. Acarol.* 117-133.

Huffaker, C. B., van de Vrie, M. and McMurty, J. A. (1969). *Ann. Rev. Entomol.* **14**, 125-174.

Hussey, N. W. and Parr, W. J. (1963). *Entomol. Exp. Appl.* **6**, 207-214.

Hussey, N. W., Read, W. H. and Hesling, J. J. (1969). "The Pests of Protected Cultivation." Edward Arnold, London.

Lee, B. (1969). *J. Australian Entomol. Soc.* **8**, 210.

Leigh, T. F. (1963). *Adv. Acarol.* **1**, 14-20.

Linke, W. (1953). *Hofchen-Briefe* **6,** 181-232.
McGregor, E. A. and Newcomer, E. J. (1928). *J. Agric. Res.* **36,** 157-181.
McMurtry, J. A., Huffaker, C. B. and van de Vrie, M. (1970). *Hilgardia* **40,** 331-390.
Nielsen, G. L. (1958). *J. Econ. Entomol.* **51,** 588-592.
Oatman, E. R., Fleschner, C. A. and McMurtry, J. A. (1967). *J. Econ. Entomol.* **60,** 477-480.
Penman, D. R. and Cone, W. W. (1972). *Ann. Entomol. Soc. Amer.* **65,** 1289-1293.
Penman, D. R. and Cone, W. W. (1974). *Ann. Entomol.Soc. Amer.* **67,** 179-182.
Potter, D. A., Wrensch, D. L. and Johnston, D. E. (1976). *Ann. Entomol. Soc. Amer.* **69,** 707-711.
Putman, W. L. (1955). *Can. Entomol.* **87,** 9-33.
Putman, W. L. and Herne, D. B. C. (1966). *Can. Entomol.* **98,** 808-820.
Qureshi, A. H., Oatman, E. R. and Fleschner, C. A. (1969). *Ann. Entomol. Soc. Amer.* **62,** 898-903.
Reeves, R. M. (1963). *Cornell Univ. Agric. Expt. Sta.,* Mem. **380.**
Saito, Y. (1977a). *Jap. J. Appl. Entomol. Zool.* **21,** 27-34.
Saito, Y. (1977b). *Jap. J. Appl. Entomol. Soc.* **21,** 150-157.
Schmidt, G. (1976). *Z. Angew. Entomol.* **82,** 216-218.
Sternlicht, M. (1969). *Ann. Zool. Ecol. Anim.* **1,** 127-147.
Takafugi, A. and Chant, D. A. (1976). *Res. Popul. Ecol.* **17,** 225-310.
van de Vrie, M., McMurtry, J. A. and Huffaker, C. B. (1972). *Hilgardia* **41,** 343-432.
Wene, G. P. (1956). J. Econ. Entomol. **49,** 712.
Wiesmann, R. (1968). *Z. Angew. Entomol.* **61,** 457-465.

FOOD QUALITY INFLUENCES ON A SPIDER MITE POPULATION

L. J. Jesiotr
Z. W. Suski
T. Badowska-Czubik

Department of Plant Protection
Research Institute of Pomology
Skierniewice, Poland

INTRODUCTION

Considerable knowledge has been accumulated about the influence of phytophagous mites on cultivated plants. Much less is known, however, about the reciprocal influence of the host plant on the mite. We feel that both aspects of this relationship are equally important to our understanding of agrocenoses.

There are many ways that a plant might affect a tetranychid mite population. The structure of the leaf surface, thickness of the cuticle, chemical composition of the saps, osmotic pressure within a cell, microclimate in the canopy, and most probably many others. All are factors which might make a plant a more or less suitable host for the mite.

METHODS

Laboratory and Insectary Research

Mineral fertilizers applied to a plant affect its chemical composition and certain morphological characters, thus, they are likely to influence indirectly the mite population as well. Numerous researchers have proved that such an influence exists. It is disquieting, however, that the conclusions concerning the effects of a particular nutrient are rather contradictory. Among macronutrients, nitrogen is generally believed to increase the mite fecundity and/or population density although contradictory statements can be found (Rodriguez, 1951). As to other macronutrients, phosphorus and potassium,

Vol. I: ISBN 0-12-592201-9

opinions are even more diverse since positive and negative influences are reported almost as frequently as neutral ones. Our laboratory and insectary research did not do much to clarify these contradictions (Suski *et al.*, 1975). We found the r_m ratio of *Tetranychus urticae* Koch fed on kidney beans to increase with doses of nitrogen, to decrease and increase again with doses of phosphorus, and to remain virtually unaffected, with contrasting doses of potassium (Suski and Badowska, 1975). However we found no significant correlations between the r_m ratio and leaf contents of each of the three nutrients.

This was quite in contrast in greenhouse rose experiments (Jesiotr, 1974). The r_m ratio of *T. urticae* was not correlated either with fertilizer doses applied to the host or with phosphorus and potassium contents in the leaves, but it was positively correlated ($P = 0.99$) with the content of nitrogen. Other characteristics measured such as lengths of preimaginal development and imaginal life, fecundity, etc., show significant differences between nutritional combinations, but no pattern could be traced, and only the mortality of immature stages showed a curvilinear relation to doses of nitrogen. In addition fecundity was positively correlated with potassium contents.

In the insectary, on the other hand,we found a significant increase in *Panonychus ulmi* Koch populations with nitrogen and phosphorus doses and no effect with potassium. But the population densities of three mite species, *P. ulmi, T. urticae* and *T. viennensis* Zach were positively correlated with nitrogen and negatively with phosphorus and potassium leaf contents (Suski, 1973; Suski *et al.*, 1975). Therefore we obtained a set of data which confirmed the results reported by some researchers and contradicted those of many others. The most conclusive, perhaps, were the results of the insectary experiment which showed the different responses of mites to nutrient doses in the spring and in summer and which furthermore suggested that the responses are basically curvilinear, providing that the doses of fertilizers cover a range broad enough (Table I).

TABLE I.
Summary of Laboratory and Insectary Research on the Influence of Fertilizers on the Intrinsic Rate of Increase (r_m) in the Twospotted Mite Population Showing Increase (+), Decrease (–), or No Effect (0).[a]

	Fertilizer Dose of			Leaf Contents of		
Host Plant	N	P	K	N	P	K
Beans (Laboratory)	+	±	0	0	0	0
Roses (Phytotrone)	0	0	0	+	0	0
Apples (Insectary)	(+)	(+)	(0)	(+)	(–)	(–)

[a] Effects on population level are shown in parentheses.

Field Experiments

It seemed logical to assume that if a fertilizer exercises any influence upon the mite population, then the effect should be magnified in a long term field experiment. Therefore, we selected three apple and two plum nutrition experiments and we observed mite population density during three consecutive years. These experiments were established in the randomized block design at least six years prior to beginning of our observations. Mite counts were made five to six times in a season. Nutrient contents in the leaves were analyzed once a year in mid-July. Statistical analyses of the data accumulated did not show any significant relationship between the mite numbers and fertilizer doses applied or the nutrient contents in the host plants.

Another set of observations were made in a raspberry nutrition experiment, which was established four years prior to the beginning of our observations. The plants were grown on so called "fixed fields" which have received standard doses of mineral fertilizers since 1921. Mite counts were made during three consecutive years, five to eight times in a season. The only reasonably consistent result was a significant depression of the population level in plots completely unfertilized for over 50 years. In other plots which were deficient in N, P, K or Ca, mite numbers did not differ from each other at $P = 0.95$. Neither could we show correlations between mite numbers and nutrient contents in the plant leaves. Similar results have been obtained by many other researchers with a few exceptions, notably Post (1961, 1962) but she manipulated in her experimental orchard not only fertilization, but also the pruning, soil management, and pesticide spraying.

Thus, we concluded these series of experiments being able to demonstrate in the laboratory that mineral nutrients applied to a host plant influence a spider mite population in a variety of ways, but we were not able to reproduce the same effect on a different host and not able to prove any effect in field conditions (Table I).

Multigeneration Research

In further investigations we chose three contrasting doses of fertilizers from the greenhouse rose experiment and investigated their effect on the twospotted spider mite population over 10 generations. Two of the doses chosen caused in the forementioned experiment the highest and the lowest fecundities of mites, the third dose approached the standard one applied in horticultural practice. The results show unexpectedly large differents in the responses of subsequent generations (Fig. 1). The data obtained suggest the following conclusions: (1 the populations have adapted to new, nutritional conditions; 2) it took about eight generations to have the adaptation process accomplished; and 3) this adaptation was achieved through selection of best fitted genotypes rather than through readjustment of the metabolic functions of specimens (Fig. 2).

The experiment, however, showed several drawbacks at close examination.

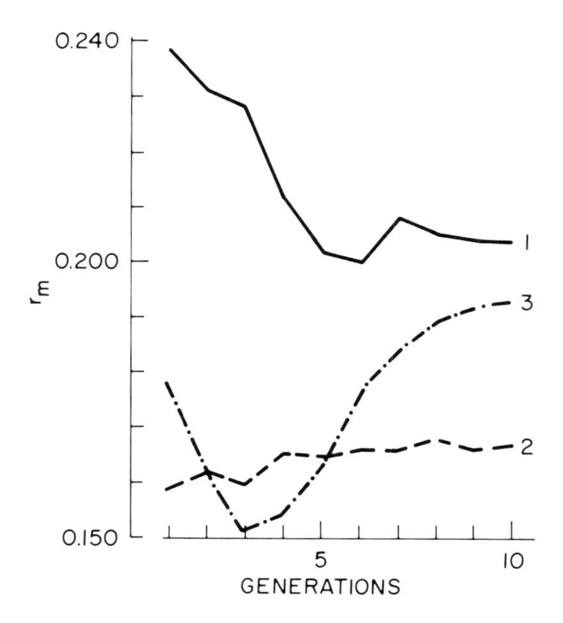

Fig. 1. The influence of various doses of N, P and K fertilizers applied to roses on the intrinsic rate of increase (r_m) of the twospotted spider mite population:

1. 600 ppm N; 37.5 ppm P; 225 ppm K
2. 300 ppm N; 37.5 ppm P; 225 ppm K
3. 300 ppm N; 75 ppm P; 400 ppm K

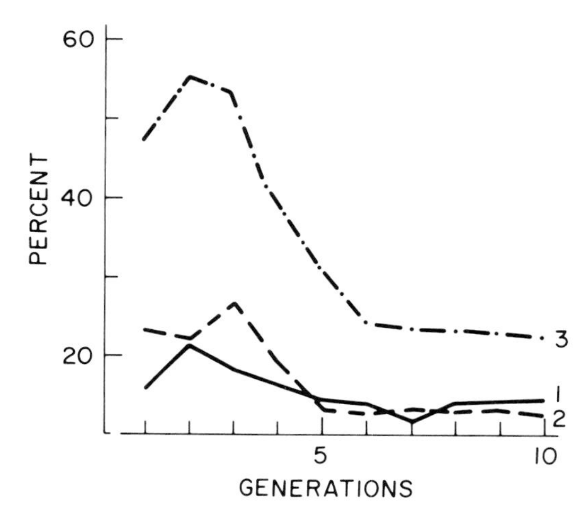

Fig. 2. The influence of various doses of N, P and K fertilizers on the mortality of preimaginal stages; for further explanation, see Fig. 1.

First, the mite population involved was a laboratory one maintained on kidney beans for about 10 years and most probably it was strongly inbred. Second, no parallel observations were made on the original host as a reference line. Therefore, it was not completely clear whether or not the responses observed were characteristic perhaps for this single population. Furthermore, it was not clear what part of the effect might be attributed to fertilizer doses and what to change in the host plant species.

For various reasons we chose to investigate better the first alternative. We selected for these purposes the following three populations of the twospotted spider mite: our standard laboratory population maintained on kidney beans since 1963; a greenhouse population from roses, maintained on roses (cv. 'Baccara') for a few generations; and a greenhouse population from carnation to be maintained on carnation (cv. 'Scania 3C'). Next, we made reciprocal transfer of each population from one host to two others and we intended to keep each population on its original host to develop a reference line. Observations were made for 20 generations in order to enable us to compute the r_m ratio. However, after 10 generations part of the mites were retransferred from the alternate host to the original one and observations were made in the same way as for other mites for the next 10 generations. While kidney beans and roses produced no particular difficulties as hosts it was quite different with carnation. We were neither able to establish the base line nor to propagate mites transferred from beans or roses on this host. Additional experiments suggested that the heavy wax layer on the leaves of the plant is partially responsible for this. It appears that wax creeps into the egg chorion, and suffocates the embryo. However, our data suggest that some other factors also play an important role, one of them being an additional stress caused to the mites by encaging and handling. Thus, we had to assume the $r_m 0$ for both the baseline and the mites transferred from other hosts to carnations.

The results of these experiments are illustrated in Figs. 3, 4 and 5. The curves developed are essentially similar. The r_m ratio of the population transferred to a new host showed quite dramatic changes during the initial six or eight generations and later it stabilized at the same or a similar level later. The extent of this suppression was larger with the laboratory population which was probably more homogenous and it was smaller with the field one which was more heterogeneous.

Retransferring of mites to the original host in both cases resulted in a depression of the r_m ratio which was comparable to that at the beginning of the experiment and in its fast increase subsequently approaching the level maintained on the original host within two or three generations. For the mites transferred from carnation to both beans and roses, the r_m ratio increased steeply during the initial two generations and it decreased gradually in the following four or five generations.

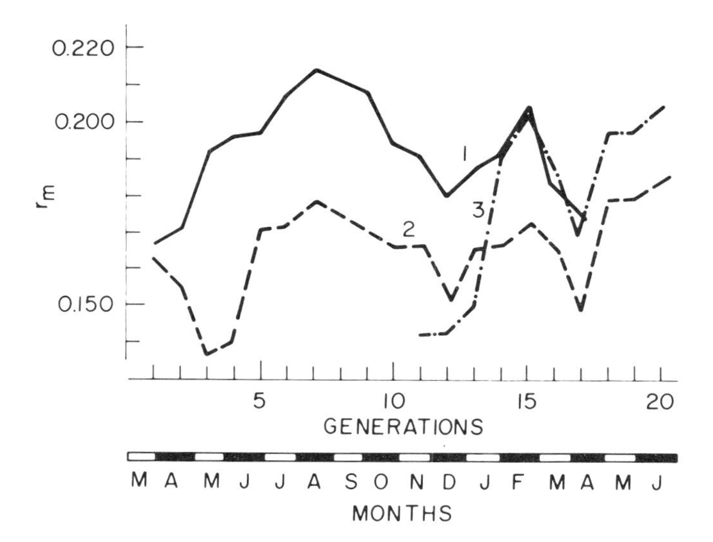

Fig. 3. The influence of the host plant on the intrinsic rate of increase (r_m) of the twospotted spider mite.

Laboratory population from beans:

1. Population maintained on the original host (beans).
2. Population transferred to alternate host (roses).
3. Population retransferred from the alternate host to the original one.

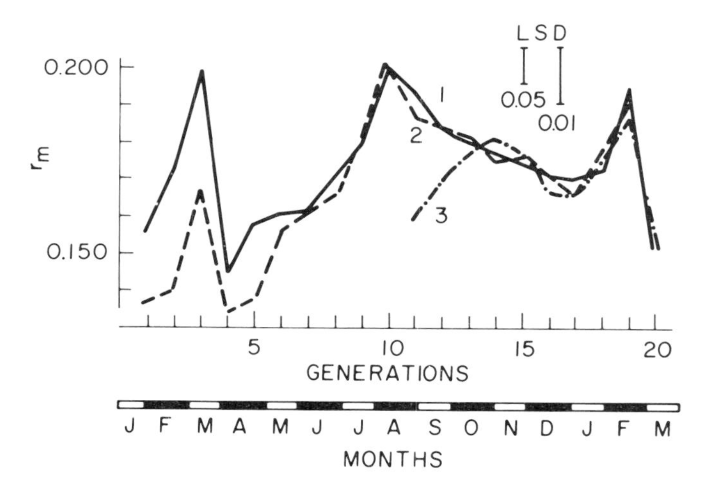

Fig. 4. The influence of the host plant on the intrinsic rate of increase (r_m) of the twospotted spider mite.

Field population from roses:

1. Population maintained on the original host (roses).
2. Population transferred to the alternate host (beans).
3. Population retransferred from the alternate host to the original one.

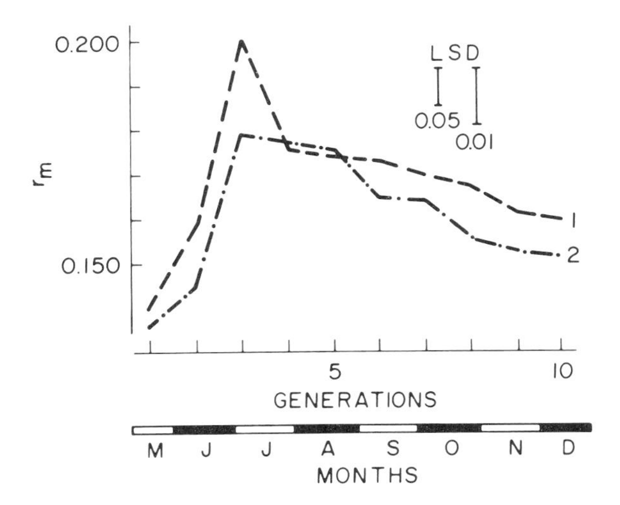

Fig. 5. The influence of the host plant on the intrinsic rate of increase (r_m) of the twospotted spider mite.
Field population from carnation.
1. Population transferred to beans.
2. Population transferred to roses.

SUMMARY

It was demonstrated in laboratory and insectary research that mineral fertilizers applied to a host plant affect the reproduction potential of tetranychid mites. However, it was impossible to demonstrate comparable effects in long term field experiments with orchard trees and raspberries. Multi-generation studies in the laboratory revealed certain temporal patterns in the response of the twospotted spider mite to altered nutritional conditions of the host. It is concluded that the adaptation of the mite population to these new conditions requires a period of several generations.

The authors suggest that any change in food quality causes environmental stress to the mite population. This results in a remodeling of the genetic makeup of the population, possibly by the suppression of certain genotypes and the favoring of other, better fitted ones. It is suggested, furthermore, that the duration of the adaptation period and the degree of change in the r_m ratio and other pertinent measurements, is relative to the extent of change of food quality and to the genetic structure of the population prior to this change. The authors believe that these findings explain certain diverse opinions concerning the influence of mineral fertilizers on the spider mite populations.

CONCLUSIONS

The results presented suggest very strongly that the adaptation of the twospotted spider mite population to a new environmental condition including food quality, takes more than one generation. In the conditions of these experiments the adaptation period lasted for three to eight generations. It was longer for populations well established on the host rather recently which were probably heterogeneous. It is premature to draw firm conclusions about the mechanisms underlying these phenomena, but we hypothesize that there occurs a remodeling of the genetic makeup of the mite population. This is possibly achieved by the suppression of some genotypes in favor of other, better fitted ones, as a result of environmental stress.

These results, we think, provide grounds for explanation of some diverse opinions concerning the mite responses to mineral fertilizers applied to the host plant. We believe that any change in food quality causes certain stresses to the mite population, regardless of whether it results from transferring the mites to another host or from alternating the fertilization regime of the same plant. The size of the stress, and the resulting shift in genetic structure of the mite population, would relate to both the degree by which the food quality has been changed, and the genetic structure of the population prior to the change. Most probably, each scientist has worked with a different population and varied the nutritional conditions in his own way. Furthermore, they have, as a rule, observed the effect in the first generation which is subject to the most dramatic changes. This is best illustrated in Henneberry's (1962) work where he observed contradictory effects of phosphorus and potassium on the OP resistant and on nonresistant populations. Whenever later generations were observed, as in our field experiments, the differences were less dramatic because populations involved were already adapted to the existing nutritional condition. Additionally, the differences were masked due to stress caused by weather conditions to both the phytophage and the host, predation, and competition with other organisms.

REFERENCES

Badowska, T., Suski, Z. W. and Mercik, T. (1977). *Fruit Science Reports* **4**, 45-49.

Henneberry, T. J. (1962). *J. Econ. Entomol.* **55**, 134-137.

Jesiotr, L. J. (1974). Ph.D. diss. at Research Institute of Pomology, Skierniewice, Poland, 77 pp.

Jesiotr, L. J. and Suski, Z. W. (1975). *Zeszyty Problemowe Postepow Nauk Rolniczych* **171**, 105-144.

Jesiotr, L. J. and Suski, Z. W. (1976). *Ecol. Pol.* **24**, 407-411.

Jesiotr, L. J. and Suski, Z. W. (1978). *Ecol. Pol.* (in press).

Post, A. (1961). *Staat Gent.* **26**, 1098-1103.

Post, A. (1962). *Entomophago* **7**, 257-262.

Rodriguez, J. G. (1951). *Ann. Entomol. Soc. Amer.* **44**, 511-546.

Suski, Z. W. (1973) Final Report of research under PI-480 grant FG-Po-256: project E 21-ENT-25 submitted to USDA, ARS, IPD: 103 pp.

Suski, Z. W. and Badowska, T. (1975). *Ecol. Pol.* **23**, 185-209.

Suski, Z. W., Badowska, T. and Olszak, M. (1975). *Fruit Science Reports* **2**, 85-94.

3.
Stored Product Acarology

Section Editors

W. A. Bruce
M. D. Delfinado

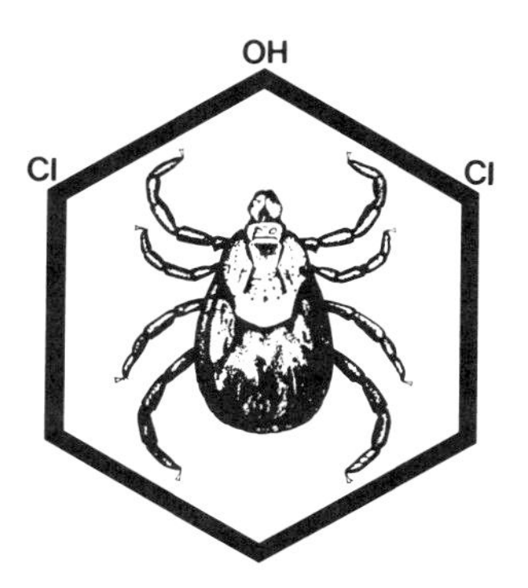

THE MORPHO-SPECIES AND ITS RELATIONSHIP TO THE BIOLOGICAL SPECIES IN THE GENUS *TYROPHAGUS* (ACARIDAE, ACARINA)

D. A. Griffiths

Pest Infestation Control Laboratory
Ministry of Agriculture, Fisheries and Food
London Road, Slough, England

INTRODUCTION

Since its inception by Oudemans in 1924, the genus *Tyrophagus* has received more attention from acarine taxonomists than most other genera of the free-living Astigmata. It has been the subject of three major revisions (Zachvatkin, 1941; Robertson, 1959; Samsinak, 1962) and, because certain species are economically important pests, there is a considerable literature on their geographic and ecological distribution. In addition, there are at least six different keys (embracing four languages) to species identification, all aimed at assisting ecologists and other non-taxonomists to determine their material.

Despite such an apparent fortunate situation it has been my experience over many years, that specific identification of *Tyrophagus* specimens invariably presents severe difficulties. Where ecologists are concerned it may be that the difference between marker characters which form the basis of specific determinations is too subtle to be appreciated by an inexperienced eye, often peering through a microscope inadequate for the task. However, many of the persons who approach me for advice are taxonomists who have already made a specific determination but are conscious that their decision is not absolute, a situation in which I find myself frequently. It suggests, therefore, that the main problem may stem from the fact that characters used to separate species exhibit continuous variation. This may be due to the occurrence of hybrid forms in nature since frequently more than one species is found sharing the same habitat. Equally, it may be due to the existence of closely similar taxa centered around, and therefore misidentified as one of the 'established' species, or maybe a combination of both phenomena is in operation. This paper examines, by means of hybridization experiments and traditional taxonomic techniques, the discreteness of *Tyrophagus* morpho-species and their value in defining biological entities.

Vol. I: ISBN 0-12-592201-9

MATERIALS AND METHODS

Preserved material representing some hundreds of populations from most of the major countries of the world have been examined using phase contrast or interference phase contrast microscopy. Details of the 30 living populations in the hybridization trials are given in Table I. Specimens from these populations have been examined by conventional and scanning electron microscopy.

TABLE I.
Details of the Experimental Cultures.

Species	Country of Origin	Habitat	Species	Country of Origin	Habitat
T. putrescentiae			*T. neiswanderi*		
Group A			T6	England	Strawberries growing crop
T10	Solomon Isles	Copra, lying beneath trees			
T13	S. Africa	Groundnuts in soil	*T. longior*		
T22	Sarawak	Illipe nuts	T4	England	Cucumber plant
T20	England	Chocolate truffle	T40	England	Mouse nest in garden
T30	England	Fungal culture	T41	Israel	Bulbs
T31	England	Fungal culture	*T. sp. H*		
T27	England	Cheese	T11	England	Packaged biscuits
T36	England	Tulip bulbs	T12	England	Nest of the bee *O. coreulesceas*
T58	England	*Ceratocystis ulmi* in galleries of the Elm bark beetle	T24	U. S. A. Carolina	Soil sample
			T. palmarum		
Group B			Group A		
T2	England	Chocolate bar	T17	England	Dead bird on ground
T9	England	Blue Tit's nest			
T37	Cyprus	Mouldy beans	T29	England	Soil of barley field
T52	Holland	Tulip bulbs			
T45	England	Rolled oats			
			Group B		
T. perniciosus			T32	England	Cheese store
Group A			T42	England	Private house
T8	England	Chaff pile on farm			
			Group C		
Group B			T7	England	Thrush (*Turdus*) nest
T38	England	Barley residues farm grain bin			
			T. similis		
			Group A		
			T44	England	Tomato plant
			Group B		
			T21	England	Beech (*Fagus* sp.) litter

Stock cultures are reared on a fungus-free mixture of wheat germ and yeast powder at 20°C and 90% relative humidity. All crosses are made with virgin adults, obtained by isolating resting tritonymphs taken from stock cultures. These are placed singly in a small length of glass tubing plugged at each end with a distinctively colored non-absorbent cottonwool. The tube contains a little food and the emerging adult remains in the tube for two to three days before it is sexed and paired. Pairing takes place in small perspex cells kept under the same conditions as the stock cultures. Single pairs are observed daily for 15 days, after which they are preserved on a microscope slide. Daily records are kept of egg output and hatch. Each cross consists of 20 pairs plus a reciprocal cross of a further 20 pairs. In many instances crosses have been repeated with the same population or different populations of the same species. When a positive result is obtained the cross is repeated in order to produce realistic percentage figures. Hybrid F_1 individuals are isolated as resting tritonymphs and sibling crosses are made using the standard technique. The female reproductive apparatus is examined microscopically to determine whether mating has been completed. This decision is based on the difference in size of the receptaculum siminis which exists between virgin and mated females.

NUMBER OF SPECIES AND FREQUENCY OF RECORDING

The number of specific names attributed to the genus *Tyrophagus* stands at around 50 although, through synonymy, the number of discrete taxa which these names represent may be as low as 20. Based on criteria currently used for specific recognition the following list represents those taxa which in my opinion can be considered to be discrete. The taxonomic arguments on which this opinion is based will appear elsewhere.

Tyrophagus Species List

T. putrescentiae (Schrank, 1781)
T. longior (Gervais, 1844)
T. cocciphilus (Banks, 1906)
T. javensis (Oudemans, 1916)
T. deliensis (Oudemans, 1923)
T. palmarum Oudemans, 1924
T. perniciousus Zachvatkin, 1941
T. molitor Zachvatkin, 1941
T. silvester Zachvatkin, 1941
T. mixtus Volgin, 1948
T. zachvatkini Volgin, 1948
T. formicetorum Volgin, 1948
T. similis Volgin, 1949

T. brevicrinatus Robertson, 1959
T. tropicus Robertson, 1959
T. neiswanderi Johnston & Bruce, 1965
T. africanus Meyer & Rodrigues, 1966
T. cucumeris Karg, 1970
T. paulensis Fain, 1976
T. litoralis Fain, 1976
T. similis Kergulensis Fain, 1976

The above list has 17 rare or very rare species known from one or few recordings, usually with a restricted geographic distribution. The remaining six are common with a wide or cosmopolitan distribution. This is undoubtedly due to the fact that all six are economically important pests which attack growing plants or infest stored foodstuffs and domestic dwellings. Some of them are able to exist successfully in all three of these environments for example, *T. putrescentiae, T. longior* and *T. palmarum.* Two species, *T. perniciosus* and *T. similis,* appear to be inhabitants of the humus layer and invade foodstuffs, particularly stored grain, whilst *T. neiswanderi* seems to be confined to plants and is a pest of crops grown under glass.

THE MORPHO-SPECIES

Specific rank is based upon the degree of variability exhibited by a number of characters, all of which have been intuitively selected by a taxonomist. The characters most commonly considered are as follows:

1) the length of dorsal setae d_1 and d_2 and lateral seta l_2 expressed as a ratio, usually $d_1 : d_2$ or $d_1 : l_2$

2) the form of solenidion omega$_1$ of tarsus I or II when viewed laterally (Figs. 1 to 4).

3) the shape of the supracoxal seta, particularly the number and length of its pectinations or side branches. (Figs. 5 to 10)

4) the form of the penis, viewed laterally. (Figs. 11 and 12)

5) the presence or absence of eye spots on the mid-lateral border of the prodorsal shield. This is a character not previously considered.

Table II illustrates the number of different forms of each character to be found when the six common economic species are compared one with another. Thus, for six species, there are three forms of supracoxal seta, four different setal ratios, and five forms each of the penis and solenidion. However, in Table II, only where numbers are divided by a thickened line do I consider that variation is discontinuous. The table shows that variability of dorsal setal ratios is rather a poor character on which to identify species, a point endorsed by Johnston and Bruce (1965). It also reveals how similar are *T. putrescentiae* and *T. neiswanderi* since the difference in penis form is the only character on

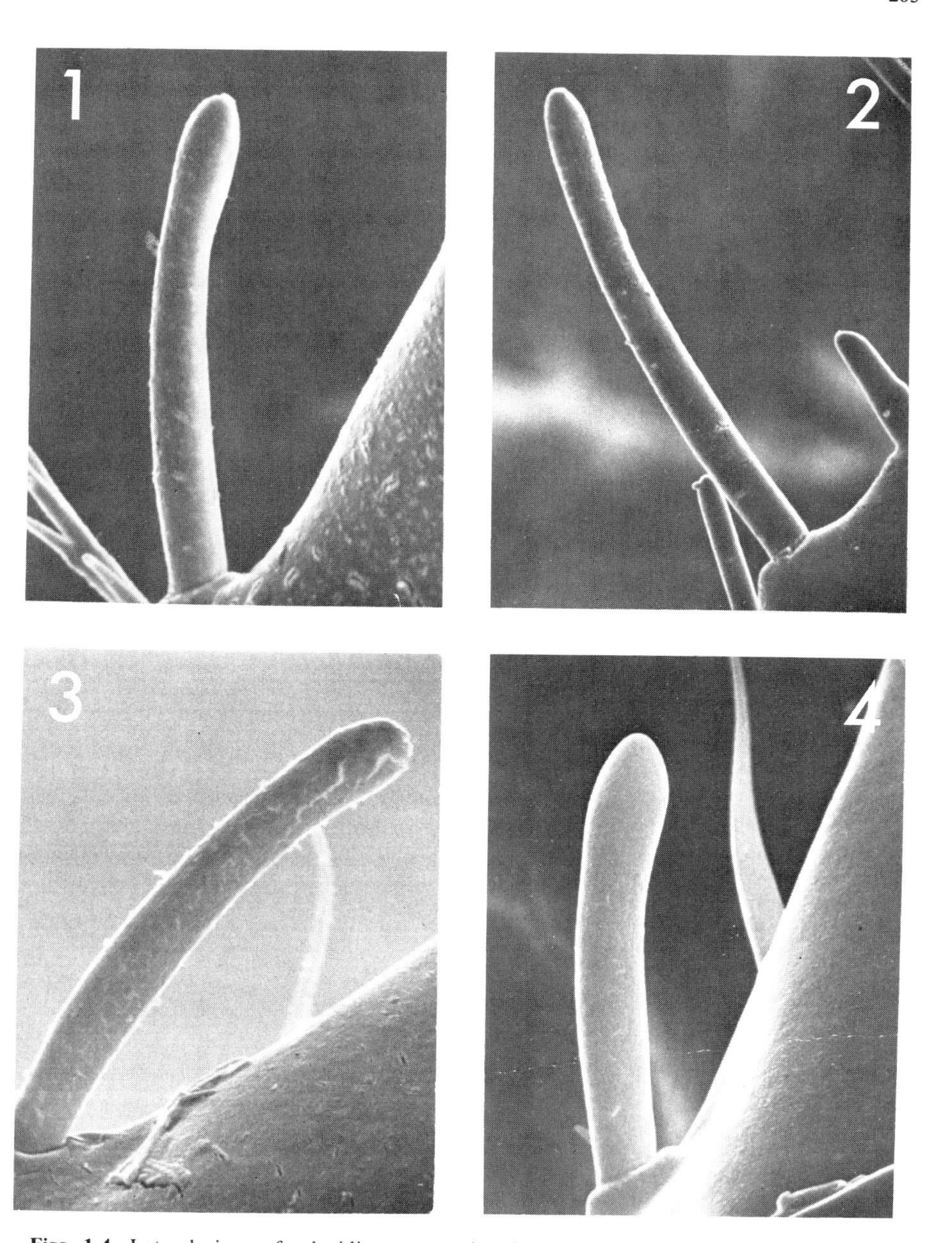

Figs. 1-4. Lateral views of solenidion omega$_1$ situation on tarsus II. Fig. 1. *T. putrescentiae* (Schrank); Fig. 2. *T. longior* (Gervais); Fig. 3. *T. palmarum* (Oudemans); Fig. 4. *Tyrophagus* sp. close to *T. perniciosus* Zachvatkin. All magnifications about 2,000 X.

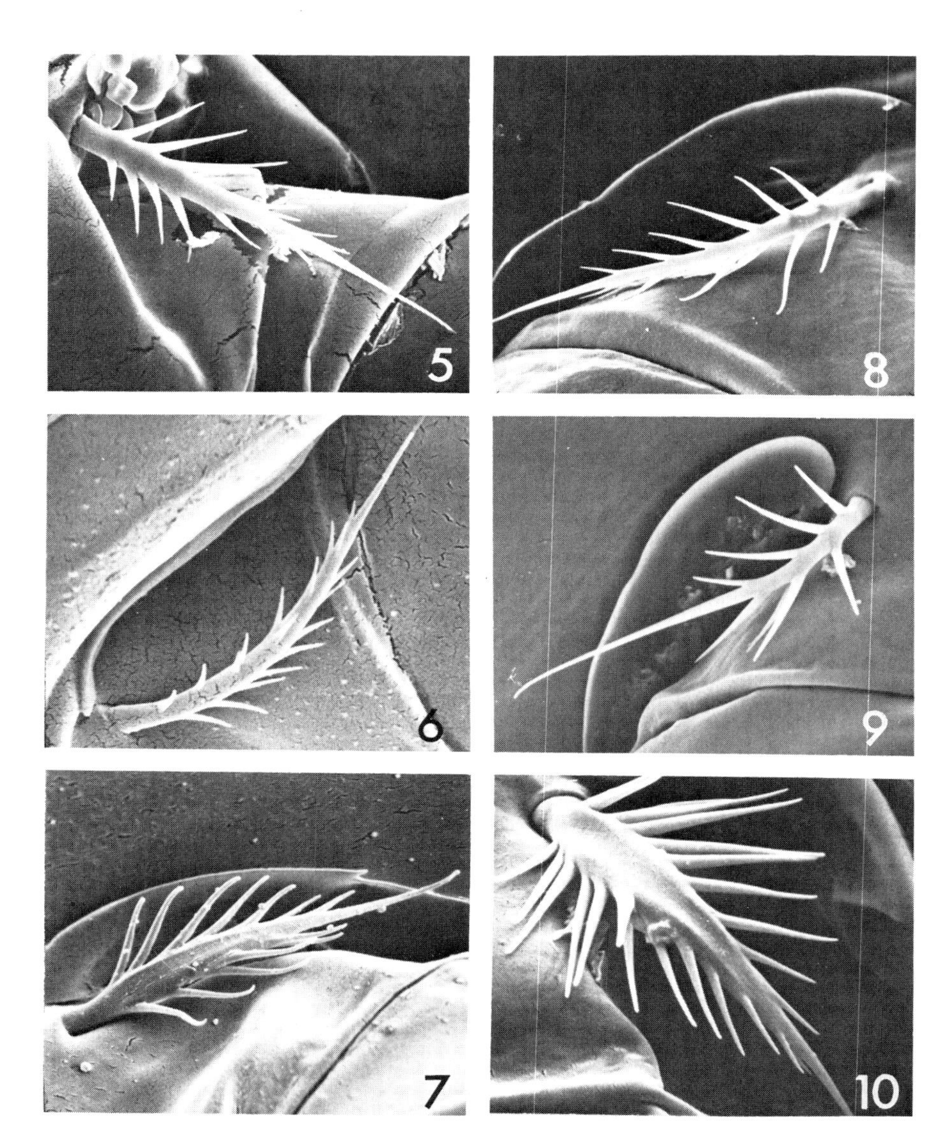

Figs. 5-10. Supracoxal setae of *Tyrophagus*. Figs. 5 and 6. *T. palmarum* viewed laterally and dorsally; Fig. 7. *T. neiswanderi*; Fig. 8. *T. similis*; Fig. 9. *T. longior*; Fig. 10. *Tyrophagus* sp. (Population T.8). Magnifications 1,700 to 2,000X.

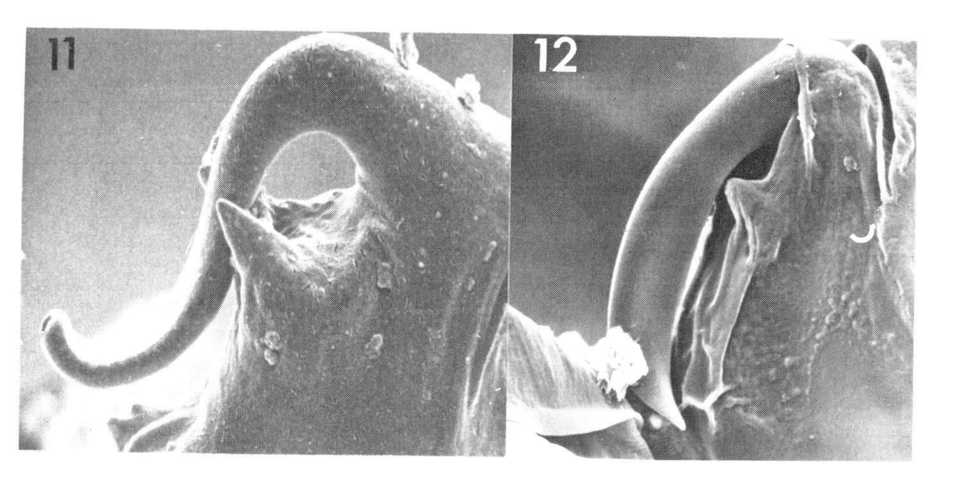

Figs. 11-12. Lateral view of penis. Fig. 11. *T. putrescentiae*; Fig. 12. *Tyrophagus* sp. (Population T.8). Magnification 3,000 X.

TABLE II.
Key Characters for Specific Determination in the Genus *Tyrophagus*.

Character	*putres-centiae*	*neis-wanderi*	*pal-marum*	*pern-iciosus*	*similis*	*longior*
D_1/d_2 setal ratio	1	2	1	3	4	4
$Omega_1$	1	1	2	3	4	5
Supracoxal seta	1	1	2	3	3	3
Penis	1	2	3	4	4	5
'Eye' spots	+	+	–	–	–	–

which these species can be separated with certainty. It is the form of the penis and that of solenidion $omega_1$ which are apparently the most reliable for species determination. Unfortunately, specimens have to be prepared so as to obtain lateral views of these organs and, even when this is done, it requires a very good microscope and considerable experience to be successful. Thus, by a judicious use of the information given in Table II it is possible to produce an identification key to the six common species. The reason why keys published on this basis appear to be inadequate will now be considered.

CAN HYBRID FORMS EXIST IN NATURAL POPULATIONS?

In order to answer this question inter- and intra-specific crosses have been completed using the 30 populations listed in Table I. They represent more than 10 years work and the results have been condensed into Figure 13. In this table certain species have been divided into groups A, B or C. The number of populations contained in each group is shown in Table I. In the case of *T. putrescentiae* and *T. palmarum* the division into groups has been made retrospective to the crossing experiments on the basis of their results. However, the division of *T. perniciosus* and *T. similis* each into two groups was made before conducting the experiments on grounds of slight morphological differences detected when populations of these species were examined following their collection. The taxon labelled species 'H' is almost indistinguishable from *T. putrescentiae* and its specific status is yet to be decided. A square without a symbol indicates that the particular cross has not been made.

Inter-specific Crosses—Results

The most significant result shown in Figure 13 is that all but one of the interspecific crosses failed to yield F_1hybrids (indicated by enclosed black squares). The single exception (*T. neiswanderi* x 'H' indicated by an open triangle) yielded a small number of F_1 adults but F_1 sibling crosses and the back cross did not produce eggs. Thus, the six economic species are reproductively isolated. Results of examination of the females' receptaculum seminis together with daily records of matings show that pre- and post-mating reproductive isolating mechanisms are operating. For example, no female of *T. palmarum* was successfully fertilized by a male of any of the remaining five species. Analysis is yet to be completed to determine whether these failures were due to the lack of attraction or a mechanical impasse. On the other hand, *T. similis* A females were fertilized by males of the other five species but in each case gametic isolation must be operating since no eggs were produced.

Intra-specific Crosses—Results

Figure 13 also shows that within a species each of the population groups are reproductively isolated from all other groups (closed circles). No F_1 hybrids were obtained between A x B crosses of *T. putrescentiae, T. perniciosus* and *T. similis,* nor between A x C and B x C crosses of *T. palmarus.* The cross *T. palmarum* A♂ x B♀ (open circle) yielded a total of three eggs, two of which hatched but the offspring died. In the reciprocal cross 40% of the females yielded an average of 18 eggs with a 30% hatch. The F_1 sibling crosses were sterile.

An unusual phenomenon was observed to take place during the period of these crosses. After about five days the posterior portion of the female's body became discolored internally, taking up a dark brown appearance. Shortly afterwards the body swelled up and eventually exploded. A similar isolating mechanism has been recorded in *Drosophila* (Mayr, 1948). The presence of reproductive isolation between populations of the same species was examined in detail employing 14 populations of *T. putrescentiae* collected in the

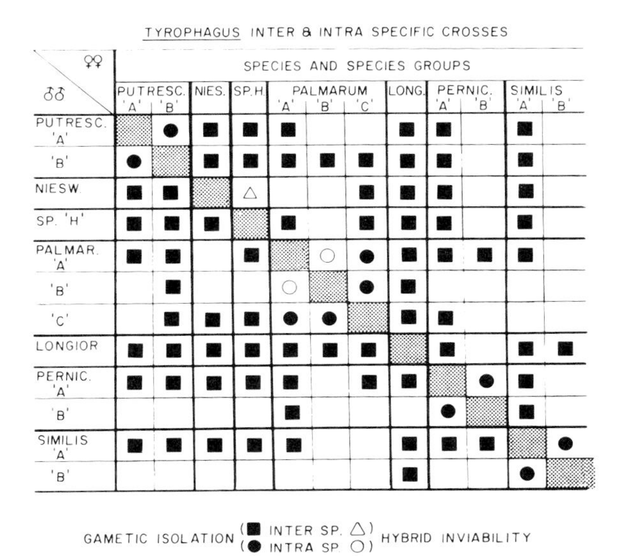

Fig. 13. Illustrating reproductive isolation between groups of populations identified on current taxonomic information as *T. putrescentiae, T. neiswanderi, T. palmarum, T. longior, T. perniciosus* and *T. similis.*

TYROPHAGUS INTRASPECIFIC CROSSES

♂♂ \ ♀♀	GROUP 'A'						GROUP 'B'				
POP. NO.	T10	T13	T20	T22	T30	T31	T2	T9	T37	T45	T52
T10		○	△	○	△	△	●	●			
T13	○		○	○	○	○	●	●	●	●	●
T20	○	○		○	○	○	●	●			
T22	○	○	○		○	○	●	●			
T30	○	△	◇	◇			●	●			
T31	○	○	○	○			●	●			
T2	●	●	●	●	●	●		○			
T9	●	●	●	●	●	●	○		○		○
T37		●						○			
T45		●									
T52		●						○			

GAMETIC ISOLATION - ● ZYGOTIC MORTALITY MEDIUM - △
NORMAL CROSS - ○ ZYGOTIC MORTALITY LOW - ◇

Fig. 14. Illustrating the degree of reproductive isolation existing between two groups of *T. putrescentiae* populations, labeled A and B.

Solomon Islands, Japan, Sarawak, India, East and West Africa, Middle East and Europe (see Table I for habitat details). The results of the majority of these hybridization experiments appear in Fig. 14. This figure is divided vertically and horizontally by a thickened black line. The rectangles at the top left and bottom right represent crosses made between populations withi group A and within group B respectively (open symbols). The remainin rectangles represent inter-group crosses (closed circles).

Crosses made within a group are for the most part normal, i.e., between 8 and 100% of the pairs produced eggs which hatched. The average number o eggs per pair was not significantly different from self crosses obtained fron these populations and the egg hatch was in excess of 90% (open circles). Onl two populations, both within group A, showed signs of some reproductiv incompatibility. The Solomon Islands population (T10) when crossed wit three English populations (T20, T30, T31) paired and produced eggs suc cessfully but the egg hatch was reduced to 19, 32 and 16%, respectively. A similar pattern was recorded when population T30 (a fungal culture take from wood, England) and population T36 (ex bulbs, England) were crosse with the South African and Sarawak populations. Such reductions i reproductive viability between allopatric populations are not uncommon. Th F_1 hybrids obtained from these crosses have yet to be paired and further inter group crosses are required to be done in group B.

The most significant result shown in Figure 14 is, of course, the discovery o two groups of interbreeding populations which are reproductively isolated on from another. The results for the *T. palmarum* crosses in Table III indicat that populations of this species may exhibit a similar phenomenon since thre discrete groups have been discovered. But, because insufficient population

TABLE III.
The Number of Different Forms of Key Characters Shared by the *T. putresentiae* and *T. neiswanderi* Species Groups.

	Tyrophagus Species Groups							
	putrescentiae 'Taxa'					*neiswanderi* 'Taxa'		
Character	A	B	H	C	D	A	B	C
Setal ratios	1	2	1	1	3	2	?	?
$Omega_1$	1	1	1	2	3	1	1	3
Penis	1	1	2	?	1	3	4	?
Supracoxal seta	1	1	1	2	1	1	1	3
Reproductive Isolation	yes	yes	yes	?	?	yes	?	?

TABLE IV.
The Number of Different Forms of Key Characters Found within the *T. perniciosus* and *T. similis* Species Groups, Respectively.

	Tyrophagus Species Groups					
	perniciosus 'Taxa'				*similis* 'Taxa'	
Character	A	B	C	D	A	B
Setal ratios	1	1	4	5	6	6
$Omega_1$	4	4	4	4	5	5
Penis	4	4	4	4	5	6
Supracoxal seta	4	5	5A	4	6	6
Reproductive Isolation	?	yes	yes	?	yes	yes

have been considered, further analysis with respect to *T. palmarum* and *T. longior* is not attempted in this paper. Thus, excluding *T. palmarum*, Figures 13 and 14 demonstrate the existence of nine genotypes. The question must now be asked do these genotypes reflect differences in their phenotypes which can be revealed by microscopic techniques?

THE PHENOTYPES

Tables III and IV depict the phenotypic appearance of the nine genotypes in respect of the four key characters presented in Table II which are used to distinguish the six common species. 'Eye spots' have been excluded since this character is common to all phenotypes apeparing in Table III and absent in those of Table IV. In these tables species 'H' has been included in the *T. putrescentiae* complex and an additional two taxa C and D have been added. Two taxa B and C and two taxa C and D have been added to the *T. neiswanderi* and *T. perniciosus* complexes, respectively. These six additional taxa are known from museum material only and represent some of the taxa in my collection which do not fit reasonably into a known species and which from inference we now know are highly unlikely to be hybrid forms.

When Tables III and IV are compared with Table II it shows immediately why the identification of *Tyrophagus* material can cause trouble. The number of different forms exhibited by the four key characters has barely increased (from 4: 5: 3: 5, in Table II to 6: 5: 6: 6, in Tables III and IV), whilst the number of taxa has more than doubled (from 6 to 14), eight of which are known to be reproductively isolated one from another. The problem of distinguishing taxa which belong to the same species group is even more daunting. Often, separation can only be decided on the basis of a subtle difference in the form of a single character (compare *T. perniciosus* B with C, Table IV). A further particular problem is the recognition on morphological grounds of *T. putrescentiae* A and B. Until recently these groups were indistinguishable, representing one morpho-species processing two biological forms. However, a careful study of the setal ratio character, known to be unreliable for species determination, has provided a method for identifying and placing populations into group A or B without recourse to crossing trials. The lengths of setae d_1, d_2, and l_2 of 50 females are measured under standard conditions and the ratio of d_1 to d_2 is plotted against d_1 to l_2. The ratio is obtained by multiplying the length of the seta considered to be always or most frequently the shortest by 100 and dividing by the length of the longest.

All 24 of the *T. putrescentiae* populations in culture together with museum material representing another 55 populations have been so treated. The results for the 24 living populations have then been compared by means of the 'Coefficient of Difference' statistic (CD) (Mayr, 1969). The results are given in Table V. It is considered that where the CD between two populations is 1.5 or more (a joint non-overlap of about 94%) a specific distinction can be made.

TABLE V.
Setal Variation Exhibited by Groups A and B as a Criterion for Specific or Sub-specific Recognition.

Length of seta d_1 to d_2 or l_2	Groups compared	No. of paired populations	Coefficients of difference			
			<1.0	1-1.49	1.5-1.99	>2.0
l_2	A x A	171	170	1	0	0
d_2	A x A	171	160	11	0	0
l_2	B x B	10	10	0	0	0
d_2	B x B	10	10	0	0	0
l_2	A x B	95	0	8	44	43
d_2	A x B	95	14	37	28	16

Table V shows that for both ratios a total of 171 paired comparisons gave a CD of less than 1.5, the vast majority had a CD of less than one. All comparison with the B group (five populations) gave a CD of less than one. Between groups the ratio d_1 to l_2 gave a CD of over 1.5 in 87 out of 95 paired populations. Out of the eight which fell between 1 and 1.49 only two were below the 75% non-overlap level. Also, all eight involved the same population, namely T52. The CD figures for the ratio d_1 to d_2 is far less useful in that more than half fell below the 1.5 level. It should be noted that it is this ratio which is employed most frequently as a character for specific determination.

A CD analysis was made on the closely similar A and B populations of *T. similis*. Again, the d_1 to l_1 ratio gave a non-overlap percentage better than 94%.

CONCLUSIONS

1. Much of the difficulty inherent in making specific determination of *Tyrophagus* material stems from the fact that too often poorly prepared material is viewed through inadequate microscopes. Key characters such as the penis, solenidion and supracoxal seta must be viewed laterally. (Compare Fig. 5 with Fig. 6.)

2. However, even when technical difficulties have been eliminated the apparent absence of a discontinuity in the variation of key characters still remains a problem.

3. This study eliminates the possibility that interspecific hybrids are responsible since the presence of such hybrids under natural conditions is considered to be extremely unlikely. Especially since the reproductive isolation mechanisms (RIM's) operating within the genus are almost entirely pre-mating RIM's or failure by the gametes to produce zygotes.

4. The rare appearance of F_1 hybrids and a complete absence of a hybrid F_2 generation is in direct contrast to that which occurs in the genus *Acarus*, Griffiths (1963). It suggests that speciation has developed further in *Tyrophagus*. The two genera also differ in that the male genitalia of *Acarus* species are morphologically undifferentiated one from another.

5. The investigation has revealed that the number of phenotypes which can be readily distinguished using contemporary taxonomic parameters. The two groups A and B of *T. putrescentiae* is a valid example. It would appear, therefore, that when we make a specific determination it often means no more than placing a taxon within a species group. Very careful consideration of its morphology may help to distinguish it further. (A paper is being prepared in which the taxa discussed here will be fully illustrated.)

6. Consideration should be given to using the setal ratio d_1 to l_1 as a taxonomic character. Indications are that it may be more reliable than the traditionally used ratio d_1 to d_2. Excluding this character separation of taxa may have to be based upon subtle differences of a single character such as solendion omega$_1$ or the supracoxal seta.

7. A study of Table I shows no evidence of habitat differences between the different genotypes. It does show, however, that all of them can occur in stored product or growing crop environments or in both situations. Their pest status and physiological propensities have yet to be considered.

SUMMARY

The experimental material consisted of living and preserved populations identified as, or considered to be very similar to, the following six economically important taxa: *T. putrescentiae, T. neiswanderi, T. palmarum, T. longior, T. similis, T. perniciosus.* Extensive hybridization trials showed that lack of discontinuity in the variation exhibited by key characters was not due to the occurrence of hybrid forms since hybrids resulting from any combination of these six species is a very rare phenomenon. However, the hybridization trials revealed the existence of genotypes the phenotypes of which are almost if not impossible to distinguish using current taxonomic keys. Consideration is given to the value of key taxonomic characters as a means of separating taxa within species group.

ACKNOWLEDGEMENTS

This study would have been impossible if my many friends in acarology had not so willingly provided me with living and preserved materials. Amongst the many have been Dr. A. M. Hughes and Miss B. John, England; Dr. Eva Zdarkova and Dr. K. Samsinak, Czechoslovakia; Dr. E. Baker and Dr. D. Coleman, U. S. A.; Dr. J. van Bronswijk and Dr. F. S. Lukoschus, Holland. Equally important is the painstaking and exacting work involved in the crossing experiments. I am greatly indebted to the Acarology Section in Slough, in particular to Miss S. Lynch who has had the responsibility of planning the work and collating the data, and to Mrs. S. Webb, Mrs. C. Singh, Miss C. George, Miss C. Ferguson and Miss S. Cooper. The scanning electron micrographs are the work of Ms. V. Cowper also of this laboratory. The measurements and analysis for the CD statistics were carried out by my 'Sandwich' student Andy Owen.

REFERENCES

Griffiths, D. A. (1964). *Proc. 1st Intern. Congr. Acarol., Fort Collins, 1963. Acarologia* (Suppl.) 101-116.
Johnston, D. E. and Bruce, W. A. (1965). *Res. Bull. Ohio Agric. Exp. Sta.* **977**, 1-17.
Mayr, E. (1948). *Adv. Genet.* **2**, 205-237.
Mayr, E. (1969). "Principles of Systematic Zoology." McGraw-Hill, New York.
Oudemans, A. C. (1924). *Tijdscher. Ent.* **67**, 22-28.
Robertson, P. L. (1959). *Aust. J. Zool.* **7**, 146-181.
Samsinak, K. (1962). *Act. Soc. Ent. Czech.* **59**, 266-280.
Zachvatkin, A. A. (1941). *Inst. Zool. Acad. Sci. Moscow N. S.* **28, 1-475.**

PYEMOTES TRITICI: POTENTIAL BIOLOGICAL CONTROL AGENT OF STORED-PRODUCT INSECTS

W. A. Bruce and G. L. LeCato

Stored-Product Insects Research and Development Laboratory
AR-SEA-USDA
Savannah, Georgia

INTRODUCTION

The straw itch mite, *Pyemotes tritici* (Lagrèze-Fossat and Montague), is a parasitic mite that long has been known to attack a large variety of stored-product insects. The excellent work of Moser and Raton (1971), Moser (1975), Cross and Moser (1971, 1975), and others (Scott and Fine, 1967; Smiley and Moser, 1976) has dealt primarily with the systematics, medical aspects, or non-stored-product pyemotid species. In addition, *P. tritici* has been virtually ignored as a biological control agent (Butler, 1972) by stored-product entomologists and acarologists primarily because it has always been regarded more as a pest species (Ghai, 1976). This is understandable, for at times infestations by this mite have reached the point of exterminating colonies of certain species of stored-product insects reared at the Stored-Product Insects Research and Development Laboratory, Savannah, Georgia. Therefore, we undertook an investigation of this mite as a potential biological control agent of stored-product insects.

PREDATION ON THE INDIAN MEAL MOTH: A PRELIMINARY TEST

Procedures

All stages of the Indian meal moth, *Plodia interpunctella* (Hübner), were placed in separate 3.78-liter jars containing whole corn (*ca.* 14% moisture) with 10% dockage. Prior to the introduction of the insects, either 5, 1, or 0.5 ml of pupae of the cigarette bettle, *Lasioderma serricorne* (F.), with attached

Vol. I: ISBN 0-12-592201-9

gravid female mites were added to the corn. One-half of the mites were added after one-third of the corn was placed into the jar, and the other half of the mites were added after two-thirds of the corn had been added. About 200 each of mid-instar larvae, pupae, adults, and 2000 eggs were added after all the corn had been placed in the jar. The cultures were observed throughout the test, and mortality was recorded after two to eight weeks. Mite-free control cultures were evaluated in the same way as the treatments. The jars which were capped with perforated lids and no. 1 filter paper were kept in pans half filled with mineral oil. The test was replicated three times.

Mites used in this test and subsequent tests were reared on pupae of *L. serricorne*. The pupae were washed free of their cocoons, dried on absorbent paper, and infested with offspring from gravid female mites. Within seven days after the mites had attached themselves to the pupae, the females had become gravid and were producing large numbers of offspring.

Results

Each density of *P. tritici* parsitized and subsequently caused 100% mortality of all stages of the Indian meal moth, whereas control moths had essentially no mortality. This preliminary test showed the need for specific information regarding predation by *P. tritici* on various species and stages of prey, the influence of the commodity (habitat) infested by the prey, and the density of *P. tritici* that is necessary to provide effective control of prey.

SPECIES AND STAGES OF PREY

Procedures

Ten each of eggs, early and late instar larvae, pupae, and adults of each of five economically important species of stored-product insects were placed in 20-dram vials, 60 x 40 mm, and covered with Parafilm® and a plastic cap. The merchant grain beetle, *Oryzaephilus mercator* (Fauvel), cigarette beetle, *L. serricorne,* red flour beetle, *Tribolium castaneum* (Herbst), almond moth, *Ephestia cautella* (Walker), and Indian meal moth, *P. interpunctella,* were studied. Newly emerged offspring ($\bar{x}$ = 11,441) of *P. tritici* placed on the tip of a no. 10/128 brush were introduced into each vial. Control vials were simultaneously established and a modification of Abbott's formula was used to adjust for natural mortality (Healy, 1952). The test lasted for 24 hr after which mortality was recorded. The temperature and relative humidity were maintained at 24 ± 1°C and 60 ± 15 RH.

Results

P. tritici caused a high mortality of all stages of *O. mercator* (Table I). Mortality of the eggs of *L. serricorne* was reduced to 78%, probably because

TABLE I.
Mortality of Stages of Three Species of Stored-Product Beetles after 24-hr Exposure to *P. tritici.*

	% Mortality				
	Eggs	Early Instars	Late Instars	Pupae	Adults
O. mercator	100	100	100	100	97
L. serricorne	78	100	4	32	44
T. castaneum	40	100	74	24	0

the eggs had large quantities of medium adhering to them, thus making it more difficult for the mites to prey upon them. Only 4 and 32% of the larvae and pupae of *L. serricorne*, respectively, were parasitized. This reduction in mortality probably occurred because the larvae have thick setaceous cuticles and the pupae are enclosed within a tightly knit cocoon that was covered with medium. Pupae with breaks in their cocoons were highly susceptible to attack by the mites. Forty-four percent of adult *L. serricorne* were parasitized. Although *P. tritici* could readily get beneath the elytra of *O. mercator*, the mites had difficulty getting beneath the elytra of *L. serricorne* and thus the latter species was not as readily parasitized.

Stages of *T. castaneum* also were resistant to attack by *P. tritici.* Only 40, 24, and 0% of the eggs, pupae, and adults, respectively, were parasitized. The eggs of *T. castaneum* are covered with a sticky fluid secreted by the accessory glands and are consequently covered with a dense layer of medium. The pupae are very hard and resist penetration by the mites that are capable of attacking only under the forming wings of the pupae.

Adult *T. castaneum* are completely impervious to attack because the mites are apparently unable to insert their mouthparts at any point on the body. However, 100% of the early instars (1-3) were parasitized, and this stage represented the most susceptible stage of *T. castaneum.* Seventy-four percent of the late instars were killed, and this reduction in mortality probably was caused by the late instars being more sclerotized and thus more resistant to penetration by the mites.

Pyemotes tritici caused considerable mortality of eggs, early instars, and adults of both *E. cautella* and *P. interpunctella* (Table II). About 50% of the pupae of these species were parasitized, and the mortality seemed to depend on whether the mites located a vulnerable spot for penetration. The data for the mortality of pupae were quite variable. On the other hand no late instars were parasitized. The mites apparently were unable to penetrate the thickened cuticle of the late instars of the moths. These findings agree with those of Hughes (1961).

The ability of the mites to parasitize and cause a high degree of mortality of both stored-product beetles and moths indicated a real potential for *P. tritici* as a biological control agent for stored-product insects.

TABLE II.
Mortality of Stages of Two Species of Stored-Product Moths after 24-hr Exposure to *P. tritici*.

	% Mortality				
	Eggs	Early Instars	Late Instars	Pupae	Adults
E. cautella	100	100	0	*ca* 50[a]	98
P. interpunctella	91	97	0	*ca* 50[a]	97

[a] Excessive variation existed in mortality of pupae.

INFLUENCE OF COMMODITY (HABITAT) ON PREDATION BY *P. TRITICI*

Procedures

This test was conducted in a manner similar to those in test 1 except for the following differences. About 12 g of flour or 18 g of whole wheat plus 10% dockage (cracked wheat) were used to half fill the respective vials. Adults and larvae (mobile prey) were placed into the vials 24 hr in advance of the mites which were placed on the top of the media. Eggs and pupae (immobile prey) were positioned in the media at approximately the depth at which they would normally occur for a given species. The tops of the vials were covered with 60-mesh Nytex® nylon cloth. All phases of this test were conducted for 24, 48, and 72 hr. However, because the data did not differ for the three durations, only data for the 24 hr test were used.

Results

In whole wheat *P. tritici* parasitized 100% of the adult and 80% of the mid-instar *O. mercator* (Table III). However, there was a highly significant reduction in mortality of the pupae. No data are shown for the eggs of the beetles or moths because these species are bran insects (not primary feeders) and consequently the early instars have difficulty feeding on whole wheat.

TABLE III.
Mortality of Stages of Three Species of Stored-Product Beetles after 24-hr Exposure to *P. tritici* in Wheat.

	% Mortality		
	Mid-instars	Pupae	Adults
O. mercator	80	4	100
L. serricorne	92	3	24
T. castaneum	94	10	0

Therefore, the diet accounted for mortality more than did the mites. Mortality of adult *L. serricorne* was reduced from 44% when the beetles had no diet in which to hide from the mites to 24% when the beetles were in whole wheat.

Ninety-two percent of the mid-instars were killed in whole wheat, but only 3% of the pupae were parasitized. Only 10% of the pupae of *T. castaneum* were parasitized. Mid-instars of *T. castaneum* had high mortality; adults had no mortality. Whole wheat afforded the mites ample space to search for prey and consequently was no impediment to them.

Virtually all mid-instars and adults of *E. cautella* and *P. interpunctella* were parasitized in whole wheat whereas less than 50% of the pupae were parasitized (Table IV).

TABLE IV.
Mortality of Stages of Two Species of Stored-Product Moths after 24-hr Exposure to *P. tritici* in Wheat.

	% Mortality		
	Mid-instars	Pupae	Adults
E. cautella	99	47	100
P. interpunctella	100	41	100

Seventy-six percent of *O. mercator* adults were killed in flour (a dense diet that would have presumably retarded the searching ability of the mites) (Table V). No eggs, mid-instars, or pupae were found to be alive in flour. This occurred not because of the presence of mites but because flour is a poor diet for *O. mercator*. No eggs and essentially no pupae of either *L. serricorne* or *T. castaneum* were parasitized in flour. A low percentage of mid-instars and adults of *L. serricorne* and no adults of *T. castaneum* in flour were parasitized, but 90% of *T. castaneum* mid-instars were parasitized.

TABLE V.
Mortality of Stages of Three Species of Stored-Product Beetles after 24-hr Exposure to *P. tritici* in Flour.

	% Mortality		
	Mid-instars	Pupae	Adults
O. mercator[a]	100	100	76
L. serricorne	6	1	21
T. castaneum	90	0	0

[a] Flour is a very adverse diet for this species.

TABLE VI.
Mortality of Stages of Two Species of Stored-Product Moths after 24-hr Exposure to *P. tritici* in Flour.

	% Mortality			
	Eggs	Mid-instars	Pupae	Adults
E. cautella	74	97	77	100
P. interpunctella	61	97	87	100

In flour all stages of *E. cautella* and *P. interpunctella* had high mortality (Table VI). The mites effectively preyed on the eggs, mid-instars, pupae, and adults of the moths because these stages occurred near the top of the flour. Consequently, the mites did not have to penetrate deep into the flour to parasitize these stages.

DENSITY OF *P. TRITICI* NECESSARY TO PROVIDE CONTROL OF *T. CASTANEUM*

Procedures

Twenty-dram translucent plastic vials were completely filled with 57.3 g of wheat. Various numbers of mid-instar *T. castaneum* were added to the wheat and allowed 24 hr for dispersal before various numbers of unattached *P. tritici* were added by means of a vacuum needle to the top of the wheat. A plastic top with a 2.5 cm hole in the center then was snapped over Nytex to cover the top of the vial which was then inverted to allow the mites to readily move up into the wheat. Prey mortality was determined after 48 hr.

Results

When the ratio of *P. tritici* to *T. castaneum* was 1:1 (10 mites:10 beetles) in 57.3 g of wheat (*ca* 81.5 cm^3) or 4.1 cm^3 per organism (mites + beetles) for 48 hr, no prey was parasitized (Table VII). An increase in the ratio to 10:1 (100 *P. tritici* and 10 *T. castaneum*, 0.7 cm^3 per organism) also did not cause mortality in *T. castaneum*. However, when the ratio was 10:1 involving 1000 *P. tritici* and 100 *T. castaneum* (0.1 cm^3 per organism), the mortality of *T. castaneum* rose to 10.4%. When the ratio was 50:1 involving 500 *P. tritici* and 10 *T. castaneum* (0.2 cm^3 per organism), mortality of *T. castaneum* was 30%. The reason for the increase in mortality was the small initial number of *T. castaneum* present. As the ratio increased to 100:1 (1000 *P. tritici* and 10 *T. castaneum,* 0.1 cm^3 per organism), mortality increased to 63.3%. This finding

TABLE VII.
Percent Mortality of Mid-Instars of *T. castaneum* Exposed 48 hr to *P. tritici* in 57.3 g of Wheat.

P. tritici (No.)	*T. castaneum* (No.)	cm^3/Organism	% Mortality of *T. castaneum*
10	10	81.5	0
100	10	0.7	0
500	10	0.2	30
500	100	0.2	4.7[a]
1000	10	0.1	63.3
1000	100	0.1	10.4

[a] Exposed for 24 hr.

was further indication that the smaller the prey population, the higher the percentage of control. However, a ratio as low as 5:1 (500 *P. tritici*, 100 *T. castaneum*, 0.2 cm^3 per organism) caused a 4.7% mortality of *T. castaneum*, but the increased number of prey present reduced the percent parasitized. Thus, as one might expect, the key to good control appears to be a high ratio of parasite to prey with preferably a relatively low number of prey. With the great fecundity of the mite this goal indeed seems feasible.

DISCUSSION

There are a number of distinct advantages for using *P. tritici* as a biological control agent: (1) high reproductive potential, (2) short life cycle (4-7 days), (3) no intermediate host or food source necessary; all development occurs within the gravid female, (4) females represent approximately 98% of progeny produced, (5) females are mated upon emergence from gravid female and begin host seeking activity immediately, (6) populations can be synchronized and reared easily, and (7) cosmopolitan in distribution.

At 25°C the mites go through a generation in seven days; thus sufficient mites would be available for parasitizing prey. One mite typically attaches to and paralyzes a single prey, but it is possible that this mite may have bitten and consequently paralyzed several prey before it becomes firmly attached (Weiser and Slama, 1964). The mites also move rapidly through the grain and may potentially contact many prey. These characteristics make *P. tritici* a promising biological control agent capable of significantly reducing the lag between a biological control agent and its prey. Also, such a fast generation time and high biotic potential for the mites ($\bar{x}$ = 200 per female) would allow them to significantly reduce the target species and/or achieve control of the prey before the commodity is significantly damaged. The mites live no more than five days without prey and thus are self-eliminating after prey extermination. If a few prey remain, the mites may also remain as an active control agent capable of disseminating in the grain mass and rapidly diminishing a renewed infestation. While recognizing the fact that *P. tritici* will bite humans, we feel that under certain circumstances, perhaps restricted situations, the potential exists to use this species as an effective biological control agent. For example, during the course of this work it was discovered that *P. tritici* may have great potential in controlling the red imported fire ant, *Solenopsis invicta* Buren. From the point of view of stored-products and stored-product pests several possibilities may exist. One such possibility might be in restricted personnel areas where commodities are left virtually defenseless to insect attack. Such conditions exist in port warehouses in which peanuts are stored prior to export. Those peanuts destined to be processed into peanut oil would require shelling and cleaning to remove dockage, insects, mites, and other contaminants. On the other hand, in storage areas such as closed silos that would tend to exclude insect entry, it might be advantageous to introduce *P. tritici* in an attempt to eliminate any residual insect population present.

It is doubtful that the straw itch mite will ever enjoy widespread use as a biological control agent in stored-product marketing channels. However two aspects are certain: (1) *P. tritici* is an extremely efficient and effective parasite and (2) its importance in the future as a biological control agent will increase as the use of pesticides decreases.

SUMMARY

The straw itch mite, *Pyemotes tritici,* has been shown in laboratory studies to have great potential as a biological control agent for stored-product insects. At least 90% or greater of one or more life stages of the five target species was parasitized by *P. tritici.* The ease with which the mites can be reared, populations synchronized, and its fast generation time (*ca* 4-7 days) ensure that sufficient numbers of parasites can be made available to effectively reduce a pest population.

REFERENCES

Butler, L. (1972). *J. Econ. Entomol.* **65,** 702-705.
Cross, E. A. and Moser, J. C. (1971). *Acarologia* **13,** 47-64.
Cross, A. and Moser, J. C. (1975). *Ann. Entomol. Soc. Amer.* **68,** 723-732.
Ghai, S. (1976). *Bull. Grain Tech.* **14,** 134-144.
Healy, M. J. R. (1952). *Ann. App. Bio.* **39,** 211-212.
Hughes, A. M. (1961). "The mites of stored food." Tech. Bull. No. 9. Her Majesty's Stationery Office, London.
Moser, J. C. (1975). *Trans. R. Ent. Soc. Lond.* **127,** 185-191.
Moser, J. C., Cross, E. A. and Roton, L. M. (1971) *Entomophaga* **16,** 367-379.
Scott, G. and Fine, M. (1967). *Pest Control* **35,** 19-23.
Smiley, R. L. and Moser, J. C. (1976). *Beitr. Ent., Berlin* **26,** 307-322.

THE CONTROL OF MITES IN CHEESE STORES

D. R. Wilkin

Ministry of Agriculture, Fisheries and Food
Pest Infestation Control Laboratory
London Road, Slough, Berks
England

INTRODUCTION

Mites have long been associated with the storage and maturing of cheese and reference to cheese mites can be found in literature dating from the 16th century. Losses attributed to mites in the United Kingdom were first estimated by Eales (1917) who reported that 2.5% by weight of each Stilton cheese produced was consumed by mites. About 2% was also added to the cost of each cheese due to the extra labor required to remove the mites before sale. Cranfield *et al.* (1934) considered that mites were the most destructive pest of Stilton cheese and evaluated some control measures. The descriptions of *Tyrophagus siro, Tyrophagus longior* and *Aleurobius farinae* given by Eales (1917) are confusing and make it very difficult to relate these mites with current specific names. However, it is likely that *T. siro* was *Tyrolichus casei* Oudemans, and that *A. farinae* was *Acarus* sp.

A further extensive survey of cheese-mite infestations was carried out by Robertson (1952) who found that infestation was widespread throughout cheese stores with Cheddar and Stilton cheeses being particularly seriously affected. Up to seven species of mites were involved but only *Tyroglyphus farinae* (*Acarus siro* L.) and *Tyrolichus casei* were named. The author also commented that infestations were rarely made up of a single species.

Reports of mites infesting cheese have also come from many other countries including North America, Canada, New Zealand, Czechoslovakia and Italy. Once again it is not always possible to determine exactly which species of mites were involved but *A. siro, Tyrophagus putrescentiae* (Schrank), *Tyrolichus casei* and *Glycyphagus domesticus* (De Geer) are frequently quoted.

The rind of traditional cheese offers little protection from mite attack but during the late 1950s many of the producers in the United Kingdom switched

Vol. I: ISBN 0-12-592201-9

from traditional to plastic-wrapped, block cheese thus almost eliminating the mite problem. However, about 15% of producers continue to manufacture traditional cheeses, particularly Cheddar, Cheshire and Stilton types and these producers' stores remained infested with mites.

The damage caused by mites has led to considerable effort being made to develop control measures. Early methods of chemical control which included fumigation with ammonia and carbon disulphide were not particularly successful and cheesemakers had to rely on physical methods such as brushing or waxing the cheeses. At best these merely reduced the problem slightly (Robertson, 1952) and brushing frequently spread mites between cheeses. However, during the early 1950s fumigation with methyl bromide was employed and gave excellent control. Unfortunately, few stores were suitable for fumigation because it required specialist labor. Efforts were therefore made to develop a technique which could be used by the store keepers. Marzke and Dicke (1959) demonstrated that gamma-HCH (lindane) was the most effective acaricide of a number of chemicals tested against cheese mites. Although there is no published data on the effectiveness of gamma-HCH under practical conditions, it became widely used by the industry, usually as a 0.5% dust applied direct to the cheese.

The Ministry of Agriculture, Fisheries and Food in England provides advice for the cheese industry on the control of mites, and during the late 1960s and early 1970s Ministry advisers began to receive requests for advice on the control of mites on traditional cheese which persisted despite the use of gamma-HCH. An investigation was undertaken to determine the extent of infestations in British cheese stores, the reason for the failure of control by gamma-HCH, and to develop new control measures as necessary.

SURVEY OF CHEESE STORES

The principal types of English traditional cheese, Stilton, Cheshire and Cheddar are produced mainly in three areas shown in Fig. 1. Other traditional, rinded cheeses, such as Leicester, Wensleydale and Double Gloucester are made either on a small scale or by large producers who also manufacture one of the principal varieties. However, stores frequently distribute or sell a variety of cheeses which ensures a regular movement of cheeses between premises.

Twenty stores producing or maturing traditional cheese were inspected during the period 1972-74. Three stores contained mostly Cheshire cheese, six contained Stilton and the remainder Cheddar, but most also handled cheeses from other producers. All premises except one had some form of temperature control and the store temperatures ranged from about 10-14°C. Few premises had any close control over relative humidity but generally aimed to maintain a level of more than 80%. These temperatures and humidities were required for the satisfactory maturing of cheese and were particularly important in the

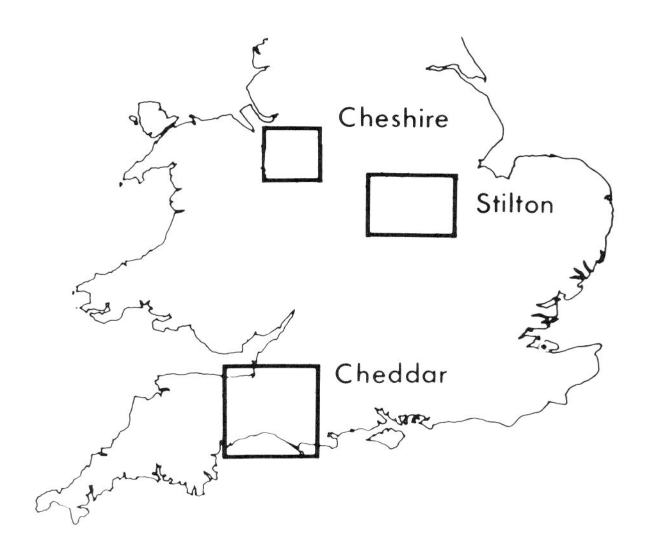

Fig. 1. The geographic location of the major producers of traditional Cheshire, Stilton and Cheddar cheese.

production of blue cheese. The size and construction of the stores varied considerably but the cheese was always stored on wooden racking or shelves. The largest store had space for 250,000 cheeses (about 7,000 tons of cheese) in a new warehouse built for the purpose but some of the smaller producers used old dairies and stores, occasionally more than 100 years old and holding a few hundred cheeses.

The stores were inspected for mites visually, and by collecting sweepings from cheeses or shelves for microscopic examination. Selected mites were mounted on slides for specific identification. Mites were found at all stores, both on cheeses and in the fabric of the building. The levels of infestation varied considerably between stores and, to some extent, between different types of cheeses. Infestation was always heaviest on the oldest cheese and careful inspection failed to detect any mites on young cheeses less than three weeks old. The surface of older, heavily infested Stilton and Cheddar cheeses was eroded by mites at many stores. These heavily infested cheeses required cleaning, and often extensive trimming before sale. Generally, Cheshire cheese was less seriously infested because of its short maturing period of about 8-10 weeks. However, one store producing blue Cheshire cheese, which was matured for at least 14 weeks, was very badly affected by mites and many of the cheeses were seriously damaged. The store manager estimated that up to 25% of each cheese was being eaten away.

The mites were identified using keys by Griffiths (1963) for the *Acarus* spp., and Hughes (1961, 1976) for the other species. The species found were *Acarus chaetoxysilos* Griffiths, *A. siro* L., *A. farris* Oudemans, *A. immobilis* Griffiths, *Tyrophagus longior* (Gervais), *T. palmarum* Oudemans, *T. putrescentiae* (Schrank), *Tyrolichus casei* Oudemans, *Glycyphagus domesticus* (De Geer), *G. destructor* (Schrank) and *Tydeus* sp. One species, *A. chaetoxysilos,* was common to all stores and was also by far the most numerous mite. Infestation in Stilton stores in particular consisted almost exclusively of this species.

During November 1973 comprehensive inspections were made at the premises of five small-scale producers of traditional Cheddar and at the large central warehouse and Cheddar grading center. Mites, collected from cheeses and the fabric of the stores, were identified to genera on the spot using a low powered microscope (x30). Representative specimens from each genus at each store were mounted for specific identification. The results are given in Table I and the species found are arranged in order of frequency of occurrence.

Several of the stores visited also handled some plastic-wrapped, block cheese. Generally this was free from infestation although small numbers of mites, mainly *T. palmarum* and *G. domesticus* were found on the outside of the wrapping. Very occasionally serious infestations were noted on individual blocks where the plastic film had been damaged.

Gamma-HCH was extensively used at many stores, both on the fabric of the building and applied directly to the cheeses. In all cases mite infestations persisted in spite of the use of this pesticide and at some stores mites could be found on cheeses under a 5 mm covering of 0.5% gamma-HCH dust.

During the inspections samples of mites from many stores were collected and cultured in the laboratory for further study.

DEVELOPMENT OF CONTROL MEASURES

Laboratory Trials

Resistance to gamma-HCH in *A. siro* from a cheese store was first detected in 1971 (Wilkin, 1973). The detection of resistance in other species of cheese mites has been hampered by the lack of known susceptible stocks. However, *A. chaetoxysilos, A. farris* and *T. longior* originating from cheese stores, were tested against gamma-HCH using the technique described by Wilkin and Hope (1973). The results, given in Table II, show that doses of between 40 and 80 ppm gamma-HCH using the technique described by Wilkin and Hope (1973). The results, given in Table II, show that doses of between 40 and 80 ppm gamma-HCH were required to give complete kill, compared to about 1 and 2.5 ppm for susceptible *A. siro* and *A. farris* respectively. Resistance to gamma-HCH has also been confirmed in *G. destructor* (Wilkin, 1975).

Laboratory trials using a range of pesticides against gamma-HCH-resistant *A. siro* have been reported elsewhere (Pest Infestation Control Laboratory Reports, 1971-73 and 1974-76). Little evidence of cross-resistance was found

TABLE I.
The Species of Mites Found at Five Traditional Cheddar Stores
and the Central Cheddar Grading Center.
(Arranged in Order of Frequency of Occurrence at Each Location.)

STORE:	SPECIES:	COMMENTS:
1	*A. chaetoxysilos* *T. longior* *G. domesticus* *T. palmarum* *A. farris* *Acarus* sp.	Store and cheeses treated with gamma-HCH but heavily infested.
2	*A. chaetoxysilos* *T. longior* *A. farris* *A. immobilis*	Heavily infested particularly on old cheeses.
3	*T. casei* *A. chaetoxysilos* *T. longior* *A. farris* *G. domesticus*	*T. casei* confined to one batch of old, heavily infested cheeses. Rest of store had mixed light infestation.
4	*G. domesticus* *T. palmarum* *A. chaetoxysilos* *A. farris*	Mostly young cheeses, lightly infested with *A. chaetoxysilos*.
5	*T. longior* *A. chaetoxysilos* *A. farris* *T. palmarum* *Tydeus* sp.	Store and some cheeses treated with gamma-HCH. Mostly young cheese in store. Infestation generally light.
Cheddar Grading Center	*A. chaetoxysilos* *G. domesticus* *A. immobilis* *A. farris* *T. putrescentiae* *T. palmarum*	Widespread, heavy infestation on cheese and fabric of store. *T. putrescentiae* found only on block cheese.

except to the organochlorine insecticide dieldrin, and the organophosphorus insecticide pirimiphos-methyl was particularly effective. Little laboratory work has been carried out on the susceptibility of other species of cheese mites to alternative acaricides, but in limited tests with mixed cultures of mites collected from cheese stores pirimiphos-methyl showed promise.

TABLE II.
The Effect of Gamma-HCH on Three Species of Mites Collected from Cheese Stores.

	gamma-HCH (ppm)	Effect after 14 days exposure[a]
A. chaetoxysilos	20	3
	40	4
A. farris	20	3
	40	3
	80	4
T. longior	20	3
	40	4

[a] 3 = 75 to 99%
4 = 100%

Field Trials

A small store producing blue Cheshire cheese with a very serious mite infestation was chosen for the initial experimental work. Observations at this and other stores had shown that young, freshly produced cheeses (less than three weeks old) were never infested and even appeared unattractive to mites. As the cheeses matured they appeared to become infested from residual populations of mites remaining on the shelves from previous batches of infested cheeses. It seemed likely, therefore, that if this residual population could be controlled the young cheeses would remain uninfested.

An initial test was carried out in which two areas of wooden shelving were sprayed with an emulsion containing 2% pirimiphos-methyl to give a deposit of 440 mg/m^2. Two infested cheeses about 10 weeks old were placed on one shelf and two young, mite-free cheeses on the other. Four similar cheeses on untreated shelves were used as controls. Mite numbers were estimated by counting all mites seen within the field of a x30 binocular microscope. Six fields were chosen at random per cheese for each count. The results are given in Table III and show that whilst the treated shelf had little effect on an established infestation the young cheeses remained virtually mite-free, although the controls became heavily infested.

A more comprehensive trial involved the treatment of the whole store. Areas of shelving were sprayed as they became empty over a period of three months and all young cheeses were placed on treated racks. This led to a dramatic reduction in infestation, damage was eliminated and the general condition of the cheese improved enormously. Six months after treatment very few mites could be found and these were confined to the oldest cheeses. Compete eradication of mites appeared unlikely since the store received regular consignments of infested cheeses from other stores.

Similar treatments were also carried out at premises producing or maturing Stilton and Cheddar and gave extremely good control. Trials at Stilton

TABLE III.
The Degree of Infestation on Cheeses Stored on Board Treated with 440 mg/m^2 Pirimiphos-Methyl.

		Time in Days			
		1	14	28	46
		Number of Mites/6 Microscope Fields			
Young Cheese	Treated surface	0	0	0	4
		0	0	0	0
	Control	0	22	63	103
		0	5	93	229
Infested Cheese	Treated surface	270	35	253	–[a]
		27	86	202	–[a]
	Control	162	188	186	–[a]
		188	132	132	–[a]

[a] Cheeses unavailable.

producers were especially successful as, in general, they did not receive other types of cheese and thus reinfestation was not a problem.

Determination of Acaricide Residues

Treatment of the fabric of the building rather than direct application of an acaricide to the cheese reduced the risk of serious contamination. However, cheese would be in direct contact with shelves treated with pirimiphos-methyl for periods of several months. Measurements were made, therefore, during the field trials to assess the amount of contamination. Only very low levels of pirimiphos-methyl were detected in the cheese but the method used did not allow the level to be precisely monitored. An accurate determination of the uptake of pirimiphos-methyl was made in laboratory experiments in which Cheshire or Stilton cheeses were placed on wooden boards treated at 440 mg/m^2 with carbon 14-labelled pirimiphos-methyl. The cheese was stored under similar conditions to those found in commercial practice (10°C, 85% rh) for three months. Full accounts of this work have been published (Thomas and Rowlands, 1975; Thomas, in press) and the results show that penetration was confined to the outer 4 mm of the cheese in which a maximum of 12-14 ppm pirimiphos-methyl was found.

DISCUSSION

The physical conditions required during the maturing and storage of cheese, particularly the high relative humidities, favor the development of mites (Cunnington, 1963; Hilsenhoff and Dicke, 1963). These conditions are regarded as essential by the manufacturers and it seems very unlikely that they could be modified sufficiently to affect mite numbers appreciably. The large-scale change to plastic-wrapped block cheese has obviously reduced the extent of the mite problem in the industry. However, infestation has remained a serious problem for stores specializing in traditional cheese, particularly with the development of resistance to gamma-HCH. The mites not only damage the cheese but their mere presence often reduces its saleability.

Most species of mites found during the survey have already been recorded from cheese stores. The most notable exception was *A. chaetoxysilos* which was previously known only from the few type specimens collected in North American ports on cargoes from Japan (Griffiths, 1963) and has never been found as a breeding population. This species was the most frequently occurring and most numerous mite in English cheese stores. It was detected in every store visited and infestations at Stilton producers were composed almost exclusively of this mite. Nothing is known of the biology and ecology of *A. chaetoxysilos* but it is obviously well suited to the conditions found in cheese stores. These mites fed on the cheese and seemed to be responsible for much of the damage. Other species found such as *T. palmarum, T. longior* and *A. farris* are usually regarded as fungus feeders and were probably feeding mainly on the mold growth always associated with traditional cheeses. It seems surprising that *A. chaetoxysilos* should be so widespread and yet not have been recorded previously or been found on other stored foodstuffs. The widespread distribution in cheese stores suggests that it has been inhabiting them for some time and may have been misidentified as *A. siro*.

The development of resistance to gamma-HCH by mites in cheese stores is hardly surprising. Many of the stores had been relying on liberal application of gamma-HCH dust to control infestations for 20 years or more, yet no formal recommendations for the use of gamma-HCH seemed to have been prepared. Consequently treatments tended to be applied only when mites were obvious, leaving some cheeses and parts of the store untreated or with only limited doses.

The application of pirimiphos-methyl to the fabric of the store has proved a very successful method of controlling mite infestations and has been widely adopted by the industry in the UK. In stores which do not receive cheese from other sources, such as Stilton producers, fabric treatment has virtually eliminated mites. However, the regular movement of cheeses between many other stores poses serious problems. Regular treatments at about six-monthly intervals are needed to cope with the reintroduction of mites. The large number of independent producers make it impossible to ensure that all stores are treated simultaneously. However, provided that regular treatments are carried out, extremely good control can be obtained and losses of cheese prevented.

ACKNOWLEDGEMENTS

I am indebted to the traditional cheese industry and the Milk Marketing Board for allowing me to visit their stores and providing much useful information. Ministry of Agriculture, Fisheries and Food Infestation Control advisors and Dairy Husbandry Advisory Officers assisted with the execution and monitoring of field trials and Miss A. Haward collected many of the mites, prepared the mounted specimens and did much of the laboratory testing. Dr. D. A. Griffiths provided much valuable assistance with the identification of the mites.

REFERENCES

Cranfield, H. T., Roebuck, A. and Stafford, J. G. W. (1934). *J. Min. Agric.* **41**, 347-352.
Cunnington, A. M. (1976). *Ann. Appl. Biol.* **82**, 175-178.
Eales, N. B. (1917). *Ann. Appl. Bio.* **4**, 28-35.
Griffiths, D. A. (1970). *Bull. Br. Mus. Nat. Hist.* (Zool.) **19**, 85-118.
Hilsenhoff, W. L. and Dicke, R. J. (1963). *Mkg. Res. Rep. U.S. Dep. Agric., Agric. Mkg. Serv.* **599**, 46 pp.
Hughes, A. M. (1976). *Tech. Bull. Minist. Agric. Lond.* **9**, 1-400.
Hughes, A. M. (1961). *Tech. Bull. Minist. Agric. Lond.* **9**, 1-287.
Marzke, F. O. and Dicke, J. R. (1959). *J. Econ. Entomol.* **52**, 237-240.
Rep. Pest Infest. Control Lab., 1971/73 (1975).
Rep. Pest Infest. Control Lab., 1974/76 (1978).
Robertson, P. L. (1952). *J. Soc. Dairy Tech.* **5**, 86-95.
Thomas, K. P. (in press). *J. Stored Prod. Res.*
Thomas, K. P. and Rowlands, D. G. (1975). *J. Stored Prod. Res.* **11**, 53-56.
Wilkin, D. R. (1975). *Proc. 8th Br. Insectic. Fungic. Conf. Brighton 1975* **1**, 355-364.
Wilkin, D. R. (1973). *J. Stored Prod. Res.* **9**, 101-104.
Wilkin, D. R. and Hope, J. A. (1973). *J. Stored Prod. Res.* **8**, 323-327.

Recent Advances in Acarology, Volume I

THE EFFECT OF SOME JUVENILE HORMONE ANALOGS ON *TYROPHAGUS PUTRESCENTIAE*

J. Czaja-Topinska

Teachers' Education and Research Center
Cyrulików, Warsaw, Poland

Z. Stepien

Institute of Plant Protection
Warsaw Agricultural University
Nowoursynowska, Warsaw, Poland

R. Sterzycki

Institute of Organic Chemistry
PAS
Kasprzaka, Warsaw, Poland

INTRODUCTION

Since the beginning of his history man has used different methods and means to control pests of stored food. The use of chemicals, especially organic pesticides, can be considered as a great progress in food protection. Chemicals used for food protection, however, have deleterious, often toxic effects on the quality of treated products which may create a hazard to the consumers.

The search for effective, selective and safe compounds, which use the natural physiological mechanisms of insects has led to the discovery of a large group of substances called insectistatics or insect growth regulators (Levinson, 1975). Many hopes are connected with the possibility of practical use of juvenile hormone analogs (JHAs) for the control of pests in storage facilities, where specific ecological conditions suggest high probability of their effective action.

To the best of our knowledge, no information exists concerning hormonal regulation of growth and development of Acaridae. However, Roshdy *et al.* (1973) found in adults of *Argas (Persicargas) arboreus* (Acarina: Argasidae) a

Vol. I: ISBN 0-12-592201-9

neurohemal organ homologous to the corpora allata of insects, and additional observations (Bassal and Roshdy, unpub.) also indicate the existence, in adults of this species, of homologues of the prothoracic glands. There are few references concerning activities of JHAs in Ixodidae (Bassal, 1974); in Argasidae (Bassal and Roshdy, 1974); in Acaridae (Czaja-Topinska and Stepien, 1975); and in Tetranychidae (Nelson and Show, 1975; Staal *et al.*, 1975). However, because of the phylogenetic relationship between the Argasidae and Acaridae and discovery of biological activity of JHAs against *Tyrophagus putrescentiae* (Schr.) (Acarina:Acaridae) (Czaja-Topinska and Stepien, 1975) more detailed study of their effect on different stages of this species has been stimulated.

MATERIALS AND METHODS

Stock Cultures and Methods of JHAs Application

Stock cultures of *T. putrescentiae* were maintained on rye germ at 25°C and 85% R.H. according to the method of Boczek (1954). Acetonic solutions of JHAs (500-0.005 μg pure active substance) in concentrations ranging from 10-0.0001%. were applied topically in 5 μl doses to the following developmental stages of *T. putrescentiae*: 1) eggs 0-24, 24-48, 48-72 and 72-96 h old; 2) active, freshly emerged, larvae, proto- and tritonymphs; 3) young, freshly emerged adults. JHAs were applied to groups of 100 eggs, larvae and protonymphs or to 20 specimens of tritonymphal and imaginal stages.

To investigate the influence of a short- and long-term JHAs contamination of adults on their fecundity, two parallel groups of young virgin females (20 specimens each) were treated in the rearing cells. One group of females was left after treatment in the contaminated cell, and the second group was transferred after 24 hours to the new clean cell. Twenty-four hours after treatment 20 males of the same age were added to each group of females. The fecundity of both groups of females was compared during a 10-day period. One hundred eggs were collected from each group of treated females and placed into separate cells. The number of adults of the F_1 generation that developed from these eggs was evaluated. All tests were repeated five times.

To exclude the toxic action of tested JHAs on *T. putrescentiae* full controls with substances mentioned in Table I were established. In the controls, mites and eggs were treated the same way as in the tests with JHAs. Biological activity of tested substances was evaluated on the basis of the following observations: 1) total survival rate, expressed as the number of adult mites of F_1 generation, developed from treated eggs or other stages; 2) survival and fecundity of treated adults; 3) duration of generation time. The significance of differences was calculated by means of the U-test, Wilcoxon-Mann-Whitney (Siegel, 1956).

Juvenile Hormones Analogs Tested

Fifty six of the more highly active insect type JHAs of different chemical structures (Table I) were tested. These substances were obtained from: 1) Institute of Organic Chemistry, PAS, Warsaw, Poland (49-56); 2) Institute of Organic Chemistry and Biochemistry, CAS, Prague, Czechoslovakia (1, 5-32, 37-46); 3) Zoecon Co., Palo Alto, California, USA (2-4, 33-36, 47, 48). Substances from (1) and (2) were received through the courtesy of Dr F. Sehnal of the Czechoslovak Academy of Sciences.

TABLE I.

The List of JHAs, by Chemical Structure, Tested on *T. putrescentiae*.

1. 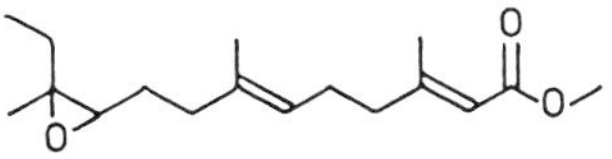 O O O
2. 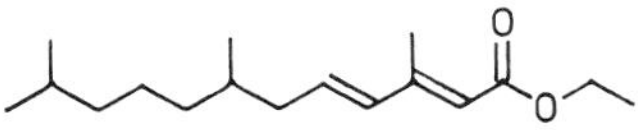 O O
3. 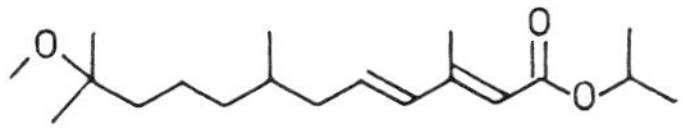 O O O
4. 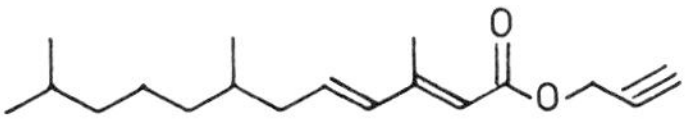 O O
5. 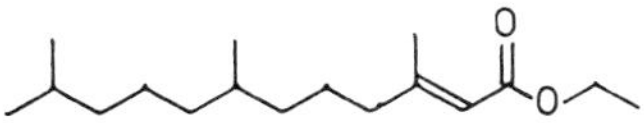 O O
6. 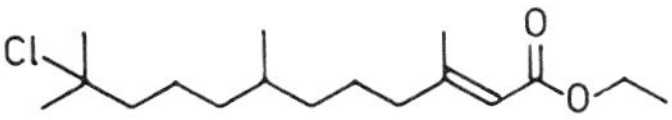 Cl O O
7. 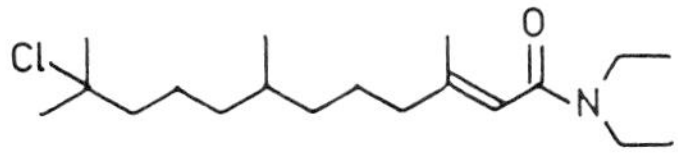 Cl O N
8. 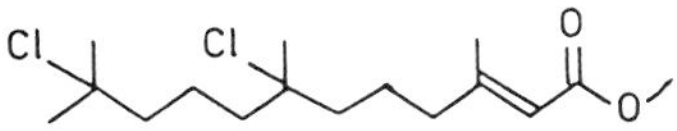 Cl Cl O O
9. 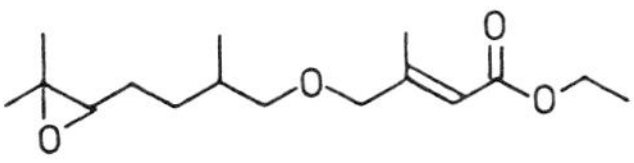 O O O O
10. 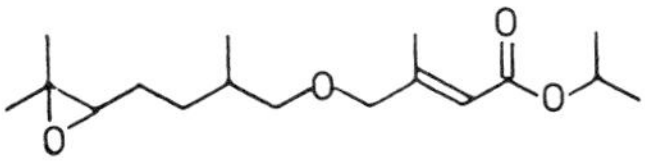 O O O O
11. 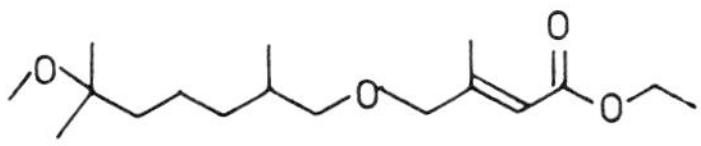 O O O O
12. 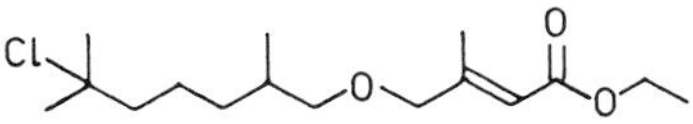 Cl O O O
13. 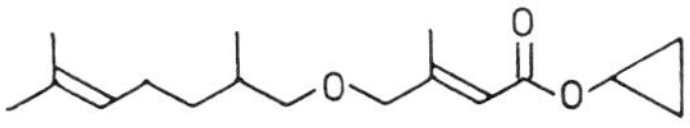 O O O
14. 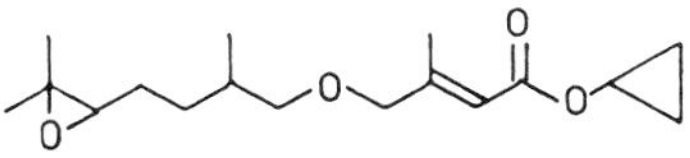 O O O O
15. 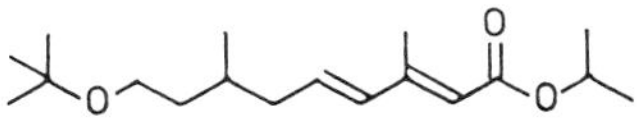 O O O
16. 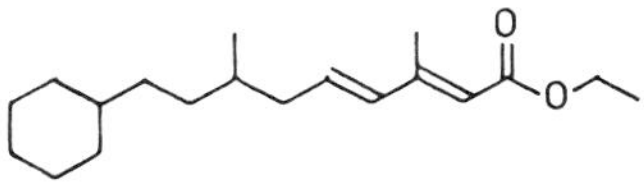 O O
17. 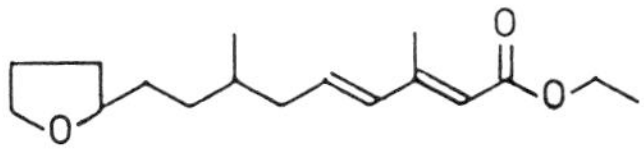 O O O
18. 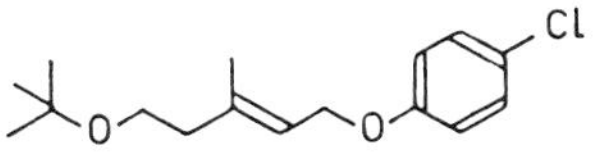 O O Cl
19. 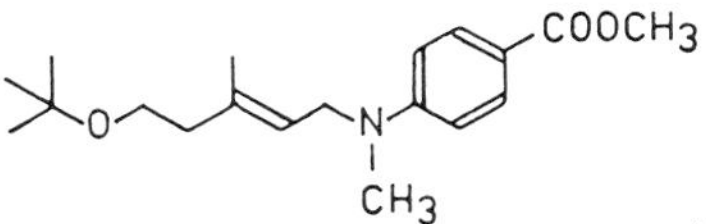 O N CH_3 $COOCH_3$
20. 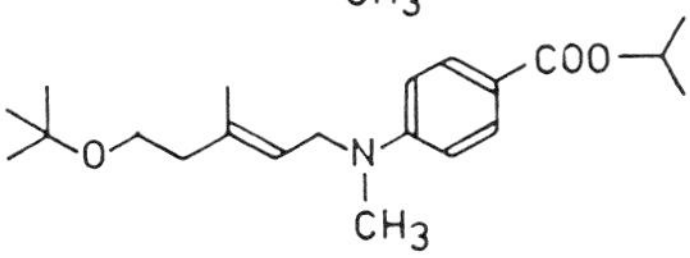 O N CH_3 COO
21. 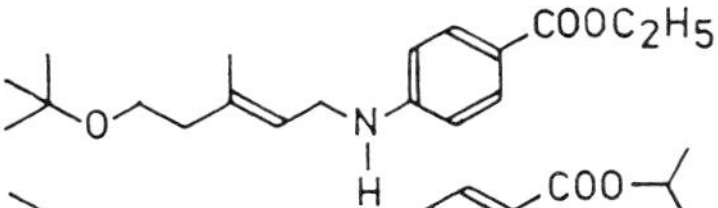 O N H $COOC_2H_5$
22. 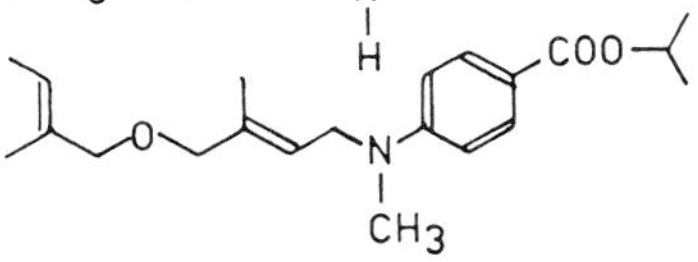 O N CH_3 COO

23. 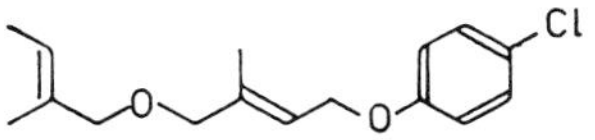 O O Cl

24. 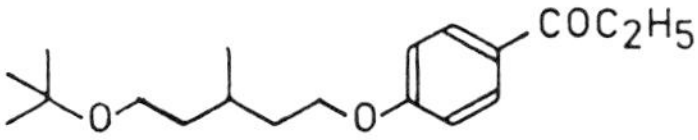 O O COC_2H_5

25. 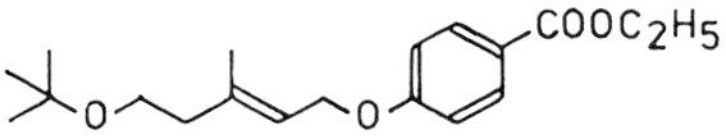 O O $COOC_2H_5$

26. 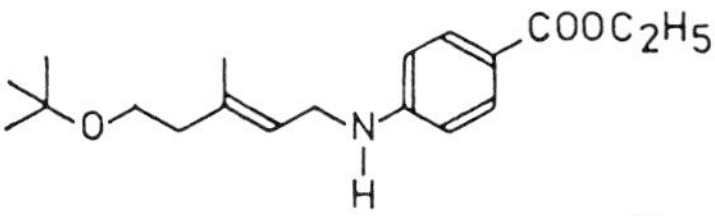 O N H $COOC_2H_5$

27. 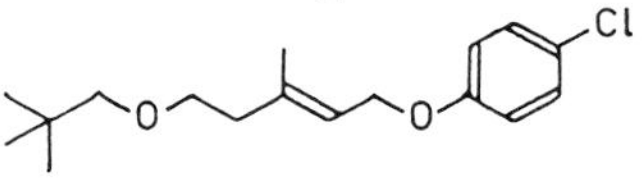 O O Cl

28. 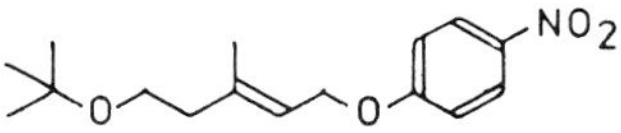 O O NO_2

29. 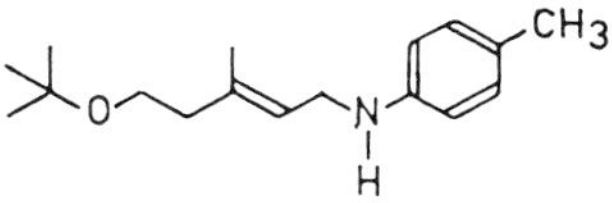 O N H CH_3

30. 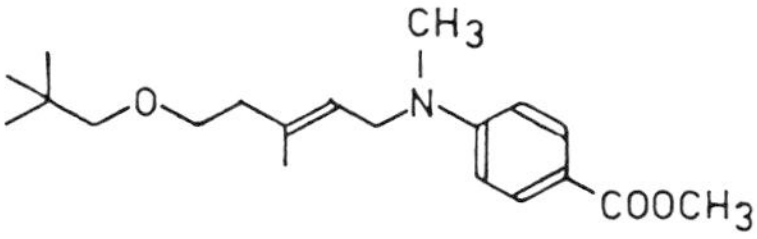 O CH_3 N $COOCH_3$

31. 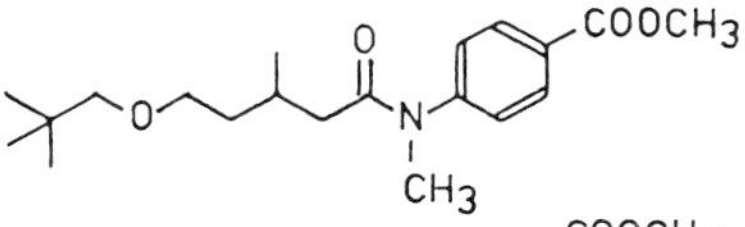 O O N CH_3 $COOCH_3$

32. 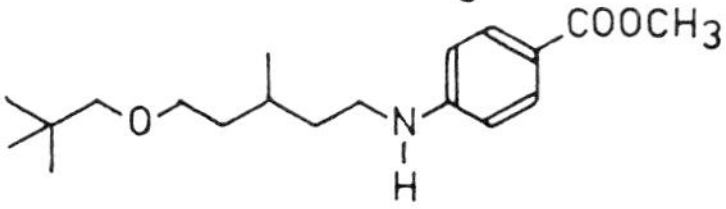 O N H $COOCH_3$

33. 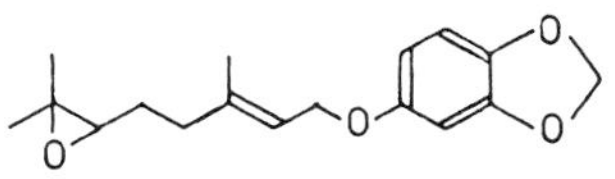 O O O O

34. 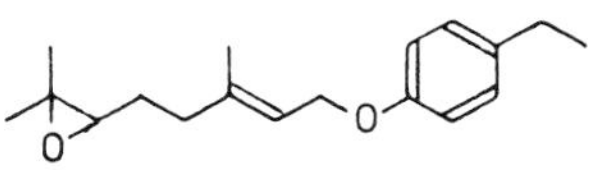 O O

35. 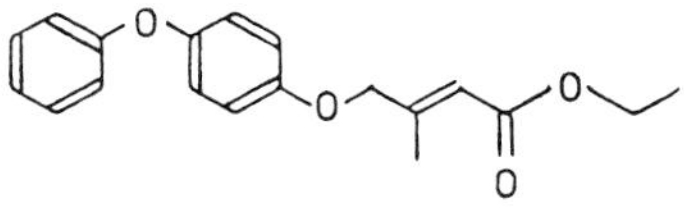 O O O O O

36. 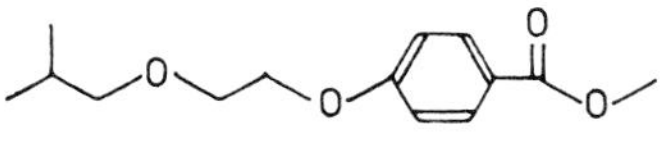 O O O O

37. 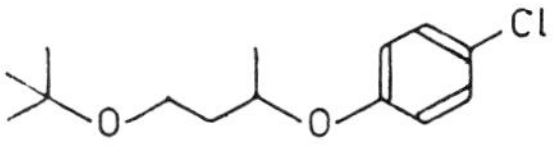 O O Cl

38. 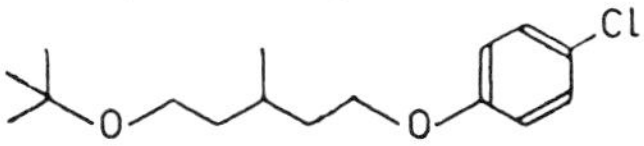 O O Cl

39. 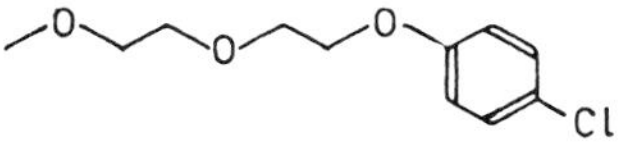 O O O Cl

40. 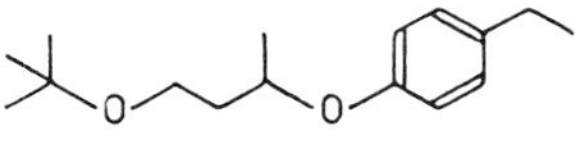 O O

41. 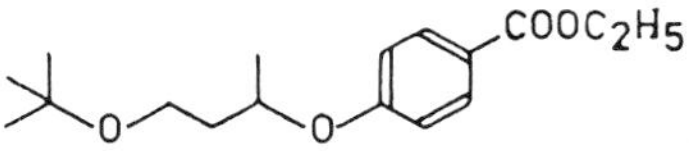 O O $COOC_2H_5$

42. 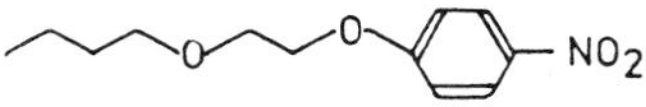 O O NO_2

43. 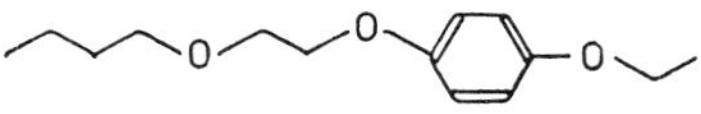 O O O

44. 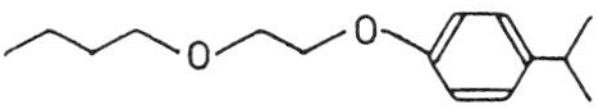 O O

45. 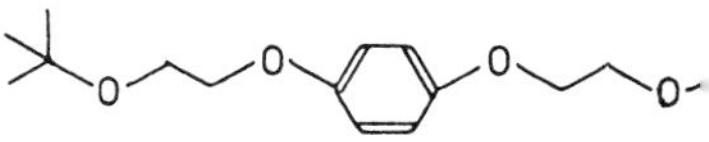 O O O O

46. 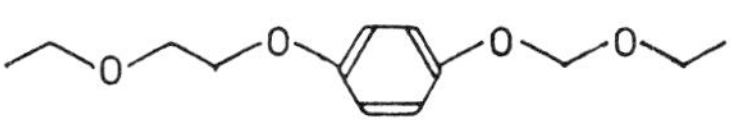 O O O O

47. 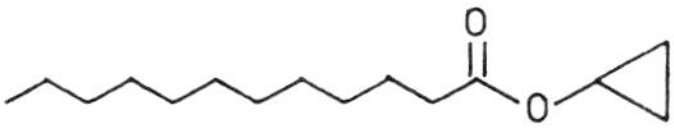 O O

48. 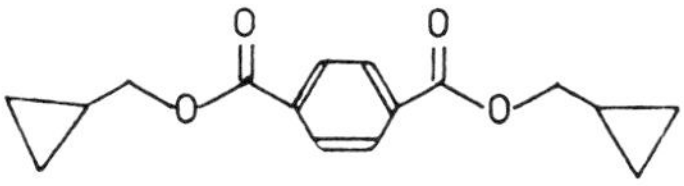 O O O O

49. 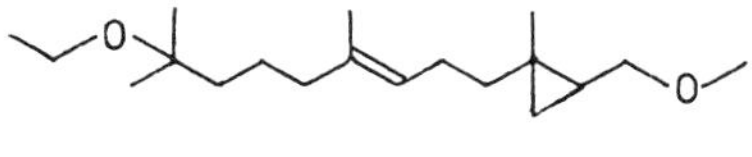 O O

50. 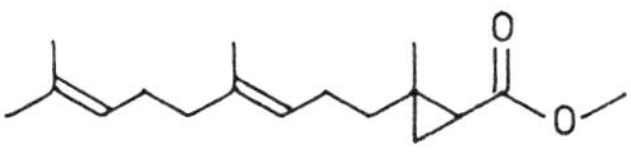 O O

51. 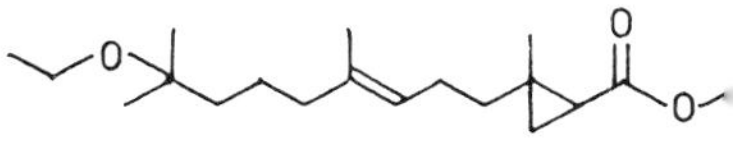 O O O

52. 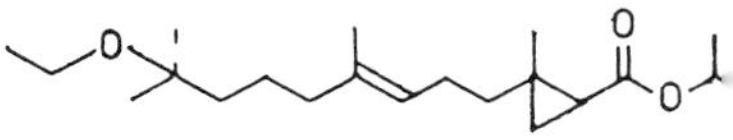 O O O

53. 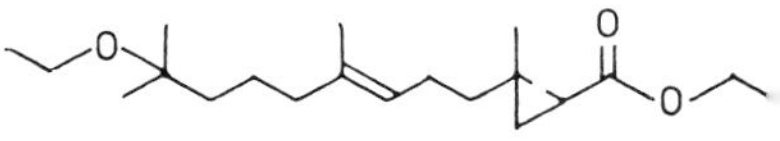 O O O

54.

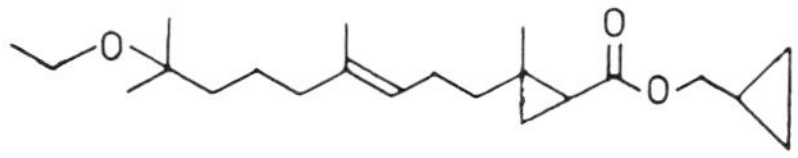

55.

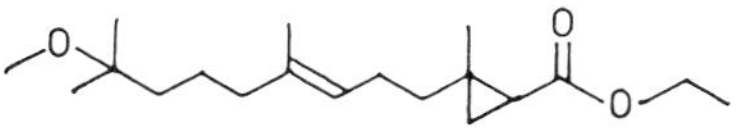

56.

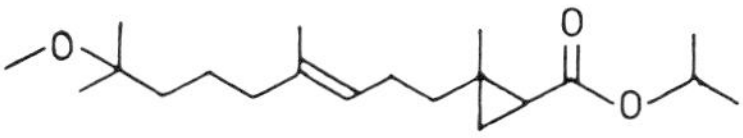

Control Substances

57. 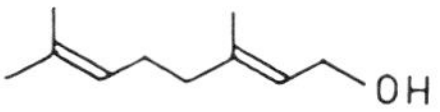 OH Geraniol

58. 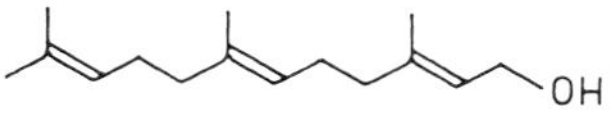 OH Farnesol

59. 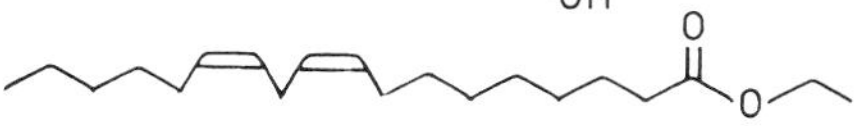 Ethyl linoleate

60. Olive Oil

61. Acetone

RESULTS

The effect of tested JHAs on the survival of *T. putrescentiae* eggs was closely correlated with the age of treated eggs. The most sensitive to JHAs were the youngest eggs (0-24 h old). Older eggs were affected slightly and only by the highest doses of tested JHAs. This discovery allowed us to focus our further study on 0-24 h old eggs.

Microscopic observations revealed that larvae emerge from treated eggs, but they do not ingest food and die a few hours thereafter.

Since JHAs showed different activity irrespective of the age of eggs, we tried to relate the chemical structure to the biological activity of the compounds against *T. putrescentiae*. Results of these studies are presented in Table II. Aliphatic sesquiterpenoidal derivatives of dodecadiene type (1-4) as well as of dodecene type (5-8) proved to be inactive against *T. putrescentiae*, despite the fact that they are considered as highly potent JHAs. The differences of activity of these compounds versus control were not significant (P is greater than 0.05, test U).

TABLE II.
Total Survival Rate of *T. putrescentiae* in Postembryonic Development, Expressed as % of Adults Developed from Eggs Treated with JHAs.

No. of JHA	Percent of Adults					
	Percent Concentration of JHA					
	10	1	0.1	0.01	0.001	0.0001
1	88	91	93	–	–	–
2	18	84	91	90	–	–
3	20	60	95	92	89	94
4	0	46	88	91	–	–
5	40	62	96	–	–	–
6	0	60	74	80	–	92
7	–	63	64	90	–	–
8	–	94	91	–	–	–
9	0	51	60	86	94	91
10	–	30	80	94	–	90
11	–	0	42	89	–	94
12	–	2	38	56	92	88
13	0	8	0	23	64	90
14	0	10	20	46	87	–
14	0	0	10	38	68	76
16	70	82	94	–	–	–
17	–	58	74	92	–	–
18	0	0	0	16	70	94
19	0	0	7	26	–	–
20	–	4	2	20	56	90
21	–	4	10	43	76	88
22	0	9	0	60	–	–
23	0	0	4	23	58	82
24	0	12	22	46	78	94
25	0	4	30	60	96	89
26	10	40	50	70	88	92
27	0	0	0	46	78	90
28	0	0	0	20	76	84
29	0	8	17	10	42	70
30	0	11	0	20	38	60
31	0	2	0	24	58	62
32	10	60	80	92	–	–
33	0	9	15	42	78	84
34	22	68	92	88	–	–
35	0	0	16	23	–	–
36	0	0	4	20	45	62
37	0	12	7	30	–	–
38	0	0	0	18	76	88
39	0	0	0	4	62	78
40	0	12	0	24	88	88
41	0	3	22	50	92	94
42	0	0	10	60	–	–
43	0	0	14	66	–	–
44	7	12	70	76	–	–
45	0	0	8	44	–	–
46	0	0	2	36	67	90

TABLE II. (Cont.)

No. of JHA	Percentage of Adults					
	Percentage Concentration of JHA					
	10	1	0.1	0.01	0.001	0.0001
47	0	0	64	72	89	94
48	12	22	26	48	64	92
49	0	50	80	78	92	96
50	0	64	68	74	91	89
51	0	45	48	64	89	92
52	11	62	78	96	91	88
53	4	64	79	96	88	92
54	0	44	58	80	91	93
55	10	67	92	88	97	90
56	19	74	93	89	95	91
Geraniol	72	84	96	–	–	–
Farnesol	92	90	–	–	–	–
Ethyl linoleate	88	91	–	–	–	–
Olive Oil	89	96	–	–	–	–
Acetone	92	–	–	–	–	–
Normal	93	–	–	–	–	–

Compounds exhibiting effects on the eggs of *T. putrescentiae* may be classified into the following groups:

1) aliphatic oxaterpenoid esters (9-15), of which the most potent at 0.01% conc. caused 80% mortality of the population developing from treated eggs. The activity is enhanced by introduction of the cyclopropyl ester moiety (13, 14)

2) 2,3-methylene farnesol derivatives (49-56), causing 40% mortality of the experimental population at 0.01% conc. (51)

3) cyclopropane carboxylic acid and cyclopropyl methanol derivatives (47, 48), active ovicides and acaricides for *Tetranychus urticae* (Nelson and Show, 1975; Staal *et al.*, 1975). They caused 30-50% reduction of *T. putrescentiae* population at 0.01% conc.

4) oxaterpenoidal *p*-substituted phenyl ethers (18-35), highly potent, reducing 90% of the experimental populations at 0.01% conc.

5) simple oxaaliphatic *p*-substituted phenyl ethers (36-46), very active, reducing 90% of the experimental populations at 0.01% conc.

Activity of JHAs against *T. putrescentiae* is connected with the presence of the cyclopropyl moiety and oxygen in active bonds. JHAs applied topically on the preimaginal stages did not show any significant effect (P is greater than 0.005, test U) in comparison with the controls. Only two JHAs (47, 51) used in 1 and 10% conc. on the tritonymphs caused elongation of this stage from 3-4 days, for the controls, to seven days for experimental populations. Similar effect was noticed with JHAs (33, 35, 36) when applied on 0-24 h old eggs.

Selected compounds (1, 2, 4, 6, 8, 9, 13, 15-18, 33-36, 47-56) and the complete controls used on adults have shown activity, demonstrated by increased mortality of both sexes, decreased fecundity of treated females and lowered survival rate during post-embryonic development. Results are presented in Table III. Compound 18 had the strongest sterilizing effect on females (inhibiting their fecundity to 10% that of the control females) and compounds 33, 47, 13, 15, 16 and 51 lowered the fecundity of treated females to 30% that of the control.

Sterilizing activity of JHAs was not always followed by the ovicidal activity. For example, compound 35, not an active sterilant, had predominant influence on survival of the F_1 generation (only 40% survived), whereas active sterilant 51 did not show any effect on survival of the F_1 generation.

TABLE III.
Mortality and Fecundity of *T. putrescentiae* Adults Treated with 1% Solutions of Selected JHAs.

No. of JHA	% mortality of adults	Fecundity[a]	Total survival of F_1[b]
49	2	227	92
50	5	196	89
51	30	52	90
52	0	250	92
53	0	242	94
54	25	184	78
55	0	238	90
56	1	243	98
1	0	260	94
6	0	198	86
9	0	144	78
15	0	94	67
13	3	108	82
18	10	30	34
8	5	180	98
16	0	240	96
17	20	105	78
33	5	90	78
2	5	250	94
36	100	-	-
47	15	90	76
4	5	245	88
48	10	123	72
35	5	218	40
34	0	245	98
Geraniol	0	247	89
Farnesol	0	262	96
Ethyl linoleate	0	240	91
Olive Oil	0	245	98
Acetone	0	263	92
Normal	0	251 ± 14	94 ± 4.2

[a] Eggs/7/10 days
[b] % of adults developed from eggs deposited by treated females.

The results of short- and long-term contamination of females with JHAs are given in Table IV. Long-term contact of females with active JHAs had significant effect on their fecundity. Females kept for 10 days in contaminated cells produced only 10% (18) to 50% (48) of the number of eggs deposited by females kept for 24 h with the same JHAs. Short-term contamination also inhibited fecundity of treated females. The survival of mites of F_1 generation in the experiments with short-term contamination did not differ from the control, and in the long-term experiments was similar to that indicated in Table III.

TABLE IV.
Fecundity of *T. putrescentiae* Females Treated with 1% Solutions of JHAs.

	Fecundity (eggs / 7 / 10 days)	
No. of JHA	24 h exposure	10-day exposure
49	230	220
50	229	235
51	120	30
52	235	251
53	248	225
54	256	180
55	230	260
56	245	242
47	170	70
48	195	81
13	200	72
18	105	10
Normal	243 ± 17	

DISCUSSION

Our work was designed to find biologically active substances that could have practical potential in the control of *T. putrescentiae,* an important pest of stored products. The literature indicates that JHAs are active not only in insects, but also in other classes of arthropods. Staal *et al.* (1975) and Nelson and Show (1975) used JHAs with cyclopropylmethyl moiety and found an ovicidal and sterilizing effect on *Tetranychus urticae*. JHAs containing this moiety were also active against *T. putrescentiae*. Acetonic solutions of cyclopropylmethyl dodecanoate (47) in 0.01% conc. applied on eggs reduced the population to 72%. Cyclopropylmethyl terephtalate (48) used in the same way reduced the population to 48% (Table II). Both JHAs in 1% conc. also reduced considerably the fecundity of treated females. The derivatives of 2,3-cyclomethylenefarnesol (51, 54) showed ovicidal as well as sterilizing activity in *T. putrescentiae*.

The highest activity in morpho- and oogenesis of *T. putrescentiae* were noticed among three groups of tested JHAs: 1) aliphatic oxaterpenoid ester (9-15); 2) oxaterpenoidal *p*-substituted phenyl ethers (18-35); 3) simple oxaaliphatic *p*-substituted pheyl ethers (36-46).

All of these compounds are highly potent insecticides. Aliphatic alkyl esters, thio- esters or amides (analogs of natural JH) exhibit gonadotropic activity, whereas aromatic JHAs have high morphogenetic and low gonadotropic activity (Sroka *et al.*, 1975).

Many aromatic JHAs (18-46) manifested high activity in *T. putrescentiae*. They reduced the total survival rate of experimental populations to 10% that in the control, as well as fecundity of treated females to 30% in the control, thereby exhibiting a sterilizing action. Moreover, sterilizing activity was closely correlated with the duration of contamination. Females kept 10 days in contaminated cells deposited only 4-30% of the number of eggs produced by control females. In addition, the sterilizing effect was considerably weaker or disappeared after removing animals from the source of contamination (Table IV). Nelson and Show (1975) observed in *T. urticae* a similar effect of transient sterilization caused by temporary contamination with JHAs.

Probably in these cases, the sterilizing effect of JHAs depends upon the stage of oogenesis (mainly vitellogenesis) at which the animals were subjected to the chemicals. Synthetic analogs used in the critical moment of vitellogenesis disturb its natural course. As a result, lowered fecundity of females may be observed (sterilizing effect).

This hypothesis is confirmed by the results of work of Bassal (1974) done on *Hyalomma (H.) dromedarii* (Acarina: Ixodidae), in which the strongest sterilizing effect was obtained within four or more days after the JHAs were applied on young females in the first day of oviposition. The author suggests that it was a result of action of JHAs applied on females prior to ovulation when the vitellogenesis was not yet completed. In the related species, *H. anatolicum excavatum*, ovulation takes place 1-3 days before oviposition (Khalil, 1970). This is in agreement with the results obtained for *H. dromedarii*, where the eggs deposited 1-3 days after JHAs application (vitellogenesis was completed) showed an 80% hatch; the biological effect of JHAs appeared later on, causing 100% mortality of larvae of F_1 generation. The author suggests that this situation results from the disturbance of the permeability of embryonic membranes caused by JHAs. We suppose that *T. putrescentiae* larvae that hatch from contaminated eggs and subsequently perish result from a similar physiological mechanism.

Anatomical and physiological studies of Argasidae and Ixodidae (Aeshlimann, 1968; Wright, 1969; Kitaoka, 1972; Bassal and Roshdy, 1974) indicate a similar mechanism of hormonal regulation of their development. It is possible that as in some insects, juvenile hormones activate the superior neurosecretory system of Acarina to secrete substances stimulating oogenesis and oviposition (Coons *et al.*, 1974), and that the endocrinal regulation is similar among all arthropods.

Biologically active JHAs for different arthropods have a great potential to be used as selective and nontoxic pesticides for the control of stored product pests. The possibility of use of JHAs in combination with conventional pesticides to enhance their activity should also be considered.

SUMMARY

The juvenile hormone analogs tested on *Tyrophagus putrescentiae* had biological but not toxic action. Their activity was demonstrated by ovicidal, morphogenetic and sterilizing action and was closely correlated with their chemical structures.

REFERENCES

Aeshlimann, A. (1968). *Revue Suisse de Zoologie* **75**, 1033-1039.
Bassal, T. T. M. (1974). *Z. Parasitenk.* **45**, 85-89.
Bassal, T. T. M. and Roshdy, M. A. (1974). *Exp. Parasitol.* **36**, 34-39.
Boczek, J. (1954). *Ekol. Pol.* **2**, 473-476.
Coons, L. B., Roshdy, M. A. and Axtell, R. C. (1974). *J. Parasitol.* **60**, 687-698.
Czaja-Topinska, J. and Stepien, Z. (1975). *Zesz. Problem. Postepów Nauk Roln.* **171**, 269-276.
Khalil, G. M. (1970). *J. Parasitol.* **56**, 596-610.
Kitaoka, S. (1972). *Abstr. 14th Intern. Congr. of Entomol.*, Canberra, August 22-30, 1972, 272.
Levinson, H. Z. (1975). *Naturwissenschaften* **62**, 272-282.
Nelson, R. D. and Show, E. D. (1975). *J. Econ. Entomol.* **68**, 261-266.
Roshdy, M. A., Shoukrey, N. M. K. and Coons, L. B. (1973). *J. Parasitol.* **59**, 540-544.
Siegel, S. (1956). "Nonparametric statistics for the behavioral sciences." McGraw-Hill, New York.
Sroka, P., Barth, R. H., Gilbert, L. J. and Staal, G. B. (1975). *J. Insect, Physiol.* **21**, 463-469.
Staal, G. B., Ludvik, G. F., Nassar, S. G., Henrick, C. A. and Willy, W. F. (1975). *J. Econ. Entomol.* **69**, 91-95.
Wright, J. E. (1969). *Science* **163**, 390-391

THE ALARM PHEROMONE OF GRAIN MITES AND ITS ANTIFUNGAL EFFECT

Katsuhiko Matsumoto, Yoshitake Wada
and Masako Okamoto

Department of Parasitology
Tokyo Women's Medical College
Tokyo, Japan

INTRODUCTION

Kuwahara *et al.* (1975) found an alarm pheromone in the body fluid of crushed cheese mites, *Tyrophagus putrescentiae,* which acted as a natural repellent, and identified this substance as neryl formate. Though some species of bees and ants are noted for having an alarm pheromone, this finding is the first case in grain mites.

It was important to investigate other mites for the alarm pheromone and to clarify its biological significance. Here, we will discuss the alarm pheromone in other grain mites, their sensitivity to it, and its effect.

MATERIALS AND METHODS

The Repellent Action of the Alarm Pheromone in Several Species of Grain Mites

The following five species of mites were used: *Tyrophagus putrescentiae, Aleuroglyphus ovatus, Lardoglyphus konoi, Carpoglyphus lactis,* and *Dermatophagoides farinae.* All of these were bred in our laboratory. Each species of mites was isolated from the culture medium by the saturated saline floatation method (Matsumoto, 1965), washed thoroughly with tap water, and kept at room temperature at least two hours to remove moisture from body surface.

Neryl formate was kindly supplied by Dr. Y. Kuwahara, Tsukuba University. Citral, composed of neral (35%) and geranial (63%), was commercially available (Tokyo Kasei).

Vol. I: ISBN 0-12-592201-9

The bioassay of sensitivity of the mites to the alarm pheromone was as follows. An appropriate number of mites was transfered and scattered evenly on a black cover slip (24x36mm). After 30 minutes when the mites were evenly distributed on the surface, a small piece of filter paper (5x5mm), impregnated with the body fluid from about 100 crushed mites, was placed in the center of the plate. Observations of the mites moving away from the filter paper were made for about 10 minutes. To observe the sensitivity of mites to neryl formate, a small piece of filter paper impregnated with 0.05 ml of each of the five concentrations of hexane solution used in this study was placed on the cover slip covered with mites. The experiments were carried out in this manner in room temperature.

The Antifungal Effect of the Alarm Pheromone, Neryl Formate and Citral

Aspergillus fumigatus, supplied by Dr. H. Nakano, Department of Microbiology, Tokyo Women's Medical College, was used in these experiments. *A. fumigatus*, cultured in a Sabouraud medium for more than 10 days at 28°C and suspended in physiological saline, was used for assay purposes.

The alarm pheromone was extracted with hexane from 20 gm wet weight of *C. lactis,* stored in -80°C and concentrated up to 1 ml.

Bioassay of antifungal effect:

A. *So-Called "Gas Method." A. fumigatus* was inoculated over the whole surface of the medium in petri dishes (90mmx20mm). These petri dishes were turned upside down on paper discs impregnated with 0.04 ml of citral hexane solution placed in the center of the lid of each dish. The growth rate for this fungus was observed at 28°C every 24 hrs for seven days.

B. *So-Called "Modified Gas Method." A. fumigatus* was inoculated in the center of the medium plates the size of a pin point. These petri dishes were turned upside down on paper discs impregnated with 0.04 ml of each alarm pheromone, citral and neryl formate, in hexane solution placed in the center of each lid as in the "Gas Method." The growth rate of the fungus at 28°C was observed by measuring the diameter of the colony every 24 hrs for three days.

RESULTS AND DISCUSSION

The Alarm Pheromone in Five Species of Grain Mites

The reactivity to neryl formate is shown in Table I. Three species of mites, *L. konoi, T. putrescentiae,* and *C. lactis* had a high sensitivity to neryl for-

TABLE I.
Repellent Reaction of Grain Mites to Neryl Formate.

Mite Species	Concentration of Neryl Formate (ppm)				
	1	10	100	1,000	10,000
Tyrophagus putrescentiae	–	–	+	+	
Aleuroglyphus ovatus	–	–	–	+	
Lardoglyphus konoi	+	+	+ +	+ +	
Carpoglyphus lactis	–	–	±	+	
Dermatophagoides farinae	–	–	–	–	+

+ + totally repelled; + almost totally repelled
± partially repelled; – unaffected

TABLE II.
Repellent Reaction of Grain Mites *T. putrescentiae, A. ovatus, L. konoi, C. lactis,* and *D. farinae* to the Body Fluid of Crushed Mites (about 100 per paper) and a Hexane Extract (1 gm of mites per 2 ml).

Reactant	Reactor				
	T.p.	*A. o.*	*L. k.*	*C. l.*	*D. f.*
T. p.	+(+)	–	+	±	–
A. o.	–(–)	–(–)	+	+(±)	–
L. k	–(–)	–(–)	+	–(–)	–
C. l.	–(±)	–(±)	+	+(+)	–
D. f.	–	–	+	+	–

( ) Results with hexane extract
+ + totally repelled; + almost totally repelled
± partially repelled; – unaffected

mate; *L. konoi* was even repelled from the filter paper impregnated with 1 ppm of neryl formate. The sensitivity of *A. ovatus* was low but it did react to 1,000 ppm. *D. farinae* had a very low sensitivity and only reacted to 10,000 ppm of neryl formate. The sensitivity of the mites to neryl formate decreased in the following order: *L. konoi, T. putrescentiae, C. lactis, A. ovatus,* and *D. farinae.*

Table II shows the repellent reaction of mites to the body fluid of each species. Three species, *L. konoi, C. lactis,* and *T. putrescentiae* reacted sensitively to their own species fluid but the other two, *A. ovatus* and *D. farinae,* did not. *L. konoi* reacted to the fluid of all five species. Even though *D. farinae* did not react to the same species, it was a repellent for other species. We therefore conclude that *D. farinae* has the alarm pheromone. According to the number of reactants from which each species were repelled, we believe the order of sensitivity decreases as follows: *L. konoi, C. lactis, T. putrescentiae, A. ovatus,* and *D. farinae.* However, the quality of the alarm pheromone contained in each species was not considered. Comparing the order of sensitivity to neryl formate with that of crushed mites, there was a contradiction in the order between *T. putrescentiae* and *C. lactis.* To find the reason for this

TABLE III.
Repellant Reaction of Grain Mites to Citral in Hexane.

	Concentration of Citral (ppm)				
Mite Species	0	1	10	100	1,000
Tyrophagus putrescentiae	–	–	–	±	+
Aleuroglyphus ovatus	–	–	–	–	+
Lardoglyphus konoi	±	+	+	+ +	+ +
Carpoglyphus lactis	–	–	–	±	+
Dermatophagoides farinae	–	–	–	–	+

+ + totally repelled; + almost totally repelled;
± partially repelled; – unaffected.

contradiction, each hexane extract of the mites was gas chromatographed by Kuwahara (1976). From this study, it was concluded that four species had citral as the alarm pheromone whereas *T. putrescentiae* did not.

Table III shows the results of the bioassay of the commercially available citral. Among the five species of mites, *L. konoi* was most sensitive to both citral and neryl formate. *D. farinae* and *A. ovatus* were so insensitive that they responded only to 1000 ppm. *T. putrescentiae* and *C. lactis* each showed the same level of sensitivity to 100 ppm of citral. Neryl formate and citral acted as an alarm for *L. konoi, T. putrescentiae,* and *C. lactis* as they responded highly to them. On the other hand, in the case of *A. ovatus* and *D. farinae*, citral is more likely to be a defensive substance as stated by Wilson (1971).

The Antifungal Effect of the Alarm Pheromones

The effect of the alarm pheromones, neryl formate and citral, on grain mites and their ecosystem must be discussed. Cole *et al.* (1975) mentioned the antifungal property of citral. Since then, we have observed that when a large population of mites was present, there was no growth of fungi in the mites' diet, but that there were fungi growing when mites were absent. There are a few reports that some grain mites eat fungi (Solomon, 1943; Nakano and Matsumura, 1935). This phenomenon may be one of the reasons that fungi could not develop in the diet with large numbers of mites present. But the antifungal effects of the alarm pheromone in the mite must not be neglected.

The diet (dried yeast powder) with a large number of *A. ovatus* and the diet without *A. ovatus*, were each inoculated with *A. fumigatus* and were observed for three days of incubation at 25°C. In the diet without mites *A. fumigatus* developed, but in the diet with mites it did not. Ten grams of *C. lactis* were crushed with 10 ml of water and were mixed into a Sabouraud medium. The final concentration of this medium was regulated as a standard. *A. fumigatus* growth in the medium with the mites was inhibited as compared with that in the medium without mites.

"The Cup Method," which is a common method to test the antimocrobial effect, was used at the beginning of this study. The results did not reveal the linear proportional relations between the concentration of citral and the diameter of the inhibition zone. It might be suspected that the gas from citral had affected the fungus rather than the liquid in the medium.

The antifungal effect of evaporated gas from citral was observed by the so-called "Gas Method." The results are shown in Table IV. In the control medium, *A. fumigatus* grew on the whole surface after 24 hrs of incubation, while with 10 mg/ml and 25 mg/ml of citral, growth of the fungus was significantly inhibited in the center of the plate. With 50 mg/ml, growth was inhibited completely for two days and even up to six days with 100 mg/ml. These inhibitions of fungal growth were caused by the evaporated gas from citral impregnated in the paper disc. When the paper disc was removed from the petri dish, *A. fumigatus* started growing within 24 hrs. This revealed that the antifungal effect of citral in these concentrations was fungistatic.

TABLE IV.
The Antifungal Effect of Citral on *Aspergillus fumigatus* Assayed by the "Gas Method" (0.04 ml of Citral Hexane Solution Impregnated in a Paper Disk).

Concentration of Citral	Incubation Period			
(mg/ml of Hexane)	1	3	5	7
0	-	-	-	-
10	+	-	-	-
25	+	-	-	-
50	+ +	+	-	-
100	+ +	+ +	+ +	-
Control	-	-	-	-

+ + total inhibition; + partial inhibition; - no inhibition.

In order to indicate the antifungal effect of lower concentration of citral quantitatively, we developed a new method and called it "The Modified Gas Method." As shown in Fig. 1, when a high concentration of citral was impregnated in the paper disc, a slow growth rate of the fungus colony was clearly observed.

We observed the effect of neryl formate to *A. fumigatus* by the same method. Neryl formate had an antifungal effect because of the gas. But this effect was much lower than that of citral.

As shown in Fig. 2, the antifungal effect of *C. lactis* was tested in the same manner. This extract revealed the same potency of antifungal effect as 4.8 mg/ml of citral.

SUMMARY

Kuwahara *et al.* (1975) reported that cheese mites, *T. putrescentiae,* had the alarm pheromone, neryl formate. We made it clear that other related mites, *C. lactis, L. konoi, A. ovatus,* and *D. farinae* were responsive to neryl formate and that they too had an alarm pheromone. From the reaction of other related

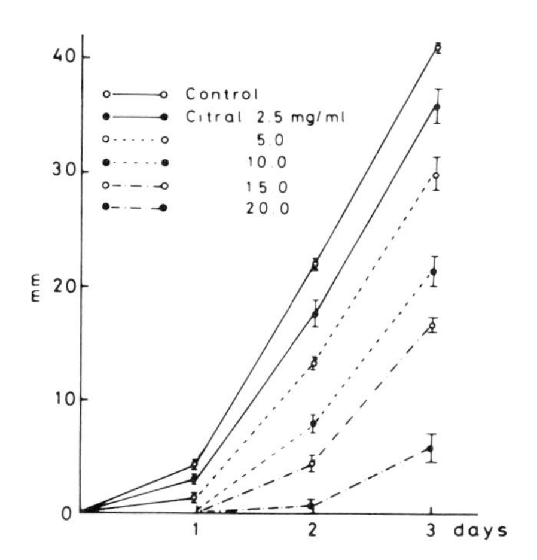

Fig. 1. Growth inhibition of *Aspergillus fumigatus* by citral with "Modified Gas Method." Diameter in mm of colonies developing from spot inoculation, incubation at 28°C. (90% confidence intervals indicated).

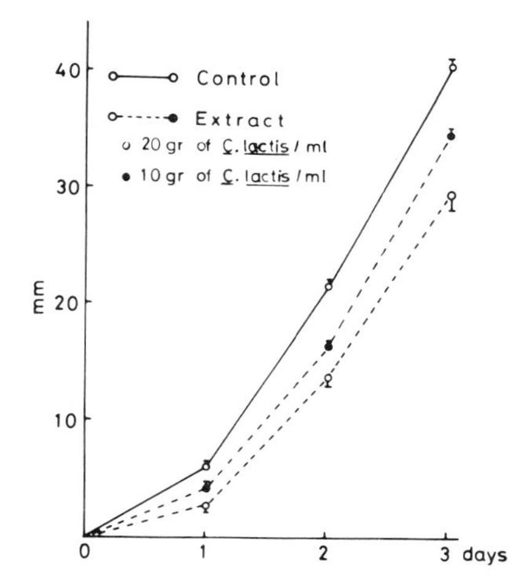

Fig. 2. Growth inhibition of *Aspergillus fumigatus* by hexane extract of *Carpoglyphus lactis* with "Modified Gas Method." Diameter in mm of colonies developing from spot inoculation, incubation at 28°C. (90% confidence intervals indicated).

mites to neryl formate and to the body fluid of each other species, we suspected this alarm pheromone to be another substance. Kuwahara proved this alarm pheromone was citral (1976).

In order to find the role citral and neryl formate play in the ecology of mites, their antifungal effects were observed. Citral inhibited the growth of *Aspergillus fumigatus* by its gas; the effect was fungistatic. Neryl formate also had a fungistatic effect but it was much less than citral.

We developed a new technique called "The Modified Gas Method" which may be used for the bioassay of antifungal effects of volatile substances. The hexane extract of 20 gm of *C. lactis* had the same antifungal effect as 4.8 mg. of citral.

REFERENCES

Cole, L., Blum, M. and Roncadori. (1975). *Mycologia* **67**, 701-708.

Kuwahara, Y. (1976). *Proceeding of a Symposium on Insect Pheromones and Their Applications, p. 65-75.* Agriculture, Forestry and Fisheries Research Council, Ministry of Agriculture and Forestry, Japan.

Kuwahara, Y., Ishii, S. and Fukami, H. (1975). *Experientia* **31/10,** 1151-1161.

Matsumoto, K. (1965). *Jap. J. Sanit. Zool.* **16**(1), 86-89.

Nakano, M. and Matsumura, S. (1953). *Shokuryoken Hōkoku* **8,** 155-162.

Solomon, M. E. (1943). "Tyroglyphoid mites in stored products. 1. A survey of published information." Dept. of Scientific and Industrial Research, London. p. 36.

Wilson, E. O. (1971). "The Insect Societies." The Belknap Press of Harvard University Press, Cambridge. p. 234.

ALLELOCHEMIC EFFECTS OF SOME FLAVORING COMPONENTS ON THE ACARID, *TYROPHAGUS PUTRESCENTIAE*

J. G.Rodriguez, M. F. Potts, and C. G. Patterson

Department of Entomology
University of Kentucky
Lexington, Kentucky

INTRODUCTION

The acarid mite, *Tyrophagus putrescentiae* (Schrank) is a cosmopolitan species encountered infesting many food stuffs and stored products. It is a pest in food or feed processing plants, and in some cases it reaches the consumer's food stuffs via the grocery store. Dried fruits and fish, cheeses, livestock feed concentrates, pet foods, nut and seeds are some examples of commodities that may be infested by this species. It is a mite that can live in association with some fungi and actually showed some tolerance to such mycotoxins as aflatoxin B_1, citrinin, ochratoxin A, and penicillic acid when these were incorporated into an axenic diet (Rodriguez, unpublished data). In fact, it was the observation that in some cases the fungi ostensibly interacted to produce a phagostimulatory (kairomone) effect while in other cases the interaction produced a deterrent (allomone) effect that led us to the question of allelochemics influencing the physiology/behavior of *T. putrescentiae*. It should be noted that in our continuing nutritional work with this species we have always been cognizant of the possibilities for a nutritional approach to control of this potential pest in so many constituted foods. It is precisely this type of food/pest interaction that lends itself to innovative control strategies of the pest (Pratt *et al.*, 1972; Rodriguez, 1972). If one accepts the fact that *T. putrescentiae* is a species that can feed/infest many foods of plant origin or plant parts, then it is reasonable to expect plant-produced allelochemics to function as kairomones or allomones to this mite. For the purpose of this study, however, we were mainly interested in focusing on the allelochemics that function as allomones either as feeding deterrents or as general inhibitors of growth/development.

Hence, the general objectives before us were to investigate the possible allelochemic effects, both in feeding and non-feeding experiments, of dried

Vol. I: ISBN 0-12-592201-9

spices and flavoring oils on *T. putrescentiae.*

INHIBITION OF *T. PUTRESCENTIAE* IN PET FOODS FORMULATED WITH DRIED SPICES AND HERBS

Procedures

Thirty experimental formulations of soft moist pet food, with commercially available spices and herbs added at 2%, were offered to *T. putrescentiae* in preference tests in an incompletely randomized arrangement. Free choice of 3C samples simultaneously was not practical in our experimental set-up; therefore, the samples were presented to the mites in groups of five, with controls. The groups of five were obtained by first randomizing the 30 samples, then arbitrarily counting one through five down the list. In the first test, six different randomizations were thus divided. Small trays containing si: micro-cups were filled with the pet food formulations, each cup containing a different one. The trays were ringed with Tree Tanglefoot® at the outside edge to confine the mites, which were released at the center of the tray and allowed free choice of the six formulations. Fifty adult females were released in each micro-tray and the resulting progeny counted after a 28-day developmental period. The results indicate not only the mites' preference within the six choices offered in each group, but also any chronic effect, i.e. either beneficial or deleterious to mite growth.

From the results of the first test, nine "preferred" and nine "non preferred" formulations were selected for further testing.

Results

The nine preferred spices and herbs selected were cayenne, cumin, dillweed ginger, marjoram, ground mustard, paprika, poppy seed, and sesame seed When these were evaluated against each other, large populations develope (Table I). Most of these spice and herb formulations contained large populations than did the standard pet food. When the counts in the individua micro-trays were examined, it was seen that the standard pet food was in variably among the least favored choices, indicating either that the spice an herb formulations were more freely chosen, or permitted better developmer either directly or through a phagostimulatory effect.

The non-preferred spices and herbs from the first test were curry powde lemon and pepper seasoning, mace, mint, oregano, poultry seasoning pumpkin pie spice, sage and thyme. In feeding tests against each other of th group, very small mite populations developed (Table I). In the 12 individua micro-trays, the standard was the first choice 11 times.

A test in which the combined preferred and non-preferred spices and herb were tested together resulted in the data shown in Table II. On a tota

TABLE I.
Progeny of *T. putrescentiae* Developing in 28 Days on 2% Spice Pet Food Formulations in a Free-Choice Test. Preferred and Non-Preferred Spice Groups Were Offered Separately.

Preferred Spice	Total Progeny	Non-Preferred Spice	Total Progeny
Cayenne	10,850	Curry powder	76
Cumin	13,750	Lemon and pepper seasoning	69
Dillweed	16,250	Mace	74
Ginger	9,650	Mint	221
Marjoram	18,000	Oregano	240
Ground mustard	5,750	Poultry seasoning	73
Paprika	7,650	Pumpkin pie spice	379
Poppy seed	14,950	Sage	113
Sesame seed	16,050	Thyme	316
TOTAL	112,900	TOTAL	1,484
Standard pet food	6,700	Standard pet food	1,105

population basis, 2.5 times as many mites developed on the preferred as on the non-preferred formulations. On an individual basis, there was some overlapping in population sizes between the preferred and non-preferred groups.

The total populations developing on the preferred formulations tested against each other (Table I) in 28 days were much higher than those developing on the preferred formulations tested against the non-preferred in 21 days.

INHIBITION OF *T. PUTRESCENTIAE* BY FLAVORING OILS INCORPORATED INTO CASEIN-WHEAT GERM-AGAR DIET

Procedures

Flavoring oils, all permitted in food products under current FDA regulations, were incorporated into the standard casein-wheat germ-agar diet used to rear acarid mites in our laboratory (Rodriguez, 1972). The oils (furnished by Fritzsche, Dodge and Olcott, Inc., New York) were: Oil Almond Bitter F. C. C. FFPA Imported; Oil Clove F. C. C.; Oil Mace F. C. C. East Indian Select; Oil Pepper Black F. C. C. Extra; Oleoresin Black Pepper Indian; Oil Onion F. C. C. Imported Extra; Oil Sage Dalmatian F. C. C. Extra; Vaniprox® Base; Oil Sassafras Artificial 29125 (contains 42.16% methyl salicylate); Oil Orange F. C. C. Sweet Valencia; Oil Lemon Cold Pressed WONF 703920; Oil Lemon F. C. C. Citrus Valley® Brand.

The oils were dissolved in alcohol (except for onion, which was dissolved in ethyl ether) and emulsified with Tween 80 and added to the sterilized diet as it was cooling, at percentages ranging downward from 0.5% to 0.001% (v/v). Each feeding vial contained 3 ml of diet and was inoculated with 100 *T. putrescentiae* larvae, and five vials (replicates) per treatment. Progeny were counted at 7, 14, 21 and 28 days and the number of eggs estimated.

Results

Initial tests made with the flavoring oils at 0.5%, 0.25% and 0.1% resulted in total inhibition of mite populations by all the oils except the orange and the two formulations of lemon oil. These permitted limited reproduction.

In the second test the oils were added at 0.1%, 0.01% and 0.001% (Fig. 1).

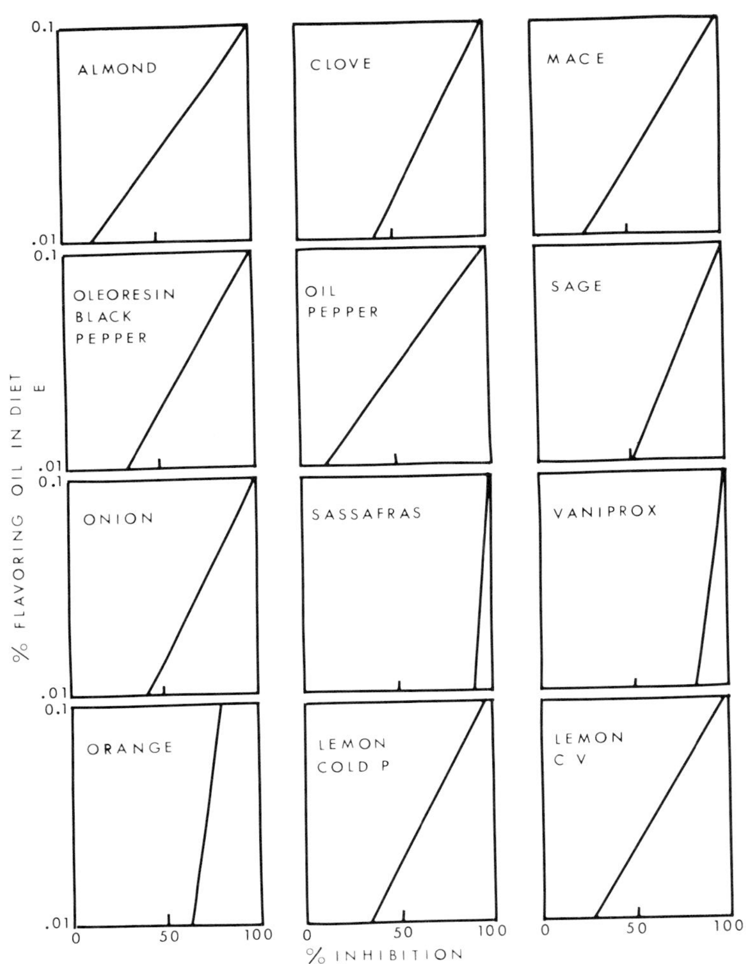

Fig. 1. Effect of flavoring oils incorporated into casein-wheat germ-agar diet on *Tyrophagus putrescentiae*. Results reflect chronic effect of oils, at 0.01% and 0.1%, as percent inhibition of mite population after 28 days.

Again, the 0.1% level resulted in total inhibition by all except the citrus oils, while at 0.01% the inhibition was less than 50% in most cases, the exceptions being sage, sassafras, vaniprox and orange. The 0.001% level, not shown on Fig. 1, resulted in partial inhibition, ranging from 0% for clove to 95.7% for vaniprox.

EFFECTS OF FLAVORING OILS ADDED TO PET FOOD

Procedures

Flavoring oils, minus the citrus oils, in a 1:1 mixture of alcohol and water were added to soft moist pet food already formulated, at levels from 0.1% down to 0.00001%. Preference tests (as described earlier) were done, presenting the various oils against each other at the same level. No-choice feeding tests were also conducted, in order to study possible chronic effects in the mites. For these, small plastic microcups were filled with the food, 10 female mites introduced into each cup and the whole covered with Parafilm M®, with 10 cups per treatment. Progeny were counted at 21 days.

Results

In the feeding tests, all the spice oils inhibited the mite populations. At 0.1%, inhibition ranged from 76.0% for oleoresin black pepper to 92% for clove; at 0.01%, from 82.4% for almond to 96.0% for sassafras; at 0.001%, from 74.2% for oil of black pepper to 94.2% for sassafras. Inhibitory effects were also found at levels of oils ranging downward from 0.001%, with no real differences among the oils.

In the preference tests, the only significant result was that at 0.001% almond oil was strikingly preferred over all the other spice oils.

PET FOOD PATTY PROTECTION THROUGH ALLOMONES

Assuming that even the best patty wrap of commercial pet foods would allow some mites to find entry, it might be technically possible to protect the pet food by providing a deterrent against entry. A paper disk, treated with the allomone, placed next to pet food interfacing with the patty seal would be an ideal method to provide such protection.

The specific objective of this study was to evaluate such possible allomone effects from natural flavoring oils. This was done by utilizing a laboratory formulated pet food patty and by simulation of a patty technique to be described later.

TABLE II.
Progeny of *T. putrescentiae* Developing in 21 Days on 2% Spice Pet Food Formulations in a Free-Choice Test. Preferred and Non-Preferred Spices Were Offered at Random in the Same Test.

Preferred Spice	Total Progeny	Non-Preferred Spice	Total Progeny
Cayenne	260	Curry powder	184
Cumin	589	Lemon and pepper seasoning	108
Dillweed	445	Mace	50
Ginger	314	Mint	288
Marjoram	279	Oregano	118
Ground mustard	160	Poultry seasoning	215
Paprika	64	Pumpkin pie spice	91
Poppy seed	654	Sage	10
Sesame seed	272	Thyme	164
TOTAL	3,047	TOTAL	1,228
Standard pet food	457		(457)

Procedures

Patty Protection Tests. An extruded soft moist dog food product in patty form, wrapped in cellophane, was made available to us by a pet food manufacturer. The following natural oils and related allomones were evaluated: Oleoresin Black Pepper Not Decolorized; Superesin Clove and Superesin Cinnamon (each containing both Oleoresin and Oil); Superesin Sage (containing Oleoresin Sage and Oil Sage Dalmatian); Oil Nutmeg F. C. C. East Indian Extra; Oil Peppermint Naural F. B.; Oil Peppermint F. C. C. Twice Rectified F. S. 7 Co.™; Oil Spearmint F. C. C. Natural; Oil Lemon F. C. C. Citrus Valley® Brand; Oil Orange F. C. C. Citrus Valley® Brand; Citral; and Neryl Formate (the latter two were purchased from ICN Pharmaceuticals, Inc., Plainview, N. Y.). The oils were applied, in 95% ethanol solution, to filter paper disks 70 mm in dia., in two concentrations, at 0.6 ml per disk. After evaporation of the solution, the disks were individually quick dipped in paraffin heated to 135°C (Histowax® Granular, MP 50-52°C, Matheson, Coleman and Bell, Norwood, Ohio). The treated disks were then inserted into the top edge of the patty by slitting halfway around the wrap, after which the cellophane was sealed with transparent tape. A slit *ca* 1 cm long was made on the top center of the patty and these were then placed (3 x 3) on the floor of a refrigerator crisper (25.4 x 35.6 cm) ringed on the inside top edge with Tanglefoot. Nine patties, each treated with a different flavoring oil, were thus tested, in each of five crispers (replicates). *T. putrescentiae* were introduced in inoculum in two dishes (35 mm dia.) placed on a wire mesh over the patties. The crispers, the tops covered with plastic film, were maintained at 27°C and 80% r.h.

Petri Dish Assay. Extruded pet food, 3.2 g, was placed in plastic petri dishes, 35 x 10 mm, and a 35 mm filter paper disk, treated in the same manner as previously described (except that 0.5 ml of solution was applied to each disk) was placed on top of the extruded food. The petri dishes were covered and randomized in a circle around a central source of mite inoculum. Two experimental designs were employed: 1) to test the flavoring oils against each other at the same concentration, and 2) to test each oil alone in a range of concentrations.

Results

Patty Protection Test. This experiment was terminated five days after it was set up. The results are given in Table III. Essentially the same strong allomone effects were demonstrated by citral and neryl formate, and to a lesser extent by spearmint. The remainder of the oils did not deter the mites from entering the patty wrap. It should be noted here that the patty wrap was not entry-proof from the sealed bottom.

Petri Dish Assay. The results of this study were quite similar to those of the patty protection test (Table IV). After seven days of population pressure the petri dishes were examined for mites. The results indicted that citral and neryl formate are relatively effective allomones at 0.1% and increased in effectiveness to 0.5%, but the differences were not significant. Spearmint and peppermint were effective only fleetingly; after 24 hr they lost their allomone effect.

GENERAL DISCUSSION

It should be noted that various types of products are encountered in the flavor industry, e.g., whole spice, ground spice, tinctures, extracts, oleoresins, essential oils and synthetics (as opposed to naturals) are the main types. The basic material from which the flavoring originates is the plant or plant part

TABLE III.
Pet Food Patty Protection Effect of Natural Flavoring Oils and Related Allomones Introduced via Treated Filter Paper Disks against *T. putrescentiae.*[a]

	Mite Nos. / % Concentration	
Oil	0.25	0.5
Blank	67 a	102 a
Oleoresin pepper	65 a	71 a
Clove	54 a	100 a
Nutmeg	53 a	92 a
Lemon C V	38 a	60 a
Cinnamon	36 a	94 a
Spearmint	13 b	28 ab
Citral	10 b	1 b
Neryl formate	5 b	3 b

[a] Patties with slitted wraps enclosing treated paper disks were exposed for 4 days to high mite population pressure; counts reflect mean number of mites (5 replicates) found inside patty wrap. Means followed by the same letter are not significantly different from each other.

from which the dried spice is made. The dried spice contains the flavor components, which can be volatile and nonvolatile organic compounds. Generally, volatile oils/essential oils are vital flavor components. For instance, the essential oils of cinnamon and clove buds comprise almost the total flavor of the dried spice. In those spices where the flavor components are generally nonvolatile, an oleoresin is formulated which is an extraction of the ground spice combined with a highly volatile selective solvent. This is further processed to remove the solvent, leaving mainly nonvolatiles, volatiles, and certain nonflavor components. Further purification/concentration follows (Rogers, 1966).

As was pointed out, *T. putrescentiae* is a species of varied food habits. It is its herbivore characteristics that concerned us in these studies, for in our rationale we aimed at elucidating the possible role of allelochemics in a behavioral or physiological reaction to the mite. Hence, we set out to incorporate spices or their components to food to study any chronic, beneficial/detrimental, effects to the mite. We considered a 21-28 day feeding test long enough to manifest a chronic effect in this species. In order to focus on the behavioral aspects, our procedures included "free-choice" and "no-choice" tests, as well as short term feeding tests (1-7 days). Thus, it was our intention to elucidate interactions in which the spice or component might be a phagostimulant/attractant (kairomone) or interactions in which the compound might be a feeding deterrent/repellent (allomone).

When whole and ground dried spices were incorporated into pet food and offered to mites in free-choice long-term feeding tests, it was evident that some spices produced kairomone or allomone effects, as compared with the untreated pet food (Tables I and II). When a previously tested group of spices had been categorized into preferred and non-preferred and each of these groups were tested within its group, the preferred group for the most part attracted and sustained a mite population higher than the control pet food, while in the non-preferred group, all materials showed allomone effects (Table I). Curry powder and lemon/pepper seasoning, mace and sage were particularly outstanding. Again in a free-choice test where both preferred and non-preferred groups of spices were combined and tested in a randomly distributed fashion, the results pointed out convincingly that the mites could sort out the spices producing the allomone effect (Table II). The lower populations occurring in the latter test (Table II) compared to the former (Table I) is partly explained by a shorter developmental time, i.e., 21 days vs. 28 days.

The study in which flavoring oils were added to a meridic diet, and presented to mites in a no-choice long-term feeding test, demonstrated the effectiveness of flavoring oils and the sensitivity of such a procedure. Of the 12 oils tested, all but Oil Sassafras Artificial 29125 were natural oils (Fig. 1). Vaniprox Base strongly inhibited mite growth/development at 0.001%. The 0.01% level resulted in total mite inhibition by all oils except the citrus oils. It would be reasonable to expect the agar meridic diet to produce more consistent results upon challenging the mites with flavoring oils in the diet. Another benefit from this type of procedure is that relatively small quantities of flavoring oil are used.

It follows that the next step was to ascertain the effectiveness of flavoring oil in pet food. First, this was done by challenging the mites in a long-term free-choice test that excluded the citrus oils, at a concentration range of 0.00001% to 0.1%. Results from this study were conclusive only to the extent that almond oil at 0.001% was highly preferred over the remainder of the flavoring oils. In a follow-up study, these same formulations were presented to the mites in a no-choice long-term feeding test. Results indicated that the oils were generally inhibitory. The chronic effects were deleterious at 0.001% for oil of black pepper and sassafras, giving about 74% and 94% inhibition, respectively. However, all the flavoring oils inhibited growth/development, indicating that in no-choice tests, the chronic effects from the allelochemics manifested themselves. It would be highly speculative to attempt to explain the metabolic basis for this allelochemics toxicity.

The toxic properties of black pepper extracts to the rice and cowpea weevils have been reported by Su (1977). Citrus oils were also shown to be toxic to the cowpea weevil and prevented these beetles from developing in diets containing 0.1% of peel oils from lemon or grapefruit (Su *et al.*, 1972) and the toxic component was later isolated (Su, 1976). These are but examples of allelochemic effects originating from natural products on stored product insects.

As was noted previously, *T. putrescentiae* can exist in association with fungi and some of their byproducts (Rodriguez, unpublished). In some cases we have observed unexplained allomone- or kairomone-like phenomena with fungi/mite associations. The report of Cole *et al.* (1975) detailing the antifungal properties of citral, an essential oil of citrus peel and the flavoring oil from lemon press extract, caused us to reflect on a possible mite interaction. Concurrently with this, Kuwahara *et al.* (1975) discovered neryl formate as having alarm pheromone properties on *T. putrescentiae*. According to the authors, the discovery of the natural occurrence of neryl formate in animals was new. Since citral is known to act as an alarm pheromone in a bee and an ant species (Blum, 1969) it seemed indicated to investigate the allelochemic effects of these terpenoids.

Having observed the effects of flavoring oils when incorporated into mite food, either meridic diet or pet food, it was indicated that flavoring oils, components, or related allomones be studied via another route, thus, the protection of the pet food patty evolved. The treated paper disk, protected further against quick chemical "fading time" by coating it with paraffin, provided interesting results (Table III). Under strong mite population pressure from the outside, the mites could find entry through a slit on top of the patty wrap, and through a not too secure wrap seal on the bottom. Results at five days were dramatic in that citral and neryl formate gave excellent protection to

TABLE IV.

Effect of Flavoring Oils and Related Allomones on *T. putrescentiae* in Petri Dish Pet Food Assay. Mites Had Choice of All Treatments in One Concentration.[a]

	Mite Nos. / % Concentration	
	0.25	0.5
Oleoresin pepper	585 a	655 a
Peppermint T R	515 a	175 ab
Sage	485 a	545 ab
Clove	445 a	139 ab
Lemon C V	422 b	390 ab
Orange C V	395 b	435 ab
Spearmint	385 ab	178 c
Nutmeg	372 ab	485 ab
Peppermint nat.	359 ab	385 ab
Cinnamon	340 b	495 ab
Neryl formate	67 c	25 d
Citral	39 c	15 d

[a] Oils introduced via treated filter paper disks placed on top of pet food in dishes 35 x 10 mm. Mite counts reflect mean of 5 replicates/concentration after 7 days. Means followed by same letter are not significantly different from each other.

TABLE V.

Effect of Flavoring Oils and Related Allomones on *T. putrescentiae* in Petri Dish Pet Food Assay. Mites Had Choice to One Treatment in a Range of Concentrations.[a]

	Mites Nos. / % Concentration				
	0	0.01	0.1	0.3	0.5
Spearmint[b]	188 a	113 a	144 a	131 a	75 b
Peppermint[b]	181 a	180 a	140 ab	94 ab	61 b
Citral[c]	419 a	400 a	79 b	31 b	18 b
Neryl formate[c]	600 a	475 a	64 b	48 b	23 b

[a] Technique same as described in Table IV. Means in rows, in concentration series 0-0.5%, followed by same letter are not significantly different from each other.

[b] Means represent 24 hour counts; counts made later were not significantly different.

[c] Means represents 5 day counts.

the patty at either 0.25% or 0.5% concentration. Of the natural flavoring oils, however, only spearmint showed possibilities (Table III). In a follow-up study to develop an assay for pet food patty protection, the treated paper disks were placed on top of pet food in small petri dishes (Table IV). In this test, however, the mites had choice of all treatments in each of the two concentrations. The mites could and did find entry by going under the dish cover. This test under great mite population pressure for seven days again proved the allomone effectiveness of citral and neryl formate, especially at 0.5%. This type of assay allows considerable room for mite entry and the petri dish top offers little if any seal.

To elucidate concentration effectiveness another petri-dish assay test was conducted using spearmint, peppermint, citral and neryl formate. Results

(Table V) showed that not only did the mints fade rather rapidly compared to citral and neryl formate, but also that it was further confirmed that these natural mint oils did not show the strong allomone effect of citral or neryl formate. Thus we have shown that these highly active terpenoid allomones can be used in a practical way to afford protection to foods against *T. putrescentiae.*

SUMMARY

Whole and ground spices incorporated into pet foods at 2% produced kairomone and allomone effects on *Tyrophagus putrescentiae,* when these were presented to mites in free-choice feeding tests. All flavoring oils mixed into meridic diet or pet food produced inhibition of mite growth at levels of 0.1% to 0.00001%. Pet food patties were protected against mite entry through the wrap by the introduction of a flavoring oil component, citral, to a filter paper disk placed inside the patty wrap. Another material equally effective was neryl formate, which has been described as being the alarm pheromone of *T. putrescentiae.*

REFERENCES

Blum, M. S. (1969). *Ann. Rev. Entomol.* **14,** 57-80.

Cole, L. K., Blum, M. S. and Roncardori, R. W. *Mycologia* **67,** 701-708.

Kuwahara, Y., Ishii, S. and Fukami, H. (1975). *Experientia* **31/10,** 1115-1116.

Pratt, J. J., Jr., House, H. L. and Mansingh, A. (1972). *In* "Insect and Mite Nutrition," (J. G. Rodriguez, ed.) pp. 637-650. North-Holland Publ. Co., Amsterdam.

Rodriguez, J. G. (1972). *In* "Insect and Mite Nutrition, (J. G. Rodriguez, ed.) pp. 637-650. North-Holland Publ. Co., Amsterdam.

Rogers, J. A. (1966). *In* "Flavor Chemistry," Advances in Chemistry Series, 56. (Robert F. Gould, ed.) pp. 203-224. American Chemical Society Publications, Washington, D. C.

Su, H. C. F. (1976). *J. Georgia Entomol. Soc.* **11,** 297-301.

Su, H. C. F. (1977). *J. Econ. Entomol.* **70,** 18-21.

Su. H. C. F., Speirs, R. D. and Mahany, P. G. (1972). *J. Econ. Entomol.* **65,** 1433-1436.

ROLE OF ACARINA IN THE STORED GRAIN ECOSYSTEM

R. N. Sinha

Research Station
Agriculture Canada
Winnipeg, Manitoba

INTRODUCTION

Mites of the family Acaridae have been known to be associated with stored grain and other foods for over a century. Although their destructive role has often been overshadowed by those of insect pests and microflora of stored food, serious acarologists of the past did not underestimate their economic importance. In 1906 Nathan Banks spelled out the role of stored-product mites when he considered the Tyroglyphidae (= Acaridae) as one of the most economically important groups of mites. He thought that mite damage of stored food has often been accredited to larger insects which happened to be present; mites were ignored because of their small size and pale color. By their rapidity in breeding, however, mites compensate for their minute size, so that products such as flour are often so badly infested that the whole mass of the substance appears to be in motion. Banks was convinced that it is chiefly through their ravages of stored food that mites are inimical to human effort. During the early 1930's the Soviet Union became aware of "enormous amount of damage" caused by granary mites and intensified research on these arachnids. Such vigorous national efforts resulted in the publication of monumental works on stored product mites by Zakhvatkin, Rodionov and others (Zakhvatkin, 1959). Much usful data has been compiled on taxonomy, biology, physiology and ecology of stored grain mites, mainly by researchers in Poland, Czechoslovakia, U. S. S. R., U. K., Japan, and Canada. The specific roles of the Acarina, comprising about 50 species of Astigmata, Prostigmata and Mesostigmata (Hughes, 1976) inhabiting various habitats in the stored grain ecosystems, however, are still not well-defined. Krantz (1961) made a broad attempt to define the role of various granary mites as members of the granary community from a review of biological and ecological data from various sources. During the last 20 years, our group in Winnipeg has

Vol. 1: ISBN 0-12-592201-9

developed a quantitative and "holistic" approach to understanding the collective role of mites, insects and fungi and the pathways of their infestations in stored grain and its products (Sinha, 1973).

STORED GRAIN ECOSYSTEM

Grain stored in bulks or in bags in farm granaries, boxcars, elevators or warehouses is considered as an immature, man-made ecosystem of limited energy. Abiotic and biotic agents that reduce its energy content also diminish the value of such man-made ecosystems as a source of human or animal food. An ecosystem is a dynamic unit; many changes within it often lead to grain spoilage. Changes occur evenly and primarily depend upon: the type of grain, post-harvest and initial storage condition, the quality and quantity of microflora, arthropods, birds and rodents invading it during storage, the climate and the location of the storage premises, the type of the granary structure and the volume of the grain stored in it, and the length of the storage period.

True to the characteristic of the fauna of an unstable and immature ecosystem in a granary, a mite species not only has a high intrinsic rate of increase but also a diverse feeding habit and high adaptibility to changing habitats and environments. Although sharing the same habitat and interacting from time to time with other animal and microbial communities within the same ecosystem, the acarine community usually plays an independent role, its numbers are regulated mainly by moisture, temperature, O_2-CO_2 levels and changes in the nutritional quality of the grain or related foods. Other variables affecting acarine populations are: the size and type of the grain bulk, seed crack, dockage, predators and parasites, microbial species (Sinha, 1973; Sinha and Wallace, 1973). An understanding of the acarine role in infested foodstuffs, therefore, requires a prior understanding of the nature and characteristics of the stored-grain ecosystem.

DISTRIBUTION, SOURCE OF INFESTATION AND DISPERSAL OF STORED-PRODUCT MITES

In the temperate belt of the Northern Hemisphere, cereal crops harvested and stored in the Middle Latitude Humid Lands (Kohn and Drummond, 1963) with warm to hot summers, cool to cold winters, over a 90-day frost-free period and four distinct seasons are often chronically infested by Acarid, Glycyphagid and Cheyletid mites. I have listed the major species of stored grain mites in several selected countries in Europe and Asia (Fig. 1). Each acarine species is indicated by the first letter of its generic name (capital letter) followed by the first letter of its specific name. The full names of mites not given in this paper can be found in Hughes (1976). The data and the ranking

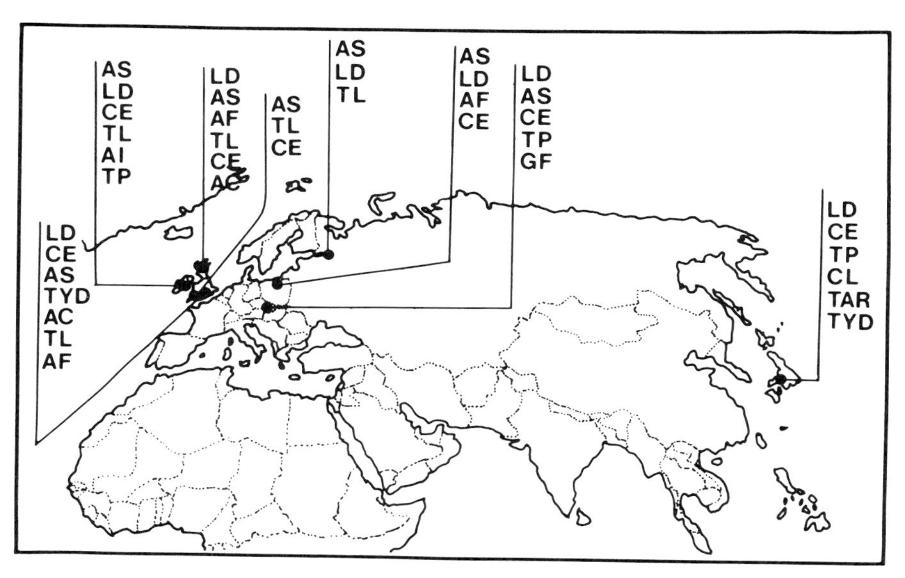

Fig. 1. A map of Europe and Asia showing major mites that infest stored grain and its products. Each species of mite is indicated by the first letter of its generic name (capital letter) and the first letter of its specific name except TYD which stands for Tydeidae. See text or Hughes (1976) for the names of the mites, and for other data source, see Griffith *et al.*, 1976; Cousal *et al.*, 1976; Jeffrey, 1976; Zabirov, 1977; Boczek and Golebiowska, Z., 1959; Zdarkova, 1967; Sinha, 1968b.

designation are based on the published data of acarologists in those countries. These data demonstrate that *Lepidoglyphus destructor* (Schrank) (= *Glycyphagus destructor*) (Ld), *Acarus siro* L. (As), *Cheyletus eruditus* (Schrank) (Ce), *A. farris* Oudemans (Af), *Tyrophagus longior* (Gervais) (Tl), *Androlaelaps casalis* (Berlese) (Ac), *T. putrescentiae* (Schrank) (Tp), Tydeidae (TYD) and *Tarsonemus granarius* (Tg) are the nine most common mites found in stored grain. The stage of the storage-transportation "pipeline" of grain movement at which mite infestation is most common, varies according to storage structure, length of storage, and the local climate. Sources of infestation also vary, although unclean storage areas with residual populations of mites are probably the main source of recurrent contamination of stored grain. Mites of stored grain often occur in nests of rodents (Krantz, 1961), birds (Woodroffe, 1953; Sander and Wasylik, 1973) and are likely to be transported by these animals when they visit granaries. Man also plays an important part in contaminating granaries by transporting mites from an infested granary to an uninfested one (Krantz, 1961).

ROLES OF ACARINA

The presence of mites makes grain and their products unacceptable or unattractive to humans and livestock. Thus in a broad sense Acarina are polluters of human and animal food. Their roles in the stored grain ecosystem may be one or more of the following: (A) energy transformer, (B) granivore, (C) herbivore, (D) fungivore, (E) predator or parasite, (F) scavenger-saprobe. It is possible to create more arbitrary groups of the acarine roles as we learn more about these mites.

Energy Transformers

Little information is available on the energy transforming role of mites in the stored grain ecosystem. Pioneering studies of Stepien and the other Polish workers of Boczek's group on the energy budgets of *Rhizoglyphus echinopus* (Stepien, 1970) and *T. putrescentiae* (Stepien *et al.*, 1973) form the sole basis for evaluating this acarine role in the basic area. These voracious acarid mites, particularly *T. putrescentiae,* are highly efficient transformers of consumed energy. In fact, both *T. putrescentiae* and *R. echinopus* act as a more efficient "energy carrier" from the producer to the 2° consumers (=predators of the next trophic level) than other mites (Berthet, 1964; Engelmann, 1961) and major granivorous stored grain beetles in the stored grain ecosystem (Singh *et al.*, 1976; Campbell *et al.*, 1976; Campbell and Sinha, 1978). Some 40-61% of food consumed by *T. putrescentiae* is transformed to body and egg production. Gross production efficiencies (production/consumption) of growth are generally higher for stored-product acarids (35-76%) (Stepien *et al.*, 1973; Stepien and Rodriguez, 1972) than five species of stored-product beetles studied (10-20%; Campbell and Sinha, 1978). The highest cumulative net production efficiency value (production/assimilation) for any stored-product beetle studied was 38%, recorded for *Rhyzopertha dominica* (Fabricius) during late larval growth. In contrast, the stored-product acarids (different life stages) have higher cumulative (K_{2c}) and/or daily (K_{2i}) net production efficiencies ranging from 77 to 89%. Beetles and particularly acarid mites seem to show a strong curvilinear relationship between their conversion efficiency of assimilated food energy into egg production and their intrinsic rate of increase. Campbell and Sinha (1978) obtained an exponential curve, $\log r = 0.879 + 0.0297\, K_{2i}$ ($R^2 = 0.928$) when they included the values of $K_{2i} = 77\%$ for *T. putrescentiae* (Stepien *et al.*, 1973) along with their energy data on beetles. When granivorous mites feed on dormant cereal seed "producers" they are in fact consuming from a nonrenewable energy source which may or may not be replenished. As a mite voraciously consumes energy from cereals, it is simultaneously altering the ecosystem in which it lives by transforming the energy into biomass, heat of respiration, egested materials, and chewed uneaten materials which accumulate and pollute the intergranular space. A most challenging area of research in stored-product acarology still remains in the exploration of the role of Acarina as energy carriers along the pathways of deterioration of stored grain.

Granivores

Of about 50 species of stored-product mites, only a few—*A. siro, T. longior* and a few other acarids—feed directly on grain kernels and can cause damage to stored grains by causing weight loss (Cusack and Brennan, 1976) and germination loss (Jeffrey, 1976). Akimov (1977) has recently found that each group has a characteristic type of structure and physiology.

Acarid mites penetrate into the hard grain and cause internal changes; whole fragments of food are bitten off by the mouth parts with the help of the chelicerae. The digestive system is adapted to the assimilation of hard particles with the aid of characteristic enzymes (Akimov, 1977). Their presence indicates impending seed spoilage and the need for immediate attention.

In contrast, Glycyphid mites cannot gnaw through hard parts of grain and feed on grain dust; only small particles of dry matter can be taken up by the mouthparts mainly with the aid of chelicerae. The digestive system and enzymes are adapted to assimilating small particles (Akimov, 1977). Their presence indicates inadequate storage facilities with the possibility of infestation by more injurious species.

Akimov (1977) claims to predict the degree of potential damage of a given species from this knowledge of the structure of the mouthparts and the digestive functions of these acarine species.

Because stored grain represents a limited, non-regenerating supply of food for grain mites, a mite infestation once established under favorable conditions is expected to reach a maximum and then decline through emigration or death, as the food supply is depleted. Such sequences of events are disrupted by predators, sudden change of environment, or by creation of a gradient of intergranular humidity and temperature that draws mites to a neighboring grain mass where conditions are more favorable. Despite several excellent laboratory studies (Boczek, 1957; Solomon, 1962, 1969a) and granary studies (Solomon, 1946; Pulpan and Verner, 1965) the precise role and the mechanism for sudden explosive increase and abrupt decline at irregular intervals of natural populations of *A. siro* (Sinha, 1973) is not well-understood. Through a schema (Fig. 2) I shall explain the role and fluctuation patterns of *A. siro* in naturally infested grain bulks. Fungal data used in this schema are taken from the reviews by Semeniuk (1954), and Martin and Gilman (1976). In grain kernels with broken seed coats, common in years of dry harvest, *A. siro* feeds exclusively on the germ. Its alternate food sources in bulk grain are: broken and shrivelled grain, seeds and flowers of weeds, and seed-borne fungi (mainly *Alternaria, Nigrospora, Hormodendrum, Mucor, Aspergillus glaucus* gr. and to a lesser extent *Penicillium* spp.). *A. siro* multiplies only within the temperature-R. H. boundary shown in Fig. 2. The critical equilibrium relative humidity of *A. siro* is 71% (Knulle, 1965). The occurrence of *A. siro* in granaries and any meaningful numerical increase of populations are possible only at humidities above the critical equlibrium humidity. Sustenance of low numbers of *A. siro* below this R. H. level and at low temperatures 0 to 4°C is

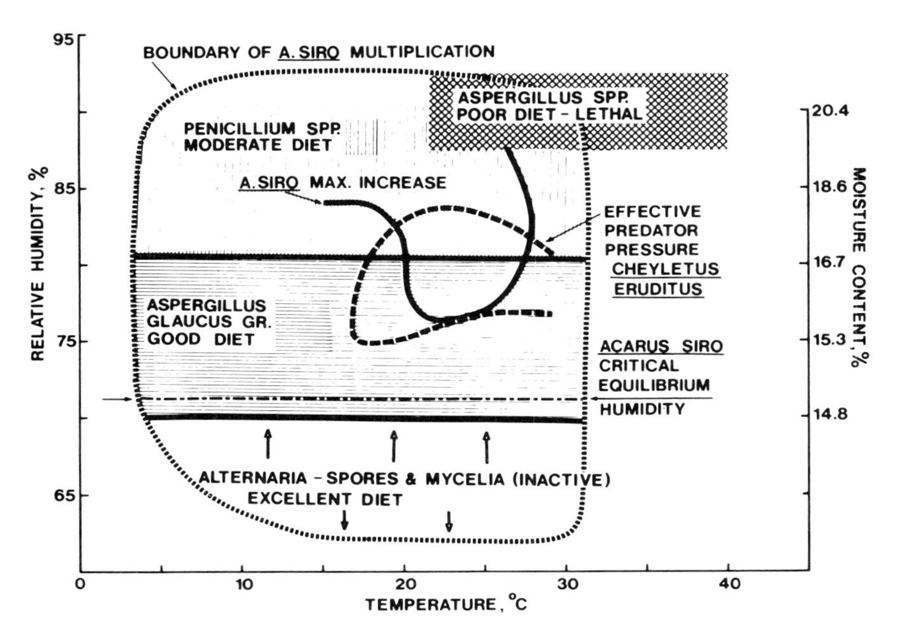

Fig. 2. A schema showing interrelations and aspects of performance by *Acarus siro* (prey) and *Cheyletus eruditus* (predator) in relation to temperature, moisture and fungal infection of stored grain. Acarine data are based on Solomon (1969), and fungal data on Semeniuk (1954), and Martin and Gilman (1976) physiological data on Knulle (1965).

possible through feeding on spores and mycelia of preharvest field fungi (*Alternaria* etc.) which contain both moisture and nutrient for the slow-moving mites living in an unfavorable physical environment. From 71 to 82% R. H. and 5 to 30°C the mites have a food choice of grain germ and xerophilic fungal species (*A. glaucus* gr.) which also flourish in these conditions. Predator pressure from *C. eruditus* is effective only within a part of this range (Fig. 2). Between 80 and 90% R. H. and 5 and 32°C *A. siro* thrives along with *Penicillium* spp. which can only be a moderately useful diet for the mite. Between 87 ad 92% R. H., *Aspergillus* spp. dominate the grain environment. These fungi are not only poor as dietary supplements but they are also polluters of the acarine environment. Successive phases in the population history of *A. siro* based on basic environmental parameters have been described by Solomon (1969) as: establishment, geometric increase, decelerating increase, the change-over from increase to decrease and the final decline in numbers. It is logical to hypothesize that in the natural grain environment the presence of *Alternaria* with relatively large spores (25μ) and other preharvest fungi and fungi of the *Aspergillus glaucus* gr. contribute in part, if not critically, to the establishment and initial geometric increase of the *A. siro* population. These fungi provide small quantities of essential nutrients and critical moisture needs. The population decline in the later stages could also be influenced, if not critically affected, by luxuriant growth of the harmful post-harvest fungi of the genus *Aspergillus,* such as *A. flavus, A. niger* and *A. fumigatus* (Sinha, 1964).

Herbivores

Several acarine species such as *G. domesticus* deGeer, *L. destructor* (Schrank), *T. putrescentiae* (Schrank), *Suidasia medanensis* Oudemans, *S. nesbitti* Hughes, *Aleuroglyphus ovatus* (Troupeau), and *Aëroglyphus robustus* (Banks) feed on a wide range of food of cereal and plant origin (Hughes, 1976). *G. domesticus* alone is found in large numbers on dried plant and animal remains in houses and stables, in flour, wheat, hay , linseed, tobacco, cheese, ham, sugarbeet seed, barn debris, seagrass fibre and on molds. *L. destructor* is probably the most common cosmopolitan species of stored grain and its products.

Fungivores

All species of Astigmata associated with stored grain and its products, whose fungal feeding habits are known, can survive and multiply exclusively on more than one species of fungi associated with stored grain (Griffiths *et al.*, 1959; Sinha, 1964, 1966, 1968a, b; Czajkowska, 1970). The ability of these mites to live and propagate on several species of seed-borne fungi—some genera of fungi are present in most stored grain—enables them to escape from extinction through starvation. Polyphagus mites seem to withstand a high level of disturbance and repeated transportation from one storage area to another by trucks, railroad cars, ships and also by conveyor belts and augers. Fungi growing on grain dust and decaying organic debris with low moisture content (12-14%) assure an adequate food supply for mites dwelling in empty granaries. Fungi of the *Aspergillus glaucus* group are particularly suitable for mites in unclean granaries without grain. *A. repens* (Cda.) Sacc., *A. echinulatus* (Delacr.) Thom and Church, *A. amstelodami* (Mang.) Thom and Church are excellent media for egg laying, development and multiplication of *A. siro* L., *A. farris* Oud., *T. putrescentiae* (Schrank) (Czajkowska, 1970). Closely related species of mites occur sympatrically in granaries. They minimize competition among themselves and thereby reduce the selection pressure by feeding not only on cereal foods of different quality and sizes, but also on characteristic groups of fungal species (Sinha, 1968a). Presence of different kinds of fungi in stored grain has undoubtedly provided stored grain mites with more choices of food and thus greater chance of survival in an unstable and immature ecosystem.

Predators and Parasites

The acarine community in the stored grain ecosystem has a well-developed predatory element. *C. eruditus* and several other cheyletid and mesostigmatid mites are the main predators in the ecyosystem which feed not only on acarids but also on eggs and immature forms of psocids and other stored grain insects (Hughes, 1976). In addition to *C. eruditus* common consumers at the second trophic level include *C. malaccensis* Oudemans, *C. trouessarti* Oudemans, *Cheletomorpha lepidopterorum* (Shaw), *A. casalis, Blattisocius Keegani* Fox, and *B. tarsalis* (Berlese).

Several in-depth studies have been made to understand the mechanism of natural control of acarid and/or glycyphagid preys in granaries (Norris, 1958; Boczek, 1959; Pulpan and Verner, 1965; Solomon, 1969b; Sinha and Wallace, 1973). Attempts to biologically control acarine pests with this method, however, have only been successful under special circumstances (Pulpan and Verner, 1965).

The parasitic role played by stored grain mites is a minor one as demonstrated by mites, such as *Acarophenax tribolii* Newstead and Duvall, and *Pyemotes ventricosus* (Newport). The former is an ectoparasite of *Tribolium* spp. whereas the latter attacks grain handlers, causing dermatitis (Hughes, 1976).

Scavenger and Saprobes

A tortoise-like unropodid mite, *Leiodynychus krameri* (Berlese) is often found in damp, moldy sprouting barley, wheat, corn in ships and warehouses and in the litter of broiler houses (Hughes, 1976). Several species of tydeid and tarsonemid mites occur in grain debris and decaying organic debris on granary floors (Sinha, 1968b; Sinha and Wallace, 1966). It seems that these and a few other mites play a scavenger and saprophagous role in the stored grain ecosystem.

OUTLOOK FOR FUTURE RESEARCH

Although considerable progress has been made on the taxonomy, life history, physical limits, feeding patterns and the community ecology of the major acarine pests of stored grain, there is a gap in our knowledge of the trophic-dynamic ecology and energy budgeting of mite-infested stored grain ecosystems. The acarine role in such ecosystems can be properly understood when system analysis can be made with reliable quantitative information, relating to involved species, on: (1) intrinsic rate of increase and physical limits, (2) energy budgets, (3) ethological and physiological survival mechanisms such as hypopus formation and critical equilibrium humidity, (4) adaptability to grain quantity and mass size, granary structure and changing environments that include both nourishing and pathogenic species of microflora, common in the movement of grain and grain products.

Research is needed in all these aforesaid areas for the so-called less important species and in areas (2), (3) and (4) for the major species. More work should be done on economic losses caused by mites in granaries, warehouses and flour mills.

SUMMARY

About 50 species of mites occur in storage premises containing grain, flour, oilseed and animal feeding material. Several countries with temperate climate lying within 40°-60° latitude in the Northern Hemisphere are most vulnerable to infestation by stored grain mites, some of which also occur in birds' or rodents' nests and agricultural fields. Although sharing the same habitat and interacting from time to time with other animals and microorganisms within the same ecosystem, the acarine community usually plays an independent role; its numbers are regulated mainly by moisture, temperature and changes in the nutritional quality of the grain or related food substrates. Other variables influencing acarine populations are: the size and the type of the grain bulk, seed cracks, dockage, predators and parasites, microflora species. Using an ecosystem approach, the specific roles of several selected mites have been described as: energy transformers, granivores, herbivores, fungivores, predators or parasites, and scavenger-saprobes. As energy transformers for grain energy, *T. putrescentiae* and *R. echinopus* are more efficient than most destructive stored grain insects. An outlook for future research emphasizing the need for more energy budget data and systems analysis of acarine populations in stored grain ecosystems is presented.

REFERENCES

Akimov, I. A. (1977). *Zashch. Rast. (Moscow) No. 2,* 44.

Banks, N. (1906). *United States Department of Agriculture, Bureau of Entomology. Technical Series No. 13,* 1-34.

Berthet, P. (1964). *Memoires de l'Institut Royal des Sciences Naturelles de Belgiques.* **152,** 152 p.

Boczek, J. (1957). *Rocz. Nauk Roln. Ser. A.* **75,** 559-644.

Boczek, J. (1959). *Pr. Nauk. Inst. Ocr. Rosl.* **1,** 175-230.

Boczek, J. and Golebiowska, Z. (1959). *Rocz. Nauk Roln. Ser. A.* **79,** 969-988.

Campbell, A., Sinha, N. B. and Sinha, R. N. (1976). *Can. J. Zool.* **54,** 786-798.

Campbell, A. and Sinha, R. N. (1978). *Can. J. Zool.* **56,** 624-633.

Cusack, P. D. and Brennan, P. A. (1976). *Sci. Proc. Roy. Dublin Soc. Ser. B (1975)* **3,** 341-353.

Czajkowska, B. (1970). *Zesz. Probl. Postepow Nauk Roln. No. 109,* 219-227.

Engelmann, M. D. (1961). *Ecol. Monogr.* **31,** 221-238.

Griffiths, D. A., Hodson, A. C. and Christensen, C. M. (1959). *J. Econ. Entomol.* **52,** 514-518.

Griffiths, D. A., Wilkin, D. R., Southgate, B. J. and Lynch, S. M. (1976). *Ann. Appl. Biol.* **82,** 180-184.

Hughes, A. M. (1976). "The Mites of Stored Food and Houses." H. M. Stat. Off. London.

Jeffrey, I. G. (1976). *J. Stored Prod. Res.* **12,** 149-156.

Knulle, W. (1965). *Z. Vergl. Physiol.* **49,** 586-604.
Kohn. C. F. and Drummond, D. W. (1963). "The World Today," McGraw-Hill, New York.
Krantz, G. W. (1961). *Ann. Entomol. Soc. Amer.* **54,** 169-174.
Martin, P. M. D. and Gilman, G. A. (1976). *Tropical Products Institute Report No. G105,* 112 p.
Norris, J. D. (1958). *Ann. Appl. Biol.* **46,** 411-422.
Pulpan, J. and Verner, P. H. (1965). *Can. J. Zool.* **43,** 417-432.
Sandner, H. and Wasylik, A. (1973). *Ekologia Polska* **21,** 323-338.
Semeniuk, G. (1954). *In* "Storage of Cereal Grains and their Products" (J. A. Anderson and A. W. Alcock, eds.), pp. 77-151.
Singh, N. B., Campbell, A. and Sinha, R. N. (1976). *Ann. Entomol. Soc. Amer.* **69,** 503-512.
Sinha, R. N. (1964). *Acarologia* **6,** 372-389.
Sinha, R. N. (1966). *J. Econ. Entomol.* **59,** 1227-1232.
Sinha, R. N. (1968a). *Evolution* **22,** 785-798.
Sinha, R. N. (1968b). *Ann. Entomol. Soc. Amer.* **61,** 938-949.
Sinha, R. N. (1973). *Ann. Technol. Agr.* **22,** 351-369.
Sinha, R. N. and Wallace, H. A. H. (1973). *Oecologia* **12,** 315-327.
Sinha, R. N. and Wallace, H. A. H. (1966). *Ann. Entomol. Soc. Amer.* **59,** 1170-1181.
Solomon, M. E. (1946). *Ann. Appl. Biol.* **33,** 82-97.
Solomon, M. E. (1962). *Ann. Appl. Biol.* **50,** 178-184.
Solomon, M. E. (1969a). *Proc. 2nd Intern. Congr. Acarology 1967, Sutton Bonington (England),* 255-260.
Solomon, M. E. (1969b). *Acarologia* **11,** 484-503.
Stepien, Z. A. (1970). "The Energy Budget of *Rhizoglyphus echinopus* (Acarina, Acaridae) During its Development." Ph.D. thesis, Warsaw Agr. Univ., 103 p.
Stepien, Z. A., Goszcynski, W. and Boczek, J. (1973). *Proc. 3rd Intern. Congr. Acarology 1971, Prague,* 373-378.
Stepien, Z. A. and Rodriguez, J. G. (1972). *In* "Insect and Mite Nutrition" (J. G. Rodriguez, ed.) pp. 127-151. North-Holland Publishing Co., Amsterdam.
Woodroffe, G. E. (1953). *Bull. Entomol. Res.* **44,** 739-772.
Zabirov, S. M. (1977). *Zashch. Rast. (Moscow) No. 2,* 45.
Zakhvatkin, A. A. (1941). "Tyroglyphoidea (Acari)." A translation of fauna of U. S. S. R. Arachoidea V., No. 1. Translated and edited by A. Ratcliffe and A. M. Hughes, 1959. American Institute of Biological Sciences, Washington, D. C.

EVOLUTIONARY ORIGINS OF ASTIGMATID MITES INHABITING STORED PRODUCTS

Barry M. OConnor

Department of Entomology
Cornell University, Ithaca, New York

INTRODUCTION

The suborder Astigmata contains the most abundant and diverse group of mites occurring in stored product habitats. Although much is known concerning the biology and control of important pest species (Hughes, 1976), very little has been published on the evolutionary origins of stored product inhabiting Astigmata. Because man has been storing food in quantity only since the advent of agriculture about 10,000 years ago, not enough time has passed to allow the evolution of such a large complex of genera and species strictly in association with man. Thus, the origin of stored product Astigmata lies in naturally occurring habitats.

Ancestrally, the Astigmata were probably fungivorous, living in specialized or ephemeral habitats and dispersing by means of a highly modified deutonymph (hypopus). From these habitats, species representing at least 34 genera in 10 families have invaded stored products and/or house dust, and it is probable that each species has invaded independently. Because only a few species in each genus have invaded stored products, the habitat preferences of the naturally occurring species may indicate the evolutionary origins of the stored product species.

The objective was to examine the natural distribution of genera that include stored product species. Analysis of the genera and criteria lead to four major categories that become apparent: first, mites that are associated with specific resources such as fruit or meat, that are not widely distributed in space or time. These species retain the ancestral astigmatid life style. The second group includes mites that have become associated with widespread field resources. The third and fourth groups include species associated with the nests of mammals and birds respectively.

Vol. I: ISBN 0-12-592201-9

MITES ASSOCIATED WITH SPECIFIC RESOURCES

Species in the first group, that are associated with uncommon or ephemeral field resources, can be subdivided into species that naturally infest the product before it enters human commerce or that utilize phoretic hosts which do, and species that infest a variety of materials in the field. Examples of the first subgroup include *Sancassania* (Acaridae) which naturally inhabits dung and rotting vegetation and often invades stables, barns and haystacks on beetle hosts. A similar case is found with *Lardoglyphus* (Lardoglyphidae) which is restricted to high protein substrates such as dried animal carcasses in the field and is phoretic on dermestid and trogid beetles that frequent such material both in natural and storage conditions. *Carpoglyphus* (Carpoglyphidae) is thought to infest rotting fruit in the field (Treat, 1975) and infests such materials in storage as well as a variety of other products with a high sugar content. The hypopus is phoretic on moths. Species of *Histiostoma* (Anoetidae) prefer very wet decaying substrates both in the field and in storage conditions. Hypopi are phoretic on a variety of insects which also frequent such wet substrates.

The remaining genera in this group are not easy to associate with a particular habitat, either because they are associated with several habitats in the field or because they are poorly known. In the Glycyphagidae, the related genera *Aeroglyphus* and *Austroglycyphagus,* are associated with a variety of specialized habitats such as bat roosts, bird nests, and the nests of social insects. These genera have evolved from ancestors that inhabited mammal nests, but the modern species have invaded a variety of natural habitats accompanied by the complete loss of the hypopus. Species in the genus *Suidasia,* which has been included in both the Acaridae and Saproglyphidae, have been associated with insects, bats, and soil, as well as stored products. A hypopus is also unknown in this genus. In the Saproglyphidae, *Nanacarus,* which occasionally infests grain, is associated with bracket fungi. The previously unrecognized hypopi of this genus are phoretic on Coleoptera and Diptera (OConnor, unpubl.). Species of *Acalvolia,* which were known only from the hypopal stage from house dust, are now known to infest stored products. All stages of the type-species, *A. squamata,* have been found infesting corn and will be described separately. The hypopi have been collected on grain-infesting Coleoptera and Psocoptera (OConnor, Delfinado and Baker, unpubl.). Species of *Acalvolia* are not yet known from natural conditions. The same is true of the genus *Procalvolia* which infests various stored products (Hughes, 1962; Fain, 1972). Species of an undescribed saproglyphid genus near *Oulenzia* are often encountered on leaves and fruit.

MITES ASSOCIATED WITH FIELD RESOURCES

The second major category contains those genera associated with widespread field resources. Few astigmatid mites are generally and widely distributed, so I consider this habitat to be derived. Among the Acaridae, the closely related genera *Tyrophagus* and *Tyrolichus* are widely distributed in a variety of natural habitats but are most abundant in grassland soil and litter. The related monotypic genus *Tyroborus* is not known from the field but probably shares a common ancestor with the preceding genera. As in *Aeroglyphus* and *Austroglycyphagus,* these genera have completely lost the hypopal stage. The wide distribution and constancy of the preferred habitat has apparently eliminated the need for a specialized dispersal stage in these genera. Species of *Mycetoglyphus* and *Rhizoglyphus* are also widespread but with a narrower range of microhabitats. *Mycetoglyphus* is related to the ant-associated genus *Forcellinia* and hypopi are carried by ants as well. Hypopi of *Rhizoglyphus* move freely through the soil although the feeding stages are restricted to bulbs, roots and tubers. Hypopi are also distributed by Diptera (Zakhvatkin, 1941; Turk and Turk, 1957).

In the family Glycyphagidae, the subgenus *Glycyphagus (Glycyphagus)* occupies the widest habitat range. Species in this subgenus occur in grassland, the foliage of trees, caves, bat roosts and rodent nests, as well as in stored products and house dust (personal observations). These species are derived from nest-inhabiting ancestors and have become widespread due to the formation of inert hypopi or the loss of the hypopus completely. (See Fain (1971) and Volgin (1971) for terminology of hypopal types.)

MITES ASSOCIATED WITH NESTS OF MAMMALS

The largest group of stored-product-inhabiting genera is derived from the third habitat category: the nests of mammals and especially rodents. The derivation of stored product mites from rodent nests is twofold. First, the nest material itself provides a substrate for the specific fungi required as food by the mites, and second, a large number of rodents store various plant parts as food in chambers within their burrow systems. Thus, in many cases, the specialization of mites for stored grain probably occurred in the burrows of rodents long before the advent of human agriculture. The nest-inhabiting Acaridae are the least specialized in this group. Species of *Acarus, Aleuroglyphus,* and *Acotyledon* inhabit a wide variety of mammal nests, including the roosts of bats. Two types of hypopi are formed in the genus *Acarus:* the normal entomophilous form that is often phoretic on fleas in natural situations (Fain and Beaucornu, 1972, 1973), and inert forms formed in species primarily associated with bat roosts (Griffiths, 1964). Both groups of species have invaded stored products. The hypopus is also reduced in *Acotyledon* and is absent altogether in *Aleuroglyphus.*

Among the Glycyphagidae, some species in the subgenus *Glycyphagus (Myacarus)* occasionally invade stored products, but most species are known only from nest-inhabiting adults or pilicolous hypopi. Formation of a hypopus is facultative in *Myacarus* (Spicka and OConnor, in press). The genus *Lepidoglyphus* especially *L. destructor,* is commonly encountered in stored products. Until recently, only inert hypopi were known in this genus; however, Philipps (pers. comm.) has succeeded in associating a pilicolous hypopus with *L. fustifer* indicating a nidicolous origin for the genus. The inert hypopus of *Lepidoglyphus* is less morphologically reduced than in *Glycyphagus (Glycyphagus)* and is independently evolved. Species of *Gohieria* and *Tropilichus* are not known to form hypopi, although species of *Gohieria* are rodent nest inhabitants (Volgin, 1961). *Tropilichus* is not yet known from natural habitats although its relationships indicate a nidicolous origin.

Ctenoglyphus (Ctenoglyphidae) contains several species inhabiting hay and straw. Most species of this genus, however, are known only from hypopi which are endofollicular parasites of rodents. These hypopi were first described under the name *Rodentopus,* but were placed in *Ctenoglyphus* by myself (OConnor, in press) following the rearing of hypopi of *C. plumiger* by Chmielewski (1975). Hypopi are rarely formed by the stored-product-infesting species. This is an unusual development because the endofollicular hypopi derive substantial nutrition from their hosts (Fain, 1971).

A similar life-cycle is found in *Chortoglyphus* (Chortoglyphidae) and *Blomia* (Echimyopidae). The hypopus of *Chortoglyphus* was formerly known as *Aplodontopus,* but studies by Spicka (1977) have led me to synonymize the two. Philipps (pers. comm.) reports the association of hypopal *Echimyopus* with adults which I consider to belong to the genus *Blomia.* The morphology of adult *Blomia* had led me to propose a new family for that genus (OConnor, in press), however the family group name Echimyopidae Fain is available and has priority.

MITES ASSOCIATED WITH NESTS OF BIRDS

The final category of natural habitats providing origins of stored product mites is the nests of birds. *Pyroglyphus* and *Euroglyphus* (Pyroglyphidae) may be derived from mammal nest inhabitants, although not enough information is yet available concerning these genera. The remaining genera in the Pyroglyphidae associated with stored products or house dust are clearly derived from species inhabiting the nests of birds. The large genus *Dermatophagoides,* and the smaller genera *Sturnophagoides, Hirstia, Malayoglyphus,* and *Guatemalichus* are known from both stored products/house dust and from bird nests (Wharton, 1976; Atyeo and Gaud, 1978).

No hypopus is formed by any species in the Pyroglyphidae.

Aside from providing insights into the natural habitats which provided stored-product species, a study of their natural distribution also indicates the geographic origins of certain groups. Many genera such as *Acarus, Tyrophagus, Sancassania, Rhizoglyphus, Glycyphagus* and *Dermatophagoides* are almost cosmopolitan in their natural distribution. However, certain genera are restricted to certain geographical regions when only naturally occurring species are considered. These are: *Gohieria,* a Palearctic genus originally associated with squirrels; *Ctenoglyphus,* a genus associated with several rodent families which probably originated in the Ethiopian region but naturally occurs in the Palearctic and Oriental as well. Single species of these two genera have been introduced into the New World and Australia. Genera with New World origins are *Chortoglyphus,* a Nearctic genus associated with several families of sciuromorph and myomorph rodents; and *Blomia,* probably originating in the Neotropical region but naturally occurring in the Nearctic as well. Natural hosts of *Blomia* include a wide variety of rodent families as well as one edentate. Apparently several species of *Blomia* have been introduced into the Old World, while only a single species of *Chortoglyphus* has been introduced.

CONCLUSIONS

After examining the natural habitats of stored product genera, some final conclusions can be drawn concerning the adaptive characteristics of the species that have been able to invade stored products. One overall adaptive syndrome involves a wider tolerance by the feeding stages of more variable environmental conditions. This, I believe, has led to a reduction in dependency on the hypopal deutonymph to survive habitat deterioration. This reduction in dependency on a hypopus for survival, accompanied by a wider range of acceptable habitats by feeding stages, has lessened the requirement for a phoretic host to distribute the species. Examples of this adaptive syndrome include *Tyrophagus, Acarus, Sancassania, Aleuroglyphus,* and *Suidasia* among the Acaridae which used insects as phoretic hosts. Among the Glycyphagidae, *Glycyphagus, Lepidoglyphus, Aeroglyphus,* and *Austroglycyphagus* exhibit the same characteristics although the original phoretic hosts were mammals. Species in these genera were able to invade new adaptive zones in field situations long before the development of human agriculture and were thus, preadapted to life in stored products.

On the other hand, other genera have been able to invade stored products only because such materials closely mimic their restricted natural environment. These include the nest-associated genera *Ctenoglyphus, Gohieria, Chortoglyphus, Blomia* and *Dermatophagoides.* These have also reduced their requirement for phoretic hosts, either permanently (*Gohieria, Dermatophagoides*) or facultatively (genera with endofollicular hypopi). Non-nidicolous genera exhibiting this same restriction include *Lardoglyphus, Carpoglyphus,* and *Acalvolia,* which only infest those products in storage which closely resemble their natural habitats. These genera often still require phoretic hosts for dispersal, even in storage situations.

SUMMARY

Of the 34 genera of Astigmata commonly encountered in stored products and house dust, six are derived from widespread field species, 12 from mammal nest inhabitants, five from bird nest inhabitants and 12 from other ephemeral or uncommon habitats. Modification of the morphology of the hypopus and/or its presence in the life-cycle are common occurrences in stored product mites. These modifications have allowed certain groups to enter new adaptive zones which preadapt the species to life in stored products. The remaining genera remain restricted to those stored products closely resembling their natural habitats.

ACKNOWLEDGEMENTS

I thank Dr. Mercedes Delfinado, New York State Museum and Science Service, Albany N.Y. and Mr. James Philipps, New York State College of Environmental Science and Forestry, Syracuse, N.Y. for permission to cite their unpublished findings. I also thank Dr. George Eickwort, Cornell University, Ithaca, N.Y. for his comments on the manuscript.

REFERENCES

Atyeo, W. T. and Gaud, J. (1977). *Steenstrupia* **4**, 121-124.
Chmielewski, W. (1975). *Zesz. Prob. Postepow Nauk Roln.* **171**, 261-268.
Fain, A. (1971). *Acarologia* **13**, 171-175.
Fain, A. (1972). *Acarologia* **15**, 225-249.
Fain, A. and Beaucournu, J. (1972). *Acarologia* **13**, 522-531.
Fain, A. and Beaucournu, J. (1973). *Acarologia* **14**, 138-143.
Griffiths, D. A. (1964). *Bull. Brit. Mus. (Nat. Hist.) Zool.* **11**, 413-464.
Hughes, A. M. (1962). *Acarologia* **4**, 48-63.
Hughes, A. M. (1976). "The Mites of Stored Food and Houses." Ministry of Agriculture, Fisheries & Food, *Tech. Bull. no. 9.* 400 pp.
O'Connor, B. M. In press, *J. Med. Entomol.*
Spicka, E. J. (1977). Ph.D. Thesis, Indiana State University, 135 pp.
Spicka, E. J. and O'Connor, B. M. In press. *Acarologia.*
Treat, A. E. (1975). "Mites of Moths and Butterflies." Comstock Publishing Assoc. Ithaca, New York. 362 pp.
Turk, E. and Turk, F. (1957) *In* "Beitrage zur Systematik und Okologie mitteleuropaischer Acarina," (Stammer, H. J., ed.) Band 1, Teil **1**, 1-231.
Volgin, V. I. (1961). *Parazit. Sborn.* **20**, 257-266 (in Russian).
Volgin, V. I. (1971). *Proc. 3d. Int. Congress of Acarology,* Prague, 381-4.
Wharton, G. W. (1976). *J. Med. Entomol.* **12**, 577-621.
Zakhvatkin, A. A. (1941). "Fauna of the U.S.S.R. Arachnoidea", Vol. VI, no. 1 Tyroglyphoidea (Acari). *Zool. Inst. Acad. Sci. U.S.S.R.* new ser. 28 (English translation, 1959). 573 pp.

SPERMATOPHORE PRODUCTION AND MATING BEHAVIOUR IN THE STORED PRODUCT MITES *ACARUS SIRO* AND *LARDOGLYPHUS KONOI*

J. Boczek

Warsaw Agricultural University
Warsaw, Ursynow, Poland

D. A. Griffiths

Pest Infestation Control Laboratory
Ministry of Agriculture, Fisheries and Food
Slough, England

INTRODUCTION

Acarus siro L. is probably the most serious pest of stored foodstuffs, attacking a very wide range of commodities over the whole of the inhabited world. *Lardoglyphus konoi,* (Sasa and Asanuma) whilst also a serious pest, is much more restricted both in its choice of habitats and distribution. It is restricted to foodstuffs containing large amounts of animal protein, especially dried fish products, and is particularly troublesome in West Africa, India and Japan.

Previously the authors (Griffiths and Boczek, 1977) showed that members of the genus *Acarus* and *Lardoglyphus* possess spermatophores. During copulation the plastic spermatophore is transferred via the male intromittent organ into the 'neck' of the bursa copulatrix. One spermatophore is transferred at each complete mating and, since these can be detected in microscopic preparations, it is possible to determine the number of matings achieved by any female. This information has been used by the authors to examine spermatophore production in both species as it occurs in single pair matings and stock cultures and to compare and contrast the results both within and between species.

Vol. 1: ISBN 0-12-592201-9

MATERIALS AND METHODS

Stock cultures and experimental cells are held at 90% R.H. and 20°C. Both species are fed on a mixture of wheat germ and yeast 3:1 but the *L. konoi* diet is enriched by adding to it 5% fish meal. Single pairs are placed together as 3-day old virgins, obtained by isolating resting tritonymphs. The females are mounted in Heinz fluid and subsequently examined by phase or interference phase contrast microscopy.

Spermatophore production in single, isolated pairs has been estimated by leaving pairs together over various periods of time. These periods have been condensed into four averages, which together with their range are as follows. Days 1 to 3; Day 8 (range 7 to 10); Day 18 (range 15 to 20); Day 27 (range 25 to 30) and Day 30. Females were extracted from stock cultures exactly on days 1, 8, 18 and 30.

RESULTS

Single Pair Matings

As one would expect, the number of spermatophores increases over time (Figs. 1 and 2). However, at all points in time the number of spermatophores produced by *A. siro* males far exceeds that produced by the *L. konoi* males. At 8 days the mean numbers of spermatophores per *A. siro* female is about 3 times greater than that for *L. konoi.* No *A. siro* female contained less than 9 spermatophores whilst only some 20% of *L. konoi* females contained 9 or more. As the exposure period gets longer (18 and 27 days) the means come closer together but the ranges about the means are quite different. In *L. konoi* the distribution resembles a normal curve whilst for *A. siro* it is skewed to the right.

Spermatophore Production in Stock Cultures

These results were obtained by picking some 200 females from a stock culture of each species at each prescribed period of time (Figs. 3 and 4). Within each species the means and distribution for the 4 time sequences are remarkably similar, indicating that the age distribution of females remains fairly constant within the culture.

The difference in distribution about the mean mentioned when single pair results were considered is even more distinct now. No *L. konoi* ♀ contained more than 20 spermatophores whereas more than 20% of the *A. siro* females contained more than this number.

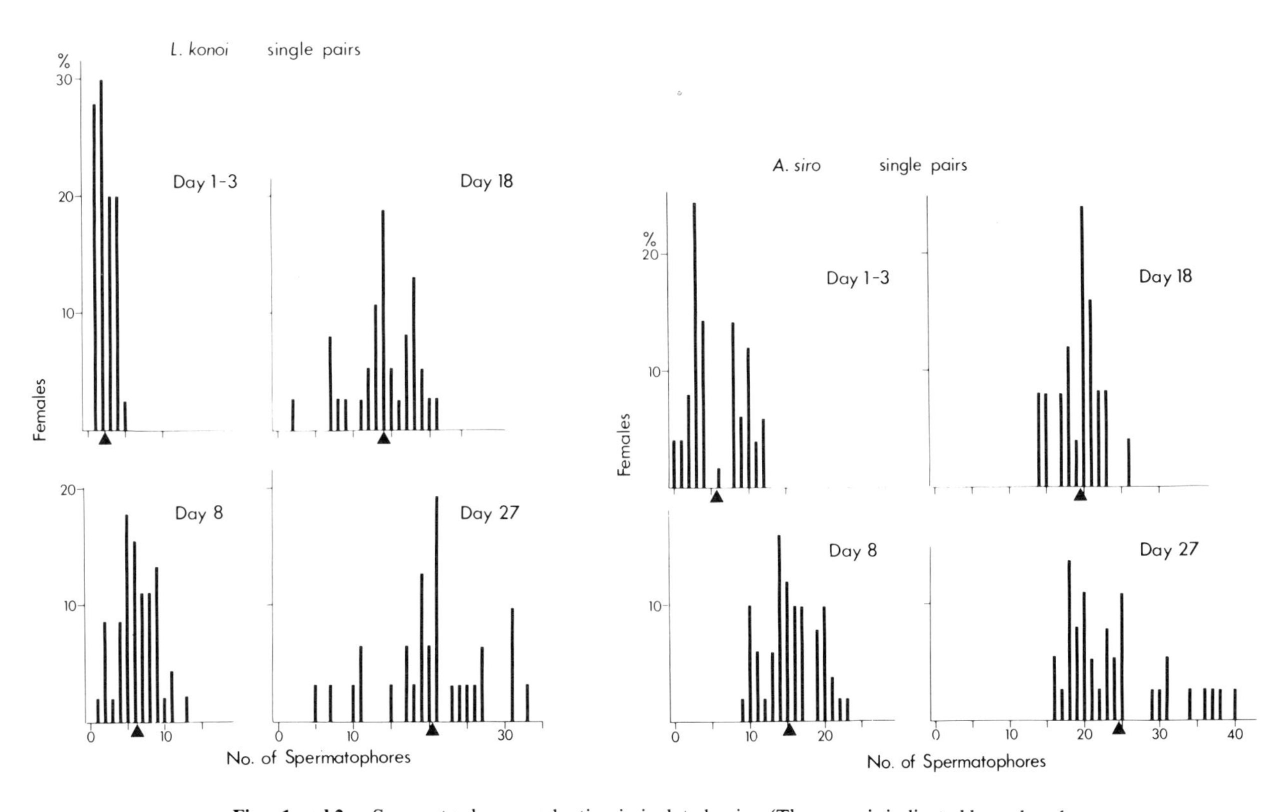

Figs. 1 and 2. Spermatophore production in isolated pairs. (The mean is indicated by a closed triangle). Fig. 1. *Lardoglyphus konoi;* Fig. 2. *Acarus siro.*

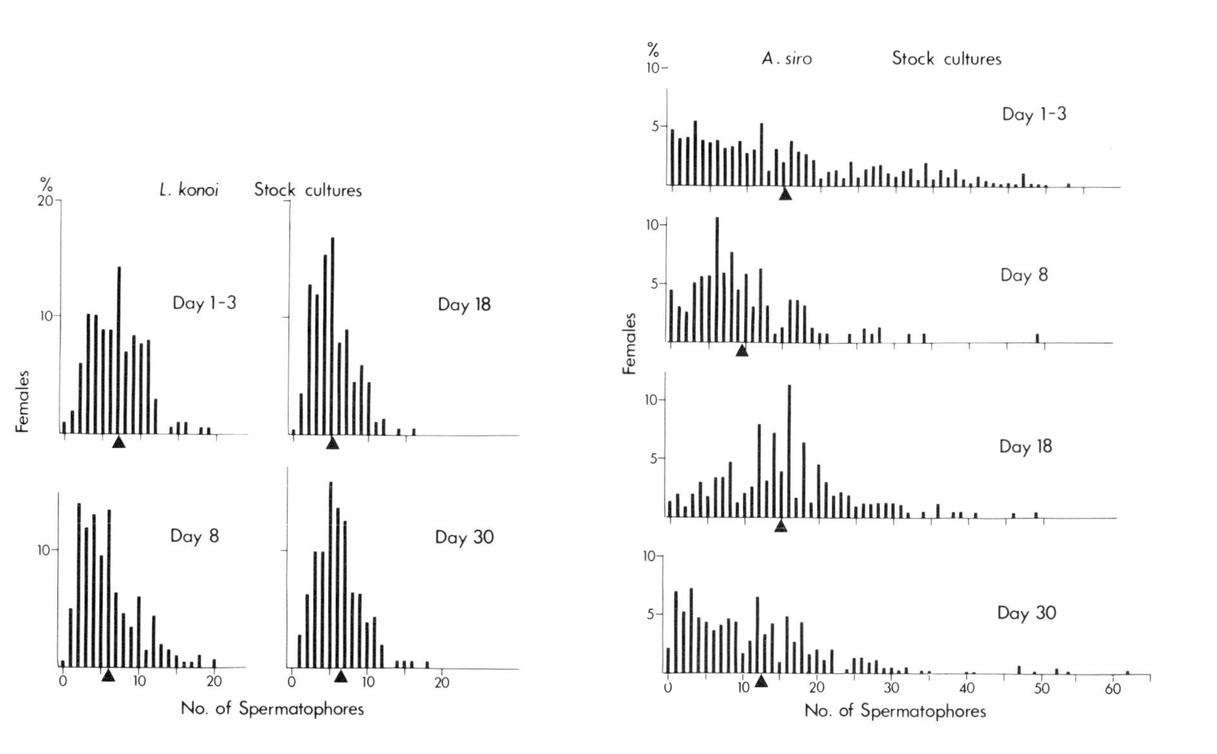

Figs. 3 and 4. The number of spermatophores contained within the receptaculum seminis of females taken, at intervals, from stock cultures. Fig. 3. *Lardoglyphus konoi;* Fig. 4. *Acarus siro.*

A Comparison of Single Pair Results with that of Cultures

L. konoi. The single pair result which most closely reflects the stock culture results is that of day 8, when the average number of spermatophores is about 7 (Figs. 1 and 3). At day 18 and 27 the average number of spermatophores per single pair female is 2 or 3 times that found in the stock cultures. At 18 days 84% of the females had been mated more than 10 times. In stock cultures the number mated more than 10 times never exceeded 15% of the female population.

A. siro. At no time do the single pair results resemble those of the cultures (Figs 2 and 4). However, in contrast with *L. konoi,* whilst *all* single pair females contained considerably more spermatophores (14 and over) than perhaps 50% of culture females, the distribution range for single pairs fell inside that for the cultures.

Performance of Males Under Different Sex Ratio Regimes (Table I)

The superiority of the *A. siro* male under single pair conditions or with 2 females is shown in Table I. At each time interval the average production of spermatophores by *A. siro* males is about 3 times that of *L. konoi* males. The interesting comparison comes at the 5 females to 1 male ratio. The addition of extra females to the cells stimulates the *L. konoi* males to increase their output of spermatophores—so that it is now about 3 times more than they produce under single pair conditions. However, *A. siro* males when faced with the same ratio of 5 females, or with 10, retain their same rate of output.

Table I.
Spermatophore Production in a Single Male
Confined for Different Periods of Time with 1, 2 or 5 Females.

Species	No. of ♀♀s	Average number of Spermatophores			
		1 day	3 days	6 days	11 days
L. konoi	1	1.7	2.6	3.8	8.5
	2	1.1	2.0	3.3	8.0
	5	2.4	6.3	12.0	22.2
A. siro	1	2.7	6.0	12.2	24.0
	2	3.0	4.1	10.0	27.3
	5	3.2	4.8	14.1	25.0
	10	3.8	5.2	13.0	24.8

DISCUSSION

Spermatophore production and, therefore, mating behaviour differs between the two species in a number of respects. Firstly, as shown by Griffiths and Boczek (1977) the spermatophore body within the receptaculum seminis of *A. siro* females breaks down within 1 to 3 days, leaving only a characteristic shaped 'tail.' Spermatophores of *L. konoi* retained their overall shape and volume throughout the life of the female. In theory, therefore, the capacity of the *A. siro* receptaculum seminis to receive spermatophores is far greater than it is in *L. konoi.*

Certainly, in single pair mating situations the *A. siro* male is able to generate spermatophores at about twice the rate of the *L. konoi* males. This suggests that the stimulus to mate must equally be greater but, at present, it is not possible to say whether the attraction performance of the female or the physiological readiness of the male provides the stimulus for mating. There is visual evidence that when virgin pairs of both species are placed together, the female is often the first to become aware of its partner. She will follow his trail around the cell and, it is not until the female's opisthosoma is close to the male gnathosoma, that the male becomes active. This indicates the release of an attractant stimulus by the female. Further evidence that such an attractant is released stems from the fact that when a female is released in the presence of a number of males, the unsuccessful males will pair with each other.

The increased performance of *L. konoi* males when confronted with 5 females and the absence of such a performance by *A. siro* males is difficult to interpret. It may reflect the nature of the sex ratio under natural conditions. For example, if the ratio of females to males is high, females may need to compete for males. If the sex ratio is close to 1 to 1 then this is no longer a requirement. Studies have shown that in cultures, the *A. siro* sex ratio is indeed 1 to 1.

The success rate of copulations between *A. siro* pairs appears to be better than for *L. konoi* pairs. For example, after 27 days as single pairs, all *A. siro* females had been mated at least 16 times, with the record standing at 40 times, almost equivalent to twice per day for a month.

The progeny results used for predicting life history tables and producing population models are always based on single pair matings. In the case of *L. konoi,* at 18 days or more, females in single pairs mate on average 2 to 3 times more frequently than do females in cultures. At 27 days they can often contain many more spermatophores than can be found in culture females. What effect these different types of behaviour have upon egg output and egg viability needs to be investigated and further work is planned in this direction.

Finally, the introduction of a technique for recording mating success will undoubtedly assist in the monitoring of performance prevailing during life history studies of these two species. In the case of *Acarus,* it can be used to predict the presence of hybrids in natural populations. Where males are given the choice of females of different species, it is now possible to record their mating preferences.

ACKNOWLEDGEMENTS

The experimental work in respect of the *Lardoglyphus* trials was carried out by Mrs. C. Singh at Slough.

REFERENCES

Griffiths, D. A., and Boczek, J. (1977). *Internat. J. Insect Morph. Embryol.* **6**, 231-238.

STERILITY INDUCED IN "COPRA MITE," *TYROPHAGUS PUTRESCENTIAE* BY IODINE SALTS

Stanislaw Ignatowicz and Jan Boczek

Agricultural University of Warsaw
Warsaw, Poland

INTRODUCTION

Some inorganic substances affect the fecundity and egg viability of vertebrates and invertebrates influencing the animal whole physiology. Among them, iodine and its compounds were found to exert strong physiological effects. Iodine induces mutations in *Drosophila melanogaster Meigen* (Ssacharov, 1936), oxidizes thiol groups (Anson, 1941), precipitates proteins (Gershenfeld and Hitlin, 1950), inhibits inorganic pyrophosphotase (Lambremont and Schrader, 1964) and carbonic anhydrase (Maren, 1967), binds to plasma proteins (Huang and Hickman, 1968) and concentrates in the ovary of vertebrates (Leloup and Fontaine, 1960). The inhibition of embryonic development was also reported for nematodes belonging to the genus, *Ascaris* (Zaman and Visuvalingam, 1967).

Crystal (1970) reported that the solutions containing I_2 and KI sterilized screw worm flies, *Cochliomyia hominivorax* (Coquerel) given multiple oral treatments and the severity of this antifertility effect was determined by the concentration. He found that reduced egg production was a more important component of sterility than reduced hatchability.

It was of interest to us to conduct the study with "copra mite," *Tyrophagus putrescentiae* (Schrank) and to compare the results obtained with those of Crystal (1970).

EXPERIMENTAL

Mineral compounds that are known to affect fecundity of various insects were chosen to study with "copra mite."

These minerals were incorporated into the mite's food as follows: Wheat

Vol. I: ISBN 0-12-592201-9

germ was ground in a porcelain mortar, and salts dissolved in distilled water were added. After thorough mixing and drying, this diet was again thoroughly ground.

Mites of one population were reared under constant temperature 25°C ±1°C and R. H. 85% in glass rearing cages on prepared diet of wheat germ. About 25 pairs of mites were observed in each test.

The initial experiments showed that among tested compounds added at 0.25% dose into wheat germ, only iodine salts caused almost complete inhibition of the egg production by *T. putrescentiae* females. Boric acid and sodium fluoride significantly affected mite's fecundity as well as longevity. Higher concentrations of these minerals were found to have strong toxic effects. The females and males lived just several days on diet with 2.0% conc. of boric acid and NaF as compared with controls, which lived 58.3 and 81.7 days, respectively. At the same time the iodine salts have no effect or moderately affect mite longevity.

The aim of the next experiment was to determine which salts cause sterility. Mites were placed on diets containing 0.25% or 2.0% of tested compounds for 0.5, 1.0, and 2.0 weeks, and then they were placed into new rearing cages with normal wheat germ diet. The data summarized in Table I show that the iodine salts—potassium iodide and iodate—induced a sterility in *T. putrescentiae;* the degree of this sterility depended upon the exposure time and the concentration of these minerals in the diet. Usually, the females laid fewer eggs feeding on wheat germ diet after 1 or 2 weeks exposure to iodine salts than they did after a half-week exposure.

TABLE I.
Effect of Salt on Fecundity of *Tyrophagus putrescentiae* Egg-Laying Females

Salt	Conc. %	Eggs Laid During Weeks of Exposure		
		0.5	1	2
KI	2.0	0.0-17.4	0.0-0.2	0.0-0.0
	0.25	0.0-267.5	0.6-171.8	1.0-27.5
KIO_3	2.0	0.0-183.6	0.6-142.2	1.2-28.6
	0.25	—	—	—
NaF	2.0	2.1-256.1	4.4-162.7	10.9-176.5
	0.25	11.8-275.7	46.2-198.7	108.1-177.4
H_3PO_3	2.0	0.0-0.0	0.0-0.0	0.0-0.0
	0.25	1.8-342.2	58.4-139.8	58.4-295.2
Control		43.7-417.2	128.1-307.9	231.5-184.7

Also fewer eggs were recorded when tested with 2.0% doses of iodine salts than in the case of diets containing 0.25% of those minerals.

The mites feeding on diet containing boric acid and sodium fluoride produced significantly fewer eggs than controls did. After transferring them to untreated food, they recovered their reproductive abilities, however, their fecundity was still lower than in the control.

This happened probably because of the accumulation of those toxic substances in their bodies. The "zero" results of the test with a 2.0% dose of boric acid did not reflect the sterilizing action of this compound, but resulted from the high mortality of the mites. Because boric acid and sodium fluoride showed simple toxic action against *T. putrescentiae,* they were eliminated from further study dealing with the sterilizing action of mineral salts.

The next experiment was conducted in order to find out how the exposure time and concentrations of potassium iodide and iodate induce sterility in "copra mite." This time, the anti-fertility effect of these minerals was measured as a percentage of non-laying females. The data of this experiment are summarized in Table II. They show that the number of sterile females was positively correlated with exposure time and concentration of iodine salts. The correlation coefficients calculated for all cases involved in this experiment varied from 0.609 to 1.000 indicating close relationships between the number of females which ceased egg production and the conditions studied. Crystal (1970) also found that the increase of concentrations of iodine compounds in the food caused the increase of sterility of *C. hominivorax.*

TABLE II.
Effect of Exposure Time and Salt Concentrations on *Tyrophagus putrescentiae* Egg-Laying Females

Salt	Conc. %	% Egg-laying Females in Weeks of Exposure		
		0.5	1	2
KI	0.25	100.0	61.0	8.0
	0.5	69.2	65.4	11.5
	1.0	43.0	24.0	0.0
	2.0	29.2	0.0	0.0
KIO_3	0.25	—	—	—
	0.5	100.0	100.0	0.0
	1.0	96.3	55.6	3.7
	2.0	88.0	65.0	8.0

T. putrescentiae females stopped egg-laying completely when they were allowed to feed for one week on diet containing 1.0% KI. After transferring to normal food, they did not recover their reproductive abilities. It was of interest to us to check if these iodine salts could inhibit production in mature egg-laying females.

Two experiments were conducted. In the first one, the mites, which had been fed on normal food for one week, were transferred onto diet with potassium iodide for second week and then again placed on wheat germ. The second experiment was done in a similar way except that the mites were allowed to feed on wheat germ for the first two weeks and then they were handled as above.

Results obtained (Tables III, IV) indicate that potassium iodide can induce a high level of sterility in one and two week old reproductive females and cause a decrease in the number of eggs laid.

TABLE III.
Anti-Fertility Effect of KI on *Tyrophagus putrescentiae* Egg-Laying Females

Salt Conc. %	1st Week: Wheat Germ Diet		2nd Week: Wheat Germ Diet		3rd-10th Weeks: Wheat Germ Diet	
	No. Eggs Laid	Egg-laying Females (%)	No. Eggs Laid	Egg-laying Females (%)	No. Eggs Laid	Egg-laying Females (%)
0	137.4	100.0	189.7	100.0	117.3	100.0
0.25	50.0	96.0	4.2	80.0	86.7	36.0
0.5	40.3	100.0	3.2	100.0	38.3	11.5
1.0	45.6	100.0	4.1	96.2	44.0	11.5

TABLE IV.
Anti-Fertility Effect of KI on *Tyrophagus putrescentiae* Egg-Laying Females

Salt Conc. %	1st and 2nd Weeks: Wheat Germ Diet		3rd Week: Wheat Germ Diet		4th-10th Weeks: Wheat Germ Diet	
	No. Eggs Laid	Egg-laying Females (%)	No. Eggs Laid	Egg-laying Females (%)	No. Eggs Laid	Egg-laying Females (%)
0	327.1	100.0	37.1	100.0	80.2	100.0
0.25	219.0	96.2	2.1	73.1	84.1	57.7
1.0	123.6	100.0	4.0	92.6	17.0	14.8

There remained another problem to solve: what sex is susceptible to sterilizing action of iodine salts—female or male or both?

To help answer this question, the following tests were conducted: 1) treated females—young females kept one week on diet containing KI were paired with untreated males—young males reared on wheat germ without any additional substances, 2) untreated females were paired with treated males, 3) treated females were paired with treated males and 4) untreated females were paired with untreated males; this served as a control.

The results are summarized in Table V. It can be seen that the highest sterilizing action of the salt were in the cases when treated females were paired with either treated or untreated males. When treated males were paired with untreated females, a very slight effect on sterility and fecundity of egg-laying females was found. The conclusion of these tests is that females are very susceptible to iodine action. The males are very slightly affected by this salt. This is supported by the next experiment (Table VI). The young males were kept 1, 4 and 8 weeks on diet and then they were paired with young females. The complete sterility of males was obtained after 8 weeks on a diet containing 1.0% of potassium iodide.

TABLE V.
Anti-Fertility Effect of 1% KI on *Tyrophagus putrescentiae*

Treatment[a] KI in Diet	Egg-laying Females (%)	Fecundity
T♀ × UT♂	26.9	178.3
UT♀ × T♂	96.2	325.7
T♀ × T♂	24.0	127.1
UT♀ × UT♂	100.0	437.8

[a]T = Treated, UT = Untreated

TABLE VI.
Effect of KI Exposure Time on *Tyrophagus putrescentiae* Male Sterility

Time (Weeks)	% of Egg-laying Females		No. Eggs Laid/Female	
	1.0% KI	Control	1.0% KI	Control
1	96.2	100.0	325.7	457.8
4	100.0	100.0	167.3	284.1
8	0.0	92.4	0.0	47.3

Also, when older developmental stages were exposed to diet with iodine salt, the females emerging from them were significantly affected (Table VII). The number of sterile females always exceeded 50%. The deutonymphal stage was very susceptible. The exposure of these mites during the duration of this form produced 73.1% sterility in females emerging from them.

TABLE VII.
Sterility of *Tyrophagus putrescentiae* Females
Induced by Feeding Developmental Stages Food with 1% KI

Developmental Stages Fed	Sterile Females (%)	No. Eggs Laid/Female
Protonymphs	50.0	215.5
Deutonymphs	73.1	77.7
Adults	100.0	0.0

CONCLUSION

It was concluded that females were more susceptible than males probably because iodine salts exerted a detrimental effect upon egg formation. There were observed a number of cases when females laid incompletely formed eggs. These eggs appeared like mere shells of the eggs, and they were, of course, not viable. On the other hand, the viability of normally formed eggs which were occasionally produced by treated females was always high, exceeding 90%. The following two points should be emphasized:

1) Iodine salts induce sterility in *T. putrescentiae* by reduction of egg production rather than by reduction of their hatchability. The same relationships were recorded by Crystal (1970).

2) Iodine salts inhibit the egg yolk production during oogenesis causing incompletely developed eggs to form. We suggest that this is caused when iodine compounds bind proteins during egg formation and inhibit their activities.

ACKNOWLEDGEMENT

This study was supported in part by a grant from the U.S.D.A. under PL-480 (FG-PO-271).

REFERENCES

Anson, M. L. (1941). *Gen. Physiol.* **24**, 399-421.
Crystal, M. M. (1970). *J. Econ. Entomol.* **63**, 1851-1853.
Gershenfeld, L., Whitlin, B. (1950). *Ann. N. Y. Acad. Sci.,* **53**, 172-182.
Huang, C. T., Hickman, C. P., Jr. (1968). *J. Fish. Res. Board Can.,* **25**, 1651-1666.
Lambremont, E. N., Schrader, R. M. (1964). *J. Insect Physiol.* **10**, 37-52.
Leloup, J. M. Fontaine. (1960). *Ann. N. Y. Acad. Sci.* **86**, 316-353.
Maren, T. H. (1967). *Physiol. Rev.* **47**, 595-781.
Ssacharov, W. W. (1936). *Genetica* **18**, 193-216.
Zaman, V., Visuvalingam, N. (1967). *Trans. Roy. Soc. Trop. Med. Hyg.* **61**, 443-444.

NATURAL REGULATION OF *TARSONEMUS GRANARIUS* NUMBERS IN STORED WHEAT ECOSYSTEMS—A MULTIVARIATE ASSESSMENT

N. D. G. White and R. N. Sinha

Agriculture Canada
Research Station
Winnipeg, Manitoba

INTRODUCTION

Tarsonemus granarius Lindquist (Tarsonemidae) has been found in association with aging stored wheat and oats in Canada (Sinha *et al.*, 1962; Sinha and Wallace, 1966), in cereal dust and spillage in England (Hughes, 1976) and on granary floors, in stored wheat and rice and on rice straw in the field in Japan (Sinha, 1968). Principal component analysis of a 95-mo study of wheat in farm granaries naturally infested with stored grain mites in Manitoba, a survey of mites and fungi in grain samples from western Canada (Sinha and Wallace, 1966) and a laboratory study of *T. granarius* propagation on seed-borne fungi (Sinha, 1964) indicate that *T. granarius* plays a fungivorous consumer role in a stored grain ecosystem. Complex interactions of mycophagous arthropod species were affected by moisture, temperature, grain depth and fungal species (Sinha *et al.*, 1969a, b).

Both preharvest and post-harvest fungi commonly found in stored cereals in Canada—*Alternaria, Penicillium, Aspergillus, Chaetomium,* and *Hormodendrum* can support thriving populations of *T. granarius* (Sinha, 1964). Microfloral activity leads to a rise in free fatty acid, changes in seed weight and volume, seed damage and discoloration, milling quality loss, moisture increase, etc., in stored grain and a deterioration in the overall quality of the grain. Because the incidence of mites and fungi are often correlated in deteriorating grain, *T. granarius* may be used as one of the biological indicators of microfloral activity and hence of overall quality of stored grain.

The purpose of this study was to determine the nature of interactions in tough wheat naturally infested with very low levels of *T. granarius* in artificial ecosystems maintained at 30°C.

Vol. I: ISBN 0-12-592201-9

MATERIALS AND METHODS

Initially, two 204 litre steel drums, standing on end, were each filled to a level *ca* 5 cm from the top with 157 kg (5.6 bu) of No. 2 Canadian Western red spring wheat (*Triticum aestivum* L., cv. Neepawa) harvested in 1976 at Glenlea, Manitoba at 15.5% moisture content and containing less than one percent dockage (weed seeds, broken kernels, etc.). Nine sample locations were selected within each drum. Holes 3.1 cm in diameter were drilled and plugged with rubber stoppers, at depths of 15, 45, and 75 cm from the surface on two sides of each drum to facilitate sample collection at regular intervals. A hole 6.3 cm in diameter at the centre-top of each drum was covered with a fine wire mesh to allow ventilation. A sliding plate covered a hole 3.1 cm in diameter at the bottom of each drum. Each 200 ml sample of wheat (*ca* 137 g) was removed by allowing the grain to flow out of the openings, except for the centre-surface and centre-middle samples which were removed with a torpedo probe. Copper-constantan thermocouples and plastic gas tubes (3.5 mm diam) were placed in positions corresponding to sample locations. The drums were held in a controlled environment room at 30 ± 2°C and *ca* 40-50% R.H. Control wheat (156 kg) was placed in plastic bags at −15°C. Samples were taken tri-weekly for 33 wk.

One day prior to each sampling date temperatures were recorded (°C) at each sample location with a Digimite® potentiometer (Thermo Electric, Saddle Brook, New Jersey) and 30 ml of gas were drawn from each sample location. The percentage of oxygen and carbon dioxide at each location was determined with a Matheson gas chromatograph, model 8430, (Matheson Gas Products, P. O. Box 85, East Rutherford, New Jersey 07073).

Once the wheat had been removed from the drums (18 jars containing 200 ml of wheat each) on each sampling date, the dust was sieved through a 40 mesh (420 μm) sieve, weighed, and the volume determined. Subsamples of 50 ml of wheat were removed from each sample jar and placed in plastic bags and refrigerated at 5°C pending further analysis. The sifted dust was returned to the initial sample (150 ml) and placed in a Berlese funnel for 24 hr to extract the mobile stages of mites (Sinha, 1961) which were preserved in 70% alcohol and counted under a dissecting microscope. Free fatty acid values (FAV) in the grain were determined in duplicate on pooled samples from each drum (AACC method 02-01, Anonymous, 1962).

The analysis of each subsample included wet kernel moisture content determination (oven-dry method, ASAE method No. S 352, Anonymous 1975), the weight, volume and damage of 100 kernels per samples. Germinability of the wheat and microfloral infection were observed from 25 randomly chosen kernels per sample by the filter paper method of Wallace and Sinha (1962) as modified by Lustig *et al.* (1977), after surface sterilzation for one minute in a one percent sodium hypochlorite solution (Tuite, 1969). Principal component analyses were done with 18 samples for each sampling date using the University of California BMDD program on an IBM 360-65 computer. Multiple linear regression was done with the Statistical Package for the Social Sciences (SPSS) program. All variables were analyzed after $\sqrt{x}$ or $\sqrt{x+1}$ transformation with the exception of mite numbers which received a $\log_{10}(x + 1)$ transformation (Goulden, 1945; Sinha *et al.*, 1969a).

Principal component analysis was undertaken on 12 variables: *Tarsonemus granarius*, FAV, CO_2, O_2, temperature, grain weight, dust weight, moisture, *Alternaria*, *Aspergillus glaucus* group, bacteria, and seed germination. Variables were omitted from analysis if they were recorded in fewer than 10% of the samples. Only the first two principal components are reported since Component 2 may generally be defined as the *T. granarius* component. The first two components together accounted for approximately 65% of the variation in the system. An arbitrary eigenvector loading cutoff level of 0.30 was used in interpretation of the analysis (Sinha, 1977).

Multiple regression analysis was undertaken on eight independent variables: FAV, CO_2, O_2, temperature moisture, *Alternaria, Aspergillus glaucus,* and bacteria, acting with the dependent variable, *T. granarius.* An analysis was done on all data from wks 0-33 with a total of 198 samples. At wk 15, both systems became lightly infested with the saw-toothed grain beetle, *Oryzaephilus surinamensis* L. However, insect free wheat simultaneously held in the same environment did not show a difference in the variables monitored for 33 wk.

RESULTS

The initial level of *T. granarius* infestation was 1-2 mites per 150 ml sample but after 9 wk the mean number of mites per sample increased dramatically to 1500/sample in drum 1 and 580 in drum 2 (Fig. 1). A sharp decline in mite numbers was evident between wk 9-15 followed by a secondary peak in numbers at wk 21. The number of *T. granarius* leveled off between wk 27 and 33 at less than 50 per sample. Generally, mite numbers were highest along the top surface and bottom centre of the drums.

The mean oxygen concentration at all levels in the two systems did not fluctuate sharply, remaining in the range of 14-18% throughout the study. The mean CO_2 concentration increased to 8-12% by wk 12 declining to 4-7% by wk 33. CO_2 was typically more concentrated in the lower levels of the drums.

The mean moisture content of the wheat in the two drums changed with time. Grain at the surface dried slightly to *ca* 14.8% moisture content by wk 33 whereas that at the bottom gradually became more moist, exceeding 17% by wk 33. Mean FAV of the two drums increased with time rising from 13 mg KOH per 100 g of wheat at wk 0 to 45 at wk 33.

Grain weight remained relatively constant throughout the study as did the grain volume. Dust weight and volume remained at low levels during 33 wk at

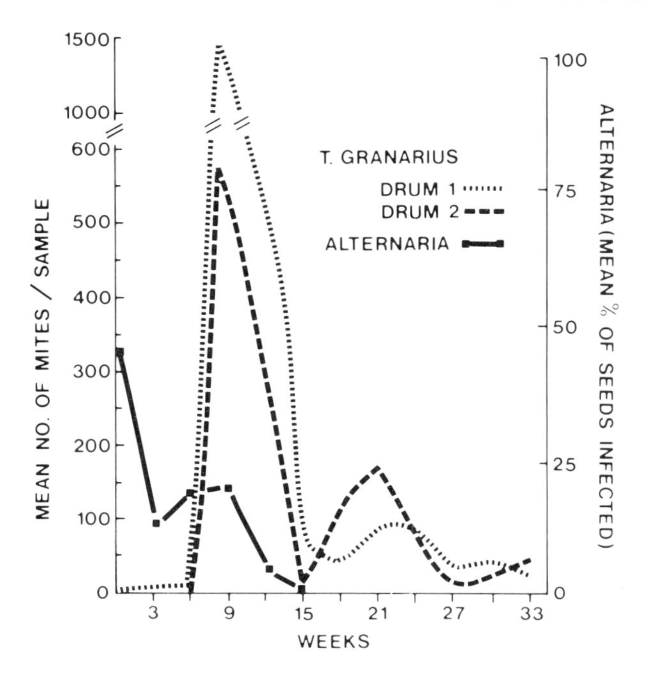

Fig. 1. Mean number of *Tarsonemus granarius* per 150 ml sample of wheat in drums 1 and 2 and the mean percent infection of seeds with the fungus *Alternaria* during 33 wk.

less than 0.05 g and 0.10 ml, respectively. No obvious kernel damage was evident during the 33 wk of study.

Slight temperature gradients were present in the grain which could have been caused by external environmental factors.

Seed germination declined sharply after wk 6 reaching 0% at wk 15 in all samples.

Alternaria spp., the most abundant seed-borne preharvest fungi, declined rapidly during the first few weeks of study (Fig. 1) and virtually disappeared after wk 15. *Helminthosporium* spp. were also relatively abundant only until wk 9.

Aspergillus glaucus group were the most abundant fungi after wk 9 increasing to 50% of the seeds infected by wk 12. This fungus sharply declined after wk 30.

Bacteria increased with time throughout the study, reaching an average infection rate of about 50% in all samples by wk 33.

MULTIVARIATE ANALYSIS

The fleeting relationships among variables in the dynamic ecosystem environment could be seen through principal components 1 and 2 from the principal component matrices. Although most variables were represented at each sampling time, usually about one-half of the total variables were statistically correlated and possibly meaningfully interacted.

Component 1: At the beginning of the study this component was defined by one biological and three non-biological variables—FAV, CO_2, O_2, and grain moisture. Several other variables such as temperature and bacteria also interacted and progressed at each time. The interacting variables ranged from 3 to 7 during the study period. They accounted for the variability in the overall system ranging from an initial level of about 25% to 45% at wk 15.

Component 2: Initially this component was defined by the variables temperature, grain weight, dust weight, and germination. The relationships encompassed by this component are somewhat less consistent than those in Component 1. However, re-occurring relationships involved FAV, temperature, moisture, *A. glaucus,* bacteria, and *T. granarius.* Three to six variables accounted for about 18% to 25% of the variability in the systems under study. *T. granarius* interacted with other variables only on wk 3, 12-15, 21 and 27.

The degree of linear dependence of *T. granarius* on the eight independent variables as revealed by multiple regression analysis is summarized in Table 1.

TABLE I.
The Multiple Regression Summary Table of the Prediction Equation for *Tarsonemus granarius* Numbers in a Model Ecosystem

Variable	Multiple R[a]	R^2[b]	Simple R[c]	B[d]	Beta[e]	Std. Error of B	F[f]	F[g]
CO_2	0.407	0.166	0.407	1.044	0.467	0.249	17.567**	42.150**
Alternaria	0.450	0.203	0.304	0.379	0.301	0.124	9.272**	9.415**
Aspergillus	0.471	0.221	0.177	0.082	0.131	0.061	1.794	4.580*
Moisture	0.481	0.231	−0.081	−1.078	−0.125	0.634	2.886	2.546
FAV	0.489	0.240	−0.164	0.140	0.156	0.141	0.986	2.290
O_2	0.505	0.255	−0.118	0.682	0.184	0.380	3.222	3.995*
Bacteria	0.506	0.256	−0.194	0.046	0.071	0.089	0.270	0.254
Temperature	0.507	0.257	−0.098	−0.329	−0.035	0.739	0.198	0.254

*Significant at the .05 level, $F_{.05}$ (1,189) = 3.84; d. f. = N − k − 1 where N = samples size, k = number of independent variables
**Significant at the .01 level, $F_{.01}$ (1,189) = 6.63
[a]Multiple correlation coefficient
[b]Coefficient of determination
[c]Correlation coefficient
[d]Unstandardized regression coefficient; constant = 0.539
[e]Standardized regression coefficient
[f]F statistic for standard regression, d. f. = 1,189
[g]F statistic for hierarchial analysis, d. f. = 1,189

The general form of the unstandardized multiple regression equation is:

$Y^1 = A + B_1X_1 + B_2X_2 + \ldots + B_kX_k$ where Y^1 represents the estimated value for Y, A is the Y intercept, and B_i are regression coefficients. Multiple regression analysis resulted in the following equation:

Y^1 (*T. granarius*) = 0.539 + 1.044 (CO_2) + 0.379 (*Alternaria*) + 0.082 (*A. glaucus*) − 1.078 (Moisture) + 0.140 (FAV) + 0.682 (O_2) + 0.046 (Bacteria) − 0.329 (Temperature). Eq. (1). Values associated with the independent variables are the unstandardized partial regression coefficients. These values were used instead of the standardized coefficients to facilitate interpretation since the data had initially been transformed. The multiple correlation coefficient (R) was 0.507 indicating that all eight independent variables accounted for 25.7% of the variation in *T. granarius* numbers. The standard error of estimate for the equation was 0.715. The F ratio for the regression equation was 8.171 indicating statistical significance ($P < .01$).

The regression coefficients (B) for CO_2 and *Alternaria* were significant ($P < .01$) using a standard regression method while those of CO_2 and *Alternaria* ($P < .01$) and *Aspergillus* and O_2 were significant ($P < .05$) using a hierarchical method (Kim and Kahout, 1975) (Table 1, col. F^f, F^g).

The control wheat stored at −15°C showed no change in FAV, moisture, or microfloral infection after 33 wk.

DISCUSSION

Principal component analysis resolves the total variation of a set of variables into linearly independent composite variables which successively account for the maximal variability in the data (Sinha, 1977). Successive and regular analysis of 12 variables including *T. granarius* revealed changing interrelationships between this mite and other variables, eight of which were then used as independent variables in a stepwise multiple linear regression using *T. granarius* as a dependent variable. The regression analysis has shown that CO_2, O_2 and the fungi *Alternaria* and *Aspergillus glaucus* were significant predictors of *T. granarius* numbers (Table I). Since multiple regression was used as an inferential tool to provide some clues about the causal factors involved in the abundance of *T. granarius,* the relatively low R^2 values were not deemed critical.

At various times in the 33 wk study, the results of principal component analyses have generally and often temporarily indicated positive relationships between *T. granarius* numbers and *Alternaria, Aspergillus glaucus,* O_2 levels, FAV, and temperature. Negative relationships were often demonstrated between *T. granarius* and bacteria, CO_2, and moisture.

Multiple linear regression of the entire data set from 33 wk has shown positive correlations between *T. granarius* and CO_2, *Alternaria, Aspergillus glaucus,* FAV, O_2, and bacteria and negative relationships with moisture and temperature.

Using the interrelations previously outlined, the following succession of events seem to have a bearing on the natural regulation of the numbers of *T. granarius.*

The rapid increase in *T. granarius* numbers between wks 6-9 and the precipitous decline between wks 9-15 may be related to the presence and decline of the fungus *Alternaria* which is one of the most favoured diets of this mite (Sinha, 1964). The decline of *Alternaria* usually corresponds to a decline in seed viability (Sinha *et al.*, 1969a). At wk 9, CO_2 levels were at the highest in the entire study, primarily due to microfloral respiration (Trisvyatskii, 1966). Accumulated CO_2 may have had a deleterious effect on *T. granarius.* As *Alternaria* infection declined, *Aspergillus glaucus* and bacteria increased. *Aspergillus* spp. are an alternative food source for this mite and probably allowed the populations to maintain themselves at lower levels. High moisture levels, also a product of microbial respiration, favour bacterial growth which inhibits the growth of *A. glaucus.*

FAV is a measure of grain deterioration which increases steadily with time primarily due to microfloral activity (Zeleny, 1954). In this study, *A. glaucus* was usually correlated with FAV and *T. granarius* numbers after several weeks of storage.

It is possible that in warm, moist grain with prior field infections of *Alternaria, T. granarius* can reach high numbers and maintain itself at lower levels in the presence of a less favorable food source such as *A. glaucus,* although further deterioration of the grain and an increase in bacterial infection leads to a decline in mite numbers.

SUMMARY

Population fluctuations of a mycophagous mite, *Tarsonemus granarius* Lindquist, were studied for over 33 weeks in two 157-kg wheat bulks of 15.5% moisture content stored at 30 ± 2°C. The causes of natural regulation of the *T. granarius* numbers were explored through simultaneous monitoring of 12 variables including several species of microorganisms, O_2, CO_2, free fatty acid, seed germination. The variables were analyzed by conventional descriptive and multivariate statistical methods including principal component analysis and multiple linear regression. The concentration of CO_2 and O_2 and fungal variables *Alternaria, Aspergillus glaucus* group appeared to have considerable influence on regulation of *Tarsonemus* numbers. An initial rapid increase in *T. granarius* numbers was associated with the presence of the food source *Alternaria* and smaller populations present after 15 wk were associated with *Aspergillus glaucus.*

ACKNOWLEDGEMENTS

We thank Drs. T. D. Galloway, N. J. Holliday, and F. L. Watters for critically reviewing the manuscript. The work was supported in part by a National Research Council of Canada operating grant to Dr. R. N. Sinha as Honorary Professor, Faculty of Graduate Studies, University of Manitoba, Winnipeg.

REFERENCES

Anonymous (1962). "Cereal Laboratory Methods." 7th ed. *Amer. Assoc. Cereal Chem.,* St. Paul.

Anonymous (1975). "Agricultural Engineer's Yearbook." *Amer. Soc. Agr. Eng.,* St. Joseph, Mich.

Goulden, C. H. (1945). *Proc. Entomol. Soc. Manit.* **1**, 29-31.

Hughes, A. M. (1976). "The Mites of Stored Food and Houses." 2D ed. *H.M.S.O.,* London.

Kim, J.-O., and Kahout, F. H. (1975). *In* "Statistical Package for the Social Sciences" (N. H. Nie, C. H. Hull, J. G. Jenkins, K. Steinbrenner, and D. Bent, eds.), pp. 320-367. McGraw-Hill, New York.

Lustig, K., White, N. D. G., and Sinha, R. N. (1977). *Environ. Entomol.* **6**, 827-832.

Sinha, R. N. (1961). *Can. Entomol.* **93**, 609-621.

Sinha, R. N. (1964). *Acarologia* **6**, 372-389.

Sinha, R. N. (1968). *Ann. Entomol. Soc. Amer.* **61**, 938-949.

Sinha, R. N. (1977). *Environ. Entomol.* **6**, 185-192.

Sinha, R. N., and Wallace, H. A. H. (1966). *Ann. Entomol. Soc. Amer.* **59**, 1170-1181.

Sinha, R. N., Liscombe, E. A. R., and Wallace, H. A. H. (1962). *Can. Entomol.* **94**, 542-555.

Sinha, R. N., Wallace, H. A. H., and Chebib, F. S. (1969a). *Ecology* **50**, 536-547.

Sinha, R. N., Wallace, H. A. H., and Chebib, F. S. (1969b). *Researches on Population Ecology* **11**, 92-104.

Trisvayatskii, L. A. (1969). "Storage of Grain." Translation of "Khranenie Zerna," by D. M. Keane. National Lending Library for Science and Technology, Boston Spa, England.

Tuite, J. (1969). "Plant Pathological Methods." Burgess, Minneapolis.

Wallace, H. A. H., and Sinha, R. N. (1962). *Can. J. Plant Sci.* **42**, 130-141.

Zeleny, L. (1954). *In* "Storage of Cereal Grains and their Products" (J. A. Anderson and A. W. Alcock, eds.) pp. 46-76. *Amer. Assoc. Cereal Chem.,* St. Paul.

SURVIVAL OF MITES ASSOCIATED WITH GROWING BARLEY THROUGH HARVEST AND INTO STORAGE

N. Emmanuel

Department of Agricultural Zoology and Entomology
Agricultural College of Athens
Athens, Greece

G. O. Evans

Department of Agricultural Biology
University College, Dublin

INTRODUCTION

A study of the Acari associated with barley during growth was undertaken at Lyons Estate, Co. Kildare, Ireland, during 1975 and 1976. Thirty-one species of mites were collected from the aerial part of the cereal. The most abundant and frequent were: *Tyrophagus longior* (Gervais), *Lupotarsonemus talpae* (Schaars.), *Siteroptes graminisugus* (Hardy), *Tarsonemus confusus* Ewing and *Tydeus* nr. *mumaei* Baker. Twenty species were found on the inflorescences and these included all the most abundant species listed above.

A comparison of the species found in the field with those recorded from stored barley in Ireland by Cusack *et al.* (1975) showed that many of the species were in common. It must be stressed, however, that *Acarus siro* L. the major pest species of stored grain was absent in all habitats—cereal, soil, weeds, organic debris, air, hedgerows—in the field. The possibility that some field species are introduced into storage at harvest was investigated.

MATERIALS AND METHODS

The crop on 4 experimental plots (I-IV) was harvested by combine harvester. Ten glass containers (2 lb. kilner jars) were used to store a sample of grain from each plot. In addition, 2 laboratory "silos" were used to store grain from 2 of the plots. Each silo was constructed from a polythene sewer pipe, 37 cm in diameter. Three holes, 3 cm in diameter, were drilled at distances of 7, 27, and 67 cm from its base. The samples from the jars were

Vol. I: ISBN 0-12-592201-9

taken on 10 occasions: the first sample was taken on the harvest day (12 September, 1975) and the last after 13 months of storage. The remaining samples were taken at irregular intervals to ascertain the survival of species during prolonged storage. The samples from the silos were taken from each hole by using a Corcoran bag sampler, on five occasions: 22 September, 7 and 22 October, 24 November in 1975 and 26 January in 1976. The grain of both silos was found to have become mouldy and compacted after 13 months storage. Samples taken from the grain at this time yielded no mites. The extraction of the mites was carried out by a modified Tullgren funnel unit (Cusack *et al.*, 1975). The temperature was obtained with a mercury thermometer while a Protimeter Grain Master Tw 70 was used to measure the moisture content of the grain.

RESULTS

Since the measurements of the temperature and of the moisture content of the grain were taken only on the sampling dates, the data give only an indication of the conditions prevailing in the samples over the period of observation. Newly harvested grain had a high moisture content; about 21% in all plots. At all other sampling dates in the jars, the moisture content had a lower value than that at the first date but it never fell below 14%. In the silos the moisture content of the grain remained relatively constant with a tendency for a slight increase in the levels of the lower sampling apertures. The temperature of the grain in the jars remained over 13°C during the experimental period while it decreased with time at all levels in silos where it reached a minimum of 6°C on 26 January. In general, from the period of harvest up to 24 November in 1975, during which frequent sampling took place the grain in the silos had a higher moisture content and lower temperature than the grain in the glass containers. The differences between the jars and silos are attributable to different storage conditions.

Species Occuring in Harvested Grain

Fourteen species were found in the experimentally stored barley throughout the sampling period. A comparison of these species with those associated with the barley and weeds at harvest showed that all species except *Tyrophagus putrescentiae* (Schrank) and *Proctolaelaps hypudaei* (Ouds.) were in common. *T. putrescentiae* was found in the samples taken 13 months after harvest and it is probable that this together with *P. hypudaei* originated from the laboratory as contaminants.

All the species found on barley inflorescences were also found in the stored grain in plots I and II while 75% and 83% of the species were present in the stored grain in plots III and IV, respectively. From the total mite species occurring on the stems of barley only 82%, 67%, 78% and 63% were recorded from the harvested grain in the plots I-IV respectively, while the comparable values for weeds were 88%, 67%, 63% and 56%. Excluding *T. putrescentiae, T. longior* was by far the most abundant species in grain from all the plots in both jars and silos. Other species occurring in numbers and with a high constancy were the Tarsonemidae and Pyemotidae. The percentage of the total population of the most abundant mites in the field at harvest day as well as in the stored grain during the first two sampling dates are shown below:

	Barley-Inflorescences	Barley-stems	Weeds	Stored grain
T. longior	50	25.8	41.8	23.2
Tarsonemidae	41.4	59.3	46.2	67.7
Pyemotidae	4.5	1.6	2	7.3

It is evident from the above that the most abundant species found in the harvested grain were also the most abundant in the field. It is also probable that the tarsonemids and pyemotids because of their smaller size and heavier sclerotization survived harvesting better than *T. longior.*

Notes on Selected Species

Tyrophagus longior: This species was introduced from the field in small numbers (1 to 29 individuals per 30 g.d.w.) of all developmental stages. The population immediately started to increase in storage and on 22 October its population reached over 600 mites per 30 g.d.w. from all plots in the jars whereas on 24 November in Silo I and on 22 October in Silo II, numbers over 1300 mites per 30 g.d.w. were found. Subsequently the population declined so that at the last sampling of the jars only 6 to 56 mites per 30 g.d.w. were recorded. The vertical distribution of this species in the silos showed a tendency toward concentration in the lower levels.

Lupotarsonemus talpae and *Tarsonemus confusus*: In the newly harvested barley these species were found mainly as females and in numbers from 12 to 49 specimens per 30 g.d.w. Although a number of larvae was recorded, they did not establish a successful breeding population. As a result they had disappeared from the jars after the 22 October, 1975 (except for one record on 12 October, 1976 in plot II) and they were virtually absent from the silos by 26 January, 1976.

Siteroptes spp.: These species were represented by a few females in newly harvested grain. They did not develop a breeding population and had disappeared from the jars and silos earlier than the tarsonemids.

All the other species considered to have been introduced in the grain from the field were recorded rarely and in very low numbers.

DISCUSSION

One of the most significant findings of the study is that the great majority of species, even those which were in relatively small numbers, found on the barley (inflorescences and stems) and weeds at harvest day were also found in the harvested grain. Although it seems that the small tarsonemids and pyemotid species are introduced in the harvested grain more readily than the other mites, relatively large species such as adult of *T. longior, Tydeus* nr. *mumaei* and *Amblyseius cucumeris* (Ouds.) were also introduced. The transference of mites from the field, however, does not necessarily lead to their establishment in the new environment.

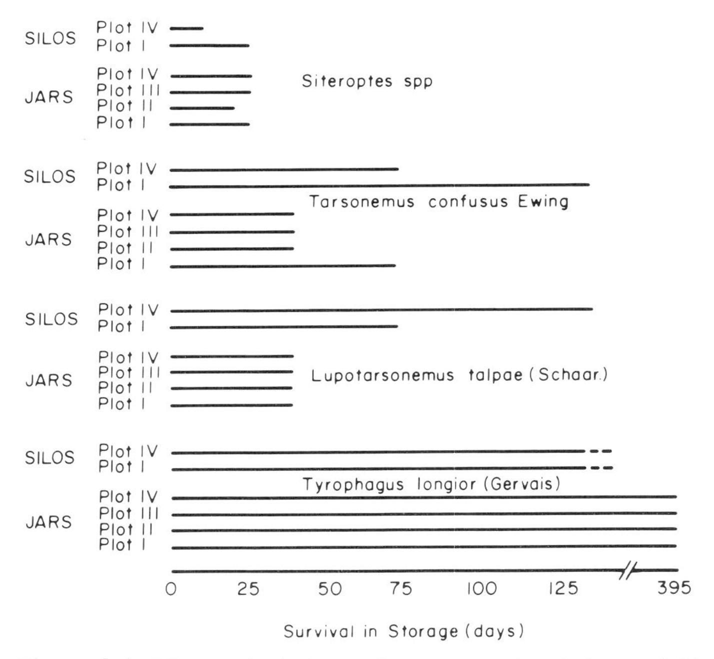

Fig. 1. The survival of the quantitatively most important species of mites carried into storage with harvested barley.

The survival of the quantitatively most important species carried into storage from the field is shown in Fig. 1. The only species which survived up to the last sampling date, that is, thirteen months after harvest, was *T. longior.* Heterostigmatic mites, especially *Siteropes* spp., by not establishing breeding populations did not survive for long. The failure of these mites to become established in grain stored under the conditions of the experiment may be attributed mainly to the lack of suitable food. According to Machacek *et al.* (1951) and Christensen and Kaufmann (1969) the major field fungi present on cereal kernels are species of *Alternaria, Cladosporium, Helminthosporium,*

Fusarium and *Nigrospora*. Although these fungi may survive for years in dry grain, they die rapidly in grain with high moisture content. Thus, fungi which apparently support populations of tarsonemids (Suski, 1972), *Siteroptes* spp. (Alfaro, 1946) and possibly *Tyrophagus* sp. (Sinha, 1964) in the field would soon disappear under the temperature and moisture regimes obtaining in the experimental jars and silos. *T. longior* possibly survived because it does not depend on a particular fungus or a group of fungi or because it may utilize grain particles as food. It is known (Sinha and Wallace, 1966) that grain spill and dust on the granary floors support a number of field and storage fungi but no such grain debris was present in the silos or glass containers used to store the harvested grain. The tendency for the mites to concentrate in the lower levels of the silos would assist their contact with the fungi which probably exist in grain debris. It is known from experiments by Cusack and Brennan (1975) and from our own experiments that clean grain is infested by mites from an infested area very rapidly. It seems, therefore, that if the introduced mites survived into the grain spill of the stores, they would disseminate with fungi through the clean material. The importance of the renewal of mite infestation of grain from the field should not, however, be overemphasized, as it seems that the main pest species have become adapted mainly or exclusively to granary habitats, for example, *A. siro* L., *Glycyphagus destructor* (Schrank) and their predator, *Cheyletus eruditus* (Schrank).

REFERENCES

Alfaro, A. (1946). *Bol. Pat. veg. Ent. agric.* **14**, 321-334.

Christensen, C. M., and Kaufmann, H. H. (1969). "Grain Storage: the Role of Fungi in Quality Loss." Univ. Minnesota Press.

Cusack, P. D., and Brennan, P. A. (1975). *Sci. Pro. R. Dublin Soc.* **3**, 341-353.

Cusack, P. D., Evans, G. O., and Brennan, P. A. (1975). *Sci. Proc. R. Dublin Soc.* **3**, 273-329.

Machacek, J. E., Cherewick, W. J., Mead, H. W., and Broadfoot, W. C. (1951). *Sci. Agr.* **31**, 193-206.

Sinha, R. N., and Wallace, H. A. H. (1966). *Ann. Entomol. Soc. Amer.* **59**, 1170-1181.

Suski, Z. W. (1972). *Zesz. Probl. Postep. Nauk. Poln.* **129**, 111-137.

STUDIES ON THE MITES OF STORED CEREALS IN YUGOSLAVIA

Neda Pagliarini

Faculty of Agricultural Science
Institute for Plant Protection
Department of Zoology
Zagreb, Yugoslavia

INTRODUCTION

So far in Yugoslavia only a few workers have been concerned with stored product mites (J. A. Freeman, B. Tomasevic, T. Stojanovic and Z. Korunic). Most of their studies were faunistic surveys undertaken while investigating stored product insects.

The present work was carried out in that part of Yugoslavia known as the Republic of Croatia, where for the first time detailed investigations on stored product mites have been made. The aim of the work was to establish which species of mites occur in our stores and in stored cereals, to determine the relations between infestation of stored cereals and quality of cereals, moisture content of grain, the method of maintenance, longevity and the method of storage. The investigations were carried out in the most important cereal production areas of Croatia, the plains.

EXPERIMENTAL

In the 4-year period from 1972 to 1976, 353 samples of wheat, maize, barley, rye, oats and refuse were collected from 63 different storage facilities at the 25 localities indicated in Fig. 1. Collections were made from the following types of storage facilities: silos built from concrete, metal or wood; different types of granaries; transit mill granaries (identified as "mills" in Table I); and temporary storage places; all with very different storage conditions. Refuse samples included dust, broken grain, shells, small pieces of straw and other impurities taken from floors, walls, windows and other hidden places in the storage facilities.

Vol. I: ISBN 0-12-592201-9

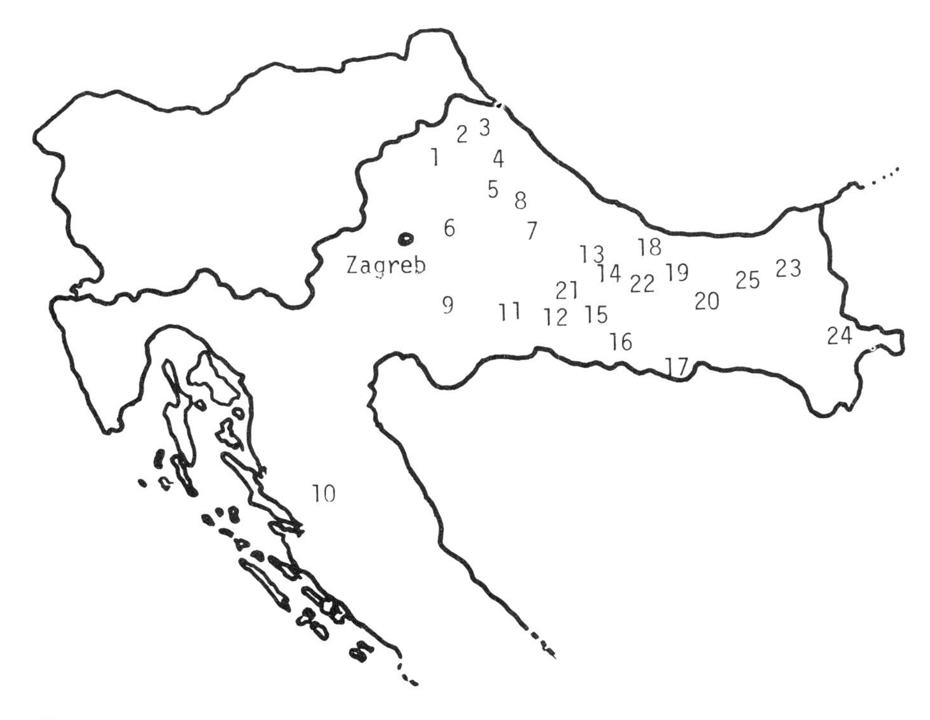

Fig. 1. Map of area in Croatia where cereal samples were collected. Numbers 1 to 25 indicate locations of storage facilities.

The material was collected under different climatic conditions ranging from an extremely humid year through standard climate to extremely dry years. Humid conditions during ripening, and particularly during harvest, in most cases caused cereals with a high moisture content to be placed into storage without drying. Such material could serve as an excellent medium for the rapid development of mites brought into the store houses directly from the field with the grain, and for mites which persist in storage facilities where hygienic measures are not satisfactory.

RESULTS AND DISCUSSION

Of the 353 samples collected, some 73.3% were infested by mites (Table I). Overall, and in the major cereals wheat and maize, the lowest percentages of infested samples were found in the silos: 66.5%, 64.1% and 59.2% respectively. This is because silos often have better conditions for storage: turning and cooling facilities and fumigation of silo bins. The moisture content of the grain in the silos was 11-13%, while that in granaries was 13-15% in bulk storage and 14-15% in bags, and in temporary storage facilities 14.5%. Grain is dried

TABLE I.
Summary of Mite Infestations in Cereal Samples Collected in Various Types of Storage Facilities in Croatia from 1972 to 1976

	Type of Storage Facility (no.)				
	Silos (16)	Granaries (26)	Mills (8)	Temporary (13)	Total (63)
Wheat					
No. samples	78	51	14	4	147
% infested	64.1	90.2	85.7	100	76.2
Maize					
No. samples	49	44	10	8	111
% infested	59.2	63.0	80.0	62.5	62.2
Barley					
No. samples	14	4	—	1	20
% infested	93.3	75.0	—	100	90.0
Rye					
No. samples	2	4	3	—	9
% infested	50.0	100	100	—	88.9
Oats					
No. samples	9	3	—	2	14
% infested	88.9	100	—	100	92.9
Refuse					
No. samples	11	21	16	4	52
% infested	63.6	72.2	81.3	75.0	75.0
TOTAL					
No. samples	164	127	43	19	353
% infested	66.5	78.0	83.7	78.9	73.3

not only to make it safe, but because grain of lower moisture content can be sold by the producer for a higher price. Grain with a moisture content higher than 12% cannot be sold.

On the other hand, the highest percentage of infested samples overall was found in grain originating in transit mill granaries (83.7%) located on small private farms, usually with very bad storage conditions.

Barley, rye and oats are produced in smaller quantities in Yugoslavia than are wheat and maize, and so fewer samples of these grains were taken. Probably a less accurate picture of mite infestation in these grains than in wheat and maize is presented here; for example, the high percentage of infested samples of barley (90.0%) was largely due to the fact that most of the samples originated from one silo with rather bad storage conditions.

The fact that 75% of the refuse samples were infested indicates that refuse is one of the main sources of infestation of the newly harvested grain brought into the storage facilities. Again, the lowest percentage of infested refuse samples (63.6%) was found in the silos.

During this investigation 23 species of mites were identified as belonging to 9 families of 3 orders: Astigmata: Acaridae, Glycyphagidae, Chortoglyphidae,

TABLE II.
Mite Species Found Most Frequently in Stored Cereals in Croatia, Sampled from 1972 to 1976

	Percent of infested samples containing listed species in						
	Wheat	Maize	Barley	Rye	Oats	Refuse	Tota
Acaridae:							
A. siro	7.1	7.2	29.4	0	0	2.6	7.4
T. putrescentiae	22.3	36.2	23.5	25.0	69.2	17.9	27.9
A. ovatus	49.1	47.8	35.3	50.0	46.2	61.5	49.6
Glycyphagidae:							
G. destructor	12.5	21.7	58.8	12.5	23.1	7.7	17.8
Chortoglyphidae:							
C. arcuatus	11.6	11.6	11.8	37.5	7.7	28.2	14.7
Cheyletidae:							
C. eruditus	29.5	15.9	35.3	37.5	38.5	5.1	25.2

and Pyroglyphidae; Prostigmata: Cheyletidae and Tydeidae; Mesostigma Ascidae, Amerosiidae, and Dermanyssidae.

The most frequently found species was *Aleuroglyphus ovatus* (Troupea which occured in half of the infested samples (Table II). It was followed *Tyrophagus putrescentiae* (Schrank) in 27.9% of the infested samples; *G cyphagus destructor* (Schrank) in 17.8%; *Chortoglyphus arcuatus* (Troupe in 14.7%; and *Acarus siro* L. in 7.4%. *Cheyletus eruditus,* a predator, w found in 25.2% of the infested samples.

Other species previously known in Yugoslavia which were found inf quently in the present work are: *Gohieria fusca* (Oudemans) (Glycyphagida *Cheyletus malaccensis* Oudemans (Cheyletidae); *Tydeus interruptus* Sig Th (Tydeidae); and *Blattisocius tarsalis* (Berlese) (Ascidae). Species here report for the first time in the mite fauna of this area of Yugoslavia are: *Suida nesbitti* Hughes (Pyroglyphidae); *Acaropis sollers* Rohdendorf a *Cheyletomorpha lepidopterorum* (Shaw) (Cheyletidae); *Lorryia bedfordien* (Tydeidae); *Blattisocius dendriticus* (Berlese) (Ascidae); *Kleemania plumo* (Oudemans) (Amerosiidae); *Androlaelaps casalis casalis* (Berlese) a *Haemogamasus pontiger* (Berlese) (Dermanyssidae). None of the new spec were found in more than 2% of the infested samples.

According to Griffiths (1964), *Acarus siro* is a major pest of stored grain the temperate regions of the world, but under our climatic conditions it w found in a rather low percentage of the samples. Cunnington (1965) stated th "its occurrence and distribution is largely confined to countries with cool a moist climates." Yugoslavia belongs to the south part of the temperate regi and this fact probably explains the high infestations of *A. ovatus* and *putrescentiae* compared to *A. siro.* According to Sinha (1968), the species m frequently found in our stored cereals, i.e., *A. ovatus,* "should be common the hottest areas." And as demonstrated by Cunnington (1967), *T. putresce tiae* is more tolerant of high temperatures (upper limit 35°C) and less tolera of low temperatures (lower limit 10°C) than *A. siro.* This would indicate that both our common species require a higher temperature than *A. siro,* and this is in accordance with our climatic conditions.

In wheat and maize, our major grains, *G. destructor* ranked third after *A. ovatus* and *T. putrescentiae,* but in barley this species was found the most often of any of the stored cereal mites.

SUMMARY

In a four-year survey, we found that 73.3% of all cereal samples collected from storage facilities were mite infested. These included wheat, maize, barley, rye, oats and refuse. The highest infestations were recorded in transit mill granaries, the lowest in silos. Twenty-three species of mites were identified from the infested samples; nine were new to the mite fauna of Yugoslavia. The most common species was generally *A. ovatus,* except in barley, where *G. destructor* was the species most commonly found. The high percentage of infested samples and the numbers of mite species found in refuse indicated that this material serves as a source for new mite infestation in grain.

ACKNOWLEDGEMENTS

This work is a part of the project done in collaboration with the U.S. Department of Agriculture, Agricultural Research Services, Foreign Research and Technical Program Division, P. L. 480, Project No. E30-MQ-8, FG-YU-223 under the title "Population of Acarina and Fungi on Stored Cereals and their Mutual Relation" for the period May 1, 1972 to December 31, 1976.

The author wishes to thank Dr. D.A. Griffiths for his valuable help in the identification and verification of the mites and for reviewing the manuscript; also to thank Miss S. Lynch for her assistance.

REFERENCES

Cunnington, A. M. (1965). *J. appl. Ecol.* **2**, 295-306.

Cunnington, A. M. (1967). *Proc. 2nd Intern. Congr. Acarol.* 241-248.

Griffiths, D. A. (1964). *C. R. ler Congr. Intern. d'Acarologie,* Fort Collins, Colorado, U.S.A. 101-116.

Sinha, R. N. (1968). *J. Econ. Entomol.* **61**, 70-75.

STUDIES ON THE MITES INFESTING STORED FOOD PRODUCTS ON TAIWAN

Yi-Hsiung Tseng

Bureau of Commodity Inspection and Quarantine
Ministry of Economic Affairs
Tainan, Taiwan, Republic of China

INTRODUCTION

Mites occur in stored food products all over the world, causing serious damage, especially in the areas where temperature and humidity are high throughout the year and large quantities of grain are grown and stored. In order to clarify the actual status of mite infestation in the stored food products on Taiwan, I conducted a series of field investigations throughout the island. This presentation reports the results of my work since 1969.

MATERIALS AND METHODS

Samples of stored food products were collected from food dealers and storage facilities of many locations throughout Taiwan. The samples were placed in plastic bags, sealed, and brought back to the laboratory in Tainan for investigation. Sample aliquots were processed by various methods such as modified Berlese funnel (Tseng, 1970), separation by using distilled water plus kerosene (Tseng, 1971) and flotation by using saturated saline water (Tseng, 1972). The mites collected by filtration of the processing liquids were transferred onto microscopic slides with a drop of either PVA or Hoyer's mounting medium. The covered mounts were gently heated to make all legs stretch naturally, and examined under a high power microscope.

RESULTS AND DISCUSSION

Mite Fauna

A total of 58 species of mites belonging to 16 families in 3 suborders (not in-

Vol. I: ISBN 0-12-592201-9

cluding several undetermined Cryptostigmata) listed in Table I were collected during the course of this investigation. The most abundant species were *Tyrophagus putrescentiae* (Schrank), *Cheyletus malaccensis* (Ouds.), *Suidasia nesbitti* Hughes, *Aleuroglyphus ovatus* (Troup.) and *Rhizoglyphus echinopus* Faumouze and Robin, whereas the rarest ones were *Cunaxa* spp. and *Bdella lignicola.* Two species, *Acarus siro* L., and *Carpoglyphus lactis* (L.) have been known to be the occasional causative agents of itch and gastrointestinal disorder of man on Taiwan (Sugimoto, 1938).

The Mites Infesting Stored Sugar

During the period between October 1969 and September 1970, a sampling survey of stored sugar (red and white) was conducted in 63 townships representing south, central, north and east sections of the island. The results for the south section revealed that 91% of red sugar and 72% of white sugar samples were positive with an average of 1914 and 1412 mites per kg sample, respectively. Four common species *Carpolglyphus lactis* (L.), *Tyrophagus longior* (Gervais), *Tyrophagus putrescentiae* (Schrank) and *Suidasia medanensis* Ouds. were found in both red and white sugar samples at all localities investigated. Other mites found were *Kleemania plumosa* (Ouds.), *Cheyletus malaccensis* (Ouds.), *Cheyletus eruditus* (Schrank), *Melichares agilis* Hering and *Melichares mali* (Ouds.) from red sugar samples, and *Cheyletus eruditus* only from white sugar samples. The results for the central section revealed that 83% of red sugar and 68% of white sugar samples were positive with an average of 262 (6 species) and 178 (4 species) mites per kg sample, respectively. The results for the north section indicated that 87% of red sugar and 66% of white sugar were infested by mites with an average of 750 and 354 mites per kg sample, respectively. The species distribution was the same as for the central section. In the east 79% of red sugar samples and 26% of white sugar samples were positive with an average of 238 and 89 mites per kg sample, respectively. Only the four commonest mite species were found in this section. *C. lactis* was more predominant compared to other species in all sections except the east where *S. medanensis* was more predominant. One sample collected from Chiayi in the south section yielded 15,124 mites per kg red sugar. Such a high density may allow stored sugar to become hemiliquidized and cause gastric troubles if infested.

The Mites Infesting Stored Grain

A total of 729 samples of stored grain (1 kg/sample) were collected from 64 townships representing south, central, north and east parts of the island during the period from July 1970 to June 1971. From these samples a total of 672,329 mites were extracted. Between 61 to 100% of the samples of various grains were infested. The important species of mites were *Tyrophagus putrescentiae* and *Lepidoglyphus destructor* for rice with husk, *Aleuroglyphus ovatus, T. putrescentiae* and *S. medanensis* for wheat and barley, *T. putrescentiae* and *S.*

Table I.
Species Distribution of Mites Infesting Various Stored Commodities in Taiwan.

Suborder/Family	Species	Host Commodity
Mesostigmata		
Ameroseiidae	*Kleemania plumigera* (Oudemans)	rice with husk, corn.
	Kleemania plumosa (Oudemans)	red sugar, rice with husk, confection, biscuit, dried minced fish, dried minced pork, dried bamboo shoot.
Ascidae	*Lasioseius allii* Chant	onion, garlic bulb.
	Lasioseius martini Tseng	onion, garlic bulb.
	Lasioseius sugawarai Ehara	pineapple fruit.
	Proctolaelaps pygmaeus (Muller)	stored food products, orange, grape.
	Melichares (Melichares) agilis Hering	red sugar, white sugar, honey, confection.
	Melichares (Blattisocius) mali (Oudemans)	red sugar.
Uropodellidae	*Uropodella laciniata* Camin	fungus-growing rice crop.
Haemogamasidae	*Eulaelaps stabularis* (Koch)	rice with husk.
Uropodidae	*Leionychus krameri* (Canestrini)	fungus-growing rice with husk, dried fish.
Prostigmata		
Cheyletidae	*Cheyletus malaccensis* (Oudemans)	red sugar, white sugar, rice with husk, dried fish, dried minced fish, dried fungus, mushroom, milk, soybean sauce, biscuit, dried meat ham etc.
	Cheyletus fortis (Oudemans)	stored food products, occasionally dried fish, soybean sauce.
	Cheyletus eruditus (Schrank)	stored food products, red sugar, white sugar, confection.
	Cheyletus trousessarti (Oudemans)	rice, kaoliang, wheat, rice with husk.
	Cheletomorpha lepidopterorum (Shaw)	rice with husk, corn, garlic bulb.
	Eucheyletia reticulata (Cunliffe)	rice with husk, wheat, kaoliang, peanut.
	Grallacheles bakeri (Deleon)	rice with husk.
	Hemicheyletia arecana Tseng	betelnut.
	Cheletomimus binus Tseng	garlic bulb, betelnut.
	Cheletomomimus bisetosa Tseng	shallot
	Cheletonella pilosa Tseng	wheat, kaoliang, millet.
	Caudacheles lieni Tseng	bean.

Table I. Continued
Species Distribution of Mites Infesting Various Stored Commodities in Taiwan.

Suborder/Family	Species	Host Commodity
Bdellidae	*Bdella lignicola* Canestrini	garlic.
Tarsonemidae	*Tarsonemus floricolus* Canestrini & Fanzago	dried fish, dried minced fish, m powder, soybean sauce, rice w husk, corn.
	Steneotarsonemus spinki Smiley	rice with husk.
Tydeidae	*Tydeus* sp.	rice with husk.
Cunaxidae	*Cunaxa taurus* (Kramer)	corn.
	Cunaxa womersleyi Baker & Hoffman	red bean.
Astigmata		
Carpoglyphidae	*Carpoglyphus lactis* (Linneaus)	red sugar, white sugar, soyb sauce, confection, dried min fish, biscuit.
Glycyphagidae	*Gohieria fusca* (Oudemans)	rice with husk, barley, red be kaoliang, rice, soybean sau dried fish, dried fungus.
	Lepidoglyphus destructor (Schrank)	rice with husk, dried sweet pot chips, green bean, wheat, ri kaoliang, imported beans.
	Glycyphagus domesticus (De Geer)	rice crop.
	Blomia freemani Hughes	dried sweet potato chips, millet.
Chortoglyphidae	*Chortoglyphus arcuatus* (Troupeau)	rice with husk, millet, dried sw potato chips, wheat.
Pyroglyphidae	*Dermatophagoides* sp.	rice, kaoliang, corn.
Anoetidae	*Histiostoma humidiatus* (Vitzthum)	rotting garlic bulb.
Acaridae	*Tyrophagus putrescentiae* (Schrank)	red sugar, white sugar, rice w husk, wheat, barley, flour, co rice, peanut, beans, kaoliang, m powder, rotting lily, tulip, na cissurs, garlic, onion.
	Tyrophagus bambusae (Tseng)	rotting bamboo shoot.
	Tyrophagus longior (Gervais)	red sugar, white sugar, ham, dri minced fish.
	Tyrophagus palmarum (Oudemans)	rice with husk, leaf litt pineapple fruit.
	Tyrophagus dimidiatus (Hering)	garlic, green bean, ham, dri minced fish.
	Acarus siro Linnaeus	rice with husk.
	Acarus gracilis Hughes	rice.
	Tyrolichus casei Oudemans	barley, rice with husk, dried mea

Table I. Continued
Species Distribution of Mites Infesting Various Stored Commodities in Taiwan.

Suborder/Family	Species	Host Commodity
	Suidasia medanensis Oudemans	red sugar, white sugar, wheat, barley, flour, corn, peanut, beans, confection, dried minced fish, milk powder, dried meat, biscuit, soybean sauce, bean sprout, ham, dried ginger, dried longan, mushroom, honey, tea, garlic, lily, tulip, onion, feather, fermenting bread, rotting mango, bacteriological culture medium, etc.
	Suidasia nesbitti Hughes	stored food products and food.
	Aleuroglyphus ovatus (Troupeau)	wheat, barley, flour.
	Aleuroglyphus formosanus Tseng	wheat.
	Thyreophagus entomophagus (Laboulbene)	garlic, onion and debris.
	Schwiebia mertzis Woodring	taro, chicken nest, feather, bird nest.
	Lardoglyphus konoi (Sasa & Asanuma)	dried fish, dried shrimp, dried minced fish.
	Caloglyphus berlesei Michael	rice with husk, banana, biscuit, dried plum, corn, orange, dried minced fish, confection.
	Caloglyphus mycophagus Oudemans	decayed garlic bulb, sugar cane.
	Caloglyphus krameri (Berlese)	rice with husk, rotting borecole, onion, corn, flour, dried fungus.
	Caloglyphus rhizoglyphoides Zakhvatkin	grape wine.
	Rhizoglyphus echinopus Faumouze & Robin	lily bulb, occasionally rice straw.
	Rhizoglyphus robini Claparede	garlic, bamboo shoot, rice with husk.

medanensis for corn, peanut and beans, *T. putrescentiae* for rice and kaoliang (tall millet). Dried sweet potato chips were infested by *L. destructor and Blomia fremani.* As regards the seasonal abundance of mites, in most cases, the number of mites for each kind of product (grain, wheat, millet, peanut and rice) reached the maximum between August and October. During peak season on an average 2,000 mites were found per kg sample.

SUMMARY

A survey on the status of mites infesting stored food commodities conducted throughout Taiwan revealed the presence of 58 species of mites belonging to 16 families and three suborders. The most abundant species were *Tyrophagus putrescentiae, Cheyletus malaccensis, Suidasia nesbitti, Aleuroglyphus ovatus* and *Rhizoglyphus echinopus*. Up to 91% of red sugar and 72% of white sugar samples collected from various regions of the island showed mite infestation. Most of the stored grain samples had positive mite infestation. The infestation was at peak between August and October.

REFERENCES

Sugimoto, M. (1938). *J. Taihoku Soc. Agric. For.* **3**, 12-28.

Tseng, Y. H. (=Tjying, I-S.) (1970). *Plant Prot. Bull.* **12**, 113-123.

Tseng, Y. H. (=Tjying, I-S.) (1971). *Plant Prot. Bull.* **13**, 147-155.

Tseng, Y. H. (=Tjying, I-S.) (1972). Bureau of Commodity Inspections and Quarantine, Ministry of Economic Affairs, Taiwan, Republic of China. 55 p. (in Chinese).

Tseng, Y. H., and Chang, A. F. (1973). *Scientific Agriculture* **21**, 172-178. (in Chinese).

THE FOOD AND DRUG ADMINISTRATION AND REGULATORY ACAROLOGY

Wynn A. Senff

Detroit District Office
U. S. Food and Drug Administration
Detroit, Michigan

J. Richard Gorham

U. S. P. H. S., U. S. Food and Drug Administration
Washington, D. C.

INTRODUCTION

Before discussing the various aspects of regulatory acarology, I would first like to introduce you to the Food and Drug Administration and briefly describe our role with specific regard to Foods.

FDA is a regulatory agency within the Public Health Services of the Department of Health, Education, and Welfare. The agency is responsible for the enforcement of the Food, Drug, and Cosmetic Act, and its amendments, as well as other laws affecting foods, drugs, cosmetics, and various radiation and medical devices, all of which must travel in interstate commerce.

Heading up our agency of some 7,700 people, of which it is estimated that 1,300 people work in Foods, is our Commissioner, Donald L. Kennedy, and his administrative staff. There are six bureaus, the large laboratory facility called the National Center for Toxicological Research, and the Executive Director of all field operations commonly referred to as "EDRO."

In function, the Bureaus serve as scientific centers, conducting studies on various products, and they serve as consultants for FDA's legal staff. They review the field's work, help in statistical analysis and help develop workplans. With regard to EDRO, or the field operations, we have 10 Regions, some of which have been further split into a total of 19 District offices, located in almost every major metropolitan area. At the District level, each office is split into four branches of service; the Investigation branch, the Laboratory branch, the Compliance branch, and the Administrative branch, while each

Vol. I: ISBN 0-12-592201-9

District also maintains a series of residence posts housing investigational staff only.

As one might expect, the Investigations Branch serves the role of examining, collecting, and documenting the various problems and situations in which the agency is involved, while the laboratory confirms the investigator's observations if necessary, and may serve as the analytical center for surveillance samples or research. Our compliance section reviews and compiles the evidence submitted by both the investigators and the analysts and determines, at least in part, what action may be taken against a firm or a product for review by FDA's legal staff and the Bureau's.

The Administrative branch serves to manage the personnel and fiscal matters of the office.

MISSION OF FDA

The mission of FDA as outlined by the law and restated in FDA's latest annual report as they relate specifically to foods follows:

1) FDA's mission is to see that foods are safe to eat, wholesome and pure.
2) That foods are produced, packed, held, and shipped under good sanitary conditions.
3) That labels are truthful and informative.
4) That pesticide residues must not exceed the safe tolerances established jointly by FDA, EPA, and USDA.

To achieve our mission, the agency has, at its disposal, two means of "**enforcement.**" The first we could refer to as "**preventative enforcement,**" which is basically an educational process, whereby through direct educational proceedings, such as schools, conferences, and publications, or through informational letters issued as a result of investigational findings, a firm or individual will take it upon himself to comply with the law. This is based on the precept of FDA enforcement policy, that, a majority of people will voluntarily comply with the law if they are made aware of it and if they understand it.

The second form of enforcement would be referred to as "**corrective enforcement,**" which includes such legal sanctions as seizure, injunction, license revocation, civil fines, and criminal prosecution. These forms of correction are, all, time consuming and expensive.

It must be pointed out, that the firm and in particular the individuals owning or operating a firm are solely responsible for the safety of the products they produce. It is for this reason, and because the Food and Drug Administration has limited resources, that FDA relies heavily upon the educational process, while not hesitating to use the formidable process of corrective enforcement, if it is deemed necessary.

As such, the agency's general regulatory goal is to be "**Anticipatory, rather than reactionary, preventative, rather than corrective.**"

SANITATION AND REGULATORY ASPECTS

Having discussed FDA, its organization and goal, let us focus on sanitation and other regulatory aspects in which the acarologist or pest management specialist may become involved.

First, we must define four basic terms: *sanitation, defect action level, avoidable* and *unavoidable* as they are related to health and the general quality of foods.

Sanitation refers to good manufacturing practices and good housekeeping. It means that the food industry must emphasize the necessity of protecting our foods to *avoid* adulteration through carelessness. It means that the first line of defense in any pest control program is cleanliness and it means that although a health hazard may not be posed, that insanitary practices can and must be eliminated without exception.

On the other hand, *Defect Action Levels* (DAL's) refer to the point or level or adulteration at which FDA will take action against a product with regard to natural or *unavoidable* defects which pose no health hazard. DAL's are not tolerances nor are they averages. They are the levels of adulteration above the product's average level in which the agency believes that undue economic cheating or loss may occur. As a result, DAL's are under constant review and will be changed as technology changes.

Again, FDA emphasizes that insanitation is inexcusable even though the products may well fall below the defect action level. It should also be noted that the agency does not allow "adulterated" foods to be mixed with "good" food in order to lower the level of adulteration, as that would render the entire lot adulterated.

Returning now to the DAL's, of the 70 products for which DAL's are established, five products specifically mention mites. In two of the five products, apple butter and druplet berries, mites are excluded from consideration as a part of the action level, because it was felt that they would not affect the appearance or wholesomeness of the product and therefore would not be of any significance to the consumer in making purchasing decisions. The three remaining products include:

Frozen Broccoli
: Where 60 aphids, thrips, and/or mites are allowed per 100 gm of product,

Frozen or Canned Spinach
: Where 50 aphids, thrips, and/or mites are allowed per 100 gm, and finally,

Canned or Dried Mushrooms
: Where 75 mites are allowed per 100 gm drained mushrooms or per 15 gm dried mushroom product.

The spinach DAL was made in 1973 after consideration of USDA's guideline in which an average of 35 aphids, thrips or mites per 100 gm was required, with a limit of 45 aphids, thrips, or mites in any one analytical unit.

The reasoning and data base for the other products, mushrooms and broccoli were not available for review, but it should be noted that new data is being reviewed on mushrooms and should be available shortly.

An informal poll of selected FDA analysts indicated that while the mushroom DAL is often applied, the broccoli and spinach DAL's are rarely needed with regard to mites. The fact however, that only these foods have DAL's does not limit FDA jurisdiction, for, though very rare, FDA has taken action against the following products to name a few: Cheese, bean sauce, dried apricots, fig paste, honey, preserved salted beans, rice powder, seaweeds, and in one situation, soda pop, based on the fact that the mites were present in excessive numbers.

As another example, with regard to sanitation, the Detroit office in August of 1976, encountered a large food warehouse which exhibited a high rodent infestation. In addition, it was quickly noted that a rodent parasite was present in high numbers in food lots which had been frequently traveled by or had become nesting sites for the rodents. A mass siezure and reconditioning of the three million dollar warehouse followed with over $55,000 dollars worth of product destroyed because of the mite and rodent adulteration. The mite was tentatively identified as *Ornithonyssus bacoti* which was confirmed by the Ohio State Acarology Laboratory.

CONCLUDING REMARKS

At FDA, information about specimens extracted from specific foods and identified to even the family level is meager. There has to this time been no legal or administrative pressure to identify the acarines and most analytical personnel have insufficient training to attempt such a task. As such, most FDA people lump the the acarines into the general taxa of "mites" with little regard for bio-economic considerations.

As a whole, regulatory acarology lags behind regulary entomology. This disparity between the two sciences has been created, at least in part, by several factors which include:

1) The lack of emphasis on acarology in university programs.
2) The general inability of many people to recognize mites as pests, or, in other words, the relative "insignificance" associated with mites as a whole.
3) The lack of understanding with regard to mites and their relationships to stored product problems and to human health.
4) The fact that in the past, little legislative, social, or pest management pressures have been brought to bear on the Agency with regard to the consideration of mites in foods.

As these differences are brought in line, it would be safe to assume that regulatory acarology within FDA will slowly develop along the same lines regulatory entomology has in the past. We expect that:

people will become available and will be hired who will have received stronger acarological training in their educational background, thus providing the needed impetus for better acarological considerations; in addition, existing personnel will have increased opportunities to obtain adequate training in regulatory acarology;

taxonomic and biological resources in acarology will become more readily available and more readily utilized in training and everyday use;

a better understanding of the health effects of mites in food, as allergens, as possible toxin producers, and as agents of gastrointestinal disorders, will lead to more emphasis on mites in the epidemiology of food associated diseases;

as the world moves toward integrated pest management, with a decreasing dependence on the use of pesticides and the probable increasing use of parasitic and predacious mites in pest management, that FDA's DAL's will be modified to specifically identify and control these areas of concern;

any attempt to precisely identify specific mite taxa and associate them with their ecological roles will require FDA analysts to have a full understanding of, rather than a superficial familiarity with, acarology;

lastly, as FDA's regulatory efforts increase and in order for FDA analysts to communicate and cooperate effectively with personnel from other programs and industry, that FDA may find it advantageous to the accomplishment of its mission to promote acarological training and research.

REFERENCES

Anon. (1974). "Enforcing the Food, Drug and Cosmetic Act," *FDA Consumer Memo,* DHEW Publication No. FDA 74-1018. Washington, D. C.

Anon. (1976). "Annual Report of The Food and Drug Administration," Washington, D. C.

Anon. (1976). "Federal Food, Drug, and Cosmetic Act, as amended." Food and Drug Administration, Rockville, MD.

Anon. (1977). "The Food Defect Action Levels." Food and Drug Administration, Washington, D. C.

Anon. (1977). "Code of Federal Regulations," Title 21, Parts 1-100, Office of the Federal register, Superintendent of Documents, Washington, D. C.

Anon. (1978). "DHEW, FDA Enforcement Policy for Certain Compliance Correspondence, Proposed Rules," *Federal Register* **43**, No. 122, June 23, 1978, pp. 27498-27502.

Boese, J. (1977). "Mites Associated with Stored Products: Selected References Published from 1960 through 1975." *FDA By-Lines* **5**, 229-261.

Gorham, J. R. (1975). *J. Milk Food Technol.* **38**, 409-418.

Terbush, L. E. (1972). *FDA By-Lines* **3**, 57-70.

4.

Physiology, Biochemistry and Toxicology of Acari

Section Editors

J. R. Sauer
L. G. Arlian

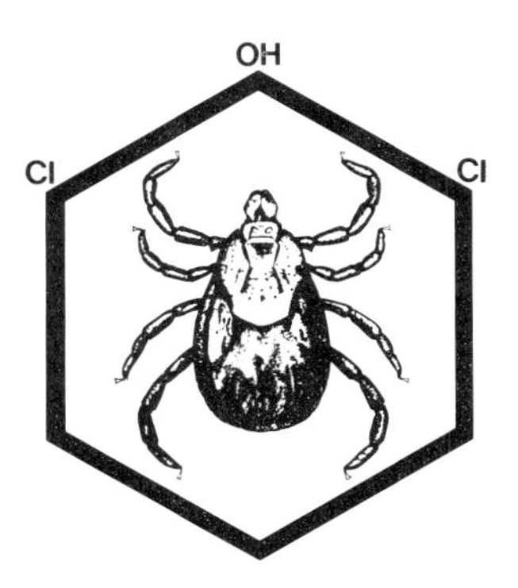

WATER AND THE PHYSIOLOGY OF HOUSE DUST MITES

G. W. Wharton

Acarology Laboratory
The Ohio State University
Columbus, Ohio

Kenneth M. Duke and Harold M. Epstein

Battelle Memorial Institute
Columbus, Ohio

INTRODUCTION

Several recent comprehensive studies or reviews have considered various aspects of water relationships of terrestrial arthropods (Wharton and Arlian, 1972; Edney, 1977; Wharton and Richards, 1978; Schmidt-Nielsen *et al.*, 1978) and one (Wharton, 1976) summarized current knowledge of house dust mites. In this paper we will consider the mechanisms responsible for the exchange of water between air and house dust mites as well as the flows and functions of water in the maintenance of the metabolic machinery of the mites. For the most part the discussion will be based on the published research of Wharton, Arlian, and their students and on previously unreported observations done cooperatively by the co-authors with partial support from NSF Grant No. DAR-76-15224ADI.

Water is the arena required for the myriads of chemical reactions that together are responsible for living phenomena. Its charges and hydrogen bonding strongly influence the three dimensional configurations of macromolecules as well as the ionization of salts and the rates of diffusion of the resultant ions. Electrical currents such as those associated with the transmission of nervous impulses are accompanied by the active and passive movement of ions though membranes and water. Bioelectric potentials that are so important in maintaining the integrity of cells are similarly generated by the movements of dissolved ions. Water is the source of O_2 used in the conversion of carbohydrates and fats into CO_2. It is also a product of such catabolism when released H^+ combines with respiratory O_2. Most metabolites

Vol. I: ISBN 0-12-592201-9

are soluble in water and can therefore readily move either by diffusion through water or as a result of bulk flow of water. Water is essentially incompressible and can be used to transmit pressure from one part of the body to another. In mites, extension of the joints of the appendages is usually accomplished by increase in the hydrostatic pressure of the hemolymph of the leg in question. Flexor muscles are antagonistic to such hydraulic extension. Water also aids in the movement of materials through the digestive, excretory, reproductive, and circulatory systems. Without water, metabolic activity ceases.

How much water is required for optimum metabolic activity? Fresh water and terrestrial organisms are usually 99% water on a mole fraction basis. That is, more than 99 out of 100 of their molecules are H_2O. A convenient unit for expressing the concentration of water is its activity, a_w. The activity of pure water is 1 and when no water is present, the a_w is 0. As a colligative property a_w values can be converted into any other such units; for example Eq. (1) relates a_w to osmoles:

$$a_w = \frac{55.508}{55.508 + \text{osmoles}} \quad (1).$$

The number of moles of water in 1000g is 55.508.

Primary marine organisms have a water concentration of a_w 0.98, the same a_w as sea water. Secondary marine organisms, those that originated in fresh water or on land and then returned to the sea, maintain the higher water concentration of a_w 0.99 even though they must expend energy to concentrate the water from a_w 0.98 to 0.99. The retention of an internal a_w of 0.99 in a marine environment of a_w 0.98 implies that a concentration of a_w 0.99 is optimum for water in living systems.

The mechanism used to concentrate sea water from a_w 0.98 to a_w 0.99 is a salt gland of some type. Salt water is imbibed but excess salt is removed from the blood before the concentration of sea water is approached. Such salt glands are found in both vertebrates and invertebrates or even in plants (Wharton, 1978).

Terrestrial organisms are exposed to air that is seldom in equilibrium with the water in their systems. The activity of water vapor in the air a_v is equal to the relative humidity divided by 100. Thus air with a relative humidity of 99 has an a_v of 0.99 and is in equilibrium with body fluids of a_w 0.99. Most terrestrial life forms lose water to air that has an a_v less than 0.99. Plants make good the loss by obtaining fresh water from the soil and most animals imbibe fresh water, a_w 0.99, or ingest it with their food. Some utilize salt water and excrete the excess salt. A few terrestrial arthropods use water vapor as a source of fresh water.

If an arthropod is exposed to water vapor labeled with tritium, it will sorb the tritiated water and in time its body water will have the same tritium concentration as the vapor to which it is exposed. If this animal is then moved to a tritium-free environment, the tritium in its system will gradually disappear and the rate of disappearance will be a measure of the transpiration of water from the animal to the air. Such observations have been made and tritium has been seen to leave as a constant % of that present.

$$\ln T_t = \ln T_O - kt \qquad (2).$$

Tritiated water is a good tracer for water; thus in completely dry air Eq. (2) also describes water loss up to the point where dehydration interferes with the normal functioning and increased water loss occurs.

In air that contains some water vapor, transpiration and sorption occur concurrently without interfering with each other. Under these conditions the instantaneous change in water mass (m) is the difference between the instantaneous rates of sorption (m_s) and transpiration (m_T):

$$\dot{m} = \dot{m}_s - \dot{m}_T \qquad (3).$$

Measurements of m_s over a period of time have demonstrated that it is constant:

$$\dot{m}_s = C \qquad (4).$$

This result is to be expected because the small amount of water sorbed compared to that in the surrounding air is insignificant and thus does not affect the overall water vapor pressure that applies the force that drives sorption. Because transpiration is a first order process and km will decrease with time as m becomes smaller and smaller and sorption proceeds at a constant rate, at some value of m, transpiration and sorption will be equal and an equilibrium weight, m_∞, will be achieved. If equilibrium is reached at a viable m, the equilibrium will be real and m_∞ can be observed. If m_∞ is not a viable water level, then the equilibrium is virtual rather than real. The higher the a_v, the more likely a real equilibrium will be reached. The lowest a_v at which a real equilibrium can be reached is the CEA (Wharton and Devine, 1968). Whether real or virtual, m_∞ = km when m = 0 and

$$m_\infty = m_s k^{-1} \qquad (5).$$

The net movement of water below the CEA is described by:

$$\ln(m_T - m_\infty) = \ln(m_0 - m_\infty) - kt \qquad (6).$$

Net loss is also at a constant percentage of the exchangeable water ($m_0 - m_\infty$) and k for net transpiration is equal to k for total transpiration (Devine and Wharton, 1973).

HOUSE DUST MITES

Many species of mites are found in houses but the term house dust mite has been restricted in its application to members of the family Pyroglyphidae that occur in houses and are commonly collected from the contents of bags from domestic vacuum cleaners (Wharton, 1976). The most common genus is *Dermatophagoides* and in most parts of the world *D. pteronyssinus* Trouessart or *D. farinae* Hughes are the most prevalent. In all, 10 species distributed in five genera are known to occur in houses. Those that have been reared in the laboratory have a CEA below a_v 0.75. The stages in their life history are egg, prelarva, larva, protonymph, tritonymph and adult male or female.

The water relations of eggs and larvae have not been studied. Because cultures thrive at a_v 0.75 it can be assumed that their CEA is below that value. It may be that the egg is almost completely impermeable because the prelarval cuticle is characterized by two sclerotized projections that are similar in appearance to the respiratory prelarval pips characteristic of the eggs of certain spider mites (Ellingsen, 1975). The prelarva, while greatly reduced, has been found in all six species recently surveyed by Fain (1977).

The water relations of tritonymphs have not been studied. Ellingsen (1974, 1975) has studied protonymphs in four physiological states above the CEA and in one instance in dry air. Above the CEA, the loss of tritium labeled water was followed. Active protonymphs lost tritium at a constant percentage of that present so that the data fit Eq. (2) with a k of $0.034h^{-1}$. Food was withheld from the protonymphs during the period of monitoring their tritium loss. After 20 h, their energy reserves were depleted and the rate of tritium loss increased. Under some conditions as yet not clearly definable, active protonymphs become immobile but do not molt. These quiescent protonymphs can maintain their water mass above the CEA for months. When they are removed from a culture and placed in a chamber for studying water loss, they lose tritiated water at a k of 0.0032 for a period of *ca* 16 days. After 16 days k is not significantly different from zero. Moving the quiescent protonymphs from the substrate damages its water proofing. Tritium loss during the first 16 days is thought to be associated with the repair of the water proofing layer. Following repair, the protonymph has a reduced O_2 consumption as well as no net water loss. It is probable though that some water loss does occur because biological systems are more permeable to water than they are to respiratory gases (Waggoner, 1967). The missing water may in fact be water produced by the low O_2 consumption (1.3×10^{-11} moles of O_2 hr^{-1}) of the undisturbed quiescent protonymph. Quiescent protonymphs killed by exposure to –18°C for 24 h lost water at a_v 0.75 more rapidly ($k = 0.004/hr^{-1}$) than their live counterparts. Disturbed quiescent protonymphs, when exposed to dry air, lost water more slowly ($k = 0.0014hr^{-1}$) than they did at a_v 0.75 until they died in about 19 days. After 19 days, their water loss was more rapid ($k = 0.0033hr^{-1}$). All of the observations of water loss by the protonymphs fit the hypothesis that water is leaving the system from a single compartment and that each molecule in the compartment has an equal opportunity of leaving. To be valid, such an assumption requires that the movement of water from one part of the compartment to another be much more rapid than the movement of water from the compartment to the air. In studies of protonymphs only 1% of the tritium content was found to remain in dried mites so that 99% of the observed tritium labeled water must remain in the water pool and thus serves as a good tracer for water.

Most studies of water relations have been done with the females of *D. farinae*. Those done with other adults (Arlian, 1975, 1977) are consistent with the interpretations of observations on *D. farinae*. Above the CEA at room temperature, females of *D. farinae* maintain a constant water mass at any a_v. Water is transpired as though the total mass were in a single compartment. Transpiration rates do change with a_v and loss through the cuticle at any a_v can be calculated.

$$k''a_v = 0.02427a_v + 0.00268h^{-1} \quad (7).$$

where k" is the rate constant for loss through the cuticle (Wharton, 1978). Above the CEA k has been measured at $0.02225h^{-1}$ at a_v 0.75 to $0.02634h^{-1}$ at a_v 1.00 (Arlian and Wharton, 1974).

Below the CEA, water seems to be lost from two compartments: a fast compartment with a rate constant $k'h^{-1}$, and a slow compartment with a rate constant $k''h^{-1}$. The fast compartment contains about 3μg of water which at a_v 0.00 is lost at a k" of $0.18h^{-1}$ (Arlian and Wharton, 1974). It is postulated that this fast compartment represents water that passes through a pump that can obtain water vapor from air above the CEA but which converts to an active mechanism that transpires water below the CEA. In the process of this active transpiration, the air pump interface must be waterproofed or enough of the body water would be lost by this mechanism to kill the mite.

In 1974 we knew that the water pump was not generally distributed over the surface of the body but little else about its location. Of the possible pump openings, that of the supracoxal gland seemed to fit the functional requirements best. The gland opening is located on the podocephalic canal that leads to the oral cavity and it is not provided with a closing mechanism. If this gland were in fact the pump, then as a mite lost water to air below the CEA its secretion should in some way close the gland opening. Examination of mites that had been killed by dessication soon led to the observation that plugs formed at the openings of the supracoxal glands. Electron probe elemental analysis demonstrated that KCl was a significant component of the plug (Wharton and Furumizo, 1977).

On the basis of the information developed up to this point it is possible to describe a water pump based on the anatomy of *D. farinae* as revealed by anatomical studies using scanning and transmission electron microscopy

(Brody and Wharton, 1971; Wharton and Brody, 1972; Brody *et al.*, 1972, 1976). Assuming that the secretion of the supracoxal gland is hygroscopic above the CEA, the secretion will flow down the podocephalic canal to the prebuccal cavity and will sorb more water from the air than it will transpire to it. This will occur as long as the secretion a_w is less than the a_v of the air. The prebuccal cavity of *D. farinae* has an enlarged surface produced by deep grooves in the dorsal wall of the hypostome and the ventral wall of the hypognathosoma as well as the walls of the medial projecting labrum. Because of its enlarged moist surface, the prebuccal cavity could serve as the major respiratory organ for the exchange of O_2 and CO_2 as well as the uptake of water vapor. These mites have neither tracheae or malpighian tubules. The prebuccal cavity opens posteriorly into a muscular pharynx that can pump the vapor enriched supracoxal fluid into the esophagus with considerable force. The empty esophagus is star-shaped in cross section and capable of expanding to a cylindrical shape as wide as the neurohaemal canal by which it makes its way through the brain to the midgut. In addition to its connection to the esophagus, the lumen of the anterior midgut connects ventrally with a pair of posteriorly directed caeca and posteriorly with the posterior midgut. The caecal and posterior midgut cells are provided with numerous microvilli that could serve to absorb salt and water and transfer it rapidly to the haemolymph. Digestion in the anterior midgut is apparently mostly intracellular because cells are seen free in the lumen as well as in the process of leaving the wall. Food passing from the anterior to posterior midgut is enclosed in a peritrophic membrance. Digestion if it occurs here is extrcellular.

When salt and water are passed to the haemolymph, the salt must be removed sufficiently rapidly to avoid intolerable decreases of its a_w. Avoidance of local buildup of a_w requires that the haemolymph be circulated. *D. farinae* has no heart but the haemocoel is so arranged that moving material through the esophagus will force blood to flow through the neurohaemal canal at the same time. The direction of flow will be from the anterior midventral region dorsally and slightly posteriorly to the dorsal region of the anterior midgut. At the same time that blood is moving from the anterior ventral region, it will be replaced by blood from lateral and dorsal areas so that as the blood moves to the dorsal area of the midgut, it will be drawn anteriorly and laterally thus completing the circular pattern.

Finally, not only must local reductions in a_w be avoided, but a general reduction as well. This requires a salt gland that can excrete the excess salt. The supracoxal gland is constructed in the manner of salt glands and in dry air its secretion was demonstrated to have a high concentration of KCl. The glands are located in the path of the blood flow from the midgut to the anterior ventral region so they have an opportunity to process the blood as it flows by. Each gland has three salt secreting units and a single unit that secretes other materials. For the secretion to work effectively, it must have a low surface tension. Perhaps a detergent is among the products of the single secretory unit.

Above the CEA, the salt secretion of the supracoxal gland flows into the mouth. Below the CEA, it loses water to the air, its solutes precipitate out as they begin to occlude the ducts and so make the gland less and less permeable to water loss until it is completely blocked. The material that blocks the opening not only serves this function but it is also hygroscopic above the CEA and when humidity again increases above this level, large amounts of water can be absorbed as the salt block deliquesces and moves with its water on into the mouth. It is because of this supply of dry salt that dehydrated animals can regain water much more rapidly than they lose it. If this mechanism were in fact in use, high concentrations of salt should be present in the ducts of the supracoxal glands in hydrated living mites.

SALT CONCENTRATIONS *IN VIVO*

To two decimal places, the a_w of *D. farinae* is expected to be 0.99 as is characteristic of most terrestrial organisms. The a_w of an optimum culture medium for the maintenance of cells of *D. farinae* was 0.988 (Wharton, 1976). We assume that this is the activity of the water in the cells and haemolymph of this mite even though direct measurements have not yet been made. Furthermore, we assume that the solute primarily responsible for this a_w is a 1:1 mix of NaCl and KCl with NaCl the chief extracellular component and KCl the primary intracellular salt. Based on these assumptions, the 9.589μg of water characteristic of the average hydrated female holds about 0.2μg of salt in solution. We have assumed that this is the quantity of salt available for the operation of the pump under discussion.

MATERIALS AND METHODS

Estimates of salt concentrations have been calculated from the data and models presented above. Observations of concentrated salts were made using laser generated X-rays (Epstein *et al.*, 1979).

The mites exposed to laser generated X-rays were females selected from laboratory cultures of *D. farinae* fed on 9:1 w/w ground Purina® lab chow: baker's yeast. The females were then held without access to food for at least 14 h prior to making contact microradiographs of them. Mites held above their CEA at a_v 0.75 were hydrated; those exposed below their CEA at a_v 0.00 were dehydrated. Air in equilibrium with saturated NaCl brine provided a_v 0.75 and air over Drierite® gave a_v 0.00. A brass exposure container was made for holding mites and the X-ray film. The film chamber was circular in outline and held the film within about 1-2 mm of the mite. Six mite chambers were formed in the cover of the exposure container. Each was separated from the film chamber by a Be window. After mites were placed in their chamber, a Be cover held them in place. The mite chambers were not hermetically sealed and mites exchanged water vapor and respiratory gases with air surrounding the exposure chamber.

Just prior to making microradiographs, film was loaded into the exposure container which was then placed in the vacuum chamber that contained the X-ray source. The system was then evacuated for about 15 min. Water loss by the mites during this procedure is negligible. After evacuation, a 5 nsec, about a 100 j pulse of 1.06μm light from a Nd doped glass laser was flashed into the chamber. This light was focused on about a 150μm spot on a Cu target. When the target is hit, a Cu plasma is formed that generates X-rays. The contact microradiographs are formed by the X-rays that pass through the mites on the way to the film. Using this system with SO424 film, resolution of about 1μm is achieved.

Quantitative analysis of the attenuation is made possible because only X-rays of about the same energy level reach the film. The characteristic energy of these essentially nonoenergetic X-rays can be calculated by a saddle point integration. Using these techniques it is possible to derive the following equation for the salt concentrations in the mites:

$$\rho = \frac{\Delta D X^{.75}}{76t} \; (0.7/fNaCl + 1.7/fKCl)^{-1} \; g/cm^3 \qquad (8).$$

Where ρ is density of salt in g/cm^3, ΔD is different in optical density, X is the thickness of the mite, t is the thickness of the salt concentration, fNaCl is the fraction of NaCl and fKCl is that of KCl.

The value of ΔD for any feature is obtained from enlargement of the microradiographs made with a light microscope. Two adjacent lines of the enlarged image are statistically scanned and the results are converted to optical density. Plots of optical density on the vertical axis against distance on the horizontal axis are used in locating and measuring areas of interest in terms of ΔD and t. The thickness of the mites, X, was taken as 0.022 cm for fully hydrated, 0.020 cm for dehydrated, and 0.018cm for rehydrated mites.

RESULTS

Exposure of *D. farinae* to a_v less than the CEA results in the waterproofing of the pump mechanism in about 14 h (Arlian and Wharton, 1974). Is the salt concentration of the mite sufficient to occlude the openings of the supracoxal glands? A minimum of 0.2μg of salt is present. The 30% of the water needed to waterproof the gland contains .06μg of salt. The mean density of NaCl and KCl is 2, thus the volume of salt available for plugging the gland ducts is about 30,000μm^3. The ducts have a diameter of about 2μm so that enough salt is available to fill 9,000μm of ducts. Only about 200 μm of ducts are present so that sufficient salt is certainly available. Salt may be precipitated on the surface in the form of a plug (Wharton and Furumizo, 1977) and no doubt some is encrusted along the podocephalic canal and in the folds of the prebuccal cavity. Build-up of salt crystals on the surfaces of the podocephalic canal and the prebuccal cavity should interfere with the normal activity of these mites. Recent observations by Arlian (1977) of the feeding behavior of *D. farinae* and *D. pteronyssinus* confirmed that feeding and movement are drastically reduced below the CEA.

When mites that are dehydrated are returned to air above the CEA, precipitated salt will deliquesce and fluid will again be available for movement through the pump. At what rate will this occur? The evaporation and condensation of water occur concurrently at rates that are determined primarily by the vapor pressure of the liquid and vapor phases. The vapor pressures on the other hand, are influenced to a great extent by temperature, a_v, and a_w. No measurements of uptake of water by house dust mites during the first few minutes of exposure to air above the CEA are available but plugs have been seen to deliquesce on dead mites. In live mites, the trochanters of legs I are held over the supracoxal plates below the CEA and maintain their position when mites are placed above the CEA. From observations of the dorsal longitudinal folds of dehydrated mites in moist air, it appears that about 3 h of exposure is required for rapid uptake to begin. In relatively still dry air (at 25°C) water will evaporate from a fairly flat surface at a rate of about 25 mg/cm^2/h or 0.00025μg μM^{-2}h^{-1} (Edney, 1957). The maximum vapor exchange of the pump of *D. farinae* at 25°C is *ca* 0.5μgh^{-1}. This rate would require a minimum surface of 2000μm^{2}which is just the area provided by the podocephalic canal. If the surface of the prebuccal cavity is added to that of the podocephalic canal, an area of over 6000μM^2 becomes available and observed losses and gains of water become feasible.

Once the solid salt is dissolved and the fluid once again begins moving through the pump, it will move fluid at a rate that will add *ca* 0.1μg of water/h to the main water pool of the mite. A flow of 1,000,000μm^{3}h^{-1} would require 16.7 turnovers/h if the depth of the fluid was 10μm. The narrowest portion of the pump is the pharynx. Figures (Brody *et al.*, 1972) can be used to estimate its cross section to be about 120,000μm^2. A flow of 1,000,000μM^3 through 120,000μm^2/h gives a rate of 8.3 μmh^{-1} which is certainly not excessive. Flows in the esophagus and midgut will be slower because of the larger cross sections of these organs.

The most concentrated salt solution moving into the midgut will have an a_w of 0.7 and it will be added to fluid of a_w 0.988. At the suggested flow rate, this would bring .4 to .5μg of salt h^{-1} that should have the same turnover rate as the rest of the cycle or 16.7 so that less than .03μg of excess salt would be in the midgut at any time. If the water content of the midgut is 2μg, then the flows suggested would change the a_w of the water in the gut from 0.988 to 0.9879. Movement of water and salt at this a_w into the haemolymph would change its a_w only slightly. If salt were concentrated by the supracoxal gland then the a_w

of the haemolymph could be held to close tolerances even while the pump delivered salt water to the midgut.

Examination of the microradiographs and analysis of their densities demonstrated concentrations of salt in the ducts of the supracoxal glands of both hydrated and dehydrated mites. A number of dehydrated mites also had concentrations of salt or plugs on the supracoxal plate at the orifice of the supracoxal glands. The mean salt concentration for measurements on 12 ducts was 0.27 ± 0.03 g/cm^3. This mean value corresponds to an a_w of 0.859 and includes both hydrated and dehydrated mites. The most concentrated salt in a duct was 0.462g/cm^3, the most dilute 0.106g/cm^3. Six plugs were also measured. Their mean salt concentration was 0.57 ± 0.10g/cm^3. The maximum concentration was 0.883 and the minimum 0.321g/cm^3. None approached the mean density of 1:1 mix of NaCl and KCl or 2 g/cm^3.

Two mites were weighed in the hydrated condition, dehydrated overnight and then X-rayed. One was then hydrated overnight while the other continued exposure below the CEA. The rehydrated mite weighed 12.3 μg originally, 9.5 μg after dehydration, and then the weight jumped to 16.7 μg following rehydration. The re-dehydrated mite weighed 15.2 μg to begin with, 11.0 μg after dehydration, and then 8.6 μg following further dessication. Both dehydrated mites had a salt concentration of 0.29 μg/cm in a duct of the supracoxal gland. In the mite that was rehydrated, the salt concentration fell to 0.25 μg/cm but increased to 0.38 μg/cm following re-dehydration. These observations demonstrate that the salt concentration in the ducts of the supracoxal glands changes from time to time. The higher salt concentrations are characteristic of the more dehydrated mites.

SUMMARY

Vapor exchange kinetics of house dust mites, especially of females of *D. farinae*, have been well studied and mechanisms have been postulated to account for the ability of these mites to use water vapor as a source of water. They drink salt water and incorporate the excess salt into a hygroscopic secretion of the supracoxal gland which then runs down the external podocephalic canal into the prebuccal cavity and becomes the salt water supply for imbibition. As the hygroscopic fluid moves to the mouth, it is enriched by water vapor so that a net gain of water results. This gain offsets the losses associated with diffusion through the cuticle and elimination of feces, secretions, and reproductive products. We have demonstrated that the amounts of water and salt, their concentrations, and estimated rates of flow, are consistent with the uptake mechanism described above. Observation of salt concentrations in the ducts of living mites was made possible by the use of low energy laser generated X-rays. Salt concentrations of the order of magnitude seen in the living mites are essential to the operation of the pump we have described.

REFERENCES

Arlian, L. G. (1975). *J. Med. Entomol.* **12,** 437-442.

Arlian, L. G. (1977). *J. Med. Entomol.* **4,** 484-488.

Arlian, L. G. and Wharton, G. W. (1974). *J. Insect Physiol.* **20,** 1063-1077.

Brody, A. R. and Wharton, G. W. (1971). *Ann. Entomol. Soc. Am.* **64,** 528-530.

Brody, A. R., McGrath, J. and Wharton, G. W. (1972). *J. N. Y. Entomol. Soc.* **80,** 152-177.

Brody, A. R., McGrath, J. and Wharton, G. W. (1976). *J. N. Y. Entomol. Soc.* **84,** 34-47.

Devine, T. L. and Wharton, G. W. (1973). *J. Insect Physiol.* **19,** 243-254.

Edney, E. B. (1957). "The Water Relations of Terrestrial Arthropods." Cambridge Univ. Press, Cambridge, England.

Edney, E. B. (1977). "Water Balance in Land Arthropods." Springer-Verlag, New York.

Ellingsen, I. J. (1974). Ph.D. dissertation, The Ohio State University, Columbus, Ohio.

Ellingsen, I. J. (1975). *Acarologia* **17,** 734-744.

Epstein, H. M., Duke, K. M. and Wharton, G. W. (1979). *Trans. Amer. Microscopical Soc.* (in press).

Fain, A. (1977). *Int. J. Acarol.* **3,** 115-116.

Schmidt-Nielsen, K., Bolis, L. and Maddrell, S. H. P. (1978). "Comparative Physiology: Water, Ions and Fluid Mechanics." Cambridge Univ. Press, Cambridge, England.

Waggoner, P. E. (1967). *Intl. J. Biometeorol.* **3 Supp. 11,** 41-52.

Wharton, G. W. (1976). *J. Med. Entomol.* **12,** 577-621.

Wharton, G. W. (1978). Uptake of water vapour by mites and mechanisms utilized by the Acaridei. *In* "Comparative Physiology: Water, Ions and Fluid Mechanics." (K. Schmidt-Nielson, L. Bolis and S. H. P. Maddrell, eds.) Cambridge Univ. Press, Cambridge, England.

Wharton, G. W. and Arlian, L. G. (1972). *In* "Insect and Mite Nutrition" (J. G. Rodriguez, ed.), pp. 153-165. North-Holland Press, Amsterdam.

Wharton, G. W. and Brody, A. R. (1972). *J. Parasitol.* **58,** 801-804.

Wharton, G. W. and Devine, T. L. (1968). *J. Insect Physiol.* **14,** 1303-1318.

Wharton, G. W. and Furumizo, R. T. (1977). *Acarologia* **19,** 112-116.

Wharton, G. W. and Richards, A. G. (1978). *Ann. Rev. Entomol.* **23,** 309-328.

METABOLISM OF BUTANEDIOL IN *TYROPHAGUS PUTRESCENTIAE*

J. G. Rodriguez, W. T. Smith, Peter Heffron and S. K. Oh

University of Kentucky
Lexington, Kentucky

INTRODUCTION

The glycols, including 1,3-butanediol, are of considerable value in food processing, in the formulation of pharmaceuticals and cosmetics, and as plasticizers and humectants. Research in synthetic sources of dietary calories was initiated in 1958 to develop high density food for extended manned space travel, and the most promising "high energy metabolite" was 1,3-butanediol, which has since been fed to various animal species at levels up to 20% without serious detrimental effects (Dymsza, 1975). In tests conducted by Tobin *et al.* (1975), 1,3-butanediol was shown to be a nontoxic metabolite providing a source of calories for human nutrition. Among other food additives, propylene glycol (propanediol) and 1,3-butanediol have been evaluated as antimicrobial agents in an intermediate moisture dog food product (Acott *et al.*, 1976).

Unpublished work in our laboratory has indicated that low dietary levels of glycols are inhibitory to acarid mites. Therefore it was deemed appropriate to study the metabolism of 1,3-butanediol in the acarid species *Tyrophagus putrescentiae*. This was accomplished by feeding 1,3-butanediol-3-^{14}C, fractionating the mite carcases and food residues and analyzing and radioassaying the fractions. In a comparison experiment, glucose-U-^{14}C was fed to mites.

EXPERIMENTAL

Rearing System

A system was devised in which feeding vials were connected to collection traps so that mites could feed and reproduce normally while their volatile

Vol. I: ISBN 0-12-592201-9

metabolic products, including CO_2, could be collected for radioassay. A gentle flow of air was filtered through glass wool, bubbled through a saturated $Ba(OH)_2$ solution to remove atmospheric CO_2 and maintained at 80% r.h. by passing over KOH. The air then passed through a tube that housed duplicate feeding vials containing a standard casein-wheat germ diet (Rodriguez, 1972) including the labeled component (50 μC glucose-U,^{14}C, sp. act. 7800 μC/mM; or 88 μC 1,3-butanediol-3-^{14}C, sp. act. 111 μC/mM, New England Nuclear, Boston, Massachusetts) and *ca* 10,000 *T. putrescentiae* eggs. The exit gases from the developing mites were then passed through an empty flask, a series of two saturated $Ba(OH)_2$ traps for precipitation of $^{14}CO_2$. The traps were 250-ml Erlenmeyer flasks, and the solutions were maintained at constant volume during the rearing period. Safety devices and procedures were used to prevent clogging of the connecting glass tubing, freezing of moisture near the dry ice-acetone trap, and the buildup of excess pressure. At the end of a 21-day rearing period, the mites were brush-collected from the walls of the feeding vials and filtered out of the food residue after suspension of both mites and food in warm water. The mites were then killed by freezing.

Analysis of Trapped Gases

The precipitated $Ba^{14}CO_3$ was filtered out of the traps, washed, air-dried, weighed and combusted in aliquots in a Parr oxygen bomb using 30 atm O_2 initially and a second increment of 10 atm to clear the bomb. The $^{14}CO_2$ was trapped in 20 ml of 2:1 2-methoxyethanol:2-aminoethanol. Duplicate 2-ml aliquots of each trap solution were radioassayed, as was the ash remaining in the combustion chamber.

The filtrates from the traps wre extracted twice with methylene chloride and once with ethyl ether, and the extracts dried, concentrated and subjected to gas chromatographic (GC) analysis for 1,3-butanediol, as were the acetone trap solutions. The analyses were performed on a 4 ft x 1/8 inch column packed with Porapak Q in an F and M Division Hewlett-Packard research chromatograph, model 5070B equipped with a hydrogen flame detector. The column temperature was 210°C and the ambient temperature 35°C with a helium flow of 25 ml/min. The injection port stood at 300°C.

Rate Studies

The relative rates of metabolism of glucose-U-^{14}C and 1,3-butanediol-3-^{14}C were studied by determining the radioactivity present in the first $Ba(OH)_2$ trap at intervals during the rearing period. For sampling, the system was briefly disconnected at the trap, the trap flask shaken to suspend evenly the $Ba^{14}CO_3$, and duplicate 0.2 ml aliquots removed for radioassay. Fresh $Ba(OH)_2$ was added to maintain constant volume, and the absolute activity on any given day was corrected for aliquots taken previously.

Fractionation Procedures

The procedures of Schneider (1945) for the separation of biological materials were used to separate the mite carcasses into four fractions: I, sugars and low molecular weight organic substances; II, phospholipids; III, nucleic acids; and IV, proteins.

Fraction V was the mite wash used to separate mites from food.

Food residues were extracted by centrifugation with 0.6% aqueous $LiCO_3$ solution to remove the uric acid, after the method of Bhattacharya and Waldbauer (1969). This extract comprised Fraction VI. The sediment was re-centrifuged with petroleum ether and methanol and the extracts combined to obtain Fraction VII. The sediment was Fraction VIII.

Analyses of the Phospholipid Fraction

Thin Layer Chromatography (TLC). Using the method of Horrocks (1963) the fractions II from both the glucose-U-^{14}C experiment and the 1,3-butanediol-3-^{14}C experiment were analyzed on a heat-activated thin layer Silica Gel GF plate (Analtech, 10 X 20 cm, 250 μ thick), using 1,3-butanediol and a mixture of known methyl esters as standards. The chromatogram was developed one-dimensionally in a mixture of chloroform, methanol and ammonium hydroxide (CMA) (65:25:4, v/v/v) and the spots visualized by short wave length UV light and I_2 vapors. Five discernible zones were scrapped off the plate and placed into scintillation vials with 0.2 ml of 95% ethanol for radioassay. Fraction II was also analyzed on TLC plates for the separation of lipid classes according to the procedures of Malins and Mangold (1960) and Malins (1966). These chromatograms were developed in a solvent system consisting of petroleum ether, ethyl ether and acetic acid (90:10:1, v/v/v). Ten discernible zones were radioassayed.

Column Chromatography. Fraction II from the 1,3-butanediol-3-^{14}C experiment was concentrated to a small volume, and taken up in a small amount of 95% ethanol and acetic acid, to be transferred to a silicic acid column (BioRad, Bio-Sil® A, 200-235 mesh, 12 X 110 mm). The material was eluted with petroleum ether-ethyl ether followed by chloroform, methanol and ammonium hydroxide (CMA) after Stein and Slawson (1966). Fractions of 5 ml were collected, an aliquot of each radioassayed, and each fraction analyzed by TLC as described above, after which the first 30 ml of eluate was condensed, taken up in petroleum ether and anhydrous ethyl ether and subjected to a second silicic acid column separation, radioassay and TLC analysis.

Characterization of Isolated Lipid Material. The most radioactive portions of the second column separation, fractions 13, 14 and 15, were combined and concentrated under nitrogen to a thick colorless and turbid oil. An IR spectrum (KBr pellet) of the material showed similarities to those of trioctanoin and of 1,3-butanediol dibutanoate. This "suspected triglyceride" was

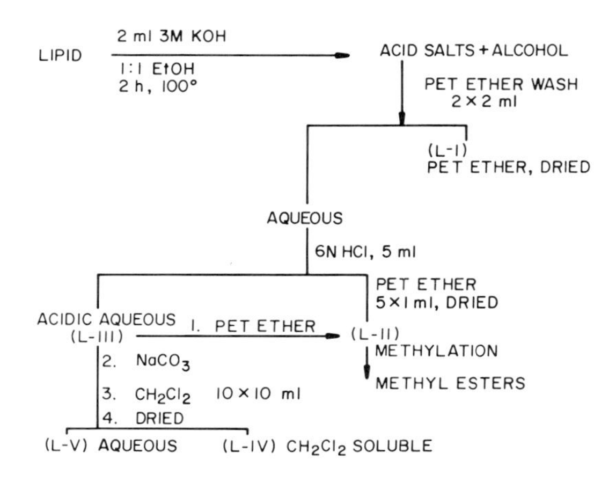

Fig. 1. Degradation of suspected triglyceride.

saponified by the procedures of Walling *et al.* (1968) with some modifications, according to the scheme shown in Fig. 1. Fractions L-I, L-IV and L-V were analyzed by gas chromatography on Porapak Q for 1,3-butanediol. Fraction L-II, containing the fatty acids, was methylated according to the procedure of Metcalfe and Schmitz (1961). The resulting methyl esters were analyzed by gas chromatography on Apiezon L columns with mixtures of known methyl esters as standards. The molar ratios of the individual fatty acids were determined by measurement of area under peaks compared with area under peaks made by known amounts of the standards, and the radioactivity under each peak was determined.

Radioassay

Scintillation counting was performed on a Packard Tri-Carb Model 3380 Scintillation Counter equipped with Model 544 Absolute Activity Analyzer. The scintillation fluid was in most cases composed of toluene, 2-methoxyethanol, Permafluor (containing PPO and POPOP) with 6% naphthalene. Occasionally a pre-blended scintillation cocktail which gave comparable results, Aquasol (New England Nuclear), was used.

The radioactivity of the soluble fractions was determined by counting aliquots of the solution, in duplicate. An aliquot of each solvent alone, of the same size as the sample, was counted and subtracted as background.

Residues were subjected to combustion in a Parr oxygen bomb, by the same procedures as those used for $Ba^{14}CO_3$.

RESULTS

The distribution of radioactivity in the fractionated mite carcasses (A), food

TABLE I.
Distribution of Radioactivity in Food Residues, Trapped Gases, and Fractionated Carcasses of *T. putrescentiae* Reared on Casein-Wheat Germ Diets Containing 50 μC Glucose-U-^{14}C or 88 μC 1,3-Butanediol-3-^{14}C.

	dpm			
	Glucose-U-^{14}C		1,3-Butanediol-3-^{14}C	
Material	% of Total in Fraction	% of Initial	% of Total in Fraction	% of Initial
A. Fractionated Mite Carcasses				
I. Sugars, low m. wt. organic cpds.	17.0	0.032	6.5	0.046
II. Phospholipids	29.5	0.055	62.4	0.442
III. Nucleic acids	7.1	0.013	1.5	0.011
IV. Proteins	46.4	0.087	29.6	0.210
Total	100.0	0.187	100.0	0.705
B. Food Residue				
V. Mite wash	9.1	0.886	5.7	3.671
VI. Uric acid and food wash	87.4	8.438	92.3	59.129
VII. ether, alcohol soluble fraction	0.7	0.069	1.0	0.637
VIII. Insoluble food residue	2.8	0.261	1.0	0.614
Total	100.0	9.654	100.0	64.051
C. Trapped Gases				
$^{14}CO_2$ as $Ba^{14}CO_3$	95.8	16.531	94.6	7.736
$Ba(OH)_2$ trap filtrates				
1st trap	2.6	0.455	3.4	0.277
2nd trap	1.3	0.219	1.7	0.144
Acetone trap	0.3	0.053	0.3	0.024
	100.0	17.258	100.0	8.181
Total Recovered		27.099		72.937

residues (B), and trapped gases (C) is given in Table I. The dpm values for the various fractions are expressed as percent of the total radioactivity in A, B, or C that appeared in the designated fraction, and also as percent of the initial radioactivity in the diet. The total dpm recovered in the glucose-U-^{14} experiment was 27.1% and in the 1,3-butanediol-3-^{14}C experiment was 72.9%.

Analyses of Trapped Gases

In both the glucose and butanediol experiments, about 95% of the radioactivity in the trapped gases was accountable as $^{14}CO_2$ precipitated as

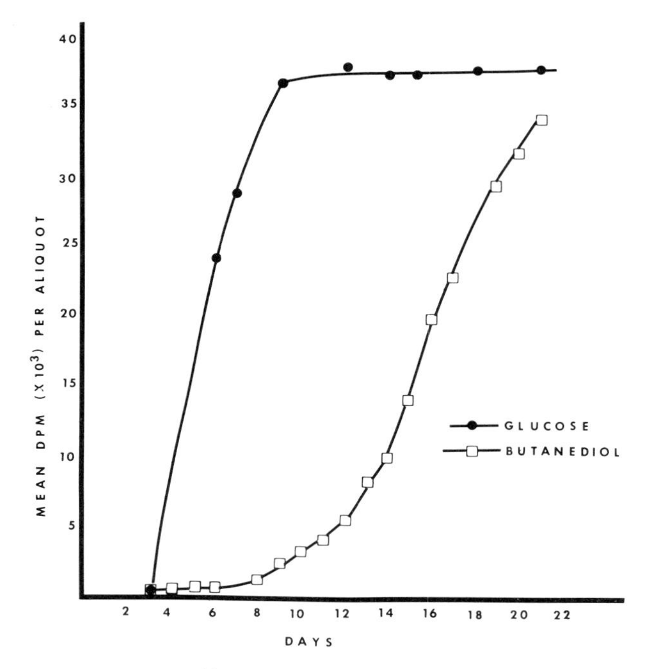

Fig. 2. Rate of production of $^{14}CO_2$ from *T. putrescentiae* reared on diets containing 50 μC glucose-U-^{14}C OR 88 μC 1,3-butanediol-3-^{14}C. Two 2-ml aliquots taken from first $Ba(OH)_2$ trap.

$Ba^{14}CO_3$. Attempts to analyze the contents of the acetone traps, as well as the methylene chloride extracts of the filtrates from the $Ba(OH)_2$ traps, were fruitless.

Rate Study

Fig. 2 shows that *T. putrescentiae* fed on glucose-U-^{14}C produced a relatively high level of $^{14}CO_2$ early in the rearing period. Production leveled off by the 10th day on the diet. The rate of production of $^{14}CO_2$ from 1,3-butanediol-3-^{14}C followed an entirely different pattern, starting out slowly and beginning to climb sharply at about the 10th day. By the end of the rearing period the total amount of $^{14}CO_2$ produced was nearly the same on the two diets.

Fractionation of Mites and Food Residues

In the glucose experiment, 0.1270 g of mite tissue was fractionated; in the butanediol experiment, 0.0369 g. Table I shows that the radioactivity in the carcasses of *T. putrescentiae* reared on 1,3-butanediol-3-^{14}C was distributed differently from that in the carcasses reared on glucose-U-$_{14}$C; 62.4% of the total radiocarbon in the diol-fed carcasses appeared in fraction II, the lipids.

The carcasses fed on glucose contained relatively more radiocarbon as proteins, sugars and nucleic acids.

When the uneaten food residue was fractionated, the greatest amount of radioactivity was found in Fraction VI, the uric acid and food wash, with a sizable amount in the mite wash (Table I). Fraction VI would predictably contain unused starting radioactive materials, since both glucose and butanediol are water soluble, and in the butanediol experiment, the presence of the parent compound was verified by GC analysis on a Porapak Q column of a methylene chloride extract of Fraction VI.

Analyses of the Phospholipid Fraction

In the first TLC analysis of fraction II, lipids, using a solvent system of chloroform-methanol-acetic acid (65:25:4) Horrocks, 1963), 68.3% of the radioactivity on the TLC plate was in a zone corresponding to neutral lipids in the carcasses of mites fed 1,3-butanediol-3-^{14}C; 0.6% had an R_f value corresponding to the parent compound; the remaining radiocarbon was found at or near the origin and was thought to consist of phospholipids. In further TLC analysis in a solvent system of petroleum ether-ethyl ether-acetic acid (90:10:1) (Malins and Mangold, 1960) (Table II) ten discernible zones were

TABLE II.

R_f Values of Lipid Standards Compared with Radioactivity in Zones Separated from Lipid Fraction II of *T. putrescentiae* Reared on Glucose-U-^{14}C or 1,3-Butanediol-3-^{14}C. (Silica Gel GF Developed in Petroleum Ether, Ethyl Ether and Acetic Acid (90:10:1).

Lipid Standard R_f		R_f Range of Zone Assayed	% of dpm on TLC Plate in Zone	
			Glucose-U-^{14}C	1,3-Butane-diol-3-^{14}C
Methyl esters, sat.	.850	.80-.99[a]	0[b]	0
Methyl esters, unsat.	.763	.64-.80	0	0
Methyl esters, low mol. wt.	.631	.60-.64	34.5	0.5
Stearic acid	.481	.46-.60	0	1.4
Trioctanoin	.419	} .35-.46	0	56.4
1,3-butanediol dibutanoate	.415			
1,3-butanediol dipropanoate	.388			
- - - - -		.30-.35	25.0	8.0
1,3-butanediol diethanoate	.245	.12-.30	0	10.5
- - - - -		.50-.12	0	4.0
1,3-butanediol and phospholipids	0	.0-.05	40.5	19.2

[a] two zones combined

[b] no radioactivity found

radioassayed, and 56.4% of the radiocarbon on the plate was in a zone comprising the R_f values .350 to .460. This zone included the R_f values of trioctanoin, 1,3-butanediol dibutanoate, and 1,3-butanediol dipropanoate. Of the remaining radioactivity, 19.2% remained at the origin where the parent compound, 1,3-butanediol, and the phospholipids would be found in this solvent system; 25.5% lay between the origin and an R_f value of 0.350, and could have been either other triglycerides or other diesters e.g., 1,3-butanediol diethanoate. This distribution pattern differed from that of the radioactivity in fraction II of mites reared on glucose-U-^{14}C, where 34.5% of the radioactivity migrated with the low molecular weight methyl esters, 40.5% remained at the origin, and 25.0% in an uncharacterized zone.

Silicic acid column separation of the lipid fraction II from the diol-fed mites, performed twice, gave a pure material the IR spectrum of which was very similar to a spectrum for a triglyceride. Upon saponification and the usual workup (Fig. 2), the radioactivity was almost evenly distributed in the aqueous and organic layers. GC analysis for possible 1,3-butanediol in the aqueous phase did not find this compound. GC analysis of the methylated fatty acid portion (organic extract) indicated that the "pure" material may have been a mixture of triglycerides containing many fatty acids. Of the fatty acids separated, negligible radioactivity appeared in those having chain lengths of C_{12} and under. The most radioactivity appeared in C_{18}, stearic, and the unsaturated C_{18} acids, oleic, linoleic and linolenic. These three were not totally distinguishable in the separation system used. Smaller amounts of radioactivity appeared in myristic, palmitic, palmitoleic and C_{20} acids.

DISCUSSION

Dymsza (1975) has suggested that "if the unpleasant taste problem can be overcome and if given FDA approval, 1,3-butanediol may have an increased role in our food supply as a functional food additive, preservative and source of calories for man and animals." Animal species are apparently able to utilize safely dietary percentages of 1,3-butanediol that in our laboratory have inhibited populations of acarid mites, including *T. putrescentiae,* a cosmopolitan pest capable of attacking many food products. Two implications may be drawn: (1) a low percentage of 1,3-butanediol in a foodstuff could protect a foodstuff from acarid mite infestation, and (2) 1,3-butanediol may play a different role in the nutrition of mites than it does in the nutrition of mammalian species. A study of the metabolism of butanediol in *T. putrescentiae,* then, has both basic and practical value.

We attempted to compare the metabolism of butanediol with that of glucose, the fundamental energy source, and some valid comparisons can be made from our results. The low recovery of label in the glucose experiment was unfortunate. It was undoubtedly procedural; initially there was some difficulty in maintaining "gas-tightness" in the rearing apparatus, and some refinements were made in the system prior to the butanediol experiment. Also, the glucose had a higher specific activity, so a smaller amount was used to attain a workable range of radioactivity, slightly more than 1 mg, as compared to 50 mg of butanediol. The unavoidable manipulative loss of label due to diet transfer, etc., would thus be intensified in the glucose experiment.

While this large loss of the glucose label limits in some respects the scope of the comparisons between the two nutrients, it in no way detracts from the significance of the findings in the butanediol experiment. Our results indicate that butanediol was metabolised in a different fashion than was glucose. This was shown first in the rate study, where the glucose label appeared as $^{14}CO_2$ at a much higher level early in the rearing period (Fig. 2). Even considering the fact that the glucose had a much higher specific activity, it was still utilized faster. Secondly, the distribution patterns of radiocarbon in the tissue fractions differed markedly in the mites reared on the two nutrients.

In research with other animal species, 1,3-butanediol has generally been administered as a replacement for dietary carbohydrate. However, the simple fact that it gives 6 Kcal/g (Dymsza, 1975) should suggest that it does not function purely as a carbohydrate. Romsos *et al.* (1975) have recently reviewed the published research on butanediol in lipid metabolism. Our results indicate that in *T. putrescentiae* the predominant, though not exclusive, routes taken by butanediol led into the lipid fraction of the mite bodies. This is consistent with the implications of the rate study that butanediol or its metabolites were stored longer and discharged more slowly as CO_2 than was glucose. Butanediol was converted to β-hydroxybutyrate in chicks, rats, and pigs (Romsos *et al.*, 1975) and possibly could enter directly into fatty acid synthesis.

An analysis of the isolated major component of the lipid fraction II of mites fed butanediol revealed a mixture of triglycerides which when saponified yielded radiolabeled fatty acids, both saturated and unsaturated, and having chain lengths of C_{14} or greater. The aqueous layer from the saponification also contained radioactivity, but too little material was left at this point to characterize the radioactive compound. One-dimensional TLC analysis of fraction II had indicated that possibly diesters of butanediol were present along with triglycerides. It has been speculated that the inhibiting effects of the diols toward mite growth/development is due to the formation of "pseudo" triglycerides in which the glycerol is replaced with a glycol. Further work is needed to clarify that point.

SUMMARY

Tyrophagus putrescentiae was reared on a standard casein-wheat germ diet to which was added either glucose-U-^{14}C or 1,3-butanediol-3-^{14}C. Vials with diet, containing *ca* 10,000 mite eggs were placed in a feeding chamber of an apparatus designed to collect trapped volatile metabolites while allowing normal rearing conditions for 21 days. Comparison of the relative rates of CO_2 production, as measured by $^{14}CO_2$ trapped in $Ba(OH)_2$, showed that glucose was metabolized at a faster rate than was butanediol. Of the total radioactivity in the trapped volatile metabolites, about 95% was identified as $^{14}CO_2$ on both nutrients. The mites were separated from the food residues, killed, and the carcasses fractionated into carbohydrate, lipid, nucleic acid and protein fractions. In mites fed butanediol, 62.4% of the recovered label appeared in the lipids, 29.6% in protein, 6.5% in carbohydrates, and 1.5% in nucleic acids. The corresponding values for mites fed glucose were 29.5%, 46.5%, 17.0% and 7.1%. Chromatographic analyses indicated that the major radioactive components of the lipid fraction of mites fed butanediol were triglycerides, which when saponified yielded labelled myristic, palmitic, stearic and one or more C_{18} unsaturated fatty acids. The results indicated that butanediol was metabolized in *T. putrescentiae* mainly through lipid pathways.

ACKNOWLEDGEMENT

We gratefully acknowledge partial funding of this study by a grant from the Pet Foods Division, General Foods Corporation.

REFERENCES

Acott, K., Sloan, A. E. and Labuza, T. P. *J. Food Sci.* **41**, 541-546.
Bhattacharya, A. K. and Waldbauer, G. P. (1969). *J. Insect Physiol.* **15**, 1129-1135.
Dymsza, H. A. (1975). *Federation Proc.* **34**, 2167-2170.
Horrocks, L. A. (1963). *J. Am. Oil Chemists' Soc.* **40**, 235-236.
Malins, D. C. and Mangold, H. K. (1960). *J. Am. Oil Chemists' Soc.* **37**, 576-578.
Malins, D. C. (1966). *Prog. Chem. Fats Lipids* **8**, 301-373.
Metcalfe, L. D. and Schmitz, A. A. (1961). *Anal. Chem.* **33**, 363-364.
Rodriguez, J. G. (1972). *In* "Insect and Mite Nutrition" (J. G. Rodriguez, ed.). pp. 637-650. North-Holland, Amsterdam.
Romsos, D. R., Belo, P. S. and Leveille, G. A. (1975). *Federation Proc.* **34**, 2186-2190.
Schneider, W. C. (1945). *J. Biol. Chem.* **161**, 293-303.
Stein, R. A. and Slawson, V. (1966) *Prog. Chem. Fats Lipids* **8**, 373-420.
Tobin, R. B., Mehlman, M. A., Kies, C., Fox, H. M. and Soeldner, J. S. (1975). *Federation Proc.* **34**, 2171-2176.
Walling, M. V., White, D. C. and Rodriguez, J. G. (1968). *J. Insect Physiol.* **14**, 1445-1458.

TOXINS OF THE AUSTRALIAN PARALYSIS TICK *IXODES HOLOCYCLUS*

B. F. Stone, B. M. Doube, K. C. Binnington

Division of Entomology

B. V. Goodger

Division of Animal Health
Commonwealth Scientific and Industrial Research Organization
Long Pocket Laboratories
Indooroopilly, Queensland, Australia

INTRODUCTION

The paralysis tick, *Ixodes holocyclus* is probably the most serious indigenous parasite affecting man and animals in Australia. It occurs along the eastern seaboard from Cairns in North Queensland to Lakes Entrance in Victoria in moist, densely vegetated areas. The principal natural hosts are the short-nosed bandicoot, *Isoodon macrourus* and the long-nosed bandicoot, *Perameles nasuta* (Doube, 1975). *I. holocyclus* often paralyzes and kills domestic pets, livestock and occasionally children. Tick toxicoses are not confined to Australia (Neitz, 1962) and paralysis, the most spectacular toxicosis, has been attributed to 31 species of seven ixodid genera and six species of three argasid genera listed for about 18 different countries (Murnaghan and O'Rourke, 1978). In this review *Dermacentor andersoni* in North America and *I. holocyclus* in Australia emerge as the most important species. *I. holocyclus* is the more consistently virulent and probably the more potent tick (Ross, 1935; Gregson, 1973; Doube *et al.*, 1977).

The symptoms caused by *D. andersoni* and *I. holocyclus* are essentially similar and in animals are loss of appetite and voice, incoordination, ascending flaccid paralysis, ocular irritation, excess salivation and vomiting, respiratory distress, asymmetric pupillary dilatation and frequently, death. All of these symptoms do not necessarily occur in every case. In Australia, if hyperimmune serum is used soon after the onset of symptoms, the chance of recovery is good but it diminishes rapidly with increasing severity of symptoms. Cardiovascular studies on *I. holocyclus*-paralyzed dogs revealed an

Vol. I: ISBN 0-12-592201-9

increase in blood pressure with a decrease in cardiac output (Cooper *et al.*, 1976) but *D. andersoni*-paralyzed dogs have normal blood pressure (Murnaghan, 1958). One respect in which these ticks differ from one another is that removal of *D. andersoni* usually leads to an immediate improvement whereas this does not necessarily occur in *I. holocyclus* and the animal may subsequently die (Gregson, 1973; Doube and Kemp, 1975).

Most reports of the physiology and pharmacology of tick paralysis have been based on studies of animals paralyzed by *D. andersoni*. This syndrome appears to be due to presynaptic effects at neuromuscular junctions caused by a disturbance in conduction in nerve fibres and from failure in the liberation, but not in the synthesis, of acetylcholine (Murnaghan, 1955, 1958, 1960; Rose and Gregson, 1956). It seems that motor paralysis induced by *I. holocyclus* also, may result from action at the neuromuscular junction but interference with conduction was not apparent in isolated hemidiaphragm preparations from mice with respiratory paralysis induced by nymphal feeding (Cooper and Spence, 1976). A direct, temperature-dependent inhibition of transmitter release occurred at neuromuscular junctions in these preparations.

Attempts to extract, isolate and characterize paralysis toxins have been reported for *I. holocyclus* rather than for *D. andersoni* ticks. Ross (1926, 1935) demonstrated toxic effects in mice and dogs injected with extracts of salivary glands taken from *I. holocyclus* when the ticks were at their most toxic stage (5th day of feeding, Goodrich, and Murray, 1978). Homogenates of salivary glands from *D. andersoni* females at a similar or later stage of feeding did not cause paralysis when injected into susceptible animals (Gregson, 1973). It is accepted that the toxins responsible for paralysis are secreted in tick saliva. Nothing is known, however, of their chemical nature except that they appear to be associated with proteins in *I. holocyclus* (Kaire, 1966; Goodrich in Sutherland, 1974). Kaire (1966) isolated from homogenates of replete females a paralyzing fraction, with marked immunological similarity to toxins secreted into hyperimmune dogs by feeding ticks. Similar fractionation of extracts of paralyzing *D. andersoni* failed to produce any characteristic symptoms in the marmot (Gregson, 1973). Toxicity was retained after digestion with proteases (pepsin, trypsin and papain) of Kaire's active *I. holocyclus* fraction. This suggested that the toxin was non-proteinaceous or that the toxic protein was resistant to digestion. On dialysis of the papain digest, toxin appeared in the dialysate and retentate. Therefore toxins were associated with extracted or digested components of differing molecular size. The principal active fraction was detoxified at 100°C but was generally stable at temperatures up to 75°C for short periods. Activity was unaffected by moderate changes in pH (Kaire, 1966). An anticoagulant isolated by Kaire from the same material was not a toxin.

Immunity in dogs to paralyzing toxins of *I. holocyclus* has been clearly established (Ross, 1935; Oxer and Ricardo, 1942) and this has resulted in the commercial production of an effective hyperimmune serum which is widely used in Australia for the treatment of animals and occasionally of man (Pearn, 1977). No corresponding development of immunity to *D. andersoni* toxins has been reported and *I. holocyclus* hyperimmune serum does not appear to relieve *D. andersoni* paralysis (Gregson, 1973).

The program at the Long Pocket Laboratories is primarily aimed at developing a vaccine against the paralysis toxin. Practical problems such as defining the best source of the toxin, developing assay techniques and separating and characterizing toxins are our major immediate concerns, but other topics, related to vaccine production and tick physiology and biochemistry, are under consideration. These include investigation of the function, origin and site of production of toxin and its role in tick physiology. Under consideration also are possible variations in tick toxicity with the physiological condition and geographic origin of the tick and with the immune status of the host. The remainder of this paper reports on some recent progress in isolating and assaying toxins produced by *I. holocyclus*, and draws to a large extent on work not yet published.

EXPERIMENTAL AND RESULTS

Assay of toxins of *I. holocyclus*

Preliminary experiments showed that adult mice (20-25 g) generally failed to show typical ascending limb paralysis when injected with paralyzing extracts. Suckling mice (4-5 g) which had been separated from their mothers were found to be much more sensitive and were used exclusively in routine bioassay. These were injected intraperitoneally (3-5 mice per dose) with 100 μl of serial dilutions of supernatants or of aseptically prepared homogenates of salivary glands in 0.9% ice cold saline. The mice were held at 30°C for 24 h. Responses were then scored on a paralysis index (P. I.) scale from 0-100 according to the severity of symptoms (Stone *et al.*, in prep.). The graphically estimated dose producing a P. I. of 50 was called the standard paralyzing dose (SPD). For fractions which were toxic but did not induce paralysis a dose producing 50% mortality was called the standard lethal dose (SLD). All doses were expressed as gland equivalent (g.e.)/g or μg Folin reactives (F.R.)/g. F.R. includes proteins, free aromatic amino acids and other phenolic compounds reacting with the Folin reagent but was converted to apparent "protein" content (Lowry *et al.*, 1951). Lyophilized supernatants were about 72% F.R. and 58% TCA-precipitable protein.

Production of toxin

Collection of ticks and salivary glands. Variation in the toxicity of ticks from different geographic regions has been reported for *D. andersoni* and *D.*

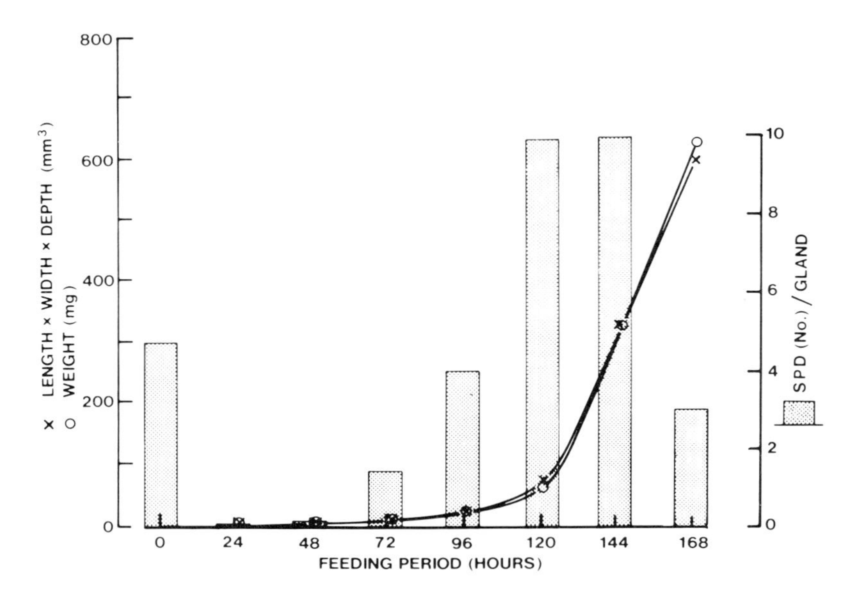

Fig. 1. Sequential changes in toxicity of homogenates of salivary glands from *I. holocyclus* females fed for varying periods on rats. Toxicity expressed in standard paralyzing doses (SPD's) per gland. Growth curve expressed as weight (mg) or dimensions (length x width x depth) plotted against time.

variabilis (Gregson, 1973) but this has not been observed in *I. holocyclus*. However, the regions sampled represent only a small proportion of the geographic range of the tick in Australia. We have found no difference in toxicity between laboratory-reared and field-collected ticks.

Unfed females are obtained by dragging woolen blankets over vegetation or by trapping infested bandicoots from which engorged nymphs are collected; the respective merits of the two methods largely depend on season. A laboratory culture of *I. holocyclus* has been established but so far has provided only a small proportion (*ca* 15%) of the unfed females needed. Females have usually been fed on mice for 96 h, at which time the hosts show terminal symptoms. Rats are now being used so that females can feed to 120-144 h and this gives a higher yield of toxin (Fig. 1). The toxicity of extracts of glands from ticks which have fed for 120 h on field-collected, tick-infested, short-nosed bandicoots and on highly paralysis-immune laboratory dogs was similar to the toxicity of those from ticks which had fed for 120 h on rats (Stone *et al.*, in prep.). This finding may conflict with that of Goodrich and Murray (1978) who reported that toxin production may be suppressed in hosts with a previous experience of tick feeding.

Sequential changes in gland toxin level and histology. Homogenates of glands from unfed female ticks were found to be quite toxic (SPD = 0.04 g.e./g). Gland toxicity declined sharply after commencement of feeding on rats, reaching a minimum after 24 h. It then increased from 48 h reaching a

TABLE I.
Relationship between Toxicity of Homogenates of Salivary Glands of *I. holocyclus* Females, Density of Cellular Granules in Glands and Esterase Activity. Ticks Fed for Various Periods on Mice; Ratings of Granular Density and Esterase Activity Based on Mean Values Obtained with 2-3 Glands per Feeding Period (1 per tick).

Feeding Time (h)	Toxicity of Gland Homogenates (SPD's/gland)	Granular Density Acinus II Acinus III Cell type a	b	d	e	Esterase Activity Acinus II Acinus III Cell type a	b	d	e
0	4.8	2+	1+	10+	10+	3	--	--	--
12	--	3+	1+	5+	3+	3	--	--	--
24	O.5	4+	1+	3+	±	4	--	--	--
48	0.3	4+	1+	3+	±	4	--	--	--
72	0.7	4+	5+	2+	±	2	--	--	--
96	2.0	4+	4+	1+	±	1	--	--	--
120[a]	10	7+	8+	2+	±	1	--	--	--
ca 168[b]	1.7	10+	3+	2+	±	1	--	--	--

[a] Ticks rat-fed or transferred from dying mouse after 96 h to healthy mouse.
[b] Engorged females from shortnosed bandicoot.

maximum at 120 and 144 h (SPD = 0.02 g.e./g), after which activity again declined to a relatively low value at 168 h (Fig. 1), which usually corresponded to the time of full engorgement and detachment. Ticks fed normally on rats, and development generally followed the expected growth curve for weight or approximated "volume" (length x width x depth). The time changes in toxicity of homogenates of glands from ticks fed on mice were very similar to those already demonstrated for mice by Goodrich and Murray (1978) and for rats in this paper.

Salivary glands were removed from *I. holocyclus* females which had fed for 0, 12, 24 48, 72 and 96 h on mice or for 120 h on rats. Glands were processed (Binnington, in press) and individual cells rated for granular content and non-specific esterase activity (Table I). The best correlations appear to be between the normal late-developing toxicity (96-120 h) and granular content in cell b and between the toxicity of unfed ticks and granular content in cells d and e. Sequential changes in salivary gland morphology during feeding of *I. holocyclus* have been studied and have been related in more detail to toxicity of gland homogenates (Binnington and Stone, in prep.).

Immunological similarity between naturally secreted toxins and gland extracts. One of the most convincing demonstrations of the similarity between secreted toxin and extract of glands is the ability of antibodies from naturally-immunized animals (exposed to feeding ticks) to neutralize the toxic effects of gland extracts. Lyophilized supernatants from centrifugation of gland homogenates were reconstituted in 0.9% saline and serially diluted to give a range of SPD's. A standard volume (100 μl) of commercial (therapeutic) canine anti-tick serum was added to 40 μl of each concentration. The mixture was incubated at 37°C for 45 min and injected into 4-5 g mice; 3-5 mice were used per dose (Stone *et al.*, in prep.). The results of the neutralization test indicated a normal toxin/antitoxin reaction. This suggested the presence, in the antiserum, of antibodies specific for the toxins extracted from salivary glands. An IgG fraction separated from commercial canine anti-tick serum and showing maximum *in vivo* neutralization of salivary gland toxins, was covalently bound to CNBr treated Sepharose 4B on an affinity chromatography column. Antigens from supernatants when passed through this column, bound to the antibodies on the matrix and were subsequently eluted with glycine-HCl buffer (pH 2.2) and shown to be toxic (Doube, unpubl. data). These results confirm that homogenates of salivary glands taken at the peak of toxicity of *I. holocyclus* contains toxins immunologically similar to toxins secreted into the host.

Separation of Salivary Gland Components

Gel Filtration. Lyophilized supernatants were reconstituted in 0.9% saline and applied to Sephadex G 200, G 100 or G 75 columns from which fractions (referred to as SX I, II, III, etc.) were eluted using 0.15M ammonium acetate or ammonium bicarbonate solutions (Goodger, 1971; Fohlman *et al.*, 1976). A number of separations on G 200 had similar profiles and resulted in fractions SX vv (void volume) I, II, and III (typical separation in Fig. 2) which were lyophilized, reconstituted and biologically assayed (typical assay in Fig. 3). SX vv, I and II all caused paralysis (and death at high doses) but SX II was the

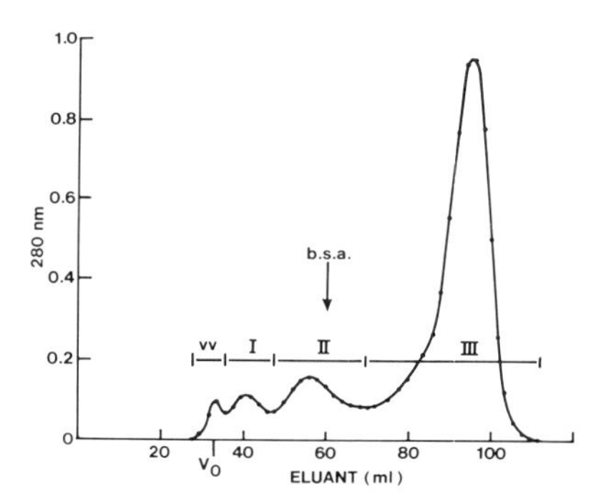

Fig. 2. Typical gel filtration of supernatant derived from 163 salivary gland equivalents of *I. holocyclus* females fed 96 h on mice. Separated on a 50 x 2.5 cm column of Sephadex G 200 equilibrated with 0.85% NaCl solution. Flow rate = 5 mlxh^{-1}, 2.5 ml fractions. Arrow indicates elution position of bovine serum albumen (MW 68,000).

principal paralyzing fraction; SX III (approx. mol. wt. range is less than 20,000) was lethal but non-paralyzing even at doses just below the lethal threshold. SX II (mol. wt. 60-100,000) appeared to be the fraction demanding further attention as the proportion of SPD's in SX II (76-84% overall) remained highest.

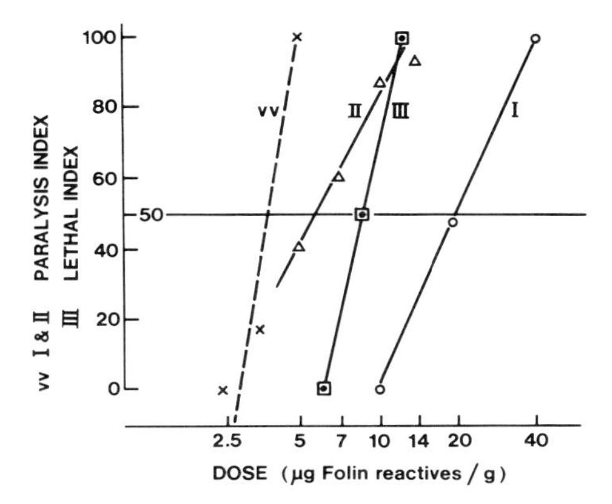

Fig. 3. Effect of intraperitoneal injection into 4-5 g mice, of fractions from Sephadex G 200 separation of *I. holocyclus* salivary gland components (3-5 mice per dose). Paralysis (toxicity) index is based on quantitative assessment of paralysis (or general toxicity) on 0-100 scale. FR = Folin Reactives, i.e., phenolic compounds including proteins and aromatic amino acids reacting with the Folin reagent.

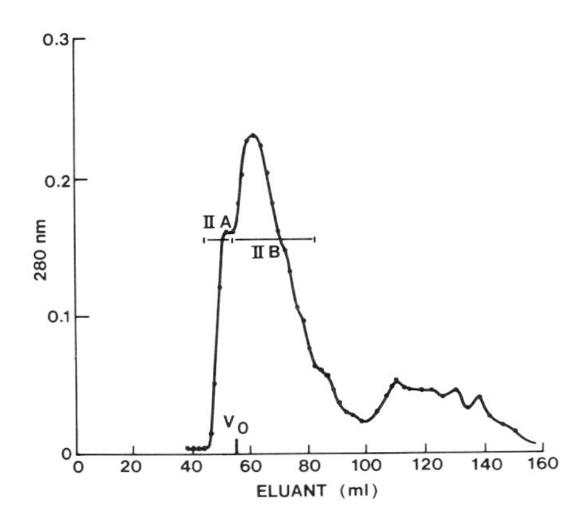

Fig. 4. Gel filtration on Sephadex G 100 of combined fraction II components derived from four G 200 separations of supernatant from *I. holocyclus* salivary homogenates. Column = 50 x 2.5 cm, equilibrated and eluted with 0.85% NaCl solution at 10 $mlxh^{-1}$.

The mean SPD values in μg/g (range in parentheses) obtained from these separations were SX vv 6(4-10), I 24(12-40), II 15(6-20) and for III the SLD was 14(8-20). SX II from one separation was resolved further on Sephadex G 75 producing a number of sub-fractions; one of these formed a major peak (mol. wt. 60-80,000) and contained 83% of the paralyzing doses. This fraction was four times more potent than the original fraction II (SPD decreased from a mean of 15 μg/g to 4 μg/g).

The combined SX II components from four runs were separated further on Sephadex G 100 resulting in a number of subfractions (Fig. 4) of which SX IIA/IIB combined, constituted the only major peak and the only toxic fractions (SPD is less than 5 μg/g). SX IIA/IIB was rechromatographed on Sephadex G 200 and four fractions resulted; one of these (SX IIY) showed the highest level of paralyzing activity so far separated (SPD = 0.5 μg/g).

Affinity Chromatography. A paralyzing fraction was separated from gland extracts by affinity chromatography as outlined previously. This is encouraging but further examination of toxic yield is required to determine whether the method should be adopted.

DISCUSSION

Clearly progress has been made towards our aim of producing large amounts of purified toxin. The collection and extraction of salivary glands containing toxins has been streamlined and there is a better understanding of the changes in toxicity during feeding. The unexpected discovery that the salivary glands of unfed females contain substantial quantities of the toxin may further simplify its collection but there is some evidence that the toxic components may be different. A simple, precise and sensitive assay for testing the toxicity of extracts has been developed which has permitted analysis of toxin present in a wide variety of extracts. Fractionation of toxic extracts by use of gel filtration and affinity chromatography has given rise to a number of toxic fractions, one of which is specifically paralyzing. Hence we are well on the way towards our first goal, purification of the toxin.

The paralysis toxin is associated with a wide range of fractions but in particular with proteins in the molecular weight range 60,000-80,000. It is not known if these proteins are toxic in themselves or are simply carrier molecules for toxins of lower molecular weight. Protease digestion studies in progress may resolve this. The presence of a toxic fraction (SX III) of lower molecular weight (MW is less than 20,000) which failed to cause paralysis is of interest because its presence may account for variations in observed symptoms. Individuals may have quite different susceptibility to the two toxins and it is possible that this latter toxin causes some symptoms, e.g. cardiovascular failure, which has previously been attributed to the paralysis toxin.

The relationship between changes in salivary gland morphology and toxicity during feeding suggests that the toxin may be contained in granules. The toxin may be a major or minor component of granular material which could have any of a number of functions in feeding. The finding (Koch, 1967; Doube and Goodger, unpubl. data) that dermal glands of engorged nymphs and females of *I. holocyclus* secrete toxins which may have similar characteristics to those in the saliva, suggests that toxins may be present in the haemolymph and fortuitously secreted by either the salivary or dermal glands. That the highest level of paralyzing toxin occurs towards the end of feeding could be a consequence of the known increase in extracellular space which occurs in the "water cells" of ixodid salivary gland at this time (Kaufmann, 1976).

There is also a relationship between toxin level in salivary glands and feeding activity. *I. holocyclus* females feeding on bandicoots produce intense capillary dilatation near the feeding lesion at 5 h; this dilatation is progressively diminished at 24 h and at 48 h (Stone and Schleger, unpubl. data). During this slow phase of engorgement, secretory activity remains at a low level, increasing again after 72 h during the rapid and more toxic phase of feeding (Fig. 1). However, further speculation on the role of toxins in the physiology of the tick must await more definitive information on their source, synthesis and chemical nature.

SUMMARY

1. Published reports on toxins of *I. holocyclus* are briefly reviewed and compared with those of the toxins of *D. andersoni.*

2. A more sensitive and precise assay involving quantitative measure of the severity of symptoms in suckling mice has been developed.

3. The paralyzing toxin is best obtained from females which have fed on rats for 120-144 h. Immunity of the host did not affect toxicity of salivary glands.

4. Gland homogenates from unfed ticks were also toxic and there were correlations between changes in morphology of particular salivary gland cells suggesting that secretory granules may contain the toxin. The possibility that toxicity of the salivary gland may be related to its role in water excretion is also discussed.

5. Neutralization test, affinity chromatography and biological assay show that secreted toxins and toxins extracted from salivary gland are very similar and possibly identical.

6. A fraction (mol. wt. 60,000-80,000) with a high level of paralyzing activity, has been isolated by gel filtration as well as a lethal fraction (mol. wt. is less than 20,000) with no paralyzing activity.

ACKNOWLEDGEMENTS

Thanks are due to the following staff members at CSIRO, Long Pocket Laboratories: Dr. James Nolan for many of the preliminary chromatographic separations; to Margaret Cowie and Alan Neish for toxicology and chromatography and to Stan Fiske for illustrations. Other colleagues assisted in the preparation of the manuscript. Special mention should be made of extended but unpublished investigations into the nature of these toxins by Dr. Ben Goodrich, CSIRO, Division of Animal Health (now of Division of Wildlife Research), Sydney and for valuable discussions with him.

REFERENCES

Binnington, K. C. (1978). *Int. J. Parasitol.* **8,** 97-116.
Cooper, B. J. and Spence, I. (1976). *Nature* **263,** 693-695.
Cooper, B. J., Cooper, H. L., Ilkew, J. E. and Kelly, J. D. (1976). Sydney University Postgraduate Committee in Veterinary Science **30,** 57-61.
Doube, B. M. (1975). *Aust. Vet. J.* **51,** 511-515.
Doube, B. M. and Kemp, D. H. (1975). *Aust. J. Agric. Res.* **26,** 635-640.
Doube, B. M., Kemp, D. H. and Bird, P. E. (1977). *Aust. Vet. J.* **53,** 39-43.
Fohlman, J., Eaker, D., Karlsson, E. and Thesleff, S. (1976). *Eur. J. Biochem.* **68,** 457-469.
Goodger, B. V. (1971). *Res. Vet. Sci.* **12,** 465-468.
Goodrich, B. S. and Murray, M. D. (1978). *Int. J. Parasitol.* (in press).
Gregson, J. D. (1973). "Tick paralysis: an appraisal of natural and experimental data." Canad. Dept. Agric. Monog. 9.
Koch, J. H. (1967). *NSW Veterinary Proceedings,* **3,** 34.
Kaire, G. H. (1966). *Toxicon* **4,** 91-97.
Kaufmann, W. R. (1976). *J. Exp. Biol.* **64,** 727-742.
Lowry, O. H., Rosebrough, W. H., Faw, L. A. and Randall, R. J. (1951). *J. Biol. Chem.* **193,** 265-275.
Murnaghan, M. F. (1955). *Rev. Canad. Biol.* **14,** 273-274.
Murnaghan, M. F. (1958). *Nature* **181,** 131.
Murnaghan, M. F. (1960). *Canad. J. Biochem. Physiol.* **38,** 287-295.
Murnaghan, M. F. and O'Rourke, F. J. (1978). *In* "Arthropod Venoms" (Sergio Bettini, ed.), Springer-Verlag, Berlin, Heidelberg, New York.
Neitz, W. O. (1962). "Tick Toxicosis." *Rept 2nd Mtg of FAO/OIE Expert Panel on Tick-borne diseases of Livestock. U. A. R.*
Oxer, D. T. and Ricardo, C. L. (1942). *Aust. Vet. J.* **18,** 194-199.
Pearn, J. (1977). *Med. J. Aust.* **2,** 313-318.
Rose, I. and Gregson, J. D. (1956). *Nature* **178,** 95-96.
Ross, I. C. (1926). *Parasitol.* **18,** 410-429.
Ross, I. C. (1935). *J. Coun. Sci. Ind. Res. Aust.* (1934). **8,** 8-13.
Sutherland, S. K. (1974). *Anaesthesia and Intensive Care* **2,** 316-328.

CONTROL OF SALIVARY FLUID SECRETION IN IXODID TICKS

William R. Kaufman

Department of Zoology
University of Alberta
Edmonton, Alberta, Canada

INTRODUCTION

The question of how blood-sucking organisms regulate their internal fluid composition has long occupied the attention of physiologists. Arthropods which exploit blood as the primary or sole source of nourishment are sporadically confronted with a glut of water which must be eliminated rapidly. Between meals, however, they share with other small terrestrial organisms the pervasive problem of retaining water reserves. It is this rather abrupt switch from a negative to positive water balance which has endowed hematophagous arthropods with some distinctive mechanisms for water economy.

Early progress in this field came from the laboratories of Wigglesworth and Ramsay, whose initial studies with insects inaugurated the conceptual framework upon which much detail is still being appended (see review by Maddrell, 1971). Less attention has been paid to ticks, which is surprising, since both major families (Ixodidae and Argasidae) have evolved fluid excretory mechanisms independent from both the insects and each other. In this paper I shall consider only ixodid ticks.

From the outset, there appeared to be a special problem associated with this family. Like most other bloodsuckers, the meal is concentrated in the gut, but contrary to the case for insects, little "urine" is secreted by the Malpighian tubules at this time. Lees (1946), working with *Ixodes ricinus,* showed that the critical temperature of the integument (that temperature at which occurs an abrupt increase in water permeability) was slightly lower than the host's skin temperature. He thus proposed that the increase in transpiration arising from the tick's sojourn on the host might account for water balance. This ingenious hypothesis is no longer accepted. Whenever water loss from the integument has been measured (Tatchell, 1967a; Kaufman and Phillips, 1973a) it accounts for only a small proportion of the total water loss.

Vol. I: ISBN 0-12-592201-9

Gregson (1967), studying feeding mechanisms in *Dermacentor andersoni,* was the first to propose that the salivary glands might serve as the major osmoregulatory system during feeding, an idea almost immediately verified for *Boophilus microplus* by Tatchell (1967a). Since first enunciation of the "salivary gland hypothesis," workers in a number of laboratories have set out to explain the physiological mechanisms employed to initiate and terminate the fluid secretory process.

PHARMACOLOGY OF SALIVARY FLUID SECRETION

Howell (1966) was probably the first to employ a pharmacological stimulant (pilocarpine) in order to elicit salivation in ticks. Since then, pilocarpine (PC) has been used extensively when collection of large quantities of saliva is desired (Tatchell, 1967b, 1969; Binnington and Schotz, 1973). The implication of these findings is that a cholinergic system controls salivary fluid secretion. Cholinomimetics are not the only drugs capable of stimulating salivation. With the introduction of an *in vitro* method for observing salivary fluid secretion (Kaufman and Phillips, 1973b; Kaufman, 1976) it became clear that isolated glands were quite insensitive to cholinomimetics, a finding confirmed in at least two other laboratories (Megaw, 1974; Needham and Sauer, 1975). The fact that cholinomimetics are effective only *in vivo*, whereas adrenergic drugs such as epinephrine, norepinephrine (NE) and dopamine (DA) are potent stimulants *in vivo* and *in vitro*, indicated that adrenergic drugs act closely to the secretory epithelium, whereas cholinergic drugs act via some intermediary. (It is possible that some drugs which stimulate secretion *in vitro* also act indirectly by stimulating the release of endogenous transmitter from the nerve terminals, but this remains to be demonstrated.) There is additional evidence to indicate direct innervation by a catecholaminergic nerve, the latter probably serving as the mediator for PC-induced salivation. 1) The salivary glands of several tick species are innervated by axons containing dense-core granules (Coons and Roshdy, 1973; Megaw, 1977), 2) NE and DA have been detected in nervous and salivary tissue of ticks (Megaw and Robertson, 1974; Binnington and Stone, 1977), 3) Although the adult nervous system is a rich source of acetylcholine (Smallman and Schuntner, 1972), cholineacetyltransferase, one of the enzymes required for acetylcholine (ACh) synthesis, appears to be present in the nervous system but not the salivary glands (Megaw, 1976) and 4) The action of cholinergic agonists *in vivo* is attenuated by a number of selective antagonists, all of which have no inhibitory effect on DA-induced salivation (Kaufman, 1978). These antagonists are: atropine (which probably blocks "central" cholinergic receptors), reserpine (which has been shown to reduce catecholamine stores in tick salivary tissue [Megaw and Robertson, 1974]) and guanethidine (a drug which in mammalian sympathetic nerves blocks the release of transmitter from the nerve terminals). That these drugs are all acting in ticks by pharmacological mechanisms similar to those demonstrated to occur in mammals should not be assumed *a priori*. For example, reserpine blocked PC-induced salivation to a similar degree regardless of whether it was administered simultaneously with, 1 h before or 17 h before PC (Kaufman, 1978). The very early onset of reserpine's effect suggests the possibility of some mechanism distinct from (or at least parallel to) depletion of nerve-terminal catecholamines. Similarly, although the classic cholinesterase inhibitor physostigmine (= eserine) induces salivation *in vivo*, it does so at a dose far below that necessary to potentiate the weak action of ACh (Kaufman, 1978), and thus the agonist effect of physostigmine in this system may not be related to inhibition of cholinesterase. Incidentally, the nicotonic blocking drugs, d-tubocurarine and toxiferine, do not block PC-induced salivation (Kaufman, 1978), a finding which further supports Megaw's (1974) hypothesis that cholinergic responses in this system are muscarinic.

One as yet unsolved question relates to the identity of the transmitter of the secretory nerve. I tested a variety of putative agonists on isolated salivary glands of *Amblyomma hebraeum* (Kaufman, 1977). Inactive were histamine, 5-hydroxytryptamine, glutamate and aspartate. Among catecholamines, DA was the most potent, but the ergot alkaloids, ergotamine and ergonovine, were even more potent than DA. It was beginning to look as if the postjunctional receptor on the salivary gland resembled a DA-receptor. DA appears to be the predominant catecholamine in many arthropod tissues (Murdoch, 1971; Gerschenfeld, 1973) and it is associated with the salivary glands of a number of insects (Klemm, 1972; House *et al.*, 1973; House and Ginsborg, 1976). However, Megaw and Robertson (1974) detected significant levels of two catecholamines (NE and DA) in the synganglion and salivary glands of *B. microplus*. (In mammalian sympathetic nerves DA is the immediate precursor of the transmitter NE. Since the enzyme which converts DA to NE is present at a high level, the normal content of DA in adrenergic nerves is vanishingly small [Iversen, 1967]. The simplest interpretation of Megaw and Robertson's results is that the salivary gland receives both dopaminergic and noradrenergic innervation. There are, however, other possible explanations for the prominence of the two amines, but further speculation at this time would be premature.) My own survey of putative antagonists for DA-induced salivation in isolated glands also proved inconclusive in terms of defining the nature of the salivary gland catecholamine receptor (Kaufman, 1977). Among four blocking agents known to be extremely potent on other dopaminergic systems (chlorpromazine, flupenthixol, spiperone, pimozide), only the second had some blocking action, but this was apparent only at excessive concentrations. The others had no blocking action whatever; on the contrary, in the presence of low concentrations of pimozide or spiperone, the maximal response to DA was increased by 100%. In some instances a similar effect was noted with chlorpromazine as well. Beta-adrenergic blocking drugs (propranolol, dichloroisoprenaline) were also ineffective except at very high concentrations as were alpha-adrenergic antagonists (phentolamine, phenoxybenzamine). In summary, pharmacological analysis of this system has done little more than to suggest a receptor which, though highly sensitive to DA, corresponds to none of the classical catecholamine receptor-types recognized in vertebrates and some invertebrates.

WHAT DOES ALL THIS PHARMACOLOGY MEAN?

Although one might gain a certain sense of satisfaction in being able to control a biological response by use of a drug, this really tells one rather little about what information is normally used to attune salivation to the tick's osmoregulatory needs. Since copious salivation occurs only during the feeding period (male ticks and unfed female ticks also secrete saliva under certain conditions [Feldman-Muhsam *et al.*, 1070; Rudolph and Knülle, 1974], but quantities are miniscule compared to those secreted by females during the feeding period), one assumes quite naturally that it is the transport of fluid

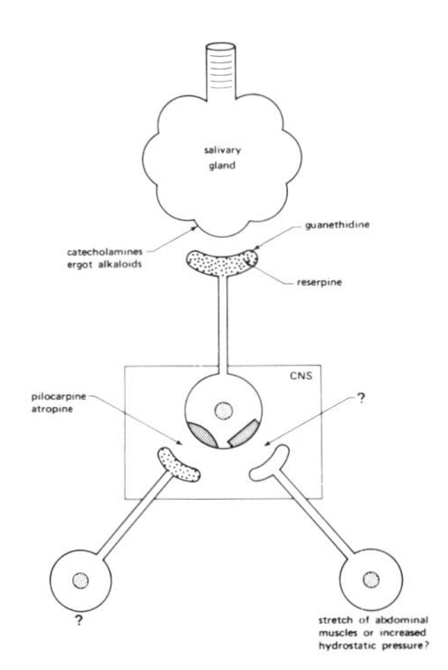

Fig. 1. A schematic representation of salivary gland innervation based on pharmacological studies. The gland is innervated by a catecholaminergic secretory nerve (various sympathomimetic drugs trigger secretion in isolated glands). The secretory nerve receives information via a cholinergic (muscarinic) nerve (cholinomimetics such as PC stimulate secretion only *in vivo,* an effect which is abolished by the muscarinic antagonist, atropine). Since atropine does not attenuate saline-induced secretion, a second sensory pathway is postulated. Possibly this nerve relays information on hemolymph volume, but the synaptic messenger is not known. For the cholinergic pathway, however, it is the physiological parameter which remains obscure. Although atropine, reserpine and guanethidine all inhibit PC-induced secretion, none of them inhibits DA-induced secretion, a finding also consistent with the present scheme.

from the gut to the hemolymph which somehow triggers salivary secretion. But is this parameter 1) increase in hemolymph volume, 2) stretch of certain mechanoreceptors, 3) decrease in hemolymph osmotic or oncotic pressure, 4) concentration change in some hemolymph constituent, or 5) some combination of the above? Recently we have been able to stimulate salivation *in vivo* by injecting large volumes of isosmotic fluids into the hemolymph (Kaufman *et al.*, 1978). At relatively low doses (5 μl/100 mg bs) solutions of NaCl, sucrose, urea and distilled water (DW) were equi-effective. At higher doses (10-25 μl/100 mg) isosmotic urea and DW were less effective, insofar as a smaller proportion of the injected load was eliminated, but NaCl was more effective. Isosmotic sucrose elicited a similar response to that of isosmotic NaCl up to 10 μl/100 mg, but at 25 μl/100 mg, the effect of sucrose was much lower.

Above I postulated that a cholinergic receptor on the secretory nerve might transmit sensory information on volume or composition of the hemolymph to the salivary gland. Since atropine is a potent antagonist of PC-induced secretion, it was of interest to test whether it could similarly interfere with saline-induced salivation. The results (Kaufman *et al.*, 1978) were contrary to expectation. Saline-induced salivation (25 μl/100mg) was essentially atropine-insensitive. Reserpine, on the other hand, did attenuate saline-induced salivation markedly. Sucrose- and urea-induced salivation were only partially blocked by atropine whereas the weak salivary response caused by DW was inhibited to the greatest extent. The results suggest that there are at least two nerve pathways bringing sensory information to the salivary gland (Fig. 1).

FUTURE WORK

We still are not at all clear as to what parameters of the hemolymph space are monitored in order to trigger salivation. Salivation does remove fluid from the hemolymph and hemolymph volume appears to be regulated by the tick (Kaufman and Phillips, 1973a). Since NaCl is the major osmolyte of hemolymph and saliva, the injection of isosmotic NaCl would have disturbed hemolymph composition less than the injection of any of the other substances we tested. Notwithstanding the latter, isosmotic saline was the most potent stimulus to secretion. Also considering that injection of very large volumes of fluid were required to elicit salivation, it seems reasonable to suppose for the moment that increase in hemolymph volume is an important parameter for triggering secretion. The physiological mechanisms by which hemolymph volume is measured by the tick have still not been elucidated.

Since mammalian blood is hyposmotic to tick tissues, one observes a fall in hemolymph osmotic concentration during the meal, an effect which is kept in check by the secretion of a hyposmotic saliva (Kaufman and Phillips, 1973a). Secretory rate *in vitro* is influenced by the ambient osmotic concentration

(Kaufman and Phillips, 1973c). Although DW administered *in vivo* is, on its own, a relatively weak stimulus to salivation, more work is needed to determine whether the normal fluctuations in hemolymph osmotic pressure play some auxiliary role in modulating secretory rate.

Megaw and Robertson (1974) demonstrated the presence of both DA and NE in tick salivary glands, but they only looked at a single stage of the feeding cycle (probably at least 24 h post-engorgement, but details are not provided). It is well known that the salivary glands are suffering autolysis at this time (Till, 1961; Kaufman, 1976), so we are currently assaying tissue from the complete feeding cycle.

Finally, there remains conflicting evidence regarding the anatomical pathway of salivary gland innervation. Binnington and Tatchell (1973) stat[e] that the salivary glands receive innervation from branches of the palpal nerves whereas Obenchain and Oliver (1975) suggest a much more heterogeneou[s] source of innervation. It may be possible to resolve this question soon by [a] combination of surgical and pharmacological techniques. It is possible to ope[n] the ventral surface of partially- or fully-engorged ticks in the region of th[e] synganglion, cut specific nerve trunks, suture the wound and then injec[t] salivary stimulants (Kaufman, 1978). By testing which nerve trunks ar[e] necessary to permit salivation after challenging ticks with PC or isosmoti[c] fluids we hope to delineate more clearly the pattern of salivary gland in[n]ervation, at least with respect to fluid secretion.

REFERENCES

Binnington, K. C. and Schotz, M. (1973). *J. Austral. Entomol. Soc.* **12,** 78-79.
Binnington, K. C. and Stone, B. F. (1977). *Comp. Biochem. Physiol.* **58C,** 21-28.
Binnington, K. C. and Tatchell, R. J. (1973). *Z. wiss Zool.* Zeipzig **185,** 193-206.
Coons, L. B. and Roshdy, M. A. (1973). *J. Parasitol.* **59,** 900-912.
Feldman-Muhsam, B., Borut, S. and Saliternik-Givant, S. (1970). *J. Insect Physiol.* **16,** 19[4]-1949.
Gerschenfeld, H. M. (1973). *Physiol. Rev.* **53,** 1-119.
Gregson, J. D. (1967). *Parasitology* **57,** 1-8.
House, C. R. and Ginsborg, B. L. (1976). *Nature* **261,** 332-333.
House, C. R., Ginsborg, B. L. and Silinsky, E. M. (1973). *Nature* **245,** 63.
Howell, C. J. (1966). *J. S. Afr. Vet. Med. Assoc.* **37,** 236-239.
Iversen, L. L. (1967). "The Uptake and Store of Noradrenaline in Sympathetic Nerves." Ca[m]bridge University Press.
Kaufman, W. (1976). *J. Exp. Biol.* **64,** 727-742.
Kaufman, W. (1977). *Eur. J. Pharmacol.* **45,** 61-68.
Kaufman, W. (1978). *Am. J. Physiol.* In press.
Kaufman, W. and Phillips, J. E. (1973a). *J. Exp. Biol.* **58,** 523-536.
Kaufman, W. and Phillips, J. E. (1973b). *J. Exp. Biol.* **58,** 537-547.
Kaufman, W. and Phillips, J. E. (1973c). *J. Exp. Biol.* **58,** 549-564.
Klemm, N. (1972). *Comp. Biochem. Physiol.* **43A,** 207-211.
Lees A. D. (1946). *Parasitology* **37,** 1-20.

Maddrell, S. H. P. (1971). *Adv. Insect Physiol.* **8**, 199-331.
Megaw, M. W. J. (1974). *Comp. Biochem. Physiol.* **48A**, 115-125.
Megaw, M. W. J. (1976). Ph.D. Thesis, Cambridge.
Megaw, M. W. J. (1977). *Cell and Tissue Res.* **184**, 551-558.
Megaw, M. W. J. and Robertson, H. A. (1974). *Experientia* **30**, 1261-1262.
Murdock, L. L. (1971). *Comp. Gen. Pharmacol.* **2**, 254-274.
Needham, G. and Sauer, J. (1975). *J. Insect Physiol.* **21**, 1893-1898.
Obenchain, F. D. and Oliver, J. H. Jr. (1975). *J. Morph.* **145**, 269-294.
Rudolph, D. and Knülle, W. (1974). *Nature* **249**, 84-85.
Smallman, B. N. and Schuntner, C. A. (1972). *Insect Biochem.* **2**, 67-77.
Tatchell, R. J. (1967a). *Nature* **213**, 940-941.
Tatchell, R. J. (1967b). *J. Parasitol.* **53**, 1106-1107.
Tatchell, R. J. (1969). *J. Insect Physiol.* **15** 1421-1430.
Till, W. M. (1961). *Mem. Ent. Soc. Afr.* **6**.

Recent Advances in Acarology, Volume I

CYCLIC NUCLEOTIDES, CALCIUM AND SALIVARY FLUID SECRETION IN IXODID TICKS

J. R. Sauer, G. R. Needham, H. L. McMullen

Department of Entomology

R. D. Morrison

Department of Statistics
Oklahoma State University
Stillwater, Oklahoma

INTRODUCTION

The salivary glands are the primary organs of osmoregulation in ixodid ticks (reviewed by Sauer, 1977). During feeding, excess fluid is extracted from the blood meal, moved across gut diverticula and returned to the host via the salivary glands. There is increasing evidence that the glands are controlled by nerves (Obenchain and Oliver, 1976; Megaw, 1977) and that the synapse at the gland is catecholaminergic-like (Kaufman, 1976).

Cyclic AMP (cAMP) is thought to mediate, in many instances, the actions of catecholamines (Robison *et al.*, 1968). In earlier communications (Needham and Sauer, 1975; Sauer *et al.*, 1976) we reported that *in vitro* salivary glands obtained from females of the lone star tick *Amblyomma americanum* secrete fluid in the presence of exogenous cAMP if the glands are prestimulated with a catecholamine or if the glands are stimulated simultaneously by cAMP and a phosphodiesterase (PDE) inhibitor (theophylline). In this communication, we provide additional evidence that cAMP is a "second messenger" during the fluid secretory process. Possible interrelationships between calcium and cAMP in controlling fluid secretion are discussed.

MATERIALS AND METHODS

Adult lone star ticks, *Amblyomma americanum,* were reared and allowed to feed on host animals (sheep) by the methods of Patrick and Hair (1975).

Vol. 1: ISBN 0-12-592201-9

Methods for dissecting salivary glands from females and subsequent measurement of fluid secretion were as described by Needham and Sauer (1975).

Support Media

The Ringer-solution used in some experiments was the same as that described by Needham and Sauer (1975). TC-199 with Hank's balanced salt solution (obtained from Difco) and as modified by Kaufman (1976) was used except that levels of streptomycin sulfate and penicillin G sodium were 0.1 g and 0.3 g/liter, respectively. Solutions were oxygenated before use and fluid secretion experiments performed at 27 ± 2°C. All drugs and chemicals were of the highest purity commercially available.

Cyclic AMP Assay

The amount of cAMP in one salivary gland was measured by the receptor protein binding displacement assay method of Gilman (1970) except that the condition of incubation was changed from pH 4.0 to 4.5 and bovine serum albumin was used to enhance cAMP binding to protein kinase (Brostrom and Kon, 1974). Prior to assay, glands were quick frozen in liquid nitrogen, purified and prepared for assay by the methods of Gilman and Murad (1974). The relationship between incubation time of glands with dopamine (10^{-5}M) and gland levels of cAMP was studied by the use of a balanced incomplete block design in which two incubation times were used in each block (the two glands from a rapidly engorging female). Other pairs were incubated at 15 and 240 sec. without dopamine to serve as controls.

RESULTS

Effects of Dopamine and Tick Weight on Maximum Fluid Secretion by *In Vitro* Glands and Assessment of Modified TC-199 to Support Secretion

Fig. 1 verifies the findings of Kaufman (1976) who found that slightly modified TC-199 is a more effective support medium than a previously used Ringer-solution for bathing *in vitro* glands. Maximum rates of secretion by glands bathed in the Ringer-solution of Needham and Sauer (1975) were much less than when glands were bathed in TC-199 with 1/10 as much catecholamine (dopamine). Fig. 1 further indicates that maximum secretion is very much dependent on the weight of the female from which the glands are obtained. Considerable variation in the maximum fluid secretory response by individual glands was observed. The dose-response of glands stimulated by varying concentrations of dopamine bathed in modified TC-199 is included in Fig. 2. Maximum secretion by glands (not shown in Fig. 2) is less than 1 nl/min (0.72 ± 0.25 ± S. E. M.; n = 4) in the absence of a stimulating drug.

Effects of Exogenous Cyclic AMP on Fluid Secretion

With the above indications of the enhancing qualities of modified TC-199, the abilities of exogenous theophylline and cAMP to stimulate fluid secretion

TABLE I.
Effect of Theophylline, cAMP, cGMP, or cGMP/Theophylline on Ability of Isolated Salivary Glands to Secrete Fluid.[a]

Wt. of Tick[b] From Which Gland was Obtained (mg)	Max. Rate of Secretion (nl/min)	Wt. of Tick[b] From Which Gland was Obtained (mg)	Max. Rate of Secretion (nl/min)
Theophylline (10^{-2}M)		cGMP (10^{-2}M)	
71	0.2	90	1.4
93	2.1	128	1.8
131	0	128	0.3
136	1.4	211	0
286	0	309	4.6
360	0.8	315	5.4
409	0.8	451	8.4
479	346.5	518	6.3
627	0	571	1.4
764	2.6	583	13.9
795	0	770	0.8
		888	0
cAMP (10^{-2}M)		cGMP (10^{-2}M) + Theophylline (10^{-2}M)	
102	2.1	68	6.3
105	0	96	4.6
127	4.6	124	0
134	10.9	133	3.2
239	31.0	165	6.3
297	21.3	232	4.6
381	98.6	324	17.3
438	111.0	409	17.3
521	21.3	426	4.6
580	295.0	568	31.0
662	58.5	787	67.2
666	0	877	0
589	8.4		
709	25.9		
856	1.4		

[a] Bathing medium, modified TC-199; Max rate during 70 min. experiment.
[b] Engorging female *A. americanum*.

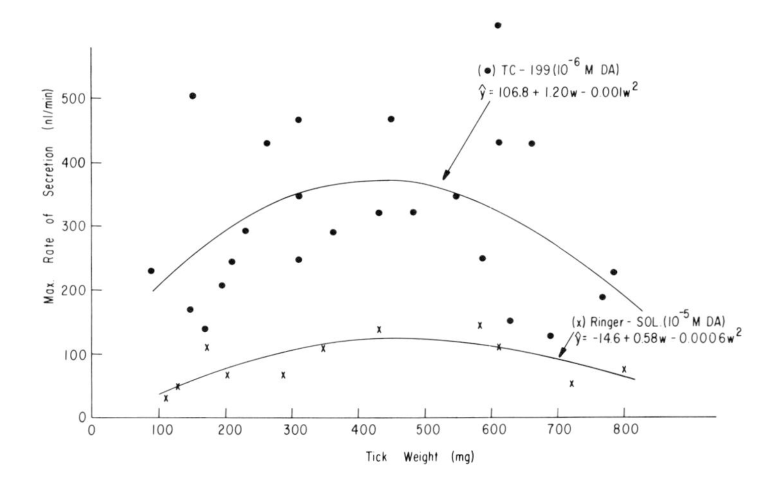

Fig. 1. Maximum rates (= y) of salivary fluid secretion (nl/min) observed in glands obtained from ticks of different weight (= w) (state of engorgement). Support medium was modified TC-199 or Ringer-solution (Needham and Sauer, 1975). Glands bathed in TC-199 were stimulated to secrete with 10^{-6}M dopamine and glands bathed in Ringer-solution 10^{-5}M dopamine. Glands were obtained from engorging *Amblyomma americanum* females. Each result represents maximum secretion attained by a gland during a 70 min. experiment.

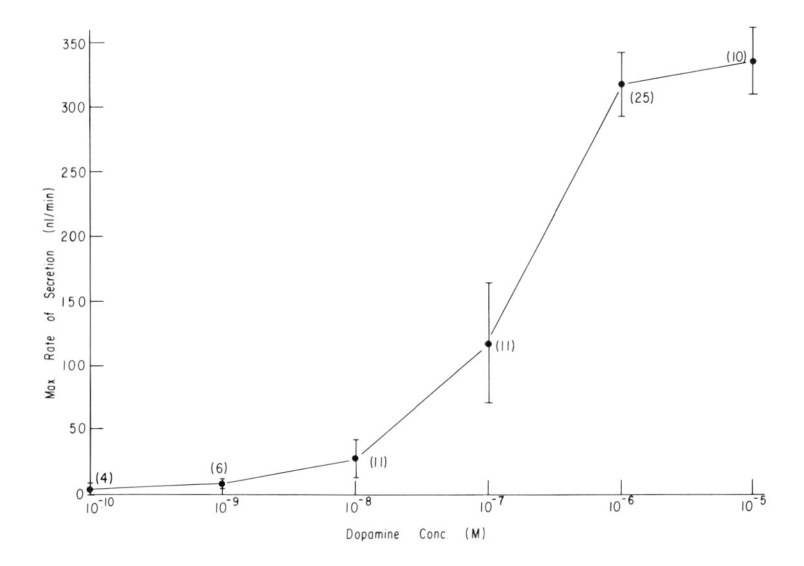

Fig. 2. Dose-response curve of maximum rates of fluid secretion observed in isolated salivary glands of engorging *A. americanum* females stimulated to secrete with dopamine. Support medium was modified TC-199. Vertical lines represent ± S. E. M. and numerals in parentheses indicate numbers of experiments.

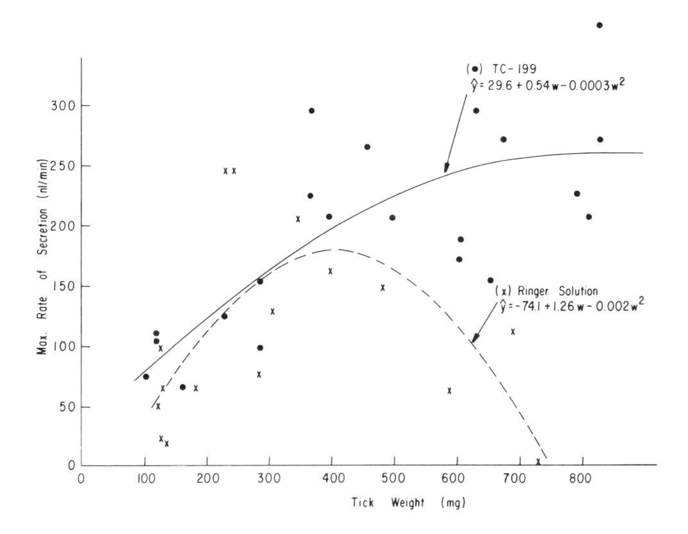

Fig. 3. Maximum rates (= y) of salivary fluid secretion (nl/min) observed in glands obtained from ticks of different weight (= w) (state of engorgement) and stimulated by cAMP (10^{-2}M)/theophylline (10^{-2}M). Support medium was modified TC-199 or Ringer-solution (Needham and Sauer, 1975). Glands were obtained from engorging *Amblyomma americanum* females. Each result represents maximum secretion attained by a gland during a 70 min. experiment.

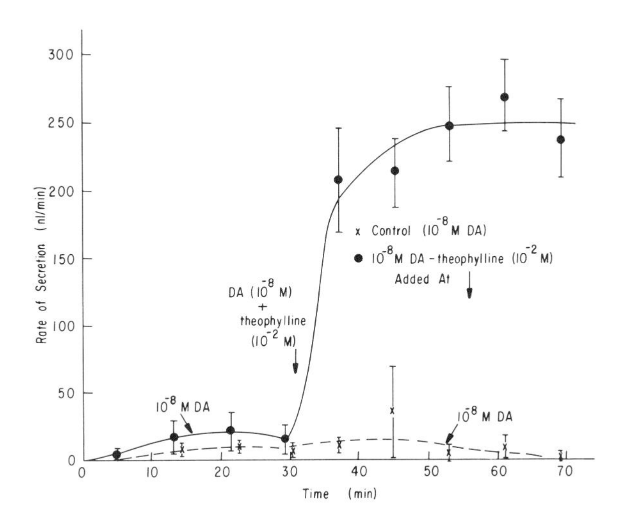

Fig. 4. Effect of theophylline (10^{-2}M) and 10^{-8} dopamine on rates of secretion (nl/min) by isolated salivary glands of rapidly engorging *A. americanum* females (wt. range 534-827 mg). Each point represents the average of five determinations and vertical lines represent ± S. E. M.

were again investigated but with modified TC-199 as the support mediu
Theophylline (10^{-2}M) did not stimulate glands to secrete substantially with o
notable exception (total of 11) (Table I). On the other hand, exogenous cAM
(10^{-2}M) stimulated all but two glands to secrete (Table I). The maximum ra
in the latter varied considerably and rates were not as high as those initiated
10^{-5} and 10^{-6}M dopamine (Figs. 1 and 2). As before (Sauer *et al.*, 1976) cAM
(10^{-2}M) stimulated glands to secrete (Fig. 3) but still not to rates as high
those initiated by 10^{-5} and 10^{-6}M dopamine with glands bathed in the sa
solution (Figs. 1 and 2). The secretory rates of glands obtained from hea
(rapidly engorging) females were enhanced by bathing glands in TC-199 a
the importance of weight of tick from which glands were taken in affecti
maximum secretion was again evident.

In tests to see if theophylline enhanced low doses of catecholamines, 10^{-2}
of the methylxanthine was added to glands that were previously stimulat
with 10^{-8}M dopamine (Fig. 4). Glands tested were obtained from rapi
engorging females (greater than 300 mg). An approximate 25-fold increase
the average rate of secretion was recorded. Results with control glan
stimulated by only 10^{-8}M dopamine throughout are included in Fig. 4 f
purposes of comparison.

Effects of Cyclic GMP

Kuo *et al.* (1971) discovered high levels of cyclic GMP (cGMP) depend
protein kinases in several arthropods and suggested that GMP may be i
portant to aspects of biological regulation in this group of animals. As can

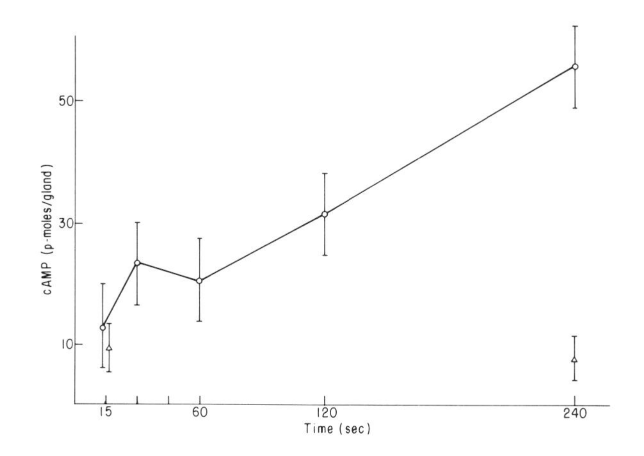

Fig. 5. Cyclic AMP (p-moles) in glands obtained from engorging *A. americanum* fema
following incubation over increasing periods of time in modified TC-199 with 10^{-5} dopami
Control glands were incubated for 15 or 240 sec. in the absence of dopamine. Each po
represents the average of at least four experiments and vertical lines represent ± S. E. M.

seen from Table I, however, neither high levels of cGMP (10^{-2}M) or cGMP (10^{-2}M)/theophylline (10^{-2}M) greatly enhanced fluid secretion by *in vitro* glands.

Levels of Cyclic AMP

The baseline control level of cAMP in salivary glands ($\bar{x}$ wet wt ± SD = 4.009 ± 2.518 mg) obtained from rapidly engorging females (greater than 300 mg) bathed for 15 sec. or 240 sec. in modified TC-199, was approximately 9.6 and 7.6 ± 3.9 p-moles/gland, respectively, in the absence of a catecholamine (Fig. 5). The concentration, on the other hand, in glands bathed in the same solution but with 10^{-5}M dopamine increased linearly with time up to 55.6 ± 6.7 p-moles/gland after 4 min.

DISCUSSION

The present results provide further proof that cAMP is involved as a "second messenger" in the salivary fluid secretory process in ixodid ticks. Exogenous cAMP mimics the effect of the primary stimulus (dopamine); an inhibitor (theophylline) of the only enzyme (PDE) known to catalyze hydrolysis of cAMP enhances the sensitivity of glands to submaximal doses of dopamine (and cAMP) and cAMP increases in glands stimulated by dopamine. At the same time, demonstration of an enzyme (adenylate cyclase) that catalyzes conversion of ATP to cAMP and an increase in its activity following incubation with a catecholamine have not yet been established. However, increases in gland levels of cAMP with increasing time of incubation with dopamine strongly suggest this possibility.

Since high concentrations of exogenous cAMP administered *in vitro* do not completely match the stimulatory effect of the primary stimulus, it is possible that salivary gland tissue is quite impermeable to the large negatively charged molecule and any that enters is rapidly degraded by PDE. The fact that theophylline greatly enhances fluid secretion brought about by 10^{-8}M dopamine suggests this possibility. In addition, the relative ineffectiveness of cAMP alone to mimic the primary stimulus but the effectiveness of theophylline and cAMP together to stimulate secretion suggests the presence of a PDE (McMullen and Sauer, 1978) that rapidly catalyzes the conversion of cAMP to 5'-AMP. Also, the general ineffectiveness of theophylline alone in stimulating secretion suggests that in the absence of a primary stimulant increased levels of cAMP are quite transient. The one exception (Table I) is difficult to explain except that gland cAMP may have been high just prior to testing the effects of the inhibitor in this instance. It is also worth noting that bathing glands in modified TC-199 enhances secretion more (compared to bathing glands in Ringer-solution) when glands are stimulated by dopamine (Fig. 1 and 2) than when glands are stimulated by cAMP/theophylline (Fig. 3). In particular, glands obtained from slowly feeding ticks (less than 300 mg) and stimulated by cAMP/theophylline secrete about the same when glands are bathed in either of the two solutions (Fig. 3). Recently, we discovered that replacement of phosphate buffer in the Ringer-solution with non-calcium chelating MOPS buffer (Morpholinopropane Sulfonic Acid) allows catecholamine-stimulated glands to secrete equally as well as when bathed in TC-199 (Needham and Sauer, 1978). With phosphate present, only micromolar unprecipitated quantities of free calcium are present in solution which contrasts to millimolar amounts when MOPS is the buffer (Needham and Sauer, 1978). However, this simple change in solutions does not increase cAMP/theophylline stimulated secretion (Needham and Sauer, 1978). Therefore, MOPS buffered Ringer-solution (and possibly TC-199) may enhance catecholamine-stimulated secretion by providing more free exogenous calcium, but the effect of TC-199 on cAMP/theophylline stimulated secretion may be to enhance cAMP/theophylline entry into fluid-secreting cells (increased permeability). It is also worth noting that enhancement of cAMP/theophylline stimulated secretion by TC-199 is best in glands obtained from rapidly engorging females (greater than 300 mg) (Fig. 3) or at times when gland PDE activity is lowest (McMullen and Sauer, 1978).

Needham and Sauer (1978) have found that removing calcium or adding the calcium antagonist verapamil to the bathing medium reduces secretion. The results are puzzling, however, because little calcium uptake can be detected in glands during stimulation by a catecholamine (Needham and Sauer, 1978) and glands will not secrete with addition of the calcium ionophore A-23187 (Needham and Sauer, 1978). The latter has often been cited (Rasmussen *et al.*, 1975) as evidence of a role for calcium in stimulus-secretion coupling in other cells. Thus, if an influx of calcium occurs during secretion the amount that enters may be small or any that enters is quickly extruded from the cell. Also, it seems that an influx of calcium alone is insufficient to trigger fluid secretion.

Depletion of calcium from the bathing medium depresses *in vitro* glandular secretion, but it does so slowly (Needham and Sauer, 1978). This could indicate that during secretion some calcium is mobilized from intracellular reservoirs. Evidence in favor of this is the relatively high level of calcium in glands (*ca* 10 mM; Needham and Sauer, 1978) and a reduction in its level during incubation with a catecholamine and an increased efflux of ^{45}Ca from pre-loaded glands when glands are stimulated by a catecholamine.

An interaction between cAMP and calcium during fluid secretion is further implicated because the sensitivity of glands to cAMP/theophylline is lost (Needham and Sauer, 1978) with calcium missing from the support medium. The same inability of cAMP/theophylline to effectively stimulate fluid secretion is seen after adding the calcium antagonist verapamil to the support medium. Thus it seems that both cAMP and calcium are involved and interrelated in some way in effecting the salivary fluid secretory process in ixodid ticks.

SUMMARY

Evidence indicates that cAMP is a "second messenger" of events that control the fluid secretory processes in ixodid female salivary glands. Exogenous cAMP stimulates glands to secrete and a phosphodiesterase (PDE) inhibitor (theophylline) enhances submaximal doses of dopamine. Theophylline also enhances the stimulatory effect of cAMP.

Stimulation of glands by dopamine causes gland levels of cAMP to increase. Glands are only slightly sensitive to cGMP / theophylline.

Additional evidence suggests a role for calcium in the fluid secretory process. A close relationship between cAMP and calcium in controlling secretion is likely.

ACKNOWLEDGEMENTS

This research was supported in part by Grant Nos. PCM 74-24140-A02, PCM 74-24140-A01, and PCM 77-23508 from the National Science Foundation. The authors are grateful to Ms. Carlene King and Kathy Dart for expert technical assistance.

REFERENCES

Brostrom, C. O. and Kon, C. (1974). *Analyt. Biochem.* **58,** 459-468.

Gilman, A. G. (1970). *Proc. Natl. Acad. Sci. U. S. A.* **67,** 305-312.

Gilman, A. G. and Murad, F. (1974). *In* "Methods in Enzymology," (J. G. Hardman and B. W. O'Malley, eds.). Vol. XXXVIII, pp. 49-61. Academic Press, New York.

Kaufman, W. (1976). *J. Exp. Biol.* **64,** 727-742.

Kuo, J. F., Wyatt, G. R. and Greengard, P. (1971). *J. Biol. Chem.* **246,** 7159-7167.

McMullen, H. L. and Sauer, J. R. (1978). *Experientia* **34,** 1030-1031.

Megaw, M. W. J. (1977). *Cell. Tiss. Res.* **184,** 551-558.

Needham, G. R. and Sauer, J. R. (1975). *J. Insect Physiol.* **21,** 1893-1898.

Needham, G. R. and Sauer, J. R. (1978). (In prep.).

Obenchain, F. D. and Oliver, J. H. (1976). *J. Parasitol.* **62,** 811-817.

Patrick, C. D. and Hair, J.A. (1975). *J. Med. Entomol.* **12,** 389-390.

Rasmussen, H., Jensen, P., Lake, W., Friedmann, N. and Goodman, D. B. P. (1975). *Adv. Cyclic Nucleotide Res.* **5,** 375-394.

Robison, G. A., Butcher, R. W. and Sutherland, E. W. (1968). *Ann. Rev. Biochem.* **37,** 149-174.

Sauer, J. R. (1977). *J. Med. Entomol.* **14,** 1-9.

Sauer, J. R., Mincolla, P. M. and Needham, G. R. (1976). *Comp. Biochem. Physiol.* **53C,** 63-66.

MECHANISMS CONTRIBUTING TO WATER BALANCE IN NON-FEEDING TICKS AND THEIR ECOLOGICAL IMPLICATIONS

D. Rudolph and W. Knulle

Institut fur Angewandte Zoologie
der Freien Universitat Berlin
West-Berlin, Germany

INTRODUCTION

During the non-feeding phases of their life cycle ticks are able to maintain their water balance in subsaturated air for considerable periods of time. This is accomplished partly by restricting discharge of feces and excreta to a rather short period after the preceeding bloodmeal. Of overriding importance, however, are mechanisms counteracting evaporative water losses, e.g., active uptake of water vapor from the atmosphere and restriction of water losses from the general body surface and from the openings of the tracheal system. Evaluation of the significance of tracheal water loss and the role of spiracular regulation for maintaining water balance is hampered by the fact that the functional morphology of the tick spiracle and its mode of action are insufficiently understood. Special emphasis is given, therefore, to this topic in the present paper.

ABSORBING WATER VAPOR

The ability to absorb water vapor from the air and to maintain water steady state has been stated for the non-parasitic phases of all instars except the eggs of more than 30 species of ticks. It seems to be poorly developed in those instars of ticks with a one-host or two-host life cycle which remain obligately on the host and in which obviously no ecological need exists; e.g. when we removed unfed nymphs and adults of *Boophilus annulatus* from the host immediately after molting and before reattachment we noticed a maximal survival time of only 14 days in high humidity of 98% R.H. at 25°C. These instars were not able even at 98% R.H. to recover a previously imposed water

Vol. I: ISBN 0-12-592201-9

deficit; only a few individuals exhibited a small uptake at 93% and 98% R.H. The lowest environmental humidity at which ticks can compensate water losses by absorption of atmospheric water vapor—the critical equilibrium humidity—ranges in ixodid and argasid ticks from approximately 75% to 94% R.H., depending on the species. Above these values ticks are able to maintain a high water level in their bodies for many weeks. In females of *Amblyomma variegatum* the body water is not delicately balanced under equilibrium conditions; instead, longer periods of 8-14 days of slight water loss alternate with shorter periods of 2-6 days of rapid uptake. However, the tick does adjust its body water from about -2% to +2% on the average of the mean water content. The amount of water in the body of the tick under equilibrium conditions is related to the level of environmental humidity, that is, higher ambient humidities correspond to higher water contents of the tick. Furthermore, the rate of uptake increases with the relative humidity, e.g. vapor uptake in partly dehydrated *A. variegatum* proceeds seven times faster at 93% R.H. than at 85% R.H.

Physiological uptake of water vapor from subsaturated atmospheres qualifies as a case of active transport. The organism overcomes a steep gradient of water activity when transferring water from the atmosphere where it occurs in low concentration to the organism with a water concentration corresponding to a relative humidity of almost 100%. In ixodid ticks water vapor uptake proceeds via the mouth (Rudoph and Knulle, 1974). There is evidence from several observations that the first step of uptake is performed by a hygroscopic salivary secretion (Rudolph and Knulle, 1978). It is likely that the saliva containing the extracted water is subsequently imbibed via the pharynx. Candidates for the production of the hygroscopic saliva are the non-granular Type I alveoli of the salivary glands. In extremely dehydrated non-feeding ticks the ejection of large quantities of a salivary secretion which accumulates as a white crystalline mass on the mouthparts is perhaps best regarded as an elimination of excess ions during periods when osmotic and ionic stress results from severe water losses. Current studies of the salivary secretion have shown so far the presence of Na, K, Cl and S. Whether the reduction of water activity in the ejected saliva to 0.85 is brought about by concentrating these ions or whether other vapor pressure lowering substances are involved is unknown.

RESTRICTING WATER LOSS DEVICES

Besides the remarkable ability to replace lost water by extracting water vapor from subsaturated atmospheres those devices which restrict water losses under dehydrating conditions are of fundamental importance for maintaining body water and thereby for the survival of ticks. Some species, when exposed to humidities below the critical equilibrium humidity, lose water very rapidly and die of dehydration within a few days, others lose water very slowly and survive for several weeks; e.g., at 0% R.H. and 25°C unfed females of *Ixodes ricinus* lose up to 50% of their original weight/day and survive for only one or two days, whereas adults of *Hyalomma asiaticum* lose 0.65% of their weight/day and survive for more than one month under these conditions.

To a high degree the striking differences among the rates of water loss can be attributed to differences in the water proofing of the cuticle. Ticks owe their resistance to desiccation primarily to waxy lipids in the epicuticle. These convey, depending on their specific nature, different degrees of water impermeability to the integument. The lipids are secreted by the epidermal cells and discharged via the pore canals and the epicuticular channels (Filshie, 1976). The waterproofing system of the cuticle can be extremely susceptible to mechanical damage, e.g. traces of cuticular lipids of larvae of many species of ixodid ticks tend to adhere to glass material of their cage leading to rapid dehydration and death (Rudolph, 1976). Removal of the lipids by various experimental procedures causes drastically increased water losses (Lees, 1947; Hafez *et al.*, 1970); the ticks are, however, capable of restoring the waterproofing lipid barrier to a high degree within a short time.

The second component of water loss to be considered is the loss of water via the tracheal system in the course of respiratory gas exchange. In a number of tick species of both families it was shown that high CO_2 concentrations in the ambient air cause drastic increases of water loss (Mellanby, 1935; Browning, 1954; Hefnawy, 1970). It has been inferred that CO_2 acts on a closing device of the spiracle and that the observed losses occured via the tracheal system. There is good evidence for the existence of a CO_2-sensitive control mechanism in the tick spiracle from Hefnawy's (1970) experiments with *Hyalomma dromedarii* and *Ornithodoros savignyi.*

That the potential water loss from the tracheal system is indeed remarkable was demonstrated in the adults of *Amblyomma variegatum.* Females exhibit a seventeen-fold increase of water loss at high CO_2-concentrations in 0% r.h. as well as in 93% r.h. This increase is completely abolished when the spiracles are blocked with paraffin wax (Fig. 1 c, d, e). We should expect that the water loss of larvae of ixodid ticks is unaffected by CO_2 because they have no tracheal system. In fact the loss rates of *A. variegatum* larvae in dry air and high CO_2-concentrations are even slightly lower than those in dry air without CO_2.

From the great difference in the rates of water loss between ticks with spiracles opened by CO_2 and ticks with functional spiracles it can be concluded that ticks keep their spiracles closed most of the time. This is true for resting ticks as can be seen from continuous weight records of a sensitive microbalance (Fig. 1a). Resting females of *A. variegatum* open their spiracles at 20°C 1-2 times/h for short periods indicated by four to seven minutes of drastically increased weight losses on the weight trace. These extra losses while the spiracles are open can be abolished by blocking the spiracles with paraffin wax.

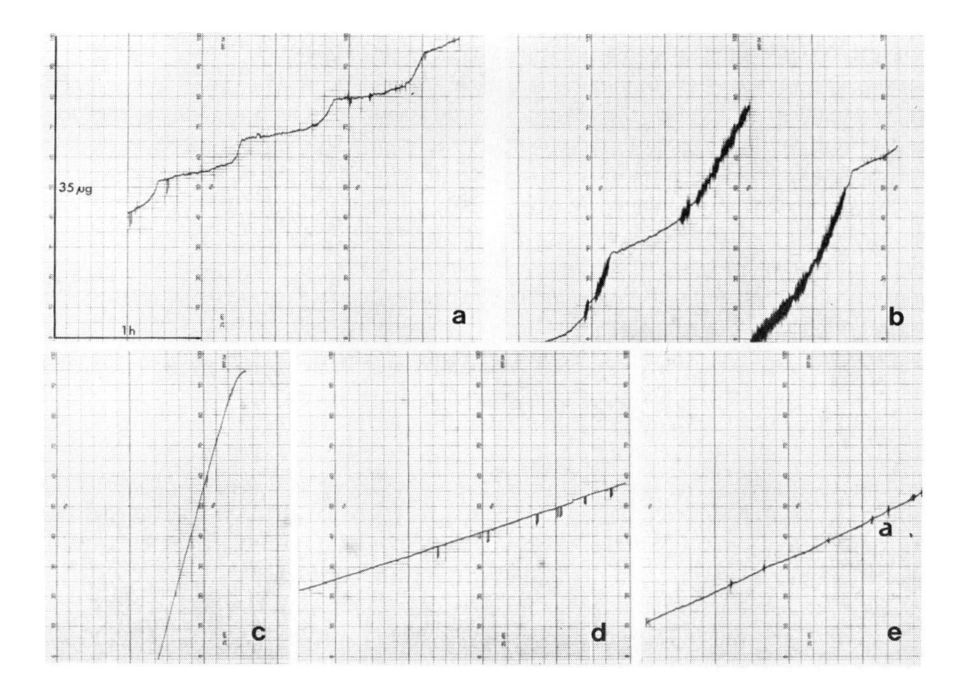

Fig. 1. Microbalance recordings of water loss of an individual female of *A. variegatum* at 20°C and 93% r.h.

a) Resting tick
b) Active tick showing locomotor activity
c) Tick exposed to a CO_2 atmosphere
d) Tick exposed to a CO_2 atmosphere, spiracles blocked with paraffin wax
e) Tick in air, spiracles blocked with paraffin wax

Low water loss from the tracheal system of the resting tick is evidently due to an effective closing device in the spiracle. However, when the unfed tick displays locomotor activity the spiracles are opened more frequently. The frequence of opening increases to *ca* 15 times/h at 20°C in *A. variegatum* females. Repeated opening of the spiracles at short time intervals leads to a marked increase of spiracular water loss as can be seen from continuous weight recordings (Fig. 1b). Locomotor activity during host-seeking periods, therefore, can be expected to stress the water reserves by increased water loss.

There has been considerable controversy concerning the functional opening and the closing device of the respiratory system in ticks. Externally the atrium of the tick spiracle is concealed by a conspicuous spiracular plate with a dark pigmented more or less central macula (Fig. 2a). The macula comprises a convex flap-like structure, the so-called ostial lip (Fig. 2c). With its crescent-shaped margin it obstructs an aperture in the spiracular plate, the so-called ostium. The part of the spiracular plate surrounding the macula represents a

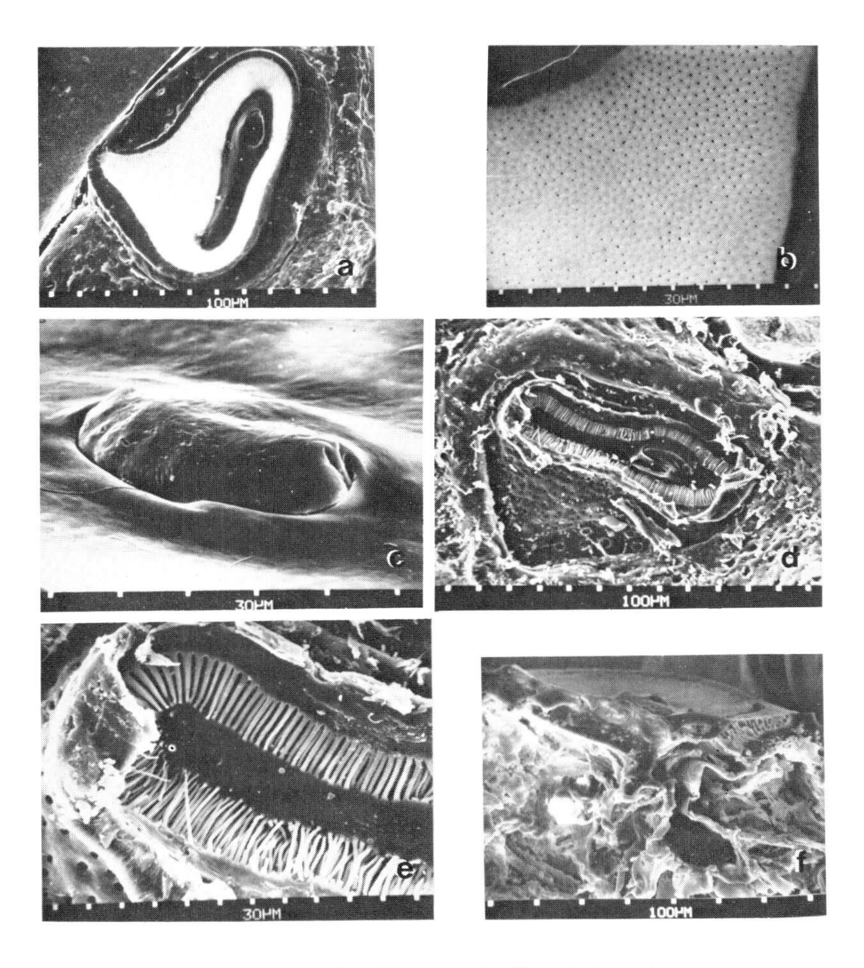

Fig. 2. Scanning electron micrographs of the spiracle of female *A. variegatum*
a) Spiracular plate
b) Aeropyles of the spiracular plate
c) So-called ostial lip
d) Spiracle as seen from the interior of the tick; tracheal trunks and atrial walls removed
e) Enlarged view of pedicellar border
f) Transversal section of spiracle

complex cuticular structure. A rather thick basal part of the cuticle is separated from an extremely thin outer part by numerous vertical rod-like cuticular pillars, the pedicels, creating an air-filled labyrinth, the so-called interpedicellar space. The thin surface cuticle is perforated by minute pores (Fig. 2b), the aeropyles, as has been clearly demonstrated by stereoscan electron microscopy (Hinton, 1967; Sixl *et al.*, 1971; Woolley, 1972; Roshdy and Hefnawy, 1973; Roshdy, 1974; Rudolph, in prep.).

The ostium has been repeatedly considered as the functional external opening of the spiracle (Falke, 1931; Browning, 1954; Roshdy and Hefnawy, 1973; Roshdy, 1974). Roshdy and Hefnawy's (1973) sections of CO_2-treated *Haemaphysalis longicornis* showed an open ostium while in ticks exposed to cyanide-gas the ostium was closed. Hinton (1967), however, considered the ostium as a non-functional collapsed ecdysial tube formed during the nymphal-adult molt and believed that the appearance of an open ostium, seen in his sections of *B. microplus,* is an artifact. By examining the late pharate phase of the adult he has observed the withdrawal of the old shedded nymphal tracheal cuticle through the ecdysial tube of the adult tick. This was also seen by Falke (1931) and Roshdy (1974).

By studying the late pharate female during molting we followed this process in *A. variegatum* and observed that the ostial aperture of the freshly molted tick is widely open. The whitish and soft ostial lip is bent down into the subostial space allowing an unobstructed view into this cavity. A few hours later the ostial lip appears inflated thereby closing the aperture tightly. Inflation of the unsclerotized ostial lip could easily be accomplished by hemolymph pressure since the macula as well as the ostial lip contain a hemolymph lumen which is continuous with the body hemocoel via the so-called stalk. Sclerotization of the ostial lip appears within 2-3 days. After sclerotization is completed the ostium remains tightly closed and no movements of the ostial lip can be observed even in a CO_2-atmosphere when increased water loss indicates that the spiracles are open. The effect of CO_2 is furthermore not cancelled when the ostium is occluded by paraffin wax indicating that the passage between the tracheal lumen and the ambient air is unimpeded. Finally, the action of the spiracular closing device as reflected by the weight change records is in no way affected by the paraffin block on the ostium.

Gas exchange between the tick and the environment evidently takes place via the aeropyles of the spiracular plate as Hinton (1967) has suggested. The aeropyles lead into the interpedicellar space and it is possible to see the pedicels by peering through the aeropyles with a stereoscan electron microscope (Hinton, 1967; Woolley, 1972; Roshdy and Hefnawy, 1973). That the interpedicellar space of the spiracular plate communicates with the underlying atrial space has been suggested by Arthur (1956) and Hinton (1967). From whole mounts of the dissected spiracle of *A. variegatum* and also by scanning electron microscopy of the dissected spiracle it can clearly be seen that the interpedicellar space is in fact along its complete circumference continuous with the atrial chamber (Fig. 1d, e). When Hinton (1967) applied oils to the surface of the spiracular plate of *Boophilus microplus* it entered the aeropyles and the atrium was flooded. In living *A. variegatum* it can be observed that paraffin oil applied to the aeropyles gradually penetrates the interpedicellar space and seeps through the pedicellar border into the atrial chamber.

The only closing device of the tick spiracle is evidently located in the atrium. The flexible ventral wall of the atrial chamber which in transverse sections of the spiracle appears to be thrown in a broad lobe (Fig. 2f) has been referred to as a possible device to occlude the atrium at this level by Falke (1931) and Arthur (1956). Roshdy and Hefnawy (1973) called the extension of the ventral atrial wall protruding into the atrial cavity "atrial valve" and inferred from serial sections of CO_2-treated and cyanide-gas-treated ticks that "the valve guarding the atrial cavity probably controls air passage to and from the tracheal trunks." The mode of action of the atrial valve is readily observed *in situ* in females of *A. variegatum* dissected transversally at the posterior level of the spiracular plates in Ringer solution (Fig. 3). In the resting tick the atrial valve is fully expanded occupying most of the atrial lumen. Its elastic wall closely joins a rigid pad on the opposing dorsal wall of the atrial cavity thereby closing up and separating the distal portion of the atrial chamber underneath the macula of the spiracular plate, the so-called subostial space, from the

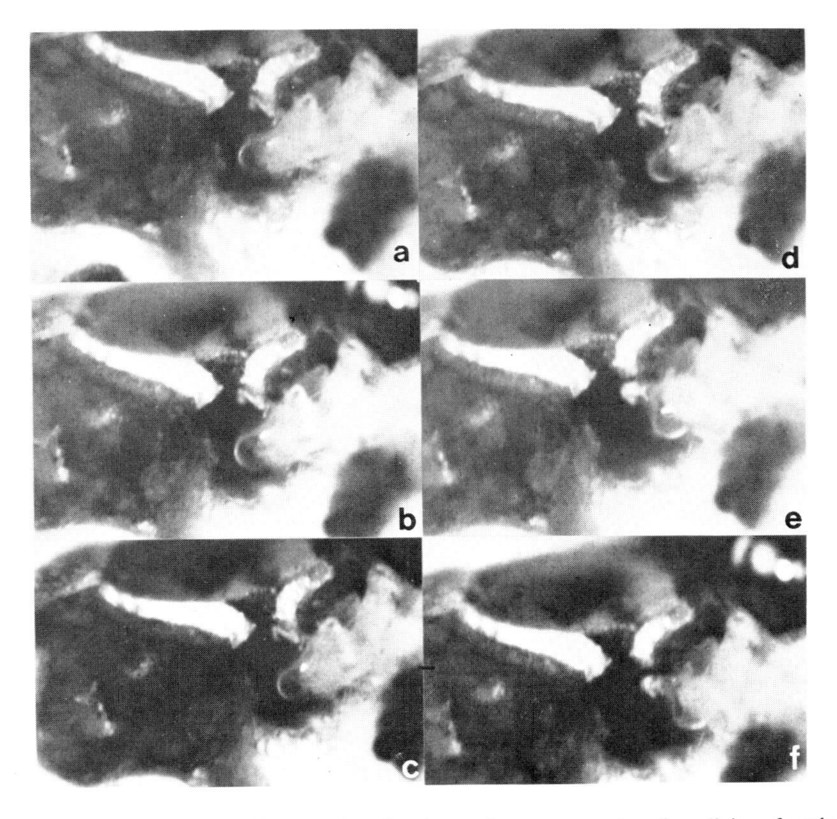

Fig. 3. Sequential microphotographs showing valve movements of a living female *A. variegatum*, dissected transversally at the posterior level of the spiracle.

proximal atrial cavity, from which the main tracheal trunks arise. When the tick becomes active the valve detaches from the opposite atrial wall and is fully retracted thereby expanding the atrial lumen to one large chamber. As long as the tick moves its legs the valve performs opening and closing movements at intervals of a few minutes. When the CO_2 concentration in the dissecting medium is increased the atrial valve comes to a standstill in the fully retracted position thereby opening the atrial chamber widely.

Doubtless the atrial valve of the tick spiracle represents an efficient closing device subject to physiological control, preventing drastic water losses from the tracheal lumen into the surrounding subsaturated atmosphere. The complicated pathway of gases between the tracheal lumen and the environment through the spiracular plate with its minute aeropyles and the underlying interpedicellar space is not yet understood. Whether it can be simply interpreted as a diffusion barrier retarding the escape of water to the environment or if it has another unknown function in connection with numerous conspicuous organs traversing the thick basal cuticle of the spiracular plate and opening into the interpedicellar space is undecided.

As regards the role of the different devices and mechanisms for maintaining water balance under natural conditions a highly impermeable lipid barrier in the epicuticle and an effective atrial valve kept closed most of the time enable the tick to spend prolonged periods on the vegetation waiting for a host animal. When a tick has suffered water loss, e.g. during increased locomotor activity in the course of host seeking or under conditions of low environmental humidities, the deficit is compensated by uptake of atmospheric water vapor whenever humidities above the critical level are available.

SUMMARY

Ticks compensate water losses by active uptake of water vapor from the atmosphere down to a relative humidity ranging from 75% to 94% depending on the species. Water vapor uptake proceeds via the mouth and there is good evidence that uptake is mediated by a hygroscopic salivary secretion produced by the Type-I alveoli of the salivary glands. Epicuticular lipids of the general body integument and a closing device in the tick spiracle under physiological control contribute essentially to the restriction of water losses. The ostium of the spiracular plate is an ecdysial opening as previously postulated by Hinton and the so-called ostial lip acts as a permanent plug of this opening after molting. Gas exchange proceeds via the aeropyles and the interpedicellar space of the spiracular plate and is controlled by an atrial valve. The valve is closed most of the time in the resting tick thus restricting water losses from the tracheal system. During locomotor activity the valve opens more frequently and water losses increase.

REFERENCES

Arthur, D. R. (1965). *Parasitology* **41,** 82-90.
Browning, T. O. (1954). *J. Exp. Biol.* **31,** 331-340.
Falke, H. (1931). *Z. Morph. Okol. Tiere* **21,** 567-607.
Filshie, B. K. (1976). *In* "The Insect Integument" (H. R. Hepburn, ed.), pp. 193-206. Elsevier, Amsterdam.
Hafez, M., El-Ziady, S. and Hefnawy, T. (1970). *J. Parasitol.* **56,** 154-168.
Hefnawy, T. (1970). *J. Parasitol.* **56,** 362-366.
Hinton, H. E. (1967). *Austral. J. Zool.* **15,** 941-945.
Lees, A. D. (1947). *J. Exp. Biol.* **23,** 379-410.
Mellanby, K. (1935). *Parasitology* **27,** 288-290.
Roshdy, M. A. (1974). *Z. Parasitenk.* **44,** 1-14.
Roshdy, M. A. and Hefnawy, T. (1973). *Z. Parasitenk.* **42,** 1-10.
Rudolph, D. (1976). Inaugural Dissertation, Freie Universität Berlin.
Rudolph, D. and Knülle, W. (1974). *Nature (London)* **249,** 84-85.
Rudolph, D. and Knülle, W. (1978). *In* "Comparative Physiology: Water, Ions and Fluid Mechanics" (K. Schmidt-Nielsen, L. Bolis and S. H. P. Maddrell, eds.), pp. 97-113. Cambridge University Press, Cambridge.
Sixl, W., Dengg, E. and Waltinger, H. (1971). *Arch. Sci. Geneve* **24,** 403-407.
Woolley, T. A. (1972). *Trans. Amer. Micros. Soc.* **91,** , 348-363.

PROTEIN DIGESTION AND SYNTHESIS IN IXODID FEMALES

Samir F. Araman

Department of Entomology
University of Hawaii
Honolulu, Hawaii

INTRODUCTION

Little work has been published on the physiology and biochemistry of protein feeding, digestion, and synthesis during the life span of ixodid females. Tatchell (1971) monitored the protein concentrations in the hemolymph of engorged female *Boophilus microplus* before and during oviposition and found that the protein concentrations increase during vitellogenesis and decrease thereafter. He also detected in the hemolymph two hemoproteins with similar electrophoretic mobilities to two egg hemoproteins but was unable to show immunological similarity between the two pairs. Ixodid hemolymph proteins were also studied electrophoretically by Dolp and Hamdy (1971) in two *Hyalomma* species. Their work also showed dissimilarity between tick hemolymph and host proteins.

The most complete information on tick proteins results from chiefly ultrastructural studies of the female argasid *Ornithodoros moubata* (Murray). Diehl (1969, 1970) found that two specific hemoglycolipoproteins in the female hemolymph were immunologically similar to two egg proteins. The uptake of the hemolymph proteins was demonstrated to occur by micropinocytosis (Aeschlimann and Hecker, 1967, 1969; Jenni, 1971). The hemolymph vitellogenins, reported to be synthesized by the gut epithelium cells, were found to constitute only a portion of the egg yolk proteins (Diehl, 1970; Jenni, 1971). The other portion is synthesized by oocytes in the ovary.

Here we provide basic information on protein feeding, digestion, and *in vivo* protein synthesis in the female ixodid *Rhipicephalus sanguineus* (Latreille). We also investigate the protein synthesis properties of the fat body, ovary, and midgut and compare these properties with those reported elsewhere for other argasid and ixodid tick species.

Vol. I: ISBN 0-12-592201-9

MATERIALS AND METHODS

R. sanguineus from a colony originating in Honolulu were used. Methods of rearing, bleeding, sample preparation, electrophoresis, and protein and hemoglobin analyses were outlined by Araman (1972, 1978).

To measure *in vivo* protein synthesis, unfed female *R. sanguineus* were each injected with 1 μliter of ^{14}C-labelled amino acid mixture (aqueous reconstituted ^{14}C protein hydrolysate 1 μCi/ul) into the hemocoel. The ticks were then allowed to feed, and hemolymph and gut fluid samples were collected at intervals during their life span and assayed for total soluble and ^{14}C-labelled proteins.

For *in vitro* protein synthesis studies, the fat bodies, midguts, and ovaries of female *R. sanguineus* on one day after detachment were used. Weight portions of each organ were separately incubated for 6 h at 30°C in 200 μliter of TRIS-buffered Ringer's solution (9.15 gm NaCl, 0.2 gm KCl, 0.4 gm $CaCl_2 \bullet 6H_2O$ in 1000 ml distilled water, buffered to pH 7.5 with 0.005 M TRIS) containing 10 μliter ^{14}C-protein hydrolysate. After incubation, aliquots from the media were removed and assayed for total and ^{14}C-labelled protein concentrations. The tissues were then rinsed and homogenized in cold non-labelled buffer, centrifuged, and the supernatant assayed for total and ^{14}C-labelled protein concentrations and subjected to electrophoresis.

RESULTS AND DISCUSSION

Feeding and Digestion

The feeding period is considered a special developmental phase of the ixodid characterized by concurrent feeding and digestion resulting in extensive body growth (Lees, 1952). A particular feature of the digestion process is the hydrolysis of hemoglobin, the major host blood protein component, into the protein, globin, and its prosthetic group, heme. Thus, hemoglobin concentrations in the tick gut, during and after feeding, are a measure of feeding activity and digestion rate. These concentrations also provide information on the rate at which amino acids and polypeptides become available for use in the biosynthesis of tick proteins.

Feeding and digestion in female *R. sanguineus* were studied by monitoring changes in body weight, midgut hemoglobin concentration and excreta during and after feeding. As with most ixodids, female *R. sanguineus* have three feeding phases during the engorgement process. The classification proposed by Balashov (1972) is used to delineate these phases.

The first feeding phase, the preparatory phase, begins immediately after attachment of the female to the rabbit host and lasts for *ca* 24 h. During this period, the body wt and size remain unchanged, but we found a slight increas

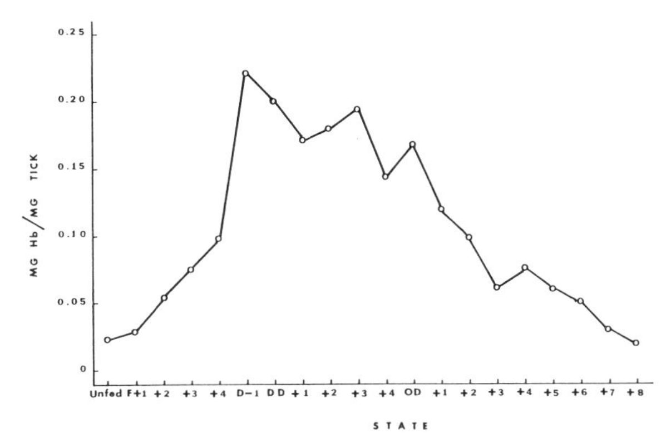

Fig. 1. Changes in the mean hemoglobin concentration in female *R. sanguineus* midgut. Samples were collected at daily intervals during feeding (F + 2 to F−1) and after detachment (DD to D + 4) and during the subsequent oviposition cycle (OD to 0 + 8).

in midgut hemoglobin concentration (Fig. 1) indicating that little feeding was taking place. During this phase the tick completes its transition from free to parasitic life and the digestive system develops (Balashov, 1961).

The second feeding phase occurs from about 24 h after attachment until 12-24 h before detachment. This phase is characterized by a gradual but continuous increase in body wt (from 3.5 mg to 50.0 mg) and size (from 2mm to 5mm). Concomitantly, there is a gradual and constant increase in the midgut hemoglobin concentration and in excreta and level of hematin in the insoluble midgut fraction, indicating high rates of feeding and digestion. Excretion during feeding, which occured mostly during this phase, was found to be constituted of 12-35% guanine, 25-31% protein, and not more than 0.3% hematin (Hamdy, 1973). Thus the second feeding phase (which lasts for 3-4 days), correctly referred to as the growth phase, is associated with active ingestion, concentration, digestion, and assimilation of host blood resulting in organ and integumental growth. Similar findings were reported by Kitaoka and Yuajima (1958) and Balashov (1972), using other parameters.

The third feeding phase, the expansion phase, takes place during the last 12-24 h of feeding and is characterized by sharp and rapid increase in body wt and size and midgut hemoglobin concentration. Nevertheless, at this time there is a sharp decrease in the amounts of hematin in the insoluble fraction and in excreta. This decrease indicates a decline in the digestion rate despite the large amount of blood ingested and may be attributed to the degeneration of the gut digestive cells. It is also associated with the finding that the proteolytic enzyme activity at D-1 (one day before detachment) decreased from its high during the growth phase to as low a level as was detected in the unfed tick (Bogin and Hadani, 1973).

The control of protease activity, and thus the digestion rate, by the female tick permits the accumulation of host blood during the expansion feeding phase for use in the vitellogenic process. Continued high protein digestion would exhaust the availability of amino acids at a time when the production of vitellogenins is critical.

After detachment, the overall rate of digestion was relatively low between detachment day (DD) and oviposition day (OD) (Fig. 1) as shown by changes in the hemoglobin concentrations. At oviposition, the rate of digestion is faster until O + 3. This may reflect the greater need for available amino acids for vitellogenins, particularly when the largest batch of eggs is laid between O + 2 and O + 3. Toward the end of the oviposition period the aqueous portion of the midgut sample contains very little heme; the transfer of most of the hematin to the insoluble fraction indicates the completion of hemoglobin digestion. The excreta collected after detachment contained no protein but 50-80% guanine and 13-22% of a second purine.

As in other hemoglobin ingesting organisms, ticks depend upon hematin for their development (Balashov, 1968). However, unlike most such organisms (Caine, 1969), ixodid ticks seem to have the ability to synthesize heme. Hamdy *et al.* (1974) found that the female *Dermacentor andersoni* is capable of *de novo* heme synthesis as demonstrated by the *in vivo* incorporation of ^{14}C of the α-carbon from glycine and carbon 4 and 5 δ-aminolaevulinic acid.

Protein Concentration and Biosynthesis in Gut Fluid

The changes in total soluble protein in cell-free gut fluid samples (Fig. 2) are closely correlated with that of the midgut hemoglobin (Fig. 1) from D-1 to oviposition day. The significant decrease in soluble protein on D + 1 coincides with the decrease in gut hemoglobin, and the rapid decrease in the concentraiton of both parameters starting from D + 3 may suggest their assimilation in the vitellogenic process.

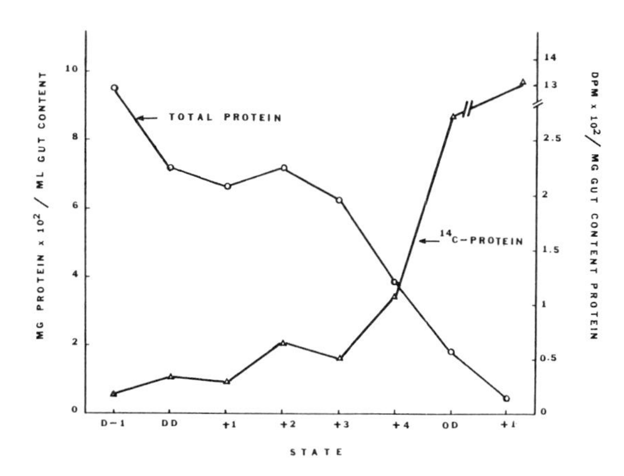

Fig. 2. The relationship of mean total soluble gut fluid protein and ^{14}C-protein specific activity as a function of development of female *R. sanguineus.*

While the concentration of soluble gut protein decreased, the ^{14}C-protein activity increased from 8 ± 1 dpm/mg gut fluid protein to 1301 ± 158 dpm/mg (Fig. 2). The increase was slow and insignificant during the first days of the experiment (D–1 to D+3), with a characteristic lag period followed by an exponential increase between D+4 and O+1. From these experiments it can be concluded that *in vivo* net protein synthesis may occur in the gut. The fluctuations in the rate of synthesis during the post-detachment period may, in part, be attributed to variation in gut water content. However, the sharp increase in gut protein ^{14}C-specific activity just prior to and during oviposition, indicates that increase in synthesis may be related to vitellogenesis. It is not known whether structural and/or yolk proteins and enzymes are produced at that time. The total soluble proteins in the gut fluid (Fig. 2) are in a state of turnover with new proteins being synthesized at a rapid rate from D+3 and 0+1 while others are being digested. In essence, then, amino acids from the hemolymph (both radioactive and non-radioactive) are taken up by the gut, and new proteins are being made and secreted into the gut as well as into the hemolymph (see Fig. 3).

Changes in Hemolymph Protein Concentration and Biosynthesis

The total soluble protein concentration and ^{14}C-protein specific activity from pooled hemolymph collected daily from female *R. sanguineus* during and after feeding are presented in Fig. 3. During feeding, the hemolymph protein concentrations may reflect the rate of bloodmeal digestion, which appears to reach a peak on F+3 (third day of feeding). They also may reflect

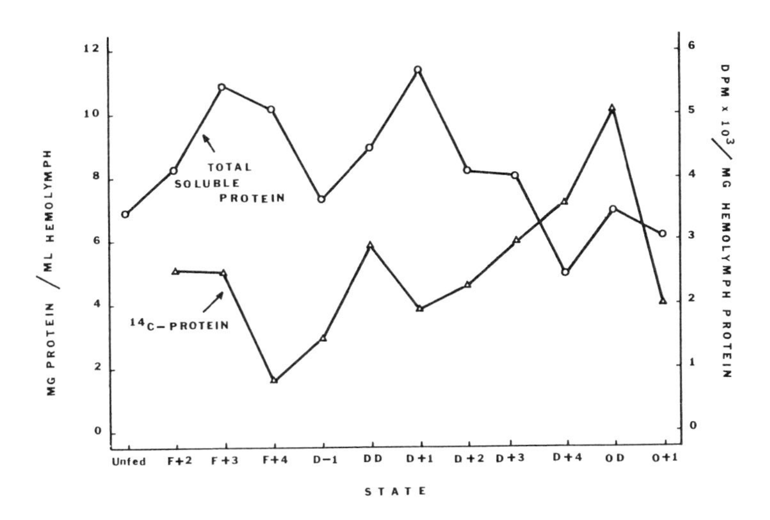

Fig. 3. Changes in the mean hemolymph total soluble protein and ^{14}C-protein concentrations at various states of development of female *R. sanguineus*.

the accumulation and utilization of hemolymph proteins as a source in the growth and enlargement of the cuticle and other body tissues. On detachment day (DD), about 60% of the recovered radioactivity was in body tissues. The relatively low hemolymph ^{14}C-protein concentrations probably do not indicate a low rate of protein synthesis but point to the influx of non-radioactive amino acids and their incorporation into hemolymph proteins. This amino acid pool becomes available from the digestion of the bloodmeal; its influx into the hemolymph results in the dilution of the radioactive amino acid pool.

A major portion of the hemolymph proteins produced during feeding is believed to go into the production of new endocuticle layers. These layers are essential if the cuticle is to expand during feeding (Hackman, 1975). In *Hyalomma dromedarii,* Bassal and Hefnawy (1972) found that the protein content was 88.4% in the cuticle of the unfed female and increased to 96.2% in the fully engorged female. During this interval, cuticular synthesis was high and 96.6% of the newly deposited material was protein.

The mechanism of transformation of tick hemolymph proteins remains conjectural because the amino acid concentrations continually change during feeding. Hemolymph proteins could enter the tissue cells intact or after partial degradation with subsequent intracellular protein synthesis. However, evidence for direct utilization of hemolymph proteins during feeding is only circumstantial, as no attempt has been made to follow their relative uptake by individual tissues.

The significant increase in hemolymph protein content on D + 1 coincides with a decrease in the concentration of midgut hemoglobin and gut fluid protein. This signifies active digestion and incorporation of the digestive products into hemolymph proteins. After detachment, biosynthesis of hemolymph proteins reaches a maximum on D + 1 which, we believe, signals the onset of vitellogenesis. This is also evident by the increase in the number, concentration, and ^{14}C-specific activity of D + 1 hemolymph proteins separated electrophoretically. Increase in protein synthesis on D + 1 is also apparent from the increase in the rate of oxygen consumption (Bassal, pers. comm.).

Ontogenetic Patterns

Seventeen protein fractions were electrophoretically separated from the gut fluid (Fig. 4). Except for G9 and G16, these fractions were weak and sometimes barely visible. The incorporation of ^{14}C-amino acids in the fractions was sporadic and no consisent pattern was detected. G9 (R_f 0.25), a complex of at least two hemoproteins, is the major fractions. It consituted 48-60% of all separated proteins. The only other prominent fraction, G16 (R_f 0.53), is also a hemoprotein and constituted 16-25% of separated proteins. G9 and G16 are not sex specific but are most probably nutrient proteins serving as a source for production of other proteins.

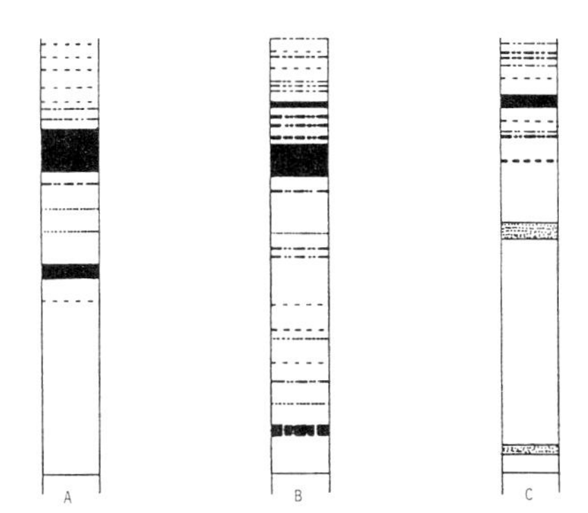

Fig. 4. Comparison between the electrophoretic protein patterns of female *R. sanguineus* (A) gut content fluid, (B) hemolymph, and (C) eggs as obtained by 6% acrylamide gels.

Electrophoretic separation of hemolymph proteins yielded 16-24 bands, with the number increasing progressively during the feeding period and up to D + 1. Bands with higher protein concentrations (H1-H13) are slow moving (R_f 0.00-0.28) while those with R_f values greater than 0.28 have much lower concentrations. On the other hand, the fast moving bands contain a higher level of ^{14}C-specific activity from the slow moving group. Fractions H6-H13 are iron containing glycolipoproteins and, except for H13, contain an appreciable amount of ^{14}C-proteins specific activity. Twelve protein fractions were visualized from egg samples. E6, a hemoglycolipoprotein, was the major fraction containing 76% of the separated egg protein.

None of the hemoproteins detected in any of the tick samples are similar to the host proteins. It was also found, both by regular acrylamide gel and SDS-gel electrophoresis, that none of the gut fluid proteins was in the eggs (Fig. 4). Similarly, Dolp and Hamdy (1971) reported that the electrophoretic patterns of hemolymph proteins from two *Argas* and two *Hyalomma* species are dissimilar to their respective host serum patterns. This dissimilarity clearly indicates that no undigested portion of the bloodmeal hemoglobin is deposited in the eggs. On the other hand, the presence of hemoproteins in ticks indicates that the heme moiety is utilized in a conjugated form. In *R. sanguineus,* 6% of the ingested heme was transferred to the eggs.

The changes in concentrations of two hemoglycolipoproteins in the hemolymph are significant in that the two faint bands in the feeding female become strongly represented prior to and during oviposition. These protein bands reach optimal concentration on the day of maximum egg production and decline thereafter at a slower rate than that of the major band, H13. The radioactivity data show that both protein bands are actively synthesized and have similar mobilities to two egg fractions, a finding confirmed by SDS-gels, suggesting that they were selectively taken up by the developing oocytes to constitute the yolk proteins. Such an uptake occurs by micropinocytosis as is shown by the work of Aeschlimann and Hecker (1967, 1969) and Diehl (1970).

In *Boophilus microplus,* Tatchell (1971) was unable to demonstrate the immunological similarity between two major hemoproteins of the eggs with those from the hemolymph, although both pairs had the same electrophoretic mobilities. In a more extensive study, Boctor and Kamel (1976) purified and characterized two lipoglycohemoproteins from *D. andersonii.* The purified proteins were shown to be immunologically identical to two hemolymph proteins but not to host hemoglobin. Similar findings were reported by Diehl (1969, 1970) for the argasid *O. moubata.*

Protein Synthesis by Tick Tissues

The results from the *in vitro* experiments indicate that in the female *R. sanguineus*, the midgut is the least active tissue in protein synthesis (Table I). Proteins released by the midgut show little similarity to those of the hemolymph or the eggs. This is in contrast to Diehl (1970) and Jenni (1971) who reported synthesis of hemolymph vitellogenins by the gut epithelium cells of the female argasid *O. moubata.* Their conclusions, however, are based on

TABLE I.
The Specific Activities of Soluble Media Released and Tissue Bound Protein after 6 h *in vitro* Incubation of Engorged Female *R. sanguineus* Fat Body, Ovary, and Midgut in Media Containing ^{14}C-amino Acids. Reported Values Are the Averages of Three Runs.

Organ	dpm x 10^3 / mg Protein Media Protein	 Tissue Protein
Fat body	572 ± 102	8246 ± 1031
Ovary	372 ± 89	136 ± 34
Midgut	151 ± 72	32 ± 9

autoradiographic and morphological evidence, which do not differentiate between sites of synthesis and sites of storage. On the other hand, the difference in results may only serve to emphasize the difference between argasid and ixodid ticks.

The ovary in *R. sanguineus* is also synthetically active, but most of the radioactivity is incorporated into tissue proteins. Analysis of the incubation medium by electrophoresis indicated that the released proteins were dissimilar to those of the hemolymph. However, the results show high radioactivity incorporation in the portions of the gel containing the hemolymph vitellogenins. This would indicate that the ovary is involved in synthesis of yolk proteins.

The fat body was the most active tick tissue in protein synthesis, incorporating 48 times more label in both tissue and media proteins than did the midgut (Table I). The proteins synthesized and released by the fat body show great similarity to those of the hemolymph. These include the major protein fraction and the vitelloenins. Synthesis of the major hemolymph proteins by the fat body suggests that the behavior of the tissue *in vitro* is representative of its behavior *in vivo*. This leads to the conclusion that in female *R. sanguineus*, the fat body is the site of synthesis of many hemolymph proteins including the vitellogenins. Such a conclusion correlates with the finding of Obenchain and Olvier (1973) that the tick fat body is metabolically active during vitellogenesis.

These results show that in *R. sanguineus,* egg yolk proteins come from two sources: the ovary and the hemolymph. These results are similar to the situation found in the argasid *O. moubata* (Diehl, 1969, 1970; Jenni, 1971).

CONCLUSION

In conclusion, the results of this study suggest the following sequence for events related to protein digestion and synthesis in the female *R. sanguineus* during development and vitellogenesis.

Ingested host proteins are digested intracellularly by digestive cells in the female midgut; limited extracellular digestion also occurs. The digestive products—smaller peptides and amino acids—are then released into the gut lumen and utilized by the tick tissues in the synthesis of enzymes, and of structural and vitellogenic proteins. Structural proteins are synthesized mostly during the *developmental* biological phase from digestive products of host blood ingested during the growth feeding phase. Vitellogenins and other proteins involved in vitellogenesis, on the other hand, are mostly synthesized during the *vitellogenic* phase utilizing bloodmeal proteins ingested during the expansion phase of feeding. A major portion of the digestive products probably enter the hemocoel where they are utilized by the fat body to produce most of the hemolymph proteins including the vitellogenins. The vitellogenins are then sequestered by the ovary and incorporated into egg yolk proteins. The hemolymph vitellogenins employ micropinocytosis to enter the ovaries, where they constitute only a portion of the yolk proteins; the other portion is produced by the oocytes in the ovary.

SUMMARY

The study of protein digestion and synthesis in the ixodid female, *R. sanguineus,* suggests the presence of two biologically and biochemically distinct phases, the *developmental* and the *vitellogenic,* during the female life span. The *developmental* phase occurs during the engorgement process and is concerned with activities related to body organ and integumental growth and development. During engorgement, a female tick goes through three feeding phases:

1. During the preparatory phase in the first 12-24 h of feeding, no changes occur in body wt, size or in gut hemoglobin indicating low rates of feeding and digestion.

2. The subsequent growth-feeding phase lasted for 3-4 days and was characterized by a gradual but constant increase in body wt, size, gut hemoglobin, excreta, and hemolymph proteins. These increases were associated with high rates of host blood ingestion, concentration, digestion, and assimilation in organ and cuticle growth.

3. The expansion phase takes place during the last 12-24 h of feeding. During this short period, there is a sharp increase in body wt, size and in midgut hemoglobin. A large amount of host blood is ingested with little digestion.

The *vitellogenic* phase represents that period from detachment through completion of oviposition. It is a period mainly concerned with biological and biochemical processes related to vitellogenesis and oviposition.

After detachment, the total soluble gut protein concentration correlates well with that of the gut hemoglobin in that both decreased. Little protein synthesis was detected in the gut fluid except just before oviposition. Separation of gut fluid proteins on 6% acrylamide gels yielded 12-14 bands. The electrophoretic pattern did not vary greatly during the various developmental states, but the concentration of the major protein bands decreased during vitellogenesis and showed a pattern similar to that of the total soluble protein. There was also little ^{14}C-protein specific radioactivity in the visualized bands.

The hemolymph total soluble protein concentration increased during the growth feeding phase but decreased during the expansion phase. The concentration reached a maximum of D+1 signaling the onset of vitellogenesis. The rate of protein synthesis in the hemolymph was apparently low during feeding due to the influx of non-radioactive amino acids resulting from bloodmeal digestion. After detachment, the synthesis rate increased during vitellogenesis, when vitellogenins, among other proteins, were actively synthesized.

Sixteen to 24 protein bands are visualized when hemolymph samples were electrophoretically separated. The number of bands increased during feeding and reached a high on D+1. Two hemolymph bands, characterized as hemoglycolipoproteins, had similar mobilities to two egg bands. Thus it appears that at least two hemolymph proteins are incorporated unchanged in the egg yolk proteins. Most of the hemolymph protein fractions are radioactively labelled indicating that the hemolymph acts as a transport medium for tick proteins. Hemolymph proteins, including the vitellogenins, are synthesized by the fat body and not by the midgut; the ovary is also capable of synthesizing the vitellogenins.

ACKNOWLEDGEMENTS

I would like to thank Drs. F. Chang and S. J. Townsley for their advice and assistance. This work was supported, in part, by a predoctoral research assistantship from the Hawaiian Fruit Flies Laboratory, USDA. This publication is Journal Series No. 2298 and is published with the approval of the Hawaii Ag. Expt. Sta.

REFERENCES

Aeschlimann, A. and Hecker, H. (1967). *Acta Trop.* **24**, 225-243.
Aeschlimann, A. and Hecker, H. (1969). *Acarologia* **11**, 180-192.
Araman, S. F. (1972). *J. Parasitol.* **58**, 354-357.
Araman, S. F. (1978). Ph.D. thesis, University of Hawii, Honolulu, Hawaii.
Balashov, Yu S. (1961). *Zool. Zh.* **40**, 1354-1363.
Balashov, Yu S. (1972). *Misc. Pub. Ento. Soc. Amer.* **8**, 161-376.
Bassal, T. T. M. and Hefnawy, T. (1972). *J. Parasitol.* **58**, 984-988.
Boctor, F. N. and Kamel, M. Y. (1976). *Insect Biochem.* **6**, 233-240.
Bogin, E. and Hadani, A. (1973). *Z. Parasitenk.* **41**, 139-146.
Caine, G. D. (1969). *J. Parasitol.* **55**, 307-310.
Diehl, P. A. (1969). *Bull. Soc. Ent. Suisse* **42**, 117-125.
Diehl, P. A. (1970). *Acta Trop.* **27**, 302-355.
Dolp, R. M. and Hamdy, B. H. (1971). *J. Med Entomol.* **8**, 636-642.
Hackmann, R. H. (1975). *J. Insect Physiol.* **21**, 1613-1623.
Hamdy, B. H. (1973). *J. Med. Entomol.* **10**, 345-348.
Hamdy, B. H., Taha, A. A. and Sidrak, W. (1974). *Insect Biochem.* **4**, 205-213.
Jenni, L. (1971). *Acta Trop* **28**, 105-163.
Kitaoka, S. and Yajima, S. (1958a). *Bull. Nat. Inst. Anim. Hlth. Tokyo* **34**, 135-147.
Lees, A. D. (1952). *Proc. Zool. Soc. Lond.* **121**, 759-772.
Obenchain, F. D. and Oliver, J. H. Jr. (1973). *J. Exp. Zool.* **186**, 217-236.
Tatchell, R. J. (1971). *Insect Biochem.* **1**, 47-55.

DEVELOPMENT OF RESISTANCE TO AZINPHOSMETHYL THROUGH GREENHOUSE SELECTION OF THE PREDATORY MITE *AMBLYSEIUS FALLACIS* AND ITS PREY *TETRANYCHUS URTICAE*

J. G. Morse and B. A. Croft

Pesticide Research Center and Department of Entomology
Michigan State University
East Lansing, Michigan

INTRODUCTION

Pest resistance is one of the main problems facing pest management, yet natural enemy resistance to pesticides is nearly unknown. Comparing pests and natural enemies, there are more than 268 cases of insecticide resistance known for pest species (Brown, 1976) but only 12 reported cases for natural enemies (Croft and Brown, 1975; Croft, 1977). Three of these 12 resistant natural enemies are parasitoids which developed resistance as a result of laboratory selections—two braconids and an aphelinid. The remaining nine are predators reported resistant in the field—seven phytoseiid mites, an anthomyiid and a cocinellid.

Many hypotheses have been presented to explain this large imbalance in reported cases of pest and natural enemy resistance. As yet it is unclear whether 1) pests and natural enemies actually differ in their ability to develop resistance to pesticides or 2) the intricate density dependence of natural enemies upon their hosts or prey limits natural enemy resistance development. It is interesting in this context, to point out that in all known cases of field natural enemy resistance, resistance in the principal prey (the pest) has preceded the resistance similarly developed by the predator (Croft, 1977).

One of the seven phytoseiid mites reported resistant in the field is *Amblyseius fallacis* (Garman), an efficient predator of several tetranychid mite species. Cases of its resistance to pesticides are widespread throughout the midwest and eastern fruit growing regions of the United States (Croft and Brown, 1975).

Vol. I: ISBN 0-12-592201-9

The two-spotted spider mite, *Tetranychus urticae* (Koch), is a common pest mite found on a variety of host plants worldwide. Its ability to develop resistance to a wide range of pesticides is well documented (Morse, 1977).

These two mite species—the predatory mite, *A. fallacis* and the two-spotted spider mite, *T. urticae* are remarkably similar in most physiological, biological and behavioral characteristics as well as in their magnitude and spectrum of resistance to organophosphates (Croft, 1977; Morse, 1977). In addition they are quite comparable in most ecological aspects, occupying nearly identical habitats throughout their life histories. They thus provide a unique opportunity to investigate resistance development in a pest—natural enemy system.

METHODS

Susceptible populations of each species were collected in 1975. Previous work by other investigators has shown that the two species are almost identical in both the mechanism and genetics of resistance to azinphosmethyl, a common orchard organophosphate (Morse, 1977).

Both species were reared on lima bean plants under similar environmental conditions in the greenhouse. An unlimiting food supply was provided to each species throughout the experiment, thus eliminating a potential bottleneck in natural enemy resistance development. Resistant prey mites (*T. urticae*) were provided as food for the *A. fallacis* population so that an unlimiting food supply would be present after chemical selection as well as before.

Both species were subjected to identical selection programs. Population sizes before selection and percent mortality levels after selection were held approximately equal.

Generations were kept distinct through the hand transfer of surviving adults from each previous generation to initiate the next. The slide-dip procedure of Anon. (1968) was used to determine resistance levels for each generation. Probit mortality was plotted versus log-dosage and the LC_{50} and slope were calculated using a computer program. Wettable powder insecticide was used in both selection and slide-dip procedures so that spray levels could be calculated to give approximately 75% selection mortality.

RESULTS

Our results show a 23.87 fold resistance development in 14 generations in the predator *A. fallacis* as compared to a 20.41 fold resistance in 22 generations for *T. urticae*. Experiments are presently underway in our laboratory to further investigate the resistance development in this pest-natural enemy system. We have nearly concluded research on the importance of the initial gene frequency of the major gene responsible for resistance in each of the two susceptible populations used in this study. A second experiment nearly completed investigates resistance development under laboratory conditions when the two species are not isolated but selected together in a more natural mode.

Our results indicate that at least in this pest-natural enemy system, the natural enemy can develop resistance at least as fast as the pest once its density dependence upon its prey is relieved through the addition of an unlimiting food supply following insecticide selection. These results have practical significance in the applied aspect of pest management. If economic thresholds for pest species were raised to maximum levels, the additional prey available for natural enemies following pesticide selection would allow for maximum resistance development in the natural enemy.

SUMMARY

For a specific pest-natural enemy system the results of a study investigating resistance development is presented. Under similar selection regimes, with unlimiting food supplies, the predatory mite, *Amblyseius fallacis* developed a 23.87 fold resistance to azinphosmethyl through 14 generations compared to a 20.41 fold resistance for the pest *Tetranychus urticae* through 22 generations. Implications to applied pest management are discussed.

REFERENCES

Anonymous. (1968). *Bull. Entomol. Soc. Amer.* **14**, 31-37.

Brown, A. W. A. (1976). *In* Amer. Chem. Soc. Symp. "Pesticide Chemistry in the Twentieth Century," ACS Washington, D. C.

Croft, B. A. (1977). *In:* "Pest Management and Insect Resistance," (D.L. Watson, A.W.A. Brown, eds) 377-393. Academic Press, New York.

Croft, B. A., Brown, A. W. A. (1975). *Ann. Rev. Entomol.* **20**, 285-335.

Morse, J. G. (1977). Masters Thesis, Michigan State Univ., E. Lansing, MI.

DOMINANT AND RECESSIVE INHERITANCE OF ORGANOPHOSPHORUS RESISTANCE IN TWO STRAINS OF *TETRANYCHUS URTICAE*

V. Dittrich, N. Luetkemeier, and G. Voss

Ciba-Geigy Ltd.
Agrochemicals Division
Basel, Switzerland

INTRODUCTION

In the past a number of investigations on the mode of inheritance of resistance (R) to organophosphate (OP) compounds in insects and mites have shown that in most cases a dominant or semidominant major gene was responsible for transmitting R. So obvious was this trend that Brown (1967) wrote in a review article on R: "OP-R has always proved to be dominant, and so has carbamate R".

In mites the dispute on the phenotypic expression of genes controlling OP-R has never been so clearly resolved among researchers as the above statement would lead one to believe. Whereas parathion R has always been found to be dominant the phenotypic expression varied from semidominant to semirecessive depending on the type of OP compound used for the tests (for a review see Dittrich, 1975). However, the work of Russian colleagues which belatedly has become available also to scientists in the West demonstrated beyond doubt that R in mites could also be a recessive character. Zilbermints (1971) reported on the XIIIth International Congress of Entomology in Moscow in 1968 that a major R gene in an experimental strain of *Tetranychus urticae* Koch from the Moscow region was completely recessive when tested with parathion and demeton. Together with coworkers Fadeev and Zhuravleva she presented this work in a more detailed form in 1969.

To us the Moscow strain containing a major recessive gene for R formed an interesting counterpart to our experimental strain which was derived from the orginal LEV-R strain. In the latter a major R gene was dominant or semidominant when tested with OP compounds. We obtained the Moscow strain from Zilbermints and compared it with LEV-R trying to define it toxicologically and biologically.

Vol. 1: ISBN 0-12-592201-9

MATERIALS AND METHODS

All toxicological experiments were carried out using the slide-dip method as described by Dittrich (1962). The mite strain designated LEV-S is the original stock of *T. urticae* from which LEV-R had been derived in the past by a long sequence of selections with demeton-type compounds. During the last few years selection pressure was reduced. The substrain used in these experiments was homozygous for a major factor, since F_1 hybrids (R♀ × S♂) crossed with S ♂♂ produced F_2 ♀♀ which could be separated in two distinct classes showing a 40/60 ratio of dead/live experimental animals over a wide range of parathion concentrations. This is good evidence for a major dominant factor for R to be present in homozygous condition. Resistance in Moscow-R is also due to a major gene (Zilbermints *et al.*, 1969).

Crosses of both Moscow-R ♀♀ and LEV-R ♀♀ were made with LEV-S♂♂ on potted Phaseolus beans in the primary leaf stage. The respective F_1 ♀♀ were analyzed toxicologically and the dominance factor D calculated according to Stone (1968). Based on the parental concentration-mortality (C.M.) lines and that of the hybrid F_1, a D factor of +1 represents complete dominance, −1 complete recessivity, and 0 stands for an intermediate expression of a gene.

RESULTS AND DISCUSSION

The LC_{50} values in Table I show that the two R strains have quite divergent reactions to the 3 OP-compounds tested. LEV-R mites are not overly resistant to monocrotophos (61 ppm), whereas the Moscow-strain can cope with 100× that concentration (6429 ppm). Against oxydemetonmethyl Moscow-R, too, is the hardier strain, whilst against parathion the LEV-R mites have the higher LC_{50} value. The extreme R of Moscow-R to monocrotophos may be explained by its having been selected with dimethoate which also contains a carbamoyl moiety.

The negative dominance factors in Table I demonstrate that Moscow-R is indeed recessive as Zilbermints and coworkers (1969) have shown, also against monocrotophos which they had not tested. The positive D-factors for LEV-R, by contrast, show a dominant or semidominant reaction of the R gene which is in strong contrast to the situation in Moscow-R.

In order to define the biochemical mechanism for R in both strains, synergists which are known to inhibit detoxifying enzymes were jointly applied with the toxicant monocrotophos. Thus piperonylbutoxide (PBO) and DEF (S,S,S-tributyl phosphorotrithioate) were used representing inhibitors of mixed function oxidases (MFO) and esterases respectively. Since monocrotophos is easily synergized by PBO an increased toxicity of its mixture with PBO could be expected if the R mechanism was based on an increased amount of, or on more efficient, mixed function oxidases. Likewise DEF as a

TABLE I.
Phenotypic Expression of 2 R-genes in 2 OP-R populations of *T. urticae*, LC_{50}, ppm.

Acaricides	Populations and R-ratios					Crosses and Dominance Factors			
	LEV-S	LEV-R	R/S	M-R	R/S	LEV-R♀ x LEV-S♂	D	M-R♀ x LEV-S♂	D
Oxydemetonmethyl	23.8	1483	62	3320	139	905	+0.77	18.1	−1.11
Monocrotophos	2.0	61	31	6429	3215	53	+0.92	1.6	−1.05
Parathion	44.3	7221	163	4289	97	6632	+0.97	110.7	−0.60

TABLE II.
Reaction of Two R-populations of *T. urticae* to Synergists

Acaricides and Synergists Tested	LC_{50}s in ppm	
	Moscow-R[a]	Leverkusen-R
Monocrotophos	579	61
Monocrotophos + PBO 1:1	294	161
Monocrotophos + DEF 1:1	252[b]	83

[a] Partly reverted from highest R-level.
[b] Significantly different from monocrotophos alone.

selective blocker of unspecific esterases would enhance toxicity if the esterases were involved in the R of both strains.

As Table II indicates DEF and PBO produced slightly lowered LC_{50} values in Moscow-R while LC_{50}s tended to be higher in LEV-R when synergists were added to monocrotophos. Even though there is a significant difference between the c.m. lines for monocrotophos and monocrotophos + DEF we do not regard this as a significant indication for esterases as being a major part of the R mechanism in Moscow-R.

In vitro studies of the biochemical mechanism of our mite strains met with an unexpected difficulty. The major gene of Moscow-R causes a considerably reduced viability in the experimental animals (Zilbermints, 1971) particularly in the homozygous condition. Therefore the strain has to be selected with an insecticide regularly in order to maintain a sufficient number of homozygotes in it which determine the R level if a major recessive gene controls R. Unless selected Moscow-R speedily reverts to a sensitive state which was unknown to us initially. Since our toxicological and biochemical tests were not carried out simultaneously reversion occurred and the biochemical tests were made on a reverted strain. When this became clear Moscow-R was reselected and toxicological and biochemical trials were made simultaneously on a strain which had regained a considerable degree of resistance. The revised toxicological data indicated that the R-mechanism in both Moscow-R and LEV-R are probably not based on metabolism of xenobiotics if such a statement is at all permissible on the basis of synergism experiments. In vitro enzyme assays on the type of acetylcholinesterase had conflicting results possibly due to a shift of our laboratory to a new location. A new environment not altogether free of e zyme inhibiting substances might have been the reason for a general decreased enzyme activity in all experimental strains. These inconsistencies a presently investigated. The curious fact remains, however, that a major dom nant factor controls R in LEV-R whereas a major recessive gene is responsib for R in Moscow-R. The implications for the development of R in populatio carrying one or the other gene are considerable particularly with respect to tl speed by which field resistance may become established under selectic pressure with OP compounds.

SUMMARY

Two resistant strains of *T. urticae* were compared toxicologically ar biochemically with a sensitive strain concerning the phenotypic expression their R-genes. In both LEV-R and Moscow-R strains a single major gene co trolled resistance. In LEV-R it was dominant or semidominant, in Moscow- it was recessive or semirecessive depending on the toxicants used. Synergis experiments with PBO and DEF did not increase the toxicity monocrotophos in both strains. Therefore it is doubtful whether metabolis forms the basis of resistance in these strains. In vitro studies of tl acetylcholinesterase in all experimental strains were inconclusive and are co tinued at present.

REFERENCES

Brown, A. W. A. (1967). *World Rev. Pest. Contr.* **6**, 104-114.

Dittrich, V. (1962). *J. Econ. Entomol.* **55**, 644-648.

Dittrich, V. (1975). *Z. ang. Ent.* **78**, 28-45.

Stone, B. F. (1968). *Bull. W.H.O.* **38**, 225-226.

Zilbermints, I. V. (1971). *Proc. 13th Internatl. Congr. Entomol. Moscow 1968.* Vol. II: 29 Leningrad, 1971.

Zilbermints, I. V., Fadeev, Y. N., and Zhuraleva, L. M. (1969). *Genetika* **5**, 96-106.

NUTRITIONAL STUDIES IN *TETRANYCHUS URTICAE* II. DEVELOPMENT OF A MERIDIC DIET

S. Kantaratanakul and J. G. Rodriguez

Department of Entomology
University of Kentucky
Lexington, Kentucky

INTRODUCTION

The two-spotted spider mite, *Tetranychus urticae* (Koch), (Acarina, Tetranychidae), is an important pest which is cosmopolitan in distribution and feeds on a wide variety of crops, including fruits, vegetable crops, forage crops and ornamentals. Studies of this economically important species have proceeded along numerous lines including fundamental nutrition. A recent publication (Kantaratanakul and Rodriguez, 1979) has reported the work done in our laboratory which was aimed at development of a chemically defined (holidic) diet for *T. urticae,* and reported the most recent improvements made to that diet. The present study reports concomitant research efforts directed toward a meridic diet.

MATERIALS AND METHODS

The materials and methods were generally the same as those used in the chemically defined diet (Kantaratanakul and Rodriguez, 1978). Briefly, the experimental animals were teneral females and males of uniform age, produced from eggs laid by females from a stock culture maintained in the laboratory on detached bean leaves. The basic axenic diet, also given (*op. cit.*) was fed in liquid form. For each diet (treatment), 10 rearing cages (replicates) were used, each containing 10 females and two males, and housed in an Aminco-Aire system isolator at 27°C and 75% R.H., under cool white fluorescent light passed through green gelatin filters. To preclude the possibility of microbial contamination during the 14-day experimental period, the mites were transferred to new rearing cages containing fresh diet every two days. The numbers of live females and the total eggs laid were recorded, and the eggs laid per female

Vol. I: ISBN 0-12-592201-9

per day was computed. The used rearing cages were kept for four more days to record percent of egg hatch.

Parameters used to measure the efficiency of the diet were 1) oviposition rate and 2) egg viability. The data were analyzed statistically by analysis of variance (the procedures of Cochran and Cox, 1957) and Duncan's new multiple range test (Duncan, 1955; Leon, 1960). An "OR × VE" value, arrived at by multiplying the oviposition rate (eggs per female per day) by the egg viability (percent hatch), made possible the comparison of the overall reproductive effectiveness of the various diets as a single figure relating the two parameters.

RESULTS AND DISCUSSION

The diet in Table I was used as a standard initially. As the work proceeded, modifications of the diet which proved to be beneficial were adopted in subsequent formulations.

It was observed that all of the female mites were actively ovipositing by the second day after they were placed on the diets, and they deposited eggs throughout the 14-day experimental period.

Effect of Bacto-agar, Gelatin and Gums

The first efforts were to promote ingestion and thus enhance the beneficial effects of the diet, by modifying the physical properties of the diet, as well as lowering the osmotic pressure created by the nutrients in the diet. This was done by adding Bacto-agar, gums and gelatin (Table II). Only the addition of 0.5% gelatin significantly increased oviposition rate and egg viability above those achieved on the standard diet. Locust bean and guar gums did not significantly affect these parameters and Bacto-agar and carragheenan adversely affected them. It was concluded that modifying the physical properties of the diet *per se* did not improve reproductive effectiveness. Since the added gelatin would contribute extra protein, its effects may have been nutritional. However, another experiment in which gelatin was added to the standard diet at levels of 0.25, 0.50, 0.75 and 1.0% indicated that 0.5% gave the best results.

Effects of Yeast Hydrolyzate and Casein

In these experiments the basic diet was that given in Table I with 0.5% gelatin added, and the yeast hydrolyzate and casein were added at levels of 0.25 and 0.5% (Table III). No significant effect was noted on the oviposition rate, but yeast hydrolyzate at both levels effected a significant increase in egg viabibility, the increase being more pronounced at 0.5%. It was thought that the negligible effect of casein might be due to insolubility in the aqueous diet and eventually the water-soluble form, sodium caseinate, was tried.

TABLE I.
Composition of Basic Synthetic Diet (Ekka *et al.*, 1971) with Modifications.

Constituent	mg/100 ml Diet	Constituent	mg/100 ml Diet
L-Amino acids			
Alanine	150	Riboflavin	0.20
Arginine	200	Thiamine HCl	0.20
Aspartic acid	200	Nicotinic acid	0.30
Cysteine	50	Biotin	0.0001
Glutamic acid	180	Ca-panotenate	0.50
Glycine	200	Choline chloride	10.00
Histidine	100	B_{12}	0.002
Isoleucine	100	Ascorbic acid	100.00
Leucine	160		118.1421
Lysine	120	Lipids	
Methionine	60	β-sitosterol	1.00
Phenylalanine	60	Stigmasterol	0.88
Proline	60	Stearic acid	0.50
Serine	80	Pamitic acid	0.25
Threonine	120	Oleic acid	0.25
Tryptophane	40	Linoleic acid	0.25
Tyrosine	50	Linolenic acid	1.50
Valine	120		
	2050	Tween 80®	5.00
RNA	100	Major salts	
		K_2HPO_4	37.50
Sugars		$Na_2HPO_4 \cdot 12H_2O$	10.60
Sucrose	3000	$MgSO_4 \cdot 7H_2O$	15.60
Glucose	500	$CaCl_2 \cdot 2H_2O$	3.62
Levulose	500	$CoCl_2 \cdot 6H_2O$	0.08
	4000		67.40
Fat-soluble vitamins		Minor Salts (Chelates)	
Vitamin A	3	NaFe	1.00
Vitamin E	10	Na_2Mn	0.50
	13	Na_2Cu	0.08
		Na_2Zn	0.12
Water-soluble vitamins			1.70
Folic acid	0.34		
Inositol	6.00		
p-Aminobenzoic acid	0.50		
Pyridoxine HCl	0.10		

Effects of Corn Starch, Soluble Starch and Pectin

Neither corn starch at levels of 0.5, 1.0 or 2.0%, nor pectin at levels of 0.1, 0.3 and 0.5%, improved the reproduction of *T. urticae* beyond that obtained on the standard diet with 0.5% gelatin. Soluble starch appeared to have a detrimental effect on reproduction.

TABLE II.
Effect of Bacto-agar, Gelatin and Gums on Oviposition Rate and Egg Viability in *T. urticae* Allowed to Develop for 14 Days on a Basic Synthetic Diet.

Dietary Component	Level (%)	Oviposition Rate[a] (Eggs/Mite/Day)	Egg Viability[a] (% Hatch)	OR × VE[b]
Bacto-agar	0.17	0.26c	9.83b	2.56
Gelatin	0.50	0.66a	18.71a	12.35
Locust bean gum	0.20	0.44b	7.27bc	3.20
Guar gum	0.15	0.52b	9.50b	4.94
Carragheenan	0.15	0.09d	5.83c	0.52
Standard (STD)[c]	—	0.43b	11.42b	4.91

[a] Numbers in columns followed by the same letter are not significantly different at the 5% level (Duncan's new multiple range test).
[b] Oviposition Rate × Viability of Eggs
[c] Standard diet as given in Table I.

TABLE III.
Effect of Casin and Yeast Hydrolyzate on Oviposition Rate and Egg Viability in *T. urticae* Allowed to Develop for 14 Days on a Basic Synthetic Diet Plus Gelatin.

Dietary Component	Level (%)	Oviposition Rate[a] (Eggs/Mite/Day)	Egg Viability[a] (% Hatch)	OR × VE[b]
Casein	0.50	1.45a	21.79c	31.60
Casein	0.25	1.47a	24.48c	35.99
Yeast hydrolyzate	0.50	1.41a	38.89a	54.83
Yeast hydrolyzate	0.25	1.39a	31.36b	43.59
STD[c] + 0.5% gelatin	—	1.24a	23.23c	28.81

[a] Numbers in columns followed by the same letter are not significantly different at the 5% level (Duncan's new multiple range test).
[b] Oviposition Rate × Viability of Eggs
[c] Standard diet as given in Table I.

Effect of Sodium Caseinate With and Without Amino Acids

Previous experiments had shown that casein had little effect on *T. urticae* reproduction, and it was believed that this was due to the relatively insoluble nature of casein in the aqueous diet. Accordingly, the soluble form sodium caseinate was tested in the diet with and without amino acids and the results are given in Table IV. The standard diet in this experiment was modified to reflect some improvements based on experimental findings: the major salts altered to 0.5x and the minor salts to 2.0x the standard in Table I, the amino acids at 0.5% and sugars at 1.0%, and without RNA, this additive having been found to have no effect on *T. urticae* reproduction. Gelatin and yeast hydrolyzate were also omitted, which would leave sodium caseinate to be the only protein source in the diets without amino acids. Table IV shows that sodium caseinate with or without amino acids dramatically increased production and viability of

TABLE IV.
Effect of Sodium Caseinate With and Without Amino Acids on Oviposition Rate and Egg Viability in *T. urticae* Allowed to Develop 14 Days on a Basic Synthetic Diet (Table I) Modified to Contain 0.5% Amino Acids and 1.0% Sugar, with Major and Minor Salts 0.5x and 2x, Resp., the Levels in Table I, and Without RNA.

Dietary Component	Level (%)	Oviposition Rate[a] (Eggs/Mite/Day)	Egg Viability[a] (% Hatch)	OR × VE[b]
With amino acids:				
STD	—	0.41e	17.83d	7.31
+ Sodium caseinate	2.0	1.91b	54.11c	103.35
	1.0	1.45c	64.15b	93.02
	0.5	1.23c	72.87a	89.63
Without amino acids:				
+ Sodium caseinate	2.0	2.31a	57.33c	132.43
	1.0	1.87b	70.66a	132.13
	0.5	0.82d	63.32b	51.92

[a] Numbers in columns followed by the same letter are not significantly different at the 5% level (Duncan's new multiple range test).
[b] Oviposition Rate × Viability of Eggs

eggs over those achieved on the standard diet (containing 0.5% amino acids). Increasing levels of sodium caseinate from 0.5% to 1.0 and 2.0%, with or without added amino acids, tended to increase oviposition but decrease egg viability; however, the overall reproductive efficiencies, as reflected in the "OR × VE" value, were highest in the diets containing 1.0 and 2.0% sodium caseinate without added amino acids. These findings were relevant to the ongoing efforts to develop a chemically defined diet as well as a meridic diet.

Effect of Varying Levels of Vitamin A and Vitamin E

When vitamin E (α-tocopherol) and vitamin A were added factorially at 1x, 2x, 5x and 10x the levels in the standard diet in Table I, the higher levels of both reduced the oviposition of *T. urticae*. It was concluded that the original levels of the two vitamins were sufficient.

Effects of Emulsifying Agents

Soybean lecithin and Tween 80 as emulsifying agents were evaluated in combination with cholesterol, cholesterol acetate and plant sterols (β-sitosterol and stigmasterol) by Kantaratanakul and Rodriguez (1979). Soybean lecithin could replace Tween 80 effectively as an emulsifying agent and gave better egg viability than Tween 80 when cholesterol served as the sterol source. Soybean lecithin has the following fatty acid percent composition: linoleic, 55.0; palmitic, 11.7; oleic, 9.8; palmitoleic, 8.6; stearic, 4.0; linolenic, 4.0; C_{20} to C_{22}, including arachidonic, 5.5. It thus has some nutritive value, and in the

TABLE V.
Composition of Final Meridic Diet

Constituent	mg/100 ml Diet	Constituent	mg/100 ml Diet
Sodium caseinate	1000	Lipids	
		β-sitosterol	1.00
Sugars		Stigmasterol	0.88
Sucrose	750	Stearic acid	0.50
Glucose	125	Palmitic acid	0.25
Levulose	125	Oleic acid	0.25
	1000	Linoleic acid	0.25
		Linolenic acid	1.50
			4.63
Fat-soluble vitamins			
Vitamin A	3		
Vitamin E	10	Soybean lecithin	5.00
	13		
		Major salts	
Water-soluble vitamins		K_2HPO_4	18.75
Folic acid	0.34	$Na_2HPO_4 \cdot 12H_2O$	5.30
Inositol	6.00	$MgSO_4 \cdot 7H_2O$	7.80
p-Aminobenzoic acid	0.50	$CaCl_2 \cdot 6H_2O$	1.81
Pyridoxine HCl	0.10	$CoCl_2 \cdot 2H_2O$	0.04
Riboflavin	0.20		33.70
Thiamine HCl	0.20		
Nicotinic acid	0.20	Minor salts (chelates)	
Biotin	0.0001	NaFe	2.00
Ca pantothenate	0.50	Na_2Mn	1.00
Choline chloride	10.00	Na_2Cu	0.16
B_{12}	0.002	Na_2Zn	0.24
Ascorbic acid	100.		3.40
	118.1412		

final meridic diet (Table V) replaced Tween 80 as the emulsifying agent. This diet gave an oviposition rate of 2.08 eggs/female/day and 69.37 percent egg hatchability.

SUMMARY

Teneral *T. urticae* females and males were reared on an axenic chemically defined diet, which initially contained 2% amino acids; 4% sugars; water-soluble vitamins, major and minor salts; and vitamins A and E, plant sterols and fatty acids with Tween 80 as an emulsifying agent. Modifications made in the diet were evaluated in terms of reproductive effectiveness as measured by oviposition rate and egg viability, which on the initial diet were about 0.4 eggs per female per day and 12 to 15% hatch, respectively. Gelatin, yeast hydrolyzate and sodium caseinate as protein sources all improved reproduction over that achieved on the chemically defined diet. Improvements were also made by altering the total amounts of major and minor salts by lowering the sugars to 1%. The best diet, containing 1% sodium caseinate as the protein source, gave an oviposition rate of 2.08 eggs per female per day and 69.37% egg hatch. Soybean lecithin replaced Tween 80 as an emulsifying agent and also served as a source of additional fatty acids in the final meridic diet.

REFERENCES

Cochran, W. G. and Cox, G. W. (1957). ''Experimental Design'' 2nd edition. John Wiley & Son, Inc., New York.

Duncan, D. B. (1955). *Biometrics* **11**, 1-42.

Ekka, I., Rodriguez, J. G., and Davis, D. L. (1971). *J. Insect Physiol.* **17**, 1393-1399.

Kantaratanakul, S., and Rodriguez, J. G. (1979). (In press, *Intern. J. Acarol.*)

Leon, H. H. (1960). *Biometrics* **16**, 671-685.

DYNAMICS OF CALCIUM, CATECHOLAMINES AND CYCLIC AMP IN CONTROL OF SALIVARY FLUID SECRETION BY FEMALE IXODID TICKS

Glen R. Needham

Acarology Laboratory
The Ohio State University
Columbus, Ohio

John R. Sauer

Department of Entomology
Oklahoma State University
Stillwater, Oklahoma

INTRODUCTION

Ixodid tick salivary glands perform important osmoregulatory functions (Sauer, 1977). In non-feeding flat ticks, they secrete a concentrated salt solution into the mouthparts to absorb water vapor hygroscopically replacing that lost during desiccation (Rudolph and Knulle, 1978; McMullen and Sauer, 1976). However, while feeding they secrete excess water and solutes from the bloodmeal back into the host. Most research relative to the latter has concerned: innervation of glands (Obenchain and Oliver, 1976; Megaw, 1977); morphological studies of alveolar types (Chinery, 1965; Till, 1961; Meredith and Kaufman, 1973); and pharmacological characteristics of the primary catecholamine-like stimulus and receptor (Kaufman and Phillips, 1973; Needham and Sauer, 1975; Kaufman, 1976, 1977).

Recently cyclic AMP (cAMP) was shown to be an intracellular messenger in the secretory process during feeding (Needham and Sauer, 1975; Sauer *et al.*, 1976, 1978). Another intracellular messenger or cofactor during stimulation of many cell types is Ca^{++} (Berridge, 1975). A localized increase in the cytosolic concentration of Ca^{++} acts as an effector of cell activity during stimulation. Various techniques were used to measure, and to alter salivary gland calcium dynamics during catecholamine and cAMP stimulation. The results of these

Vol. I: ISBN 0-12-592201-9

experiments are discussed relative to the involvement of Ca^{++}, catecholamines and cAMP in the secretory process.

MATERIALS AND METHODS

Salivary glands were dissected from female *Amblyomma americanum* undergoing rapid engorgement. Fluid secretion was monitored *in vitro* by the methods of Needham and Sauer (1975).

Tissue calcium was determined using standard atomic absorption spectrophotometric techniques. $LaCl_3$ (0.5% w/v) was added to all preparations to limit interference from extraneous substances.

Four different bathing media were used in various experiments. The support medium, tick saline with phosphate buffer (TS/PO_4), is described by Needham and Sauer (1975). Because of extensive Ca^{++} precipitation, the phosphate buffer was replaced by 10 or 20 mM Morpholinopropane Sulfonic Acid ($TS/MOPS_{10}$, $TS/MOPS_{20}$). TC 199/MOPS was used with modifications described by Kaufman (1976).

(±) Verapamil • HCl (VA) was the kind gift of Knoll Pharmaceutical Co. (Whippany, N. J.) and the acid salt of ionophore A23187 was donated by Eli Lilly and Co. (Indianapolis, Ind.). Additives were used at the following concentrations: noradrenaline (NA) and dopamine (DA), 10^{-5} M; cAMP and theophylline (C/T), 10^{-2} M; ethylene glycol bis-(amino-ethyl) tetraacetic acid (EGTA), 5×10^{-3} M.

RESULTS

Preliminary experiments to investigate the importance of exogenous Ca^{++} revealed a ⅔ reduction in NA stimulated secretion with Ca^{++} absent from the support medium. Complete recovery to control rates occurred upon its addition.

This suggests that the glands may not be responding maximally when bathed in TS/PO_4 due to a low soluble Ca^{++} concentration. Using standard formulae for calcium phosphate formation at pH 7 (Frieser and Fernando, 1966) and ignoring the influence of other ionic species, we estimated 10^{-7} to 10^{-5} M unbound Ca^{++} in TS/PO_4. Phosphate buffer was replaced with MOPS (10 mM), a non-chelating type buffer. This slight saline modification increased the Ca^{++} concentration to a millimolar quantity and the average gland response 5 to 8 fold. Four salines and other factors that may also influence DA stimulated secretion were compared (Table I). The change in Ca^{++} levels affects C/T stimulated secretaion differently than that of catecholamines. TC 199/MOPS supported the highest rate and TS/MOPS and TS/PO_4 were not significantly different (Table I). MOPS added to glands bathed in TS/PO_4 did not change the secretory response of glands to DA or C/T.

TABLE I.
Effect of Bathing Media on Dopamine and Cyclic AMP/Theophylline Stimulated Fluid Secretion by Isolated Salivary Glands

Medium	Medium Components[a]				Dopamine (10^{-5} M)	Cyclic AMP/ Theophylline (10^{-2} M)
	pH	HCO_3^-	Na^+:K^+ (mM)	Ca^{++} (mM)	Maximum Response ± SE (nl/min)	Maximum Response ± SE (nl/min)
					(n) = number of points in the mean	
TS/PO_4[b]	7.0 ± 0.1	yes	207:19	0.0001 to 0.001	110 ± 16(4)	125 ± 30(5)[f]
TC199/MOPS[c]	7.0 ± 0.1	no	137:5	1.3	327 ± 29(6)	236 ± 15(9)
$TS/MOPS_{20}$[d]	7.1 ± 0.1	yes	170:19	3.5	381 ± 53(6)	151 ± 21(5)
$TS/MOPS_{10}$[e]	7.5 ± 0.2	yes	170:19	3.5	487 ± 60(6)	

[a] See Needham, 1978 for complete listings of components.
[b] 3.5 mM Ca^{++} was added, but because of precipitation by phosphate the Ca^{++} concentration was estimated to be 10^{-7} to 10^{-6} M.
[c] Modified tissue culture medium 199 with Hank's balanced salt solution and 10 mM MOPS buffer (Kaufman, 1976).
[d] Tick saline containing 20 mM MOPS.
[e] Tick saline containing 10 mM MOPS.
[f] CyclicAMP (4×10^{-3} M).

Ca^{++} removal from $TS/MOPS_{10}$ reduced secretion whether NA (Fig. 1) or DA (unpublished) was the stimulant. Addition of 3.5 mM Ca^{++} restored rates to control values with NA, however DA stimulated recovery was over 200 nl/min greater than control rates.

To determine if C/T stimulated secretion is equally sensitive to changes in exogenous calcium, this cation was deleted from $TS/MOPS_{20}$ which also contained the calcium chelator EGTA. This completely blocked secretion and addition of Ca^{++} at 40 min had no effect (n = 3). Exposure to DA 32 min later stimulated secretion (37 ± 16 nl/min, $\bar{x}$ ± SE).

The Ca^{++} antagonist VA was used to block both DA and C/T stimulated secretion. Use of this inhibitor enables one to retain Ca^{++} in the support medium and block its entry into the cell, yet avoids possible "cytotoxic effects" caused by Ca^{++} removal or inclusion of chelators like EGTA. VA at 10^{-3} M inhibits both DA and C/T stimulation (<10 nl/min). Removal of VA allows recovery of DA stimulated glands to near control rates (DA control, 370 ± 69; treatment, 273 ± 19 nl/min; 40-45 min interval). However, during the same time interval, C/T stimulated recovery was not reversible (C/T control, 125 ± 29; treatment, 18 ± 5 nl/min). Interestingly, VA (10^{-3} M) inhibition of DA was quicker and more effective than removal of Ca^{++}. Furthermore, VA prevented recovery of DA stimulated secretion when Ca^{++} was re-added to Ca^{++} depleted glands (compare to recovery in Fig. 1, 40-45 min interval). DA and C/T stimulated secretion was depressed by 10^{-4} M VA but to lesser degree, while

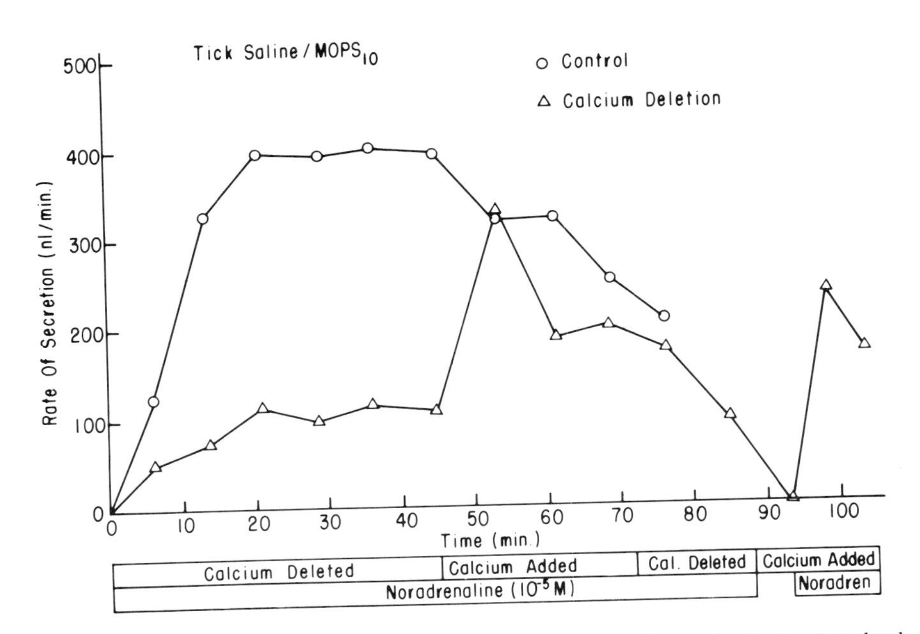

Fig. 1. Effect of Ca^{++} removal and addition on the rate of fluid secretion by *in vitro* glands. Rates of secretion by control glands where Ca^{++} was present throughout are included for purposes of comparison (0, n = 4; △, n = 5). Addition of Ca^{++} produced a significant rate increase (48-53 min interval).

10^{-5} M VA was not effective against DA (10^{-5} M VA was not tested against C/T).

To determine if increased cytosolic free Ca^{++} alone is sufficient to initiate fluid secretion, glands were incubated in TS/MOPS$_{20}$ containing ionophore A23187 (dimethylsulfoxide, DMSO, was used to solubilize A23187). This ionophore facilitates the movement of divalent cations (Mg^{++}, Ca^{++}) across plasma and mitochrondrial membranes (Babcock *et al.*, 1976, Prince *et al.*, 1973). When added to glands at concentrations of 0.1, 0.5, 1.0, 5.0 and 10.0 mM it was unable to induce secretion. Incubation in a Ca^{++}-free medium (EGTA included) with subsequent Ca^{++} introduction yielded the same lack of response.

At concentrations listed above, A23187 may cause excessive influx or release of Ca^{++} into the cytosol. Reduction of Ca^{++} to 1.0 and 0.1 mM in TS/MOPS$_{20}$ with 1.0 and 0.1 mM A23187 failed to produce secretion. The viability of each gland was tested with DA after exposure to the ionophore, and DMSO did not alter DA stimulated secretion.

Glands are able to function in the absence of exogenous Ca^{++} and the use of intracellular stores could explain this phenomenon. To investigate this possibility glands were incubated in TS/PO_4 with and without NA for 15 min.

Stimulated tissues analysed for Ca^{++} showed a reduction from 10.1 ± 1.2 (n = 24) in control glands to 5.7 ± 0.9 mM in treated.

DISCUSSION

The accumulated results indicate that at the neuroglandular junction, a primary catecholaminergic stimulus induces cAMP formation probably within the water cells of type II and III alveoli. The cytosolic Ca^{++} concentration may increase due to a release of intracellular Ca^{++} or influx across the plasma membrane or both. Together cAMP and Ca^{++} act as messengers to stimulate the fluid secretory process, probably via protein phosphorylation. Phosphodiesterase (PDE) may then be activated to catabolize cAMP to 5'AMP and reduce fluid secretion.

Although portions of this hypothesis are speculative, the following results are offered in support of the above. Dopamine stimulates increased gland cAMP with time, suggesting involvement of adenylate cyclase which catalyzes cAMP formation from ATP (Sauer *et al.*, 1978). Exogenous cAMP mimics the primary stimulus; when added with the PDE inhibitor theophylline (Sauer *et al.*, 1976); if the glands are pre-exposed to a catecholamine (Needham and Sauer, 1975); or if bathed in TC 199/MOPS support medium (Sauer *et al.*, 1978). Stimulation by near threshold concentrations of DA are enhanced by the addition of theophylline (Sauer *et al.*, 1978). Also, theophylline can maintain secretion if the gland has been pre-exposed to a catecholamine (Needham and Sauer, 1975). PDE activity has been demonstrated in tick salivary glands as high in light ticks (<300 mg) and low in heavier fast feeding ticks (McMullen and Sauer, in preparation).

Investigations into Ca^{++} dynamics of isolated glands show that the primary stimulus can induce secretion in the absence of exogenous Ca^{++} but at reduced rates (Fig. 1). Ca^{++} addition restores secretion which is suggestive that external (hemolymph) Ca^{++} is necessary for maximum functioning. However, the ability to secrete in a Ca^{++}-free solution indicates use of intracellular Ca^{++}. What are the concentration requirements for exogenous Ca^{++}? Replacement of the phosphate buffer with MOPS made more Ca^{++} (∿ millimolar) available to enhance DA stimulated secretion 8 fold, indeed, the secretion supported by low Ca^{++} (∿ micromolar; TS/PO_4; Table I) is reminiscent of the Ca^{++} deletion experiment (Fig. 1). The higher level also agrees with the hemolymph concentration in non-feeding (Shih *et al.*, 1973) and engorging adult female *A. americanum* (unpublished).

If the primary stimulus initiates Ca^{++} influx, then this should be measurable in glands incubated in $^{45}Ca^{++}$ with DA. Unstimulated control glands retained more $^{45}Ca^{++}$ at 5 and 10 min incubation (unpublished). With the available information these results are not easily explained. The result that VA blocks fluid secretion, however, suggests that Ca^{++} influx during stimulation is important. Of interest is that VA inhibition is more effective than exposure to a

Ca^{++}-free medium. One might therefore expect glands to function for a time on intracellular stores before complete inhibition occurs. However VA may be effecting secretion in other ways than Ca^{++} antagonism. The (+)-isomer of VA was found to affect Na^{+} influx in cardiac cells (Bayer *et al.*, 1975) which could be of particular interest because Na^{+} is the primary cation in the saliva produced by the salivary glands (Hsu and Sauer, 1975). Attempts to overcome VA inhibition with Ca^{++} concentrations up to 10 mM were unsuccessful so this antagonism may not be competitive with respect to Ca^{++}.

The ionophore experiments were performed in an attempt to determine if Ca^{++} influx or release from intracellular stores can by itself induce fluid secretion as this is often cited as evidence for Ca^{++} as a "second messenger." At the concentrations of A23187 and Ca^{++} used, it seems that Ca^{++} is unable to initiate the fluid secretory process.

The mobilization of intracellular sources to elevate cytosolic Ca^{++} during stimulation is supported by the 56% reduction in NA treated glands as compared to controls. Furthermore, pre-labeled glands incubated with DA for 5 or 10 min lose more $^{45}Ca^{++}$ than controls (unpublished). Generally efflux of Ca^{++} above non-stimulated rates is reflective of what is occurring in the cytoplasm (Berridge, 1975).

The inability of C/T to stimulate secretion in a Ca^{++}-free medium is significant. First, many investigators feel that this cyclic nucleotide controls release or influx of Ca^{++} in most cells (Berridge, 1975; Rasmussen and Goodman, 1977). If cAMP is penetrating the cell then one might expect release of intracellular calcium and stimulation of fluid secretion. Surprisingly glands do not secrete even when Ca^{++} is restored to the bathing medium. Possibly some imbalance within the cell has been created or the membrane permeability to cAMP is reduced and neither is rectified by the addition of Ca^{++}.

Use of VA allows Ca^{++} to remain in the medium and reportedly blocks movements of this ion into the cell. Because of VA's success in inhibiting C/T stimulated secretion and C/T's inability to initiate secretion in calcium's absence it seems that cAMP cannot by itself mimic the primary stimulus. The interaction of cAMP and Ca^{++} is apparent in tick salivary glands during secretion and many references to such involvement in other tissues are known (Berridge, 1975; Rasmussen and Goodman, 1977).

Greengard (1978) believes that protein phosphorylation may be the final common pathway for many biological regulatory agents. Both cAMP and Ca^{++} are known to affect protein phosphorylation. Information relative to this subject may clarify some of the intricacies associated with cAMP and Ca^{++} mediated processes, including that of ixodid tick salivary fluid secretion.

SUMMARY

The results indicate that intracellular and hemolymph Ca^{++} are used by ixodid tick salivary glands during stimulation. Furthermore, Ca^{++} and cAMP interact to control fluid secretion.

ACKNOWLEDGEMENTS

This research was supported by National Science Foundation grants PCM 74-24140-A02, PCM 74-24140-A01 and PCM 77-23508; and by the Agriculture Research Station, Oklahoma State University, Stillwater, Oklahoma, 74074. The authors would like to thank Ms. Kathy Dart, Kila Roggow and Peggy Bennett for their technical assistance and Mr. Clive Bowman for his helpful criticisms of the manuscript.

REFERENCES

Babcock, D. F., First, N. L., and Lardy, J. A. (1976). *J. Biol. Chem.* **251**, 3881-3886.
Bayer, R., Hennekes, R., Kaufmann, R., and Mannhold, R. (1975). *Nauyn-Schmiedeberg's Arch. Pharmacol.* **290**, 69-80.
Berridge, M. J. (1975). *In Adv. Cyclic Nucleotide Res.,* P. Greengard and G. A., eds. Raven Press, New York. **6**, 1-98.
Chinery, W. A. (1965). *Acta Tropica* **22**, 321-349.
Frieser, H. and Fernando, Q. (1966). "Ionic Equilibria in Analytical Chemistry." John Wiley and Sons, New York.
Greengard, P. (1978). *Science* (Wash., DC) **199**, 146-152.
Hsu, M. H. and Sauer, J. R. (1975). *Comp. Biochem. Physiol.* **52A**, 269-276.
Kaufman, W. (1976). *J. Exp. Biol.* **64**, 727-742.
Kaufman, W. R. (1977). *Eur. J. Pharmacol.* **45**, 61-68.
Kaufman, W. R. and Phillips, J. R. (1973). *J. Exp. Biol.* **58**, 537-547.
McMullen, H. L. and Sauer, J. R. (1976). *J. Insect Physiol.* **22**, 1281-1285.
Megaw, M. W. J. (1977). *Cell Tiss. Res.* **184**, 551-558.
Meredith, J. and Kaufman, W. R. (1973). *Parasitology* **67**, 205-217.
Needham, G. R. (1978). Ph.D. Thesis, Okla. State Univ.
Needham, G. R. and Sauer, J. R. (1975). *J. Insect Physiol.* **21**, 1893-1898.
Obenchain, F. D. and Oliver, J. H. (1976). *J. Parasitology* **62**, 811-817.
Prince, W. T., Rasmussen, H., and Berridge, M. J. (1973). *Biochem. Biophys. Acta.* **329**, 98-107.
Rasmussen, H. (1970). *Science* **170**, 404-412.
Rasmussen, H. and Goodman, D. B. P. (1977). *Physiol. Rev.* **57**, 421-509.
Rudolph, D. and Knulle, W. (1978). *In* "Comparative Physiology—Water, Ions and Fluid Mechanics," (K. Schmidt-Nielsen, *et al.,* eds.) Cambridge Univ. Press, Cambridge. p. 97-113.
Sauer, J. R. (1977). *J. Med. Entomol.* **14**, 1-9.
Sauer, J. R., Mincolla, P., and Needham, G. R. (1976). *Comp. Biochem. Physiol.* **53C**, 63-66.
Sauer, J. R., Needham, G. R., McMullen, H. L., and Morrison, R. D., (1979). *In* "Recent Advances in Acarology," (J. G. Rodriguez, ed.) Academic Press, New York.
Shih, C., Sauer, J. R., Eikenbary, P., Hair, J. A. and Frick, J. H. (1973). *J. Insect Physiol.* **19**, 505-514.
Till, W. M. (1961). *Mem. Entomol. Soc. S. Afr.,* **6**, 1-124.

Recent Advances in Acarology, Volume I

ELECTROPHYSIOLOGICAL RESPONSES OF TWO TYPES OF AMMONIA-SENSITIVE RECEPTORS ON THE FIRST TARSI OF TICKS

D. A. Haggart and E. E. Davis

SRI International
Menlo Park, California

INTRODUCTION

The host-seeking behavior of hematophagous arthropods and the role of airborne stimuli in eliciting and/or modulating this behavior pattern has been extensively investigated, especially in those organisms that are major vectors of disease (cf, Hocking, 1971). Temperature, lactic acid, and CO_2, for example, are important stimuli associated with the host-seeking activity of mosquitos; and some evidence indicates that similar kinds of stimuli are important in the host-locating behavior of ticks. Garcia (1962) suggested that the frequent passage of animals along game trails can increase the amount of CO_2 in the vicinity and that the increased CO_2 causes ticks to congregate near the trails. In field samples of ticks, both Garcia (1965) and Wilson *et al.* (1972) reported that they increased the numbers of ticks caught when they released CO_2 at their respective sampling sites. A similar rationale can be used with the NH_3 gradient along game trails. In laboratory tests, El-Ziady (1958) reported that adult *Ornithodoros erraticus* were attracted to 0.5 and 1.0% solutions of ammonia. In contrast, in their laboratory behavioral bioassay, Dukes and Rodriguez (1976) found that nymphal stages of *Amblyomma americanum, Rhipicephalus sanguineus,* and *Dermacentor variabilis* gave inconsistent responses to NH_3.

Reported here are the electrophysiological responses of a set of NH_3-sensitive neurons in the anterior pit of Haller's organ and in the medial 4-group on the first tarsi of the adult brown dog tick, *Rhipicephalus sanguineus.*

METHODS

For each experiment, an adult tick was secured to a small mount with

Vol. I: ISBN 0-12-592201-9

Scotch^{TM} tape. The first tarsi were extended and held with tacky-wax. Th mount with the tick was then placed under a compound microscope for inser tion of the electrodes. An uninsulated tungsten micro-electrode (< 1μm ti diameter) was inserted into the hemolymph space at the distal tip of the tarsu and connected to ground. A similar electrode was inserted through the cuticl at the base of a sensillum and connected to various electronic devices to detect amplify, and record the extracellular action potentials (nerve spikes) of th sensory neurons. The nerve spikes were also routed to a 3-channel amplitud discriminator, the output of which was connected to a frequency converter fo plotting the instantaneous (spike-to-spike) frequencies of up to 3 individua neurons.

Airborne chemical stimuli were generated in 1 of 2 ways. One method wa to apply neat compounds to small pieces of filter paper placed inside in dividual 10 ml syringes. The vapors of the test substances in the syringes wer expelled manually over the tarsus of the tick. The other method of stimulu generation was to pass air through a flask containing the test liquid. Thi saturates the airstream at the vapor pressure of the test substance. The flov rates of the various stimulus airstreams were controlled by individual flov meters and metering valves. Stimulation of the tick was accomplished by ac tivating a solenoid valve that routed the appropriate odor stream over th tick's tarsus. We used 21 different stimuli, which are listed in Table I.

TABLE I.

Responses of NH_3-Excited Neurons to Other Chemical Stimuli.
The Numbers are Sensilla Responding in a Particular Manner—Excited: +, ++;
Inhibited: –, ––; or No Change in Spike Frequence: 0.

	Medial (4-) Group (N = 23)					Anterior Pit (N = 11)				
Stimulant	++	+	0	–	––	++	+	0	–	––
Host										
warmth	7	·8	13				2	9		
cold		1	1							
breath		1	14				1	8		
10% relative humidity		1					1	1	1	
90% relative humidity				1		1	1		1	
lactic acid		3	17	3			2	8		
butyric acid		1	12	9	2		1	7	2	
acetic acid					3			1		1
Repellent										
DEET			2					3		
612			1	1				3		
Indalone		1					1	1		
Pheromone										
2,6-dichlorophenol		1	19	1	1		1	8		

Other stimuli:

The following failed to elicit a response in either type sensilla: methyl butyrate, ethyl and isopropyl alcohol, formic acid, eugenol, acetone, halothane, air, and carbon dioxide.

RESULTS AND DISCUSSION

The results presented below were obtained from 34 sensilla on 20 adult ticks (11 female and 9 male). Many more sensilla of different types were examined, but only those showing a response to low levels of NH_3 are included herein.

Examination of the neural discharge activity and the instantaneous frequency records revealed 2 types of neurons sensitive to NH_3 on the first tarsi of the tick: 1 in the anterior pit of Haller's organ and 1 in the medial (4) group. Both types of NH_3-sensitive neurons showed an increase in spike discharge frequency on exposure to low levels of NH_3. The magnitude of the responses was proportional to the stimulus intensity. Fig. 1 is an example of the increase in spike frequency elicited by NH_3 from an anterior pit neuron. The instantaneous frequency plots in Fig. 2, obtained from the same sensillum, show the phasic-tonic response pattern typical of sensory neurons—i. e., a sudden burst of activity that decays to a tonic, or steady-state, level of spike activity with continuous stimulation. The stimulus intensity in both Fig. 1 and Fig. 2 was 38 x 10^{-9} moles/sec NH_3 and is indicated by the bar. We can also see that the response is "slow adapting"—that is, there is little or no decrement in the steady-state firing rate over 18 min of stimulation. In addition, when a subsequent stimulus was presented 5 min later, the neuron was observed to respond with nearly the same level of spike activity as the first stimulus, indicating a relatively fast recovery time. Qualitatively similar response characteristics were observed in the NH_3-sensitive neurons from the medial sensilla.

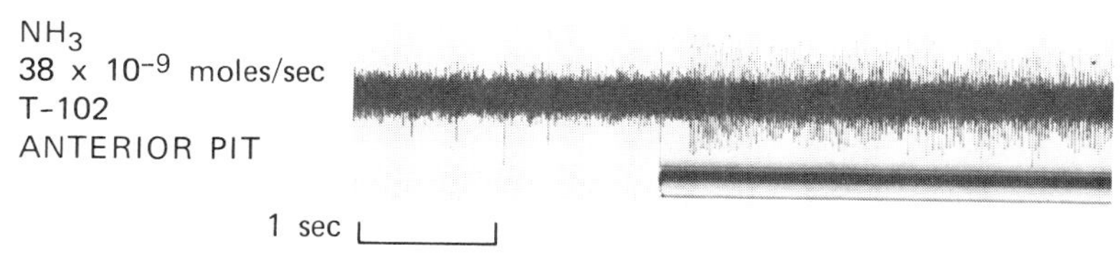

Fig. 1. Nerve impulses from anterior pit neuron.

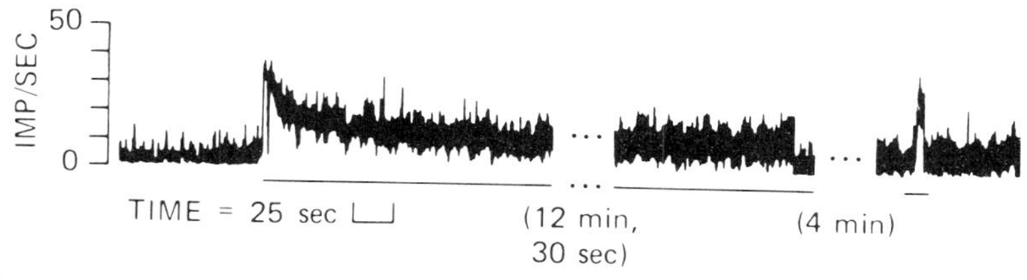

Fig. 2. Instantaneous frequency plot showing typical phasic-tonic response.

To determine how specific the responses of these neurons were to NH_3, we presented the series of compounds listed in Table I. The numbers in the table indicate how many sensilla responded in a particular manner to neat vapors presented manually via syringe.

Morphologically, the medial group sensilla are indistinguishable (Foelix and Axtell, 1971). Each sensillum as 1 or 2 bipolar neurons extending unbranched dendrites into the lumen of multiporous seta. Similarly, we have not been able to correlate the electrophysiological response with individual sensilla. This suggests that the neurons within each sensillum respond in a similar manner to the stimuli presented so far. It appears that one of the neurons may be sensitive to warmth. However, the data are not sufficient to describe this response in detail. Of the remaining stimuli, only acetic and butyric acids and, in 1 case, 2,6-dichlorophenol evoked a strong response. In these instances, the responses were characterized by a decrease in spike frequency. When the responses to these 3 substances were reexamined at lower, more physiological levels, they were either very weak or absent. Thus, at least 1 of the neurons in each of the medial group sensilla appears to be relatively specific for NH_3.

Similar qualitative results were obtained from the anterior pit sensillum, except that only acetic acid elicited a strong inhibitory response when presented neat. No response to acetic acid was observed at lower intensity levels. The anterior pit of Haller's organ contains 6 different sensilla so closely grouped that we could not be certain from which sensillum we were recording (Balashov and Leonovich, 1977).

To complete the initial description of the physiological characteristics of the NH_3-sensitive neurons in the medial group and anterior pit sensilla, we presented NH_3 in a graded intensity series over the range of 0.2 to 72 × 10^{-9} moles/sec. The stimulus was an aqueous solution of 7 mM/L NH_4OH at 36°C, the approximate level of NH_4OH in sweat (Dittmer, 1961). Air at various flow rates was passed through this solution and routed over the tick. Fig. 3 is a plot of the results of this set of experiments for neurons from both types of sensilla.

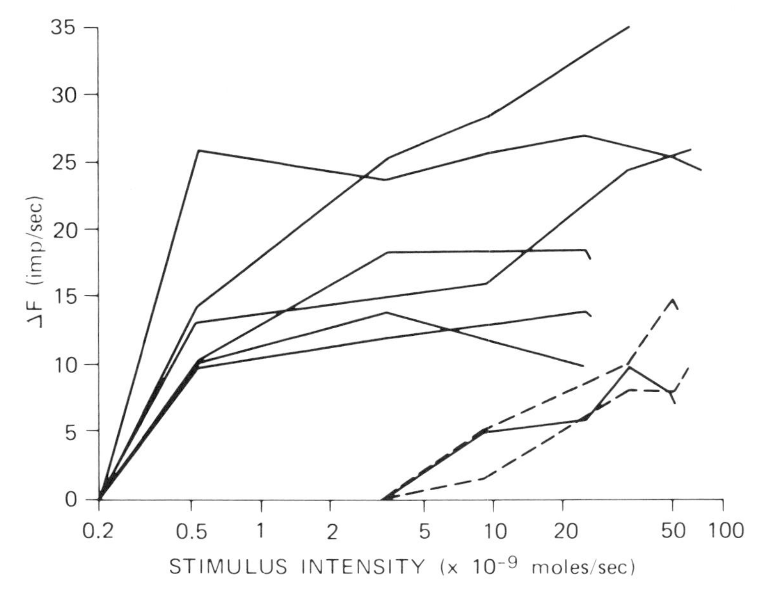

Fig. 3. Graph of stimulus intensity-response vs. response. The NH_3-sensitive neurons in the anterior pit (———) and the NH_3-sensitive neurons in the medial group (- - - -).

The NH_3-sensitive neurons in the anterior pit appear to be sensitive to lower levels of NH_3 and, over the range tested, gave a higher maximum response than the NH_3-sensitive neurons in the medial group. Because of limitations of our apparatus, we have not as yet exposed the medial group to stimulus intensities greater than 72×10^{-9} moles/sec. One exception was noted in which the response function of 1 of the anterior pit cells fell in the range of responses from the medial group NH_3-sensitive neurons. We have no explanation for the response of this cell at present. Neither type of NH_3-sensitive neuron responded to the control stimulus—water vapor at 36°C—over the same range of flow rates as NH_3.

These results indicate that with these 2 types of NH_3-sensitive neurons, which appear to differ only in their relative sensitivities to NH_3, the tick has the means of perceiving NH_3 over a broad range of intensities from very low levels such as are produced in sweat to higher levels as might be present along a well-used game trail. We must qualify any interpretation of these results by stating that we have shown only that there are primary afferents in the anterior pit of Haller's organ and in the medial group sensilla on the first tarsi of the tick *Rhipicephalus sanguineus* that are sensitive to NH_3. However, the biological relevance of this finding remains to be demonstrated. Behavioral tests must be conducted to establish the validity of our assumption that the tick actually uses NH_3 in its host-orienting or perhaps in other unspecified behavior.

SUMMARY

Using electrophysiological techniques, we have demonstrated the presence of two types of NH_3-sensitive neurons—one in the anterior pit of Haller's organ and one in the medial group of sensilla—on the first tarsi of the adult *Rhipicelphalus sanguineus.* These NH_3-sensitive neurons are relatively specific for NH_3, as determined from their reaction spectra to a series of pure compounds. Both the neurons in the anterior pit and the medial group responded to NH_3 with an increase in spike frequency that was proportional to the stimulus intensity over the range 0.2 to 72×10^{-9} mole/sec. Based on these data, we suggest that NH_3 is involved in the host-orienting behavior of the adult brown dog tick.

REFERENCES

Balashov, Yu S. and Leonovich, S. A. (1977). *Entomol. Rev.* **55**, 149-154.

Dittmer, D. S. (ed.). (1961). "Biological Handbooks: Blood and Other Body Fluids." Federation of American Societies for Experimental Biology, Washington, D. C., p. 467.

Dukes, J. C. and Rodriguez, J. G. (1976). *J. Kan. Ent. Soc.* **49**, 562-566.

El-Ziady, S. (1958). *Ann. Entomol. Soc. Am.* **51**, 317-336.

Foelix, R. F., and Axtell, R. C. (1971). *Z. Zellforsch.* **114**, 22-37.

Garcia, R. (1962). *Ann. Entomol. Soc. Am.* **55**, 605-606.

Garcia, R. (1965). *Am. J. Trop. Med. Hyg.* **15**, 1090-1093.

Hocking, B. (1971). *Ann. Rev. Entomol.* **16**, 1-26.

Wilson, J. G., Kinzer, D. R., Sauer, J. R., and Hair, J. A. (1972). *J. Med. Entomol.* **9**, 245-252.

FUNCTIONAL MORPHOLOGY AND CYTOCHEMICAL LOCALIZATION OF CHLORIDE IONS AND OUABAIN SENSITIVE PHOSPHATASE ACTIVITY IN THE SALIVARY GLAND TRANSPORT EPITHELIA OF FOUR SPECIES OF TICKS

Lewis B. Coons

Center for Electron Microscopy and Department of Biology
Memphis State University
Memphis, Tennessee

Mohamed A. Roshdy

Department of Zoology
Faculty of Science
Ain Shams University
Cairo, Egypt

INTRODUCTION

Tick salivary glands have been shown to be important in both water elimination (Tatchell, 1967) and ionic regulation (Tatchell, 1969). Electron microscopic studies of the salivary glands of several species of hard ticks (Balashov, 1968; Coons and Roshdy, 1973; Kirkland, 1971; Meredith and Kaufman, 1973) and one species of soft tick (Roshdy and Coons, 1975) show the presence of cells with ultrastructural features common to transport epithelia (Schmidt-Nielsen, 1965). These features include a highly infolded plasma membrane which forms extracellular channels and large numbers of mitochondria closely associated with these channels.

In both families of ticks, transport epithelia are located in two distinct groups within the salivary glands. The first group consists of those cells that make up the agranular alveoli. They are located mostly in the anterior region of the salivary glands adjacent to the main salivary duct. The agranular alveoli in unfed hard and soft ticks have a similar ultrastructure (Balashov, 1968; Coons and Roshdy, 1973; Meredith and Kaufman, 1973; Roshdy and Coons, 1975). The second group consists of those cells found in the granular alveoli.

Vol. 1: ISBN 0-12-592201-9

These cells are much more numerous than the first group. In *A. (P.) arboreus* the transport epithelia of the granular alveoli are present in the unfed tick (Roshdy and Coons, 1975). During feeding they expand in direct relation to the depletion of the granular secreting cells (Coons and Roshdy, unpublished data). In the hard ticks *Dermacentor variabilis* (Say), *Amblyomma americanum*, (L.) and *Rhipicephalus sanguineus* (Lat.) these cells are not recognizable as transport epithelia in the unfed ticks. During feeding they undergo a complete alteration into transport epithelia with the above ultrastructural characteristics (Coons and Roshdy, unpublished data). These cells have an ultrastructure similar to that of the "water cell" described in partially fed female *D. andersoni* (Meredith and Kaufman, 1973).

This paper presents the results of studies designed to localize chloride ions and a ouabain sensitive transport enzyme system in tick salivary glands. A chloride pump has been postulated as the force behind fluid secretion in the salivary glands (Kaufman and Phillips, 1973) and the demonstration of a transport enzyme system is a necessary prerequisite to establishing the transport function of epithelia. We also present the results of a study on the agranular alveoli of fed and unfed ticks at the light microscopic level. We use these data to discuss the functional morphology of the transport epithelia in tick salivary glands.

METHODS

Ticks were obtained from the following sources: *A. (P.) arboreus* from a NAMRU-3 Medical Zoology laboratory colony originating from the type locality of the species. *D. variabilis* and *R. sanguineus* were field collected in Shelby County Tennessee. *A. americanum* were obtained from a laboratory colony at the Department of Entomology, Oklahoma State University, Stillwater, Oklahoma. Male and female ticks were placed in a small cage and fed on either a chicken (*A. (P.) arboreus*), a rabbit (*D. variabilis*), or a dog (*R. sanguineus* and *A. americanum*). Soft ticks termed completely fed were those that had removed themselves from the host and had emitted a drop of coxal fluid. Female hard ticks termed completely fed were those observed to be with a male and were taken only as they removed themselves from the host.

We localized chloride ions using the method of Komnick and Bierther (1969). The procedure was carried out *in situ* in a photographic darkroom under a red safelight. To obtain an X-ray analysis of the deposits in the chloride ion localization experiment, gold (*ca* 1400 Å thick) sections were placed in an Hitachi HU-12A transmission electron microscope equipped with a STEM mode and a KEVEX Si(Li) dispersive X-ray spectrophotometer and the resulting energy spectrum recorded. Cytochemical localization of ouabain sensitive phosphatase was carried out according to the method of Ernst (1972).

For conventional transmission electron microscopy salivary glands were prepared as previously described (Coons and Roshdy, 1973) and examined with a Zeiss 10A transmission electron microscope. For light microscopy, 0.5 μm sections were stained as described by Roshdy and Coons (1975). Measurements were taken using a calibrated ocular micrometer mounted in an Olympus Vanox photomicroscope. Photomicrographs were recorded using the same microscope.

RESULTS

Light microscopic studies of the agranular alveoli of female and male *D. variabilis* show a significant decrease in size and a change in shape of these alveoli during feeding (Figs. 1-4). Using serial sections we measured the widest portion of a given agranular alveolus from each of 4 different ticks of the same feeding stage. The mean and standard deviation of each are: unfed female *D. variabilis* 184 μm (17.28), 48 hr fed female *D. variabilis 122* μm (10.06), completely fed female *D. variabilis* 86 μm (12.00) unfed male *D. variabilis* 152 μm (11.31), 9 days fed male *D. variabilis* 84 μm (10.32). Comparable reduction in the size of these alveoli were observed in the other two species of hard ticks used in this study. We found no reduction in the size of adult *A. (P.) arboreus* (Figs. 5-6). The mean and standard deviations are: unfed ticks 174 μm (13.66) and completely fed ticks 168 μm (6.53).

The silver chloride precipitation technique produced discontinuous deposits in vacuoles but not along membranes in the agranular alveoli of unfed female *D. variabilis* (Fig. 7), and along membranes in transport epithelia of the granular alveoli of completely fed female *D. variabilis* (Fig. 8). The chloride content of the deposits was confirmed using dispersive X-ray spectrophotometry (Fig. 9). Localization was similar in the same type of cells in *A. americanum* and *R. sanguineus.* We found no deposits in any transport epithelia from completely fed adult *A. (P.) arboreus.*

The cytochemical procedure designed to localize ouabain sensitive phosphatase activity produced consistently poor fixation as can be observed by comparing Fig. 10, a cell prepared in a conventional manner, to Fig. 15 and Fig. 16 which are similar cells from a different tick species prepared using the localization procedure. The reaction product was localized within the extracellular channels on the outer surface of the plasma membrane in the agranular alveoli of unfed adult *A. (P.) arboreus* (Fig. 11). Ouabain controls showed an absence of the reaction product in this area (Fig. 12). An identical pattern of localization was observed in the agranular alveoli of unfed *D. variabilis.* Likewise the reaction product was found between the infolded plasma membrane in transport epithelia of the granular alveoli in unfed *A. (P.) arboreus* (Fig. 13). Oubain controls showed no reaction product in this area (Fig. 14). In completely fed *A. americanum* the reaction product was localized along infolded membranes in the transport epithelia of the granular alveoli (Fig. 15). Ouabain controls showed no reaction product in this area (Fig. 16). A similar pattern of localization was observed in the same type of transport epithelia in the other two species of hard ticks.

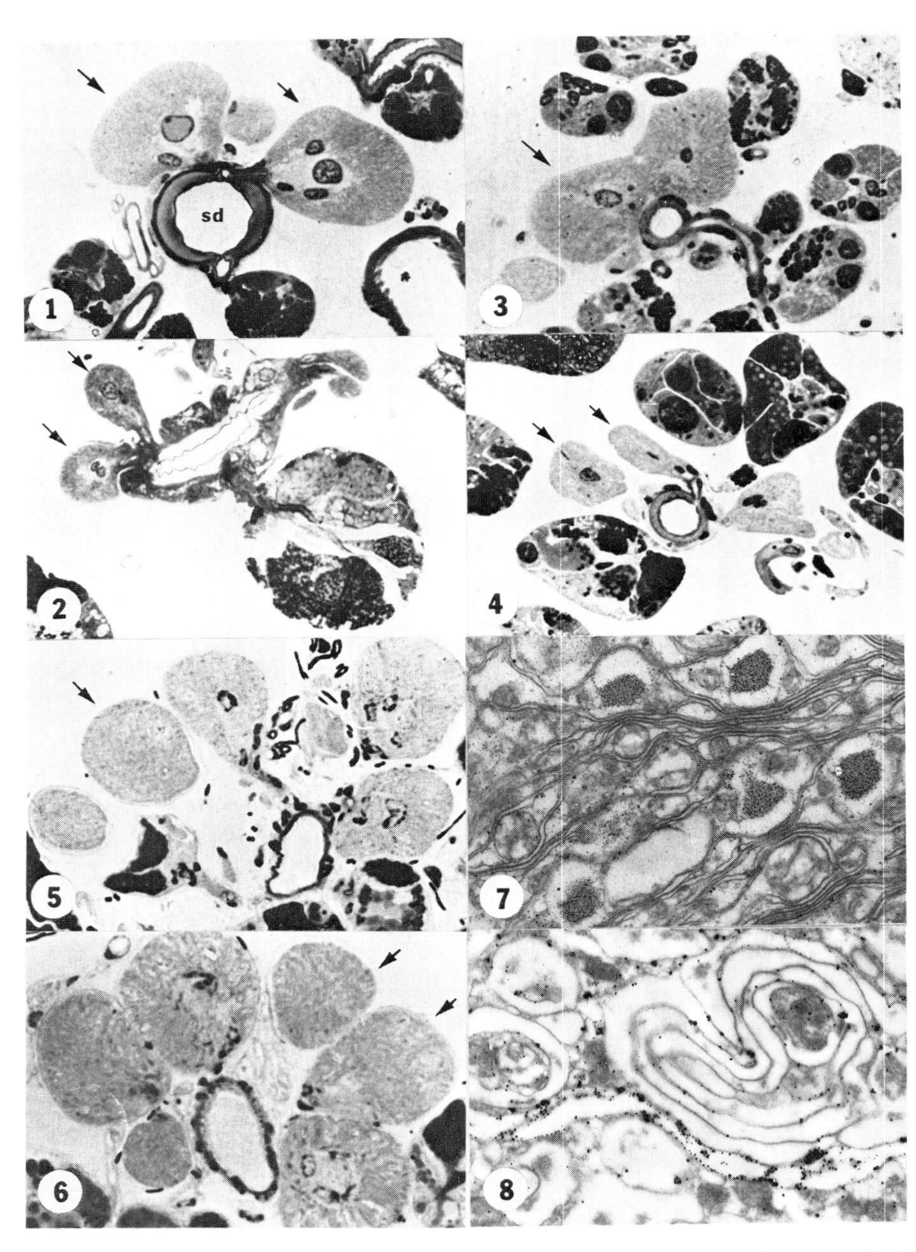

Fig. 1-6. Photomicrographs showing agranular alveoli (arrows) from the salivary glands of ticks. Fig. 1 unfed female *D. variabilis*. Fig. 2 completely fed female *D. variabilis*. Fig. 3 unfed male *D. variabilis*. Fig. 4 male *D. variabilis* fed for 9 days. Fig. 5 unfed adult *A. (P.) arboreus*. Fig. 6 completely fed adult *A. (P.) arboreus*. Note the close proximity of the agranular alveoli to the main salivary duct (sd). All figures × 348.

Figs. 7-8 Electron micrographs showing silver chloride localization in the transport epithelia of tick salivary glands. Fig. 7 Agranular alveolus of unfed female *D. variabilis*, × 14,880. Fig. 8 Transport epithelia from granular alveolus of completely fed female *D. variabilis*, × 11,720.

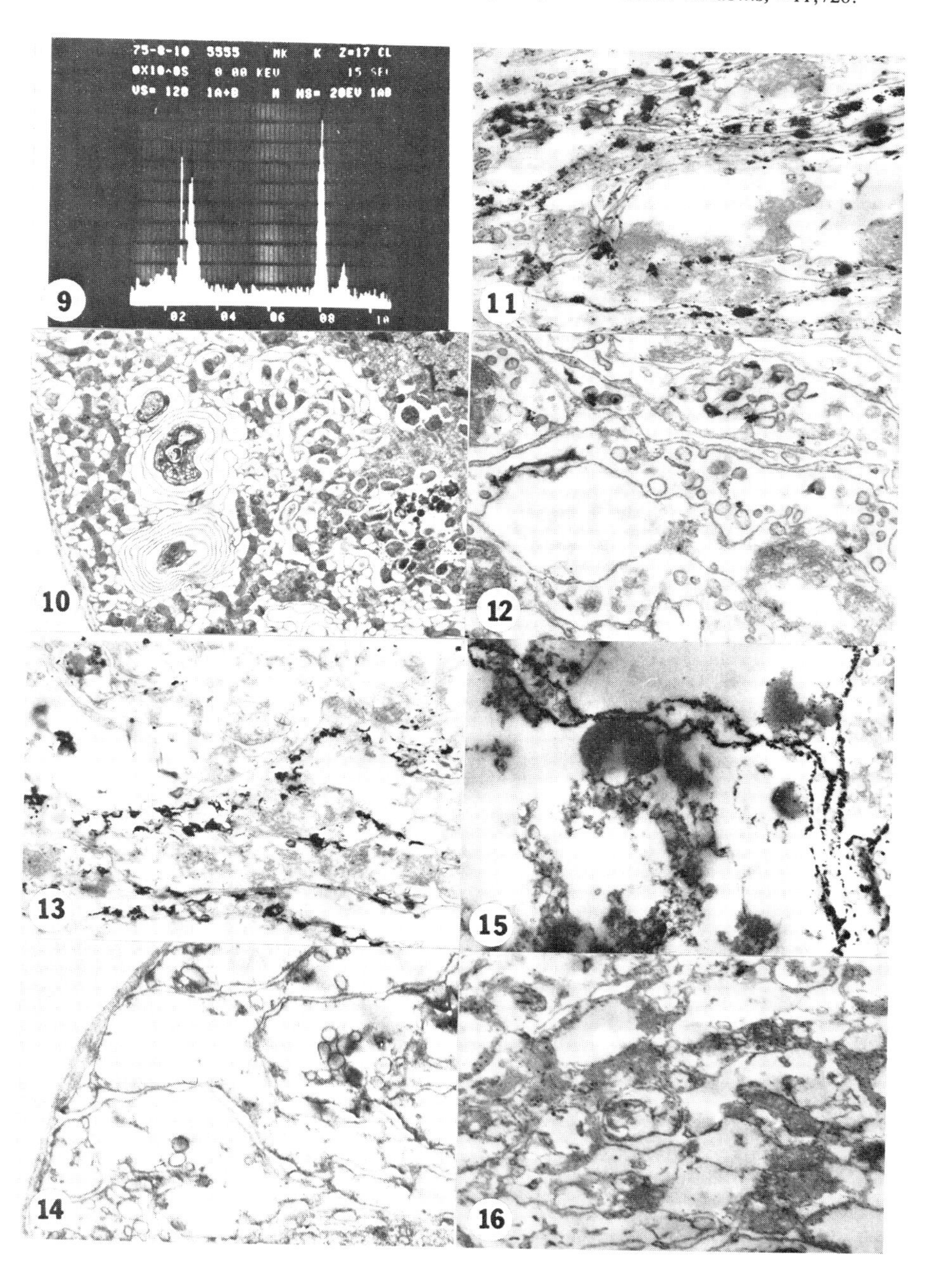

Fig. 9. Energy spectrum of deposits found in tick salivary glands after localizing chloride.

Fig. 10. Low power electron micrograph of a transport epithelial cell from a granular alveolus of a completely fed female *D. variabilis*, × 4,460.

Fig. 11. Electron micrograph of a portion of an agranular alveolus from an unfed adult *A. (P.) arboreus*. The reaction product is mostly localized within the channels formed by the infolded plasma membranes, × 18,220.

Fig. 12. Electron micrograph of the same type of cell as Fig. 11, from an unfed *A. (P.) arboreus*. Ouabain control. Note absence of the reaction product from the channels, × 18,220.

Fig. 13. Electron micrograph of a portion of a transport epithelial cell from the granular alveolus of an unfed adult *A. (P.) arboreus*. The reaction product is mostly localized between the infolded plasma membranes, × 10,880.

Fig. 14. Electron micrograph of the same type of cell as Fig. 13 from an unfed *A. (P.) arboreus*. Ouabain control. Note absence of reaction product from between the plasma membrane. × 11,710.

Fig. 15. Electron micrograph of a portion of a transport epithelial cell from a granular alveolus of a completely fed female *A. americanum*. The reaction product is mostly localized along the unfolded membrane. × 11,710.

Fig. 16. Electron micrograph of the same type of cell as Fig. 15 from a completely fed female *A. americanum*. Ouabain control. Note absence of reaction product from the infolded membrane. × 14,500.

DISCUSSION

The procedure we used to localize ouabain sensitive phosphatase activity produced consistent localization patterns of the reaction product within the extracellular channels and along membranes in the transport epithelia of tick salivary glands. This is where active transport is believed to be initiated in this type of epithelia. Although the reaction product is the result of phosphatase activity, it is held to be the site of ouabain sensitive Na, K-activated transport ATP-ase (Lewis and Knight 1977). We conclude that the localization of this enzyme system further demonstrates a transport function for the epithelia in which it was localized.

We propose that the transport epithelia of the granular alveoli in hard ticks functions to produce the watery portion of the saliva, to eliminate the excess water and to regulate the ions accumulated from the blood meal. We base these conclusions on the following facts: First, there is the presence in the epithelia of a transport enzyme as described above. Second, there is the early appearance of the epithelia during feeding and their dominance in size and number among salivary gland cells during the rapid engorging phase of feeding when water elimination and ionic regulation is most active. Third, there is the localization of chloride ions in the epithelia which is consistent with the suggestion of Kaufman and Phillips (1973) that a chloride pump is the force behind fluid secretion, and is consistent with the data of Sauer *et al.*, (1974) which showed significant uptake of ^{36}Cl in the stimulated salivary glands of adult female *A. americanum* in the rapid engorging phase. In the latter study salivary glands from non-feeding female ticks failed to take up significant amounts of ^{36}Cl and uptake in stimulated salivary glands from non-feeding female ticks was only slightly increased—this at a time when the transport epithelia of the granular alveoli are absent. Fourth, there is the position of these epithelia throughout the salivary glands and in intimate association with the granule secreting cells which we believe supports the conclusion that they produce the watery portion of the saliva.

It is well documented that the coxal glands are the means by which water is eliminated and ions regulated after the blood meal in soft ticks (Araman and Said, 1972; Balashov, 1968; Lees, 1946). However, this does not explain how the watery portion of the saliva is produced during feeding. We propose that this is the function of the transport epithelia of the granular alveoli. We base this on the presence of a transport enzyme system, the anatomical position of these cells in the salivary gland which is similar to that of the transport epithelia of the granular alveoli in hard ticks described above, and to their increase in size during feeding.

In view of their reduction in size, it is difficult to assign the agranular alveoli an active role in hard ticks during the feeding process. We support the conclusion of Rudolph and Knulle (1978) that the agranular alveoli in hard ticks function to produce the oral secretion involved in water uptake. We base our conclusion on the demonstration of a transport enzyme system in these epithelia and on the localization of chloride ions presumably associated with the sodium and potassium ions found by Rudolph and Knulle (1978) in the oral secretion. It is interesting to note the presence of the deposits within vacuoles which may be a means of accumulating chloride ions in these alveoli. We repeated the experiments of McMullen *et al.,* (1976) using the silver chloride localization procedure described in this paper. Their study showed ^{36}Cl uptake in dessicated then rehydrated salivary glands of *A. americanum.* We could not show increased deposits in the salivary glands of rehydrated *A. american.* However, our technique is poorly adapted to quantitate results and we believe the results if interpreted in this fashion are misleading.

Most likely the agranular alveoli function in a similar fashion in soft ticks but additional experimentation is needed. If this proves to be the case, the agranular alveoli appear to be able to function in *A. (P.) arboreus* after feeding. This would extend an important water regulating mechanism into a part of the life cycle during which soft ticks are susceptible to dehydration. Following the secretion of coxal fluid, sodium and chloride ions probably move into the agranular alveoli from the hemolymph. Araman and Said (1972) showed the presence of substantial amounts of these ions in the hemolymph of *A. (P.) arboreus* after coxal fluid secretion.

CONCLUSIONS

We conclude that the transport epithelia of the granular alveoli in hard and soft ticks function to produce the water portion of the saliva, and in hard ticks also function in water elimination and ion regulation especially in the latter part of feeding. We also conclude that at least in hard ticks the transport epithelia of the agranular alveoli function to produce the oral secretion involved in water uptake.

ACKNOWLEDGEMENTS

We thank Ms. N. L. Roberts for her expertise in electron microscopy, Dr. J. R. Sauer for specimens, and Dr. Harry Hoogstraal for his continued interest in our work. Research Project MR 041.09.01-0152 Naval Research and Development Command, National Naval Medical Center, Bethesda, Maryland. The opinions and assertions contained herein are the private ones of the authors and are not to be construed as official or as reflecting the view of the Department of Navy or of the naval service at large.

REFERENCES

Araman, S. F. and Said, A. (1972). *J. Parasitol.* **58**, 348-353

Balashov, Yu S. (1968). *Misc. Pub. Ent. Soc. Amer.* **8**, 161-376.

Coons, L. B. and Roshdy, M. A. (1973). *J. Parasitol.* **59**, 900-912.

Ernst, S. A. (1972). *J. Histochem. Cytochem.* **20**, 23-38.

Kaufman, W. R. and Phillips, J. E. (1973). *J. Exp. Biol.* **58**, 549-564.

Kirkland, W. L. (1971). *J. Insect Physiol.* **17**, 1933-1946.

Komnick, H. and Bierther, M. (1969). *Histochemie* **18**, 337-362.

Lees, A. D. (1946). *Parasitology* **37**, 172-184.

Lewis, P. R. and Knight, D. P. (1977). *In* "Practical Methods in Electron Microscopy." (A. M. Glauert, ed.) North-Holland Publishing Co, New York, pp. 311.

McMullen, H. L., Sauer, J. R. and Burton, R. L. (1976). *J. Insect Physiol.* **22**, 1281-1285.

Meredith, J. and Kaufman, W. R. (1973). *Parasitology* **67**, 205-217.

Roshdy, M. A. and Coons, L. B. (1975). *J. Parasitol.* **61**, 743-752.

Rudolph, D. and Knulle, W. (1978). "Comparative, Physiology-Water Ions and Fluid mechanics." (K. Schmidt-Nielsen, L. Bolis and S.H.P. Maddrell, eds) pp. 97-113.

Sauer, J. R., Frick, J. H., and Hair, J. A. (1974). *J. Insect Physiol.* **20**, 1771-1778.

Schmidt-Nielsen, B. (1965). "Ideas in Modern Biology" (J. A. Moore, Ed). The Natural History Press, Garden City, pp. 391-425.

Tatchell, R. J. (1967). Nature, Lond. **213**, 940-941.

Tatchell, R. J. (1969). *J. Insect Physiol* **15**, 1421-1430.

MOULTING HORMONE ACTIVITY IN THE FIFTH NYMPHAL INSTAR OF THE TICK *ORNITHODOROS PORCINUS PORCINUS*

Christine K. A. Mango and L. Moreka

International Centre of Insect Physiology and Ecology
Nairobi, Kenya

INTRODUCTION

Ticks like insects develop and grow through intermittent replacement of their integument. In insects, ecdysone is responsible for moulting. In ticks, however, only indirect evidences for the role of ecdysone exist (Kitaoka, 1972; Mango *et al.*, 1976; Mango 1978). The *Musca* bioassay for ecdysone has been used to test for the presence and titre of ecdysone in insects and is used here in an assay of its role in ticks.

MATERIALS AND METHODS

Synchronization of the development of *Ornithodoros porcinus porcinus* Walton 1962 is easily obtained due to the role of the bloodmeal in the initiation of the developmental cycle. Ticks used in this experiment were cultured by the *in vitro* breeding method of Youdeowei and Mango (1978) on defibrinated pig blood and were kept at 28°C and 75 to 85% R.H. between feedings. One month old fifth instar nymphs were artificially fed through bat's wing membrane on the defibrinated pig blood and were then divided into 14 groups of 50 nymphs each. Starting on day one and continuing to day 14 after feeding, samples of 60 μl of tick haemolymph were obtained from the day's group by cutting the legs and collecting the haemolymph with micro-capillary pipettes. The collected haemolymph was then extracted in a total volume of 1 ml of 10% ethanol.

The *Musca* bioassay was performed by injecting 2 μl of the extract into larvae which had been ligated 24 hrs previously by the rubber band technique. Groups of 30 *Musca* larvae were used for each experimental group and individual extracts were tested three times. Control groups were injected with

Vol. I: ISBN 0-12-592201-9

either 2 μl of 10% ethanol or 2 μl of 10% ethanol containing 0.2 ng of beta-ecdysone. All injected *Musca* abdomens were incubated at 30°C and 65% R.H. for 24 hrs, when experimental and control groups were observed and scored for % response. The equivalent beta-ecdysone titre was then calculated from the mean % response for each group of haemolymph extracts on the basis of a standard dose response curve.

RESULTS

Two peaks of activity were observed in the extracts of tick haemolymph. The first peak is relatively small and is reached on the fourth day after the bloodmeal. The second peak is higher and is reached on the ninth day. There is then a sharp drop in ecdysone-like activity prior to the moult (Fig. 1).

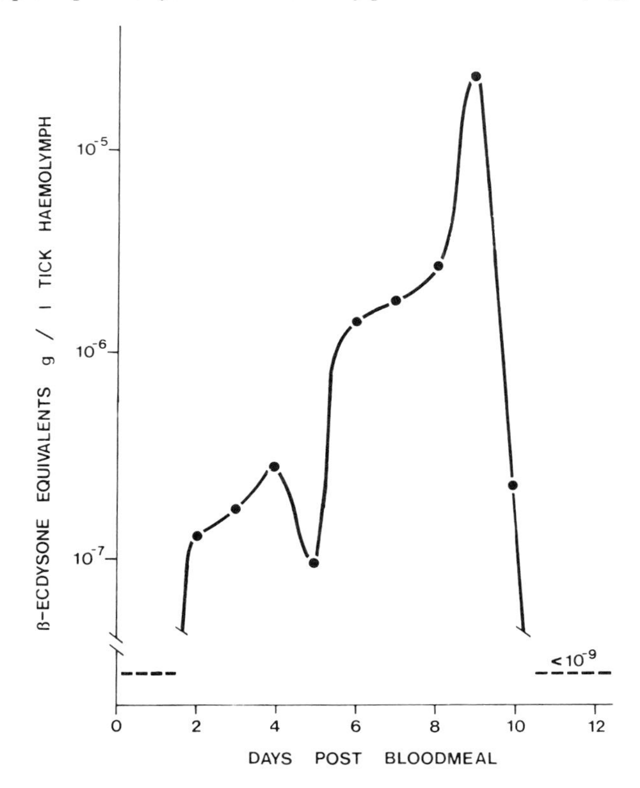

Fig. 1 Relative moulting activity in the haemolymph of fifth instar nymphs of *Ornithodoros moubata* as determined in the *Musca* bioassay and expressed as μg equivalents beta-ecdysone. Values on days 0, 1, 11 and 12 were less than 1×10^{-9}.

DISCUSSION

The data presented in Fig. 1 demonstrate the moulting hormone activity in the haemolymph of fifth instar nymphal *Ornithodoros p. porcinus* ticks. The two peaks of activity are in agreement with patterned ecdysone titres observed in the larvae of *Manduca, Galleria* and *Leptinotarsa* (Bollenbacher *et al.*, 1977, 1978; Hsiao *et al.*, 1976) prior to their moult. In *Pisaura* the moulting hormone activity was determined by the radioimmune assay for ecdysones and the level of the lowest peak at about 50 ng/g and of the main peak at 500 to 700 ng/g compare favorably with the titres determined here for *Ornithodoros* with the *Musca* bioassay. The data reported here, together with those of Bonaric and Reggi (1977) indicate that the moulting patterns and mechanisms of Arachnida are similar to those found in the Insecta.

SUMMARY

Changes in the moulting activity of haemolymph extracts of fifth nymphal *Ornithodoros p. porcinus* ticks were investigated with the *Musca* bioassay. Titre levels and the two peaks of activity on days four and nine post-feeding are similar to those reported previously from spider nymphs and insect larvae.

ACKNOWLEDGEMENTS

We are most grateful to Professors Thomas R. Odhiambo and Rachel Galun for their encouragement and guidance, to Dr. F. D. Obenchain for his help with titre calculations and to Mr. A. Bwire for technical assistance.

REFERENCES

Bollenbacher, W. E., Zvenko, H., Kumaran, A. K., and Gilbert, L. I. (1978). *Gen. Comp. Endocrinol.* **34**, 169-179.

Bollenbacher, W. E., Vendeckis, W. V., Gilbert, L. I., and O'Connor, J. D. (1975). *Develop. Biol.* **44**, 46-53.

Bonaric, J. C., and DeReggi, M. (1978). Experientia **33**, 1664-1665.

Hsiao, T. H., Hsiao, C., and Wilde, D. J. (1976). *J. Insect Physiol.* **22**, 1257-1261.

Kitaoka, S. (1972). *Proc. 14th Int. Cong. Ent.* Australia, 272.

Mango, C., Odhiambo, T. R., and Galun, R. (1976). *Nature.* **260**, (5549) 318-319.

Mango, C. K. A. (1978). *In:* "Tick-Borne Diseases and their Vectors," (J. Wilde, ed.) Univ. Edinburgh. pp. 35-37.

Youdeowei, A., and Mango, C. (1978). Proc. 4th Int. Bat Res. Conf. Kenya.

5.
Ecology, Behavior and Bionomics of Acari

Section Editors

G. W. Krantz
J. H. Camin
H. L. Cromroy

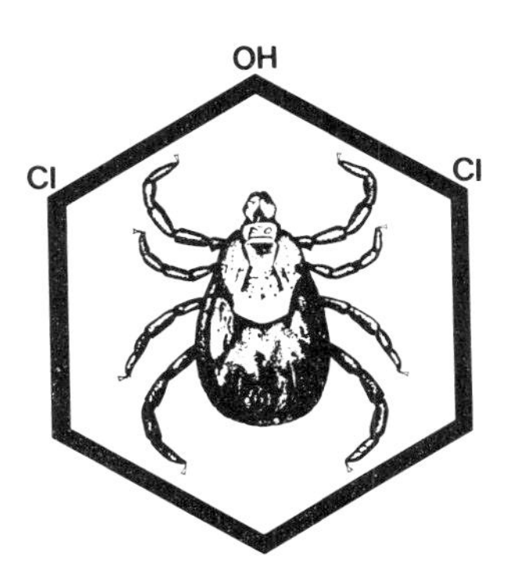

DIURNAL ACTIVITY BEHAVIOR OF *RHIPICEPHALUS APPENDICULATUS* IN THE FIELD

D. K. Punyua and R. M. Newson

*International Centre of Insect Physiology and Ecology
Nairobi, Kenya*

INTRODUCTION

Although seasonal activity of *R. appendiculatus* is fairly well studied and documented (Lewis, 1939; McCulloch *et al.,* 1969; Newson, 1977; Smith, 1969; Wilson, 1946, 1950, 1953; Yeoman, 1964, 1966, and 1967) there has been no attempt to relate this to the daily activity behavior of the population.

The blanket dragging method of Milne (1943) has proved very satisfactory for the sampling of immature stages of *R. appendiculatus,* but rather unsatisfactory in sampling adults on vegetation. For this reason cattle hosts have been used as a better index for adult tick sampling.

The present paper describes the daily activity patterns of *R. appendiculatus* both on cattle and by direct observation on vegetation.

MATERIALS AND METHODS

Using an already established tick population in a paddock at Muguga, Kenya (Purnell *et al.,* 1975), native cattle were introduced into the paddock in pairs at two hour intervals in April. A week before exposure the cattle were put in pens and treated with a non-persistent acaricide (pyrethrin). Two days before release the cattle were washed thoroughly with soap and water.

After release the cattle were allowed to graze freely in the paddock before they were driven into a crush and all the ticks collected after a search on each animal for a standard time of 30 min. In one experiment, pairs of different individual animals were used, while in another experiment the same animals were used throughout by searching and releasing back into the paddock for three days.

Although this represented a natural situation of host-tick interaction, it was indeed possible that the host stimuli like CO_2, odor and warmth could also

Vol. I: ISBN 0-12-592201-9

alter the natural tick activity patterns on the vegetation by exciting the inactive ones. To overcome this possibility, 100 ticks, marked with "Day Glo" powder, were released into tick-proof cages (Semtner *et al.,* 1973). The cages were placed in open and tree-shaded sites in the paddock occupied by the cattle. Observations were carried out simultaneously with cattle. Ambient temperature was measured using a YSI Telethermometer; relative humidity was measured using a dry and wet bulb thermometer; measurements of both parameters were made at ground level and at the level of the vegetation top during each observation.

RESULTS AND DISCUSSION

The results from the pairs of cattle showed a marked peak of activity during the 1200-1400 hours period of exposure (Table I), with a mean of 345 ticks/animal/hr. In the other experiment the cattle pick up was compared with the tick activity behavior on the vegetatin in the wire-mesh cages, and the results are shown in Fig. 1. In the latter, the cattle pick up showed two peaks of activity with a bigger peak between 0830 and 1130 hours, and a marked drop of activity in the middle of the day. There was another increase at 1530 hours which quickly dropped and remained at a lower level throughout the night. There was, however, activity at night as the cattle were picking up >40/animal/hr. The pattern of activity seems to be very much related to the daily air temperature patterns. The afternoon peak seems to have been suppressed by high temperatures (Punyua, 1978). When tick activity on the vegetation is compared with that in the tree-shaded and open sites, (Fig. 1) it can be seen that the open site also showed two peaks of activity with a marked drop at 1200 hours which corresponds with the midday drop in the cattle pick up. The tree-shaded site showed only one daily peak of activity at 1600 to 1800 hours.

When these experiments were repeated during different times of the year, the cool dry month of June (Fig. 2) showed similar results, again with the open site showing two peaks of activity with a marked drop during midday. The tree-shaded site also showed a delayed peak of activity between 1600 and 1800 hours.

TABLE I.
Mean Numbers of Adult *R. appendiculatus* Picked Up Per Animal Per Hour by Pairs of Cattle During Two Hours of Exposure in a Tick Infested Paddock.

	Time of Exposure				
	0800-1000	1000-1200	1200-1400	1400-1600	1600-1800
April	104.5	194.5	345.0	202.0	168.5
November	94.6	30.6	54.0	41.6	40.6

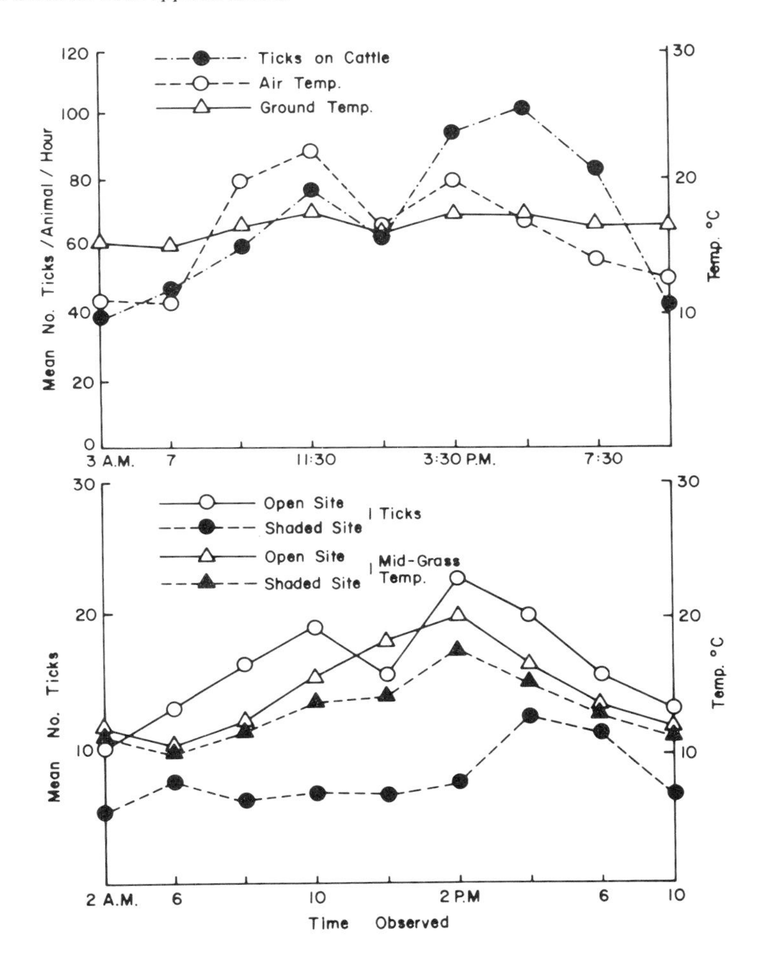

Fig. 1. Daily activity patterns of unfed *R. appendiculatus* in relation to climatic factors in the field during the month of April.

During the warm wet month of November, the open site again showed a similar pattern except that the two peaks were more separated with the morning peak at 0800 and the afternoon peak delayed until 1800 hours. The cattle released during this season also showed an early peak of activity from 0800 to 1000 hours (Table I). Although activity by the cattle is essential to enable the tick to encounter its host, the present results suggest that the level of tick activity itself during the day is not uniform, and therefore affects the number of ticks being picked up.

The magnitude and location of two peaks in the open site and one peak in the tree-shaded site, seem to vary from season to season. The delayed activity in the tree-shaded site may be due to the delayed temperature rise towards the optimum. Balashov (1972) observed that the desert ticks *Hyalomma asiaticum*

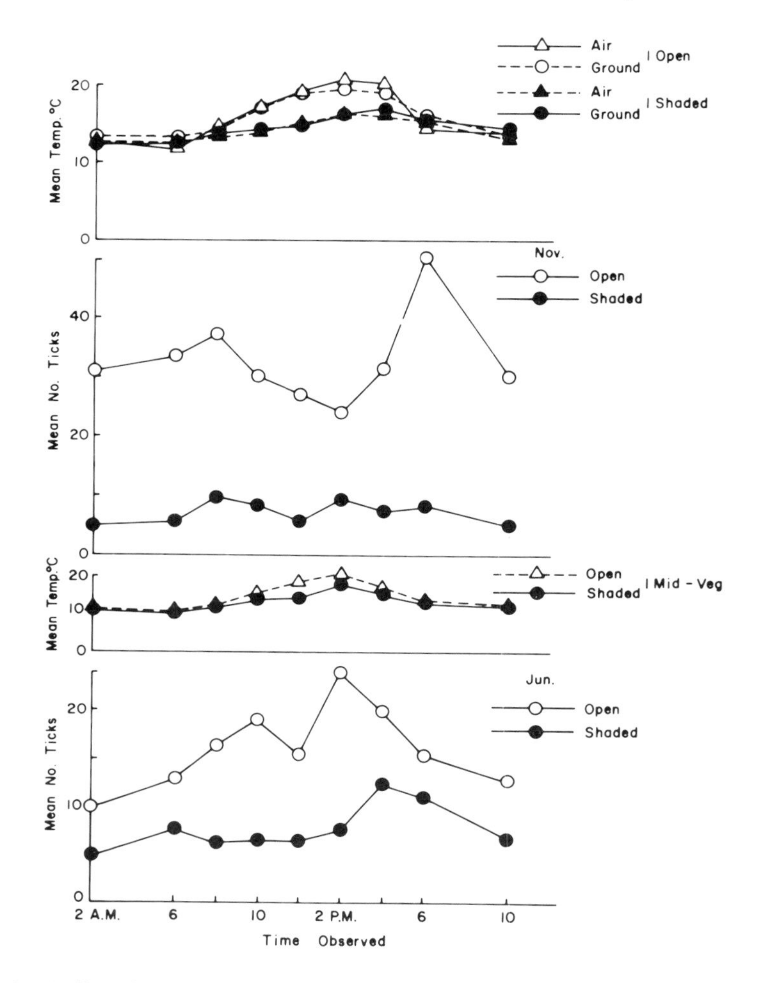

Fig. 2. Daily activity patterns of unfed *R. appendiculatus* in relation to climatic factors during the months of June and November.

were inactive before sunrise, remaining motionless on the soil surface or hiding under cover. When air and soil surface temperature increased at sunrise the ticks became active. Maximum activity occurred at temperatures between 30-40°C and 20-50% RH. As temperatures continued to increase the ticks hid again in various shelters. When temperatures began to drop towards evening the ticks again became active, but soon became motionless at nightfall.

Similarly, Semtner *et al.* (1973) reported two peaks of activity for *Amblyomma americanum,* the fewest ticks being found on the ground in the evening and the most during the morning, with movement of ticks to the ground during latter part of the day.

During the present study, ticks were also observed hiding under the leaves or the litter or even burying themselves under the grass tufts. The midday drop in activity may be due to temperatures above the tolerable limits.

SUMMARY

Daily activity patterns of *R. appendiculatus* revealed a diurnal host seeking rhythm. The cattle pick up and the direct observations of active ticks in an open site showed almost identical results with two marked peaks, one in the morning and the other in the late afternoon. The tree-shaded site, however, gave one large peak in late afternoon. It also appears that the location of these peaks and the magnitude varies according to the seasons. During the cooler month of June the morning peaks are delayed to late morning (1000-1200 hour) and late afternoon (1600-1800 hours), while during the warmer month of November, the morning peaks come at about 0800 hours. Daily air temperature fluctuations may be the important factor controlling daily activity patterns of the tick *R. appendiculatus*.

ACKNOWLEDGEMENTS

We wish to express gratitude to Dr. W. N. Masiga, Director, Veterinary Research Department for facilities provided, to Professor Thomas R. Odhiambo, Director, ICIPE for encouragement, and to Dr. M. P. Cunningham for reviewing the manuscript.

REFERENCES

Balashov, Yu, S. (1972). (Translation from Russian). *Ent. Soc. Amer. Misc. Publ.* **8**, 161-376.
Lewis, E. A. (1939). *Emp. J. Exp. Agric.* **7**, 261-270.
McCulloch, B., Kalaye, W. J., Tungaraza, R., Suda, B. Q. J. and Mbasha, E. M. S. (1968). *Bull. Epizoot. Dis. Afric.* **61**, 477-500.
Milne, A. (1943). *Ann. Appl. Biol.* **30**, 240-250.
Newson, R. M. (1977). *Proc. Conf. Tick-borne Diseases and Vectors, Edinburgh,* 1976.
Punyua, D. K. (1978). M.Sc. thesis Univ. of Nairobi.
Purnell, R. E., Cunningham, M P., Musisi, Fl L., Payne, R. C., and Punyua, D. K. (1975). *Trop. Anim. Hlth. Prod.* **7**, 133-137.
Semtner, P. J., Sauer, J. R., and Hair, J. A. (1973). *J. Med. Entomol.* **10**, 202-205.
Smith, M. W. (1969). *Bull. epizoot. Dis. Afric.* **17**, 77-105.
Wilson, S. G. (1946). *Parasitology* **37**, 118-125.
Wilson, S. G. (1950). *Bull. Ent. Res.* **41**, 415-428.
Yeoman, G. H. (1964). *Outlook Agr.* **4**, 126-135.
Yeoman, G. H. (1966). *Bull. epizoot. Dis. Afric.* **14**, 113-140.
Yeoman, G. H. (1967). *Bull. Epizoot. Dis. Afric.* **15**, 89-113.

A STUDY OF DIURNAL ACTIVITY OF LARVAE OF THE TICK, *HAEMAPHYSALIS LONGICORNIS*

T. Yoshida

Biological Institute of Liberal Arts
Shinshu University
Matsumoto, Japan

INTRODUCTION

There are numerous studies on the behavior of ticks, for example: behavior in relation to light (Belozelov, 1969), behavior in relation to CO_2 (Korenberg, 1969), rhythmicity in relation to "drop-off" of engorged ticks (George 1964, 1971; Hadani and Rechav, 1969, 1970; Hadani and Ziv, 1974; Camin *et al.*, 1971), distribution of ticks in relation to vegetation (Semtner, *et al.*, 1971), vertical distribution of ticks in relation to the period of hibernation (Dusbabek, 1971) and circadian rhythm of oviposition (Kitaoka, 1962).

Data concerning the diurnal activity of the larvae of ticks, *Haemaphysalis longicornis,* are lacking, partly because we have no measurement apparatus for recording the movement of the ticks. Therefore, it is assumed that a basic pattern of behavior of the tick larvae is to ascend on the grass and to descend to the ground. The existence or non-existence of diurnal activity in the larvae of ticks, and of the behavior in relation to CO_2 is examined in this paper by means of a new apparatus developed to record this type of activity.

MATERIALS AND METHODS

The construction for the apparatus was as follows: The experimental chamber, C, was constructed of glass sheets 30 cm × 30 cm × 40 cm. Flood light projector, A, with a red light 1 mm dia. from a tungsten lamp is projected through glass walls of the chamber onto a flood light receptor. Other flood light projectors and receptors are set out at three other locations, at heights of 10 cm, 20 cm, 30 cm and 40 cm respectively. Interruption of the light path by tick larvae ascending or descending the walls is received by the flood light receptor and the stimulus is transmitted to a control box (Figs. 1 and 2). In the

Vol. I: ISBN 0-12-592201-9

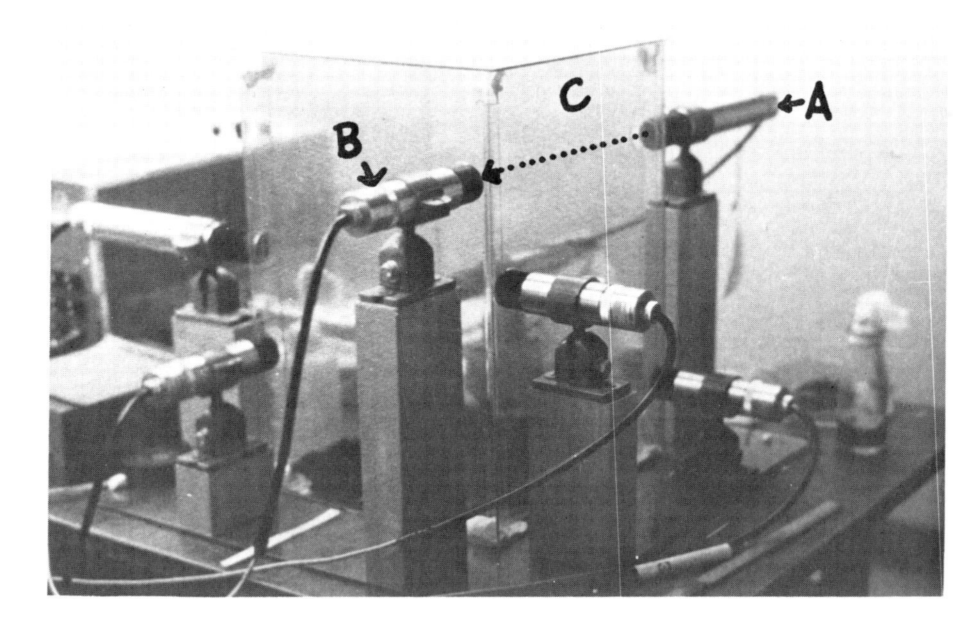

Fig. 1. A newly developed measurement apparatus for recording the movement of ticks:
(A) flood light projector
(B) flood light receptor
(C) experimental chamber

control box the stimulus is changed into an electric rectangular pulse of 500 mV which in turn is transmitted from the control box to a recorder. Signals from each of the 4 receptors are recorded with 4 pens of different colors. The recorder paper advances at the rate of 30 mm per min.

Following introduction of a small amount of mud placed on the bottom of the box to provide a relative humidity of 90% R.H., approximately 500-700 larvae were released onto the bottom of the box in each experiment. The experiments were conducted using four conditions, namely 1) 20°C and total darkness, 2) 20°C and continuous light, 3) 26°C and total darkness and 4) 26°C and continuous light. The light intensity was approximately at 800-1000 Lux (fluorescent light) at the ceiling level.

To study behavior in relation to CO_2, the experiments were conducted at 1) 26°C and total darkness and 2) 26°C and continuous light, *ca* 800-1000 Lux. CO_2 was transported from a piece of dry ice in a gas washing bottle; the CO_2 evaporated from it was transported into two gas washing bottles which were connected with rubber hoses 1 mm long and released into the environmental chamber through a glass pipe 1 mm in dia.

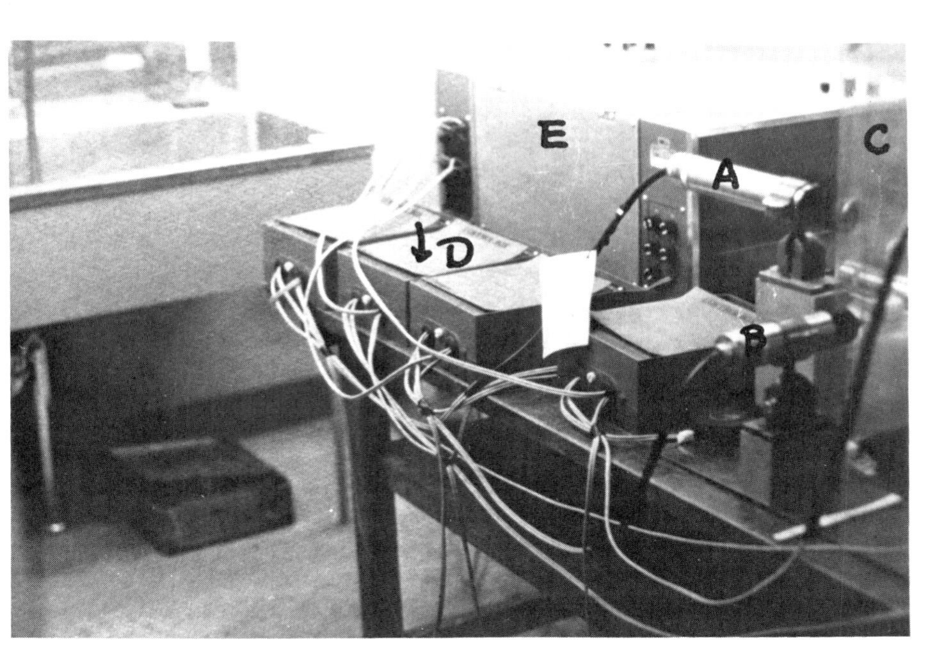

Fig. 2. Control box units:
(A) flood light projector
(B) flood light receptor
(C) experimental chamber
(D) control box
(E) recorder

RESULTS AND DISCUSSION

Two activity peaks were detected, one from 900 to 1200 hours, the other from 1700 to 2400 hours. During these activity periods, the ticks waited for the host on the grass or in the mat. Tick larvae were found to be responsive to CO_2. Following 5 minute periods of CO_2 gas release into the experimental box, the larvae became active and began to ascend or descend the glass walls. Activity diminished gradually with time and returned to the original diurnal rhythm.

Field observations demonstrated larval activity, with periods of ascending or descending to the negative mat during the hours of forenoon and evening. In addition to these data, we observed many larvae of these ticks attached to the hosts.

I assumed that this was related to the diurnal activity of the larvae which ascend or descend the vegetation and then I tried to prove it by experiments. As a result, it was found that there were two peaks in the diurnal activity of larvae tiks at the hours of forenoon and from evening to night. Especially, it was observed that larvae were most active during the evening hours. This activity of the larvae in the experiment was coincident with that observed in the field. Thus, they were found to ascend on to grass and wait for the host with many of them crowded in massive clusters from the forenoon to evening; during night hours they descended from the grass to the ground mat to await a passing host.

The activity of many wild mammals increases at night, when they begin to move about seeking food. In contrast, domestic animals are active during the daylight hours. They begin to graze in the pasture at the hours of forenoon. The hours of grazing activity of the host were coincident with those of the waiting behavior of the ticks.

REFERENCES

Belozelov, V. N. (1969). *Med. parazitologia i parazit. bolezni* **38**, 219-223.
Camin, J. H., George, J. E., and Nelson, V. E. (1971). *J. Med. Entomol.* **8**, 394-398.
Dusbabek, F., Daniel, M., and Cerny, V. (1971). *Folia Parasitol.* **18**, 261-266.
George, J. E. (1964). *Acarologia* **6**, 343-349.
George, J. E. (1971). *J. Med. Entomol.* **8**, 461-479.
Hadani, A. and Rechav, Y. (1969). *Acta Trop.* **26**, 173-179.
Hadani, A. and Rechav, Y. (1970). *Acta Trop.* **27**, 184-190
Hadani, A. and Ziv, M. (1974). *Acta Trop.* **31**, 89-94.
Kitaoka, S. (1962). *Nat. Inst. Anim. Hlth Quart.* **2**, 106-111.
Korenberg, E. I. (1969). *Med. parazitologia i parazit. bolezni* **38**, 427-431.
Semtner, P. J., Howell, D. E., and Hair, J. A. (1971). *J. Med. Entomol.* **8**, 329-335.

INTERSPECIFIC RELATIONS IN SMALL MAMMAL ECTOPARASITES

Anders Nilsson and Lars Lundqvist

Department of Animal Ecology
University of Lund
Lund, Sweden

INTRODUCTION

Generally the abundance of external parasites is low. The main part is supported by a restricted portion of the host population; the parasite population mostly fit the negative binomial distribution (Crofton, 1971; Nilsson and Lundqvist, 1978). The abundance of the parasites is restricted by factors such as immunity reactions of the host and its deparasitizing behavior (Bell and Clifford, 1964). Thus, due to this environmental stress the host body is a limited resource for the parasites. The deparasitizing behavior of the host probably also plays an important role for the distribution of the different species on the host body. On small mammals, species which anchor themselves to the skin, e.g., ticks and chiggers, are primarily found on the host's head, away from direct predation. Small and/or mobile species like lice, fleas and different mites may occur on other parts of the body, mostly, however, in rather specific areas (Nilsson, 1978).

The number of parasitic species on small mammals in Scandinavia is fairly low. At most 15 species of different taxonomic groups may be found on one small mammal host species. In other areas the number of species may be even less. In Iceland, e.g., where the common field-mouse, *Apodemus sylvaticus* L. is the only field-living small mammal, there are no ticks, no lice, only two flea and eight mite species.

Ectoparasites are usually divided into three groups according to their life strategies: permanent, nest-dwelling and "free-living" parasites (Brinck-Lindroth *et al.*, 1975). This grouping will be used here.

Interactions between endoparasites was reviewed by Halvorsen (1976). Do species within an ectoparasitic community also interact? If so, have mechanisms limiting competition evolved? Interactions within a parasitic community could be detected by restricted coexistence on host individuals, but also

Vol. I: ISBN 0-12-592201-9

along different niche parameters. In this paper joint occurrences, seasonality and spatial distribution of ectoparasites on small mammals from Iceland and southern Scandinavia will be discussed.

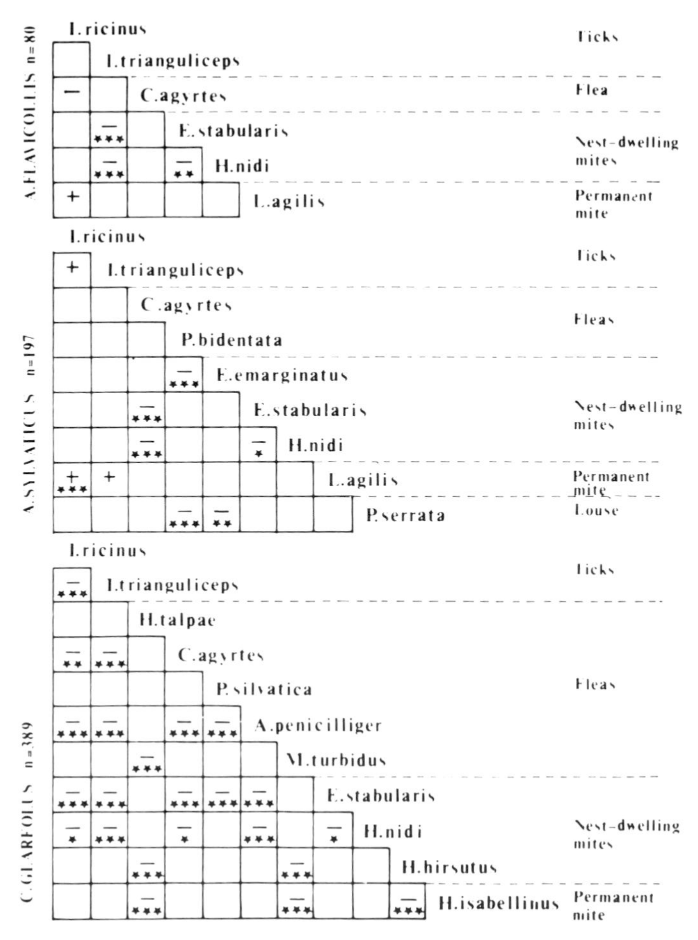

Fig. 1. Tests of differences between observed and expected number of joint occurrences using Fager's affinity index (Daniel and Holubickova, 1972; Lundqvist, 1974).—observed number < expected. + observed number >expected. $P < 0.05$; $P < 0.01$; $P < 0.0001$. Species list: Insecta, Anoplura: *Polyplax serrata.* Siphonaptera: *Hystrichopsylla talpae, Ctenophthalmus a. agyrtes, Peromyscopsylla bidentata, Peromyscopsylla silvatica, Amalaraeus penicilliger mustelae, Megabothris turbidus.* Acari, Ixodoidea: *Ixodes ricinus, Ixodes trianguliceps,* Mesostigmata: *Euryparasitus emarginatus, Laelaps agilis, Eulaelaps stabularis, Haemogamasus nidi, Haemogamasus hirsutus, Hirstionyssus isabellinus.*

JOINT OCCURRENCES

Using Fager's index of affinity (Fager, 1957) the number of joint occurrences were tested on an autumn material from southernmost Scandinavia

(Draved, Denmark). The host species were common field-mouse, *A. sylvaticus,* the yellow-necked field-mouse, *A. flavicollis* and the bank vole *Clethrionomys glareolus* (Fig. 1). In almost all testable cases there are negative correlations between the species, i.e. the observed number of joint occurrences are smaller than the expected, with different degrees of statistical significance. (Daniel and Holubikova, 1972; Lundqvist, 1974). Using a test based on probability, Lundqvist and Stenseth (unpubl.) showed negative correlations also between species not testable with Fager's affinity test. A theoretical background to the phenomenon of negative correlations is given in Smith (1974) and Lundqvist and Stenseth (unpubl.).

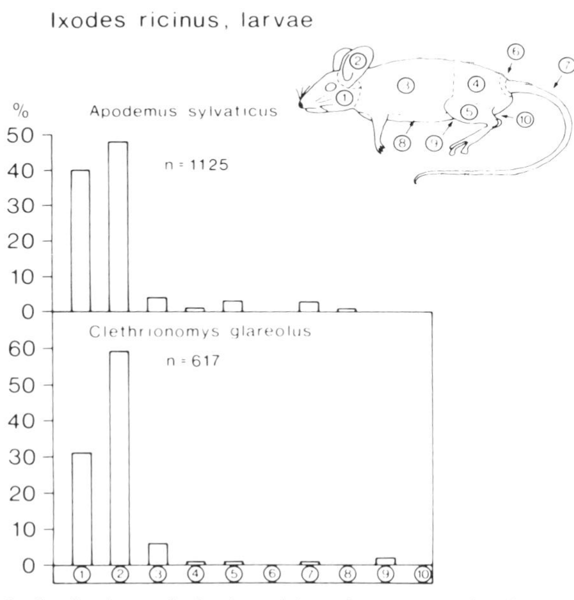

Fig. 2. Spatial distribution of *Ixodes ricinus* larvae on *Apodemus sylvaticus* and *Clethrionomys glareolus,* southern Sweden. n = number of ticks. Division in areas of the host in accordance with Nilsson (1978).

SEASONALITY AND SPATIAL DISTRIBUTIONS

Some species were studied as regards seasonality and the spatial occurrence on the host.

Ixodes ricinus-I. trainguliceps: I. trianguliceps is bound to burrowing small mammals and attaches almost only to the ears of the host. *I. ricinus* is more common and may occur on a variety of host species. The larvae, which primarily infest small mammals have a somewhat less restricted distribution on the host body than *I. trianguliceps* but also prefer the ears (Fig. 2), especially in *C. glareolus,* where almost 60% of all *I. ricinus* attach to the ears, compared with a 47-48% in the *Apodemus* species (Nilsson, 1978). Consequently it is on *C. glareolus* we may expect the strongest interaction. This is also obvious from Fig. 1. There are also certain differences in the seasonality between the two species. Their peaks of activity, however, do overlap (Nilsson, 1974; 1978).

Ixodids-*Laelaps agilis: L. agilis,* a permanent parasite of *Apodemus,* shows no negative correlation with the two tick species (Fig. 1), but at least in one case a statistically positive correlation. From the viewpoint of spatial distribution there is no reason to expect an interaction of negative character. *l. agilis* is concentrated primarily to the tail base and to the hind legs (Fig. 3) (Nilsson, 1978). There are also differences in the parasite strategies; *L. agilis* being a permanent parasite reproducing on the host, while *Ixodes* spp. are free-living parasites occupying the host for only a few days.

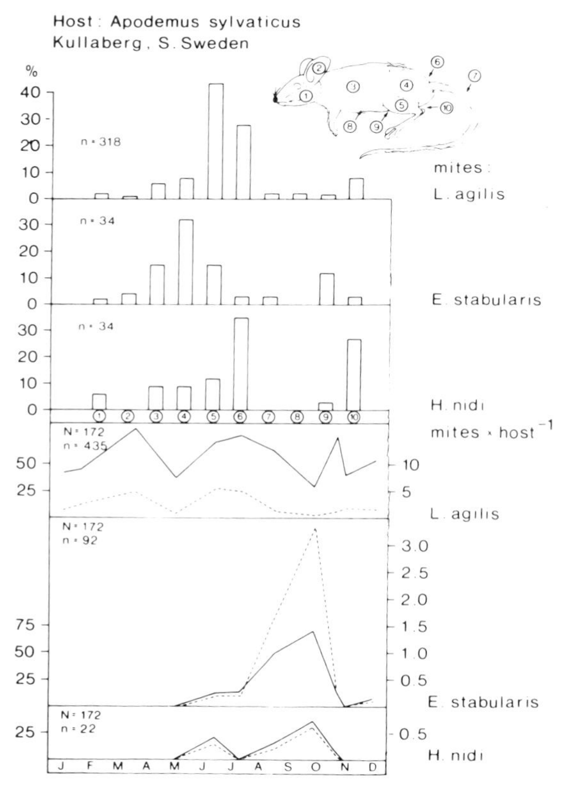

Fig. 3. *Apodemus sylvaticus* from Kullaberg, southern Sweden. Seasonal variation and spatial distribution of some parasitic mite species. ——— % of host population infested. --- number of mites × host^{-1}. N = number of hosts, n = number of mites. Division in areas of the host in accordance with Nilsson (1978).

Laelaps agilis-Eulaelaps stabularis-Haemogamasus nidi: The two nest-dwellers *E. stabularis* and *H. nidi* may in certain localities have almost the same seasonal occurrence on the host. They differ in their spatial distribution on the host, although there is an overlap (Fig. 3). *H. nidi* concentrate at the tail base and *E. stabularis* on the back a few mm in front of the former. *L. agilis* partly overlaps both, but is also abundant on the legs. The joint occurrences of *l. agilis* in relation to the nest-dwelling species could not be tested. There are, however, other facts that indicate a negative interaction between them. During the peaks of *E. stabularis* and *H. nidi* in southern Sweden the infestation frequency of *L. agilis* decreases significantly ($P < 0.001$) (Fig. 3).

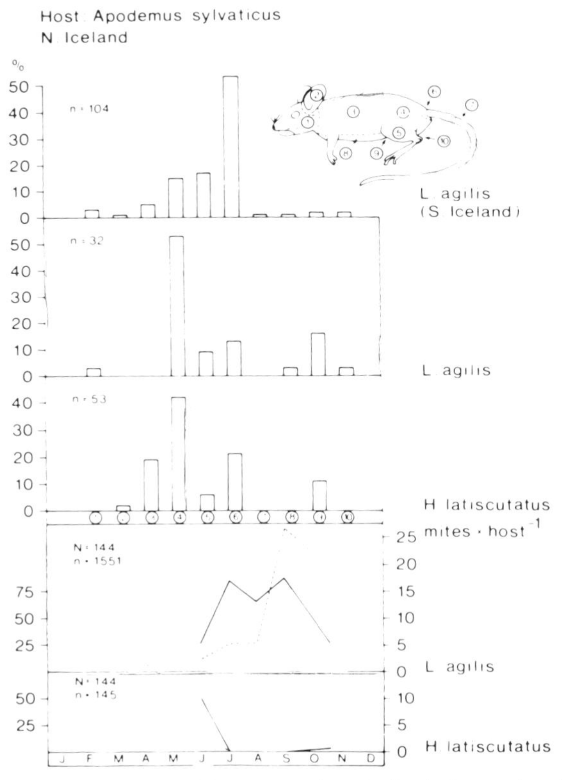

Fig. 4. *Apodemus sylvaticus*, from Myvatn, northern Iceland. Seasonal variation and spatial distribution of some parasitic mite species. — % of host population infested. --- number of mites x host $^{-1}$. N = number of hosts, n = number of mites. Division in areas of the host in accordance with Nilsson (1978).

Laelaps agilis-Hirstionyssus latiscutatus: In Scandinavia, *L. agilis* clearly dominates among permanent mites on *Apodemus* ssp. This is also the case in Iceland. In southern Iceland, where *L. agilis* is the only permanent mite, it distributes on the host's body in about the same way as in Scandinavia (Figs. 3 and 4). In the presence of another permanent mite, *H. latiscutatus* (Myvatn-area, northern Iceland), however, the spatial distribution changes. In this situation most *L. agilis* occur on the hind part of the back, in front of the tail base (Fig. 4, area 4). Furthermore, the seasonal occurrences of the two species are quite different. The highest abundance and infestation frequency of *H latisculatus* were found in June, when corresponding values for *L. agilis* were at the lowest.

CONCLUSION

There is a relation between the degree of niche overlap and the possibilities of coexistence among species. From the present results we may conclude that there is a competitive situation between certain species of ectoparasites. This results in a negative correlation as regards joint occurences on host individuals. Hence we would expect a separation along niche parameters. This was also shown for the easily measurable parameters of seasonality and spatial distribution. Such separations may be evolutionarily fixed or simply a result of interaction when species meet.

REFERENCES

Bell, J. F., and Clifford, C. (1964). *Exp. Parasitol.* **15**, 340-349.

Brinck-Lindroth, G., Edler, A., Lundqvist, L., and Nilsson, A. (1975). "Biocontrol of rodents." (Hansson, L. and Nilsson, B. eds.). *Swedish Nat. Sci. Res. Council.* 73-98.

Crofton, H. D. (1971). *Parasitol.* **62**, 179-193.

Daniel, M., and Holubickova, B. (1972). *Folia parasitol. (Praha)* **19**, 67-86.

Fager, E. W. (1957). *Ecology* **38**, 586-595.

Halvorsen, O. (1976). *In* "Ecological Aspects of Parasitology." (Kennedy, C. R., ed.) Amsterdam. 474 pp.

Lundqvist, L. (1974). *Xnt. scand.* **5**, 39-48.

Nilsson, A. (1974). *Folia parasitol. (Praha)* **21**, 233-241.

Nilsson, A. (1978). Unpubl. Ph.D. Diss., Univ. of Lund.

Smith, J. M. (1974). "Models in Ecology." Cambridge Univ. Press, Cambridge. 146 pp.

THE DEVELOPMENT OF *RHIPICEPHALUS APPENDICULATUS* POPULATIONS AT DIFFERENT HOST STOCKING DENSITIES

R. M. Newson

International Centre of Insect Physiology and Ecology
Nairobi, Kenya

INTRODUCTION

Rhipicephalus appendiculatus Neumann, the brown ear tick, is a three host African species that parasitises large herbivores. All stages can, and commonly do, subsist on cattle. The unfed larvae, nymphs and adults occur on pasture. Questing ticks ascend the grass to await contact with a passing host. Ticks that are not questing remain lower down in the vegetation, or on the surface of the soil. Climatic conditions on the study area are such that in the absence of a host, mass disappearance of the larvae occurred after 12 months, of nymphs after 15 months, and adult numbers declined to zero over a period of 25 months (Newson, unpubl.).

Sutherst *et al.* (1978) discussed host finding on the pasture by larval *Boophilus microplus* and confirmed that the rate at which they are picked up increases with the density of hosts. On the other hand, cattle also reduce the grass cover by grazing and trampling so that adverse effects on the unfed ticks are also likely to increase with the density of hosts.

The present experiment, which is still in progress, is designed to study the development of five populations of *R. appendiculatus* at three different host stocking densities and to observe the results of these interactions.

MATERIALS AND METHODS

Contiguous plots with areas of 1,000 m^2 (nos. 1 and 3): 4,000 m^2 (nos. 4 and 5): 12,000 m^2(no. 6) at the Kenya Agricultural Research Institute, Muguga, Kenya, were seeded in June 1976 with unfed larvae and nymphs at the density of 9 larvae and 4 nymphs per m^2. Two weeks later one cross-bred steer (predominantly *Bos taurus* type) was introduced into each plot where, with the

Vol. I: ISBN 0-12-592201-9

exception of plots Nos. 1 and 3 during August-December 1977, they have been kept ever since. Plot 2 was an unstocked control. Once per month all the adults on one half of the entire body were collected to idenfity sex and species (since *Rhipicephalus hurti* Wilson was also present) and thus obtained an estimate of total *R. appendiculatus* adults.

Ticks on the vegetation were sampled once per month by blanket dragging and the height of the grass was measured along 2 fixed transects on each plot. Daily rainfall, temperature and R.H. records are kept.

The study area is located at 2,100 M, 1°13′ S. lat., with mean annual rainfall approximately 1,000 mm coming in two rainy seasons. The daily temperature range varies from 9°-18°C (July) to 12°-23°C (February). The principal grasses on the study area are *Cynodon dactylon, Setaria sphacelata* and *Themeda triandra.*

RESULTS

Examples of changes in grass height in response to rainfall and grazing pressure are shown in Fig. 1. The results for plots 1 and 3 were identical and once the cattle had reduced the grass to about 50 mm in height it became necessary to feed supplementary hay. The cattle were removed in August 1977 but the recovery of the grass was so marked that they were put back again in December and it has not yet (July 1978) been necessary to resort to hand feeding, beyond a small daily ration of concentrates. The changes in the grass in plots 4 and 5 have paralled those shown for plot 6.

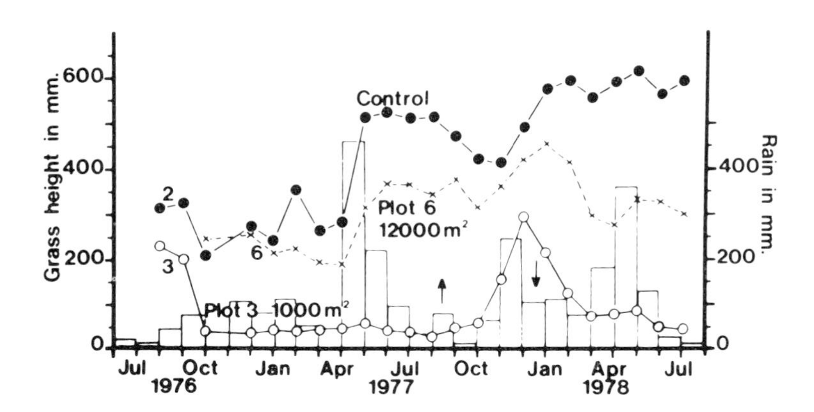

Fig. 1. Changes in grass height on a heavily stocked plot (No. 3, 1,000 m²), a lightly stocked plot (No. 6, 12,000 m²) and an unstocked control (No. 2). The bovine host was removed from No. 3 in August, 1977 and put back in December. Monthly rainfall data are also shown.

TABLE I.
Estimated Total Adult *R. appendiculatus* Burden Per Host
Based on a Single Sample Covering One Half of the Body Each Month.

Month	Plot Number and Size				
	1 1,000 m²	3 1,000 m²	4 4,000 m²	5 4,000 m²	6 12,000 m²
Aug. '76	1	1	3	1	1
Sept.	106	390	276	132	146
Oct.	70	242	92	278	306
Nov.	36	129	93	277	268
Dec.	13	45	175	605	428
Jan. '77	20	58	320	522	384
Feb.	160	1495	276	276	289
Mar.	382	3880	726	861	385
Apr.	361	3272	2849	4046	647
May	158	1489	2844	3367	885
June	32	429	1296	1346	576
July	13	147	397	744	295
Aug.	24	211	262	276	301
Sept.	—	—	164	210	252
Oct.	—	—	178	—	227
Nov.	—	—	257	151	351
Dec.	—	—	282	120	233
Jan. '78	28	1093	160	34	251
Feb.	66	930	155	31	143
Mar.	34	409	112	14	84
Apr.	22	187	111	5	35
May	0	14	39	4	21
June	6	103	12	19	12
July	15	403	12	18	7

Changes in *R. appendiculatus* adult numbers on the hosts are shown in Table I, from which it can be seen that there was an initial rapid increase as the introduced nymphs fed and moulted to adults. These in turn produced a generation which appeared on the hosts in peak numbers during March-May 1977. Since then the trend on the three larger plots has been downward, with no sign of an annual pattern in numbers as usually seen in this species. The 2 smallest plots showed peak numbers 1 month befroe the other 3 plots. Their subsequent history is less clear-cut, due to the break in adult collections when they were de-stocked. The marked and consistent differences in numbers between plots 1 and 3, though absolutely parallel, led to the suspicion that the two hosts varied in their ability to feed ticks, since the habitats are only separated by 30 m and are exactly similar. When the animals were returned to the plots in December 1977 they were switched, but the previous trends in the populations have so far been maintained.

The results of sampling larvae and nymphs on the grass are given in Table II. For the first one and a half years of the experiment there was a definite pattern of succeeding cohorts of nymphs, adults (Table I), larvae and then more

TABLE II.

Monthly Mean of Larvae (LL) and Nymphs (NN) Collected Quarterly by Blanket Dragging, Corrected to Numbers per 100 m² of Ground Covered, Results for 3 Samples Per Week, August, 1976-July, 1977, and for One Batch of 10 Samples Per Month Thereafter. [Means ($\bar{x}$) Higher than 10 are Rounded-Off to Nearest Whole Number.]

Quarter	Plot Number and Size									
	1 1,000 m²		3 1,000 m²		4 4,000 m²		5 4,000 m²		6 12,000 m²	
	LL	NN	LL	NN	LL	NN	LL	NN	LL	NN
A, S, O '76 $\bar{x}$	0.3	0.7	1.6	1.1	0.6	4.4	0.4	5.1	0.8	3.0
N, D '76 J '77, $\bar{x}$	154	3.2	322	6.3	114	1.0	74	0.5	50	0.3
F, M, A '77, $\bar{x}$	12	3.5	40	12	82	16	164	12	176	7.6
M, J, J, '77, $\bar{x}$	35	0.8	352	4.0	57	3.0	13	9.4	30	1.7
A, S, O '77, $\bar{x}$	67	14	1524	102	383	2.8	662	6.3	40	1.4
N, D '77 J '78, $\bar{x}$	17	10	216	118	77	2.7	74	2.1	8.5	0.5
F, M, A '78, $\bar{x}$	9.7	0.1	226	2.1	5.2	0.5	5.0	0.5	3.6	0.2
M, J, J '78 $\bar{x}$	0.1	0.0	70	5.4	2.6	0.1	1.0	0.0	0.8	0.1

nymphs again. The number of larvae and nymphs on plot 3 consistenly exceeded those on plot 1 by a large factor. They were numerous on the grass whilst the plots were host-free but disappeared just after the cattle returned. On plot 3 a third cohort of larvae appeared in 1978 after more adults had fed.

On all plots larvae, the offspring of the March-May adults, reached a second peak in numbers in October 1977; l on the 3 largest plots they have since declined almost to zero. Nymphal numbers have also fallen to nothing.

DISCUSSION

After nearly one year of increase, initially, by all five populations there has subsequently been a decrease which has gone beyond that of the normally expected annual cycle in *R. appendiculatus,* even though climatically it would appear to have been unusually favorable with 50% more rain in April 1977-April 1978 than on average. The grass cover on plots 4-6, at least, is also better than in sites where the tick flourishes naturally. The most obvious difference from the normal field situation is that each tick popualtion has been limited to feeding on a single individual host. Strong host resistance is known to develop to *B. microplus* (Rick, 1962) and it is postulated that during 1977 host resistance to *R. appendiculatus* also developed. In the usual field situation the tick population will be feeding on many hosts, with differing levels of resistance, including young animals which would be in the process of becoming resistant. In the present experiment, the same individuals served as the hosts throughout the experiment.

The other conclusion one can draw is that in the small, heavily grazed plots the tick/host interaction is intense and responses are rapid, whereas in the largest plot there is a notable lack of oscillation as if there were a buffer in the system. This could be provided by a substantial population of unfed ticks on the ground which can only gain access to their host at a low rate as the grazing pressure is light.

SUMMARY

Five populations of *Rhipicephalus appendiculatus* were set up in June 1976 at initial densities of nine larvae and four nymphs per m^2, on plots of 1,000 m (2), 4,000 m (2) and 12,000 m (1). One steer has been kept continuously on each plot for two years. Adults were sampled on the cattle, larvae and nymphs from the vegetation.

REFERENCES

Riek, R. F. (1962). *Aust. J. Agric. Res.* **13**, 532-550.

Sutherst,. R. W., Dallwitz, M. J., Utech, K. B. W., and Kerr, J. D. (1978). *Aust. J. Zool.* **26**, 159-174.

THE BIOLOGY OF AN ENCYRTID WASP PARASITIZING TICKS IN THE IVORY COAST

J. F. Graf

Centre Suisse de Recherches Scientifiques
Abidjan, Ivory Coast

INTRODUCTION

Encyrtid wasps make up an important family parasitizing insects and other arthropods. Among the five species known to parasitize ticks, two have been found in Africa. These are the widespread *Hunterellus hookeri* (Hoogstraal, 1956; Oliver, 1964) and the much rarer *H. theilerae,* which has been found only in South Africa (Fielder, 1953) and in Egypt on migratory birds coming from East Africa (Kaiser and Hoogstraal, 1958; Hoogstraal and Kaiser, 1961). Many attempts have been made to make use of *H. hookeri* for tick population control, but without great success (Cooley and Kohls, 1934; Smith and Cole, .943).

This paper deals with an as yet undescribed species of *Hunterellus,* which parasitizes the nymphs of *Amblyomma nuttalli* in Ivory Coast, and gives details on the biology and the breeding of this wasp. Specimens of the wasp have been deposited at the Geneva National History Museum and at the Institute of Zoology at the University of Neuchatel.

MATERIAL AND METHODS

Wild hedgehogs *Altelerix albiventris* and wild tortoises *Kinixys belliana,* were captured for use as hosts of *Amblyomma nuttalli.* The captured animals were kept in wired cages and the engorged ticks gathered every day in a water-filled tray underneath. They were then kept in standard laboratory conditions (25°C, 95% R.H.). A part of the wasps which emerged were used to study the breeding and the biology of this species.

Vol. I: ISBN 0-12-592201-9

RESULTS

Natural infestations. The wasp was obtained on four occasions from 19 nymphs of *Amblyomma nuttalli.* These nymphs were themselves obtained from wild hedgehog and tortoise during the months of August, September and December (Table I).

A. nuttalli larvae gathered on these same hosts were not parasitized, nor were the nymphs which emerged from these larvae.

Engorged nymphs of *Amblyomma variegatum* (15), *Haemaphysalis leachii* (4), *Rhipicephalus sanguineus* (4) and *Rh. senegalensis* (22) were collected at the same place and during the same period on various hosts. None of them was found to be parasitized.

TABLE I.
Occurrence of the Wasp in *Amblyomma nuttalli* Under Natural Conditions.
(LL, Larvae; NN, Nymphs).

Date	Host	No. and stage of ticks	Parasitized Nymphs	
			No.	%
Aug. 19, '76	*Atelerix albiventris*	14LL	—	—
		20 NN	12	60.0
Sept. 18, '76	*A. albiventris*	159 LL	—	—
		10 NN	1	10.0
	Kinixys belliana	4 LL	—	—
		5 NN	4	40.0
Oct. 14, '76	*A. albiventris*	278 LL	—	—
		55 NN	—	—
Mar. 23, '77	*A. albiventris*	5 NN	—	—
Jul. 22, '77	*K. belliana*	1 N	—	—
Dec. 21, '77	*A. albiventris*	57 LL	—	—
		23 NN	4	17.4
Feb. 20, '78	*A. albiventris*	7 NN	—	—

Experimental infestations, breeding. Wasps which had just emerged from ticks were placed in contact with larvae and nymphs of *Amblyomma nuttalli* before, during and after engorgement. The host used for tick feeding was the wild hedgehog *Atelerix albiventris* (Table II).

The larvae placed in contact with the wasps before or during feeding and the nymphs issuing from these larvae were not parasitized. The engorgement and molting rate for these ticks was not significantly different from the normal rate.

Unfed nymphs infested by the wasps rapidly died. Mortality began on the fifth day and reached 100% after 3 wks. Unengorged nymphs which had not been in contact with the wasps lived for more than 6 mon. under the same conditions.

Nymphs infested before feeding had a great deal of difficulty in feeding normally. A large number died and dried out on the host after one or two

TABLE II.
Experimental Infestations of Tick Immatures. Number of Larvae (LL) and Nymphs (NN).

Stage of Infestation	Total Immatures	Dead Before Engorgement	Engorged Ticks	Parasitized Ticks	Mortality
LL, unfed	166	7	159	—	17
during engorgement	323		323	—	9
NN, emerging from LL	72	—	70	—	10
NN, before engorgement					
21 days	30	30	—	—	30
15 days	30	19	—	—	30
8 days	30	9	—	—	30
5 days	30	1	12	8	29
1 day	30	—	21	16	30
NN, during engorgement					
0 days	20		19	7	15
2 days	30		26	20	26
7 days	45		43	37	45
NN, after detatching					
1 day	27		27	25	27

days. Among the nymphs which finished feeding, a certain percentage were parasitized. This percentage increased when the time of infestation got closer to the beginning of the bloodmeal.

When infestation by wasps took place on the host, during engorgement, then feeding continued normally. The highest percentage of parasitized ticks was obtained when infestation took place towards the end of the bloodmeal or immediately after detachment of the tick.

The total mortality of the nymphs which had been in contact with the wasps was always very high, significantly higher than that observed in cases of natural infestations, where it varies from 26 to 80%.

This breeding method has enabled us up to obtain five generations of wasps bred in the laboratory.

Number of parasites per tick, sex ratio. The number of parasites per tick was very high. The 262 wasps found in one nymph in the course of an experimental infestation was, as far as is known, the highest number mentioned in the literature (Table III). The difference observed between natural and experimental infestations was not seen to be significant in our sample.

The number of wasps that developed per tick depended on the engorgement weight of the nymph. On engorgement weights of 25-125 mg/nymph the relationship was constant and worked out to be an average of 1.6 wasps/mg of tick weight.

The sex ratio was very variable from one tick to another (Table III). On three occasions (once during a natural infestation and twice during experimental infestation), we found ticks containing only male wasps. Ticks containing only females were never found. The sex ratio difference between natural and experimental infestations was not significant.

TABLE III.
Number of Wasps that Emerged Per Tick and Sex-Ratio.

Infestation	No. of Ticks	No. of Wasps/Tick		Sex-Ratio (♂:♀♀)	
		Mean	Range	Mean	Range
Natural	14	64.8	35-115	1:3.6	1:0-1:22.5
Experimental	42	91.9	31-262	1:3.0	1:0-1:12.7

Development of the wasp. Ticks infested by the wasp were easily recognized by their coloring, which was initially brown and subsequently yellowish. Unparasitized ticks remained grey until molting. At the time of emergence, the wasps gnawed from 1 to 3 holes in the cuticle of the tick. These holes were not found in any particular situation.

The total duration of the wasp's development (from oviposition to emergence) varied according to the time of infestation. On the other hand, the duration of development after detachment of the tick was fairly constant (Table IV). When the wasps oviposited in engorged nymphs, the duration of development was significantly longer than when wasps oviposited in unfed or partially engorged nymphs. Two days must be added to this latter period for the difference to cease to be significant. The length of the wasp's development period was slightly longer than the tick's premolting period, which varied from 25 to 32 days in the same conditions.

The development of the wasps varied according to temperature, e.g. at 30°C, 25 to 27 days and at 20°C, it lasted more than 2 mon. When infested nymphs were placed outside under natural conditions, 21° to 32°C, 60% to 100% R.H., the duration of the wasp's development was from 28 to 36 days.

The emergence rate of the wasps also varied in relation to external conditions. This was 69.4% under laboratory conditions, 18.2% at 30°C and 70% under natural conditions.

Incubation of parasitized ticks at relative humidities up to 30% did not affect development and emergence of the wasps. After emergence, the wasps

TABLE IV.
Wasp Emergence Time (Days) Under Laboratory Conditions

Stage of Infestation	From Wasp Oviposition		From Tick Detachment	
	Mean	Range	Mean	Range
NN, before engorging				
5 days	44.5	41-49	32.0	30-25
3 days	42.6	41-46	31.6	30-34
1 day	40.0	38-44	29.6	28-32
NN, during engorging				
0 days	41.2	40-43	33.0	32-34
7 days	34.6	33-36	32.8	31-35
NN, after detaching				
1 day	35.6	34-37		

died rapidly and never survived for more than a week under standard laboratory conditions. At 30% R.H., they died in less than 3 days.

DISCUSSION

The results of studies with the new hymenopteran species resembles those obtained with *Hunterellus theilerae* (Fiedler, 1953) very closely, but differs from it in certain characteristics. Dr. C. Ferriere, to whom we have submitted our material, considers that it may be a new species (Pers. comm.). A description of this species is in preparation.

Up to now, this wasp has only been found in one species of tick, *Amblyomma nuttalli,* and this is the first time that this tick has been observed as the host of a hymenopterous parasite. Attempts to infest other tick species in the laboratory are in progress. First results show that the wasp can be bred on nymphs of *Rhicipicephalus sanguineus.* Engorged nymphs of *Amblyomma variegatum,* put in contact with the wasps, show a high rate of mortality, but development of the hymenopteran in this tick has yet never been observed.

Parasitism of tick larvae has never been observed in the field nor obtained in the laboratory. In this respect, this species differs from *Hunterellus hookeri* (Cooley and Kohls, 1934), *Ixodiphagus texanus* (Larson and Green, 1938) and also from a *Hunterellus* of an Australian species (Doube and Heath, 1975); in all instances a larval infestation has been observed, either with direct emergence or followed by a latent period up to the nymphal stage.

A. nuttalli nymphs, on the other hand, can be infested at all stages. Study of the different lengths of the development period shows, however, that the wasp cannot begin its own development until the tick has absorbed a certain quantity of blood, i.e. about two days before the end of engorgement.

After emerging, the wasp has a short life span. The females fly but very little and move mainly by jumping. The males do not fly. There must be a high degree of synchronization between the activity patterns of the host, the tick and the wasp to allow the cycle to be completed, and all the more so because *A. nuttalli* is never particularly abundant in the field.

SUMMARY

An as yet undescribed encyrtid wasp was found parasitizing nymphs of *Amblyomma nuttalli* in a savanna area of central Ivory Coast. The wasp was obtained on four occasions from natural hosts. Laboratory breeding was conducted on *A. nutalli* nymphs which had fed on wild hedgehog. The development of the wasp was studied under various conditions.

ACKNOWLEDGEMENTS

The Swiss National Fund of Scientific Research provided support for this project, Drs. C. Besuchet and C. Ferriere of the Geneva Natural History Museum identified the wasp.

REFERENCES

Cooley, R. A., and Kohls, G. M. (1934). *Proc. 5th Pacif. Sci. Congr.* **5**, 3375-3381.
Doube, B. M., and Heath, A. C. G. (1975). *J. Med Entomol.* **12**, 443-447.
Fiedler, O. G. H. (1953). *Onderstepoort J. Vet. Res.* **26**, 61-63.
Hoogstraal, H. (1956). *Res. Rep. U. S. Govt. Printing Office,* 1101 pp.
Hoogstraal, H., Kaiser, M. N. (1961). *Ann. Entomol. Soc. Amer.* **54**, 616-617.
Kaiser, M. N., Hoogstraal, H. (1958). *J. Parasitol.* **44**, 392.
Larson, C. L., Green, R. G. (1938). *J. Parasitol.* **24**, 363-368.
Oliver, J. H. Jr. (1964). *Pan-Pacif. Entomol.* **40**, 227-230.
Smith, C. N., Cole, M. M. (1943). *J. Econ. Entomol.* **36**, 569-572.

BOOPHILUS MICROPLUS AND THE FIRE ANT *SOLENOPSIS GEMINATA*

J. F. Butler

Department of Entomology and Nematology
University of Florida
Gainesville, Florida

M. L. Camino and T. O. Perez

Campana Nacional Cortrala Garaparta
Centro Nacional de Parasitologia Animal
Mexico D. F., Mexico

INTRODUCTION

The monophasic tick of bovine cattle, *Boophilus microplus* (Can.), is found to be the primary cause of great cattle losses in the State of Morelos and Mexico in general.

In order to control the transmission of the protozoan *Babesia bigemina* (Smith and Dilbourne) by the tick *B. microplus,* a national eradication campaign has been set up. Tick control in this program is based on using chemicals in dip vats.

The fire ants *Solenopsis invicta* Buren (red imported fire ant) and *S. ritcheri* Forel (black imported fire ant) were introduced into Alabama, thirty-five to fifty-six years ago, and eventually spread to nine states from California to Texas. These species are considered dangerous pests because they sting man and his animals, as well as damage pastures by building their anthills. In Mexico, *S. geminata* (Fab.) is the native fire ant and is considered a predator and is seldom seen feeding on plants and only occasionally on seeds (Whitcomb, pers. comm., 1977).

Harris and Burns (1972) and Burns and Melancon (1977) demonstrated in Louisiana that the red imported fire ant *S. invicta* caused dramatic reductions in *Amblyomma americanum* (L.) which had been attracted to dry ice.

Vol. I: ISBN 0-12-592201-9

MATERIALS AND METHODS

The Mexican experimental area of the study was located at 1,400 M above sea level in the state of Morelos, in three different sites: Cuautla, the Municipality of Yecapixtla and Mr. Juvencio Yanes's 300 ha farm with *ca* 200 domestic dairy and beef animals. The climate at this site is semi-warm and subhumid with summer rainfalls. There is little temperature oscillation with a mean annual temperature of 20°C and annual rainfall of 110 mm, with monthly precipitation ranging from 150 to 225 mm during mid-May to October.

The local grass, *Cynodon dactylon,* is extremely resistant to grazing. There are other plants present, like heath consisting of weeds and "acahual," a woody vegetation whose main plant is *Ipomoea murucoides.* An experimental area of 50 x 80 m was fenced to protect it from cattle.

Predation by ants or other insects was determined by placing gravid *B. microplus* females in exposed wire mesh cages, which prevented their escape, but allowed the entrance of smaller arthropods. The cases contained plant debris as found in the habitat being evaluated. Five replicate cages were placed in each of the three habitats (Table I). The number of exposed ticks varied from 5 to 30 per container, ticks were exposed to predation for one week. They were then examined by emptying the contents of containers over a piece of white flannel cloth. Periodical observations were made at different intervals, in order to identify the insects which had fed on them.

The following numbers of ticks were exposed to predation (Table I): 13 ticks, November 4, 1977; 11 ticks, November 9; 48 ticks, November 24; 45 ticks, December 1; 86 ticks, January 26, 1978; 44 ticks, February 22. Cattle were grazed from June to December on the farm. From January to May they were moved to an area 20 to 30 km away, and fed on residues of corn and sorghum fodder left in the fields during the dry season when no grass was available.

RESULTS AND DISCUSSION

The only natural predator of *B. microplus* detected in this study was identified as *Solenopsis geminata,* (identified by Dr. W. Buren, University of Florida, Gainesville, Florida), known as the native fire ant. This ant devours gravid female ticks, leaving no more than remnants of cuticles. We also observed that attack began on the coxae of the legs. Further observations are needed, to determine how rapidly the ticks are destroyed since the present ones were made after six days of exposure.

Table I shows the dates of exposure and observations of predation in brushy areas, grass, and tall grass; it also gives the number of exposed females and percentage of predation. Hence, predation was greatest in brushy areas, then short grass areas, followed with the least predation in areas of brush and woody vegetation.

TABLE I.
Gravid *Boophilus microplus* (Can.) Females and Their Natural Predation by *Solenopsis geminata* (Fab.) in Yecapixtla, Morelos, Mexico—1977-1978.

Date of Exposure and Vegetation[a]		No. Exposed Females	Date Observation Made	Predated Females	Predation (%)
Nov. 6	Th	6	Nov. 12	4	66
	G	3		3	100
	BW	7		1	14
Nov. 9	Th	3	Nov. 18	3	100
	G	8		1	12
	BW	4		2	50
Nov. 24	Th	6	Dec. 1	6	100
	G	6		3	50
	BW	26		0	0
Dec. 1	Th	15	Dec. 12	15	100
	G	15		2	13
	BW	15		2	13
Jan. 26	Th	20	Feb. 3	5	25
	G	20		1	5
	BW	20		1	5

[a] Th = Thicket or brush; G = Grass; BW = Brush and woody vegetation

TABLE II.
Total Predation of *B. microplus* by *S. geminata* from November 1977 to February 1978 in Cages Placed in Different Habitats.

Site	Exposed Females	Predated Females	Predation (%)
Thicket	60	38	63
Grass	63	12	19
Brush and Woody Vegetation	81	6	7

Table II shows the predation of *B. microplus* by the fire ant. Chi-square analysis of the data indicated significant differences in predation in different sites, i.e., predation is not independent of environment, in this case, the configuration of vegetation.

These data are concordant with the results obtained by Harris and Burns (1972) and Burns and Melancon (1977) for predation of gravid females of *Amblyomma americanum* ticks by the red imported fire ant, *Solenopsis invicta*, in thickets. The same phenomenon was observed and substantiated in this study for predation of *B. microplus* ticks by the native fire ant, *S. geminata*.

SUMMARY

In thickets in the municipal boundary of Yecapixtla, Morelos, Mexico, fire ants present a high percentage of predation (63%) on gravid *B. microplus* females during the months of November through February (1977-1978). Predation was significantly influenced by the type of vegetation in the different sites of study (brush or thicket, grass, and tall grass). Further studies on natural and applied biological control using this reported predation phenomenon are needed and are continuing.

REFERENCES

Burns, E. C. and Melancon, D. G. (1977). *J. Med. Entomol.*, **14**, 247-249.
Harris, W. G. and Burns, E. C. (1972). *Environ. Entomol.*, **1**, 362-365.

ERIOPHYOIDEA IN BIOLOGICAL CONTROL OF WEEDS

Harvey L. Cromroy

Department of Entomology and Nematology
University of Florida, Gainesville

The unique characteristics of the superfamily Eriophyoidea make them strong candidates in the consideration of biological control of weeds. To best explain their potential for weed control, it is prudent to summarize the characteristics of the superfamily and then illustrate these on a pest species which we have studied and finally some of the species which we consider candidates for biological control of weeds.

Eriophyoid mites are distinguished from all other members of the Acari by some obvious morphological features. Eriophyoid mites have only two pairs of legs, a worm-shaped body with apparent annular rings behind the cephalothorax (thanasomal rings). These mites are all phytophagous and many are specific to plant species within a genus or even to a single species within the genus. The mites are small, 100-400μm in length. It is the family Eriophyidae (superfamily Eriophyoidea) that has the greatest potential in weed control. This family includes mites which are gall makers, producers of erinea, and rust mites.

There is much lack of information on the basic biology of the eriophyoid mites, but some general facts are known. The eriophyoid mites have preferred site locations on plants and this applies to both the gall makers and the erinea makers. The eriophyoid cheliceral stylets can penetrate plant tissue to a depth of 25-50μm and when the gall is produced the growth modification is initiated only on embryonic plant tissue. The gall forming substance introduced by the mite into the plant is very localized in its activity. In the case of erinea-producing mites, there is a lateral dispersion of the substance produced by the mite within the leaf. Eriophyoid mites are typically pests of perennial plants but they infest annuals also. Some authors believe that the members of the family Eriophyidae have a narrow host range especially on the broad leaf plants and apparently over 95% of the eriophyoid species are restricted to a single genus of plants and within this grouping possibly 40% are restricted to a single genus and species of plant.

An additional feature of the eriophyoids is the fact that they are recognized as carriers of ten different plant viruses (Whitmoyer, 1972). The best way of

Vol. I: ISBN 0-12-592201-9

emphasizing the potential of this group of mites in weed control would be to describe the effect of one species on an economic crop since the economic species have been better studied than those on weeds. The species I will discuss in some detail is the Bermudagrass mite, *Eriophyes cynodoniensis* (Sayed) (Jeppson, *et al.*, 1975). The Bermudagrass mite is distributed throughout the entire state of Florida and has also been reported from California, Arizona, Texas and Georgia. Florida has a very large area in turf (approx. 900,000 A.) which includes golf courses, commercial areas, cemeteries and home lawns as well as sod growers (Anon., 1976). This mite when present in large numbers can cause stunting of the grass and with secondary stress will kill grass. It is found only on Bermudagrass and epecially on four varieties, namely, St. Lucie, Ormond, Common and No-Mow, which are some of the most common grasses planted in Florida. Our studies (Cromroy and Johnson, 1972; Johnson, 1975) indicated that this mite will develop only on these four strains of Bermudagrass but will not develop on any of the hybrid strains grown or developed at Tifton, GA with the mixed parentage from Africa which would include *Cynodon transvaalensis* Davy as a parent. This particular mite points out the high degree of specificity found in the eriophyoids where the mite will infest only strains of Bermudagrass which come from *Cynodon dactylon* (L.) parentage but from a mixed parentage which includes the hybrid species Tiftway (419) and Tiftgreen (328).

This specificity is further demonstrated by the eriophyid gall mite *Eriophyes chondrillae* (Can.), the one eriophyoid mite documented in the biological control of skeleton weed, *Chondrilla juncea* L. (Wapshere, 1978). *C. juncea* is an apomictic triploid with geographic clones throughout its native range. The mite apparently has several strains some of which are highly adapted to certain forms of *C. juncea* while others were ineffective against other forms of the weed. Wapshere (1978) has pointed out the problem of eco-climatic matching in predicting the results from the introduction of a biological control agent.

In Florida we have found three different species of eriophyoids on three weeds in our preliminary examination of eriophyoids for the state. One species was found on *Lantana camara,* one on *Mikania,* and one on poison ivy. The damage done to *Lantana* in the north and northcentral part of the state is to produce large galls which consist of a mass of very small green leaves, and distorted flowerbeds and flowers.

This mite, *Eriophyes lantanae* Cook, redescribed by Keifer and Denmark, (1976) damages *Mikania* in the southern portion of the state by producing a silvering and dropping of the leaves. The mites found on poison ivy are primarily makers of erinea and according to observations made by D. H. Habeck (unpubl.) will produce a general dwarfing or stunting of the plant. All three species are in the genus *Eriophyes,* family Eriophyidae. There is a high probability that *E. lantanae* and the poison ivy mite are probably quite specific because of the type of damage which they do whereas the *Mikania* mite may well be broader in its host selection. A major problem in the evaluation of these mites for selection in biological control is not only their host specificity but a measure of the decline in vigor which they will produce in the weed. This type of parameter is one of the most difficult to evaluate as to the success or failure of the eriophyid.

In any consideration of the eriophyids for biological control of weeds one can then look on the positive aspects of these mites, which would include high degree of host specificity, the mites can be wind dispersed, they have a selective site preference as to where they will find food, they generally produce a slow decline in the plant vigor, large numbers can be accommodated in minute spaces and they can be used easily in conjunction with other control agents, e.g., microbials or beneficial insects and will not be competitive with these other agents.

On the negative side would be the two items where little is known, namely, mite physiology and mass culture techniques. The latter have been done exclusively by rearing the mites on live material and then transferring live mites by grinding up the material on another medium.

SUMMARY

A number of advantages for the use of eriophyoids as an adjunctive organism in the biological control of weeds have been presented. The success of the gall mite used in Australia on skeleton weed would seem to be a strong indicator for the future use of this group of mites along with other organisms in weed control. The mites produce a gradual decline in the plant and this is cause for consideration of the mites as an adjunct to be used with any other control system being used. It may well be that certain species of eriophyoids will be the only organisms that will exert any biological control on some weeds.

REFERENCES

Anon. (1976) *Fla. Turfgrass 1974, 1976.* Fla. Dept. Agr. & Consumer Ser. Div. Marketing, 39 pp.

Cromroy, H. L., and Johnson, F. A. (1972). *Fla. Turf* **6**, 5-6.

Jeppson, L. R., Keifer, H. H., and Baker, E. W. (1975). "Mites Injurious to Economic Plants." Univ. Calif. Press. 327-394.

Johnson, F. A. (1975). Ph.D. dissert., Univ. of Fla., unpubl.

Keifer, H. H., and Denmark, H. A. (1976). *Fla. Dept. Agr. & Consumer Ser. Ent. Cir. 166.*

Wapshere, A. J. (1978). *Proc. 4th Int. Symp. Biol. Cont. Weeds.* Univ. Fla. Press. 124-127.

Whitmoyer, R. E., Noult, L. R., and Bradflute, O. E. (1972). *Ann. Ent. Soc. Amer.* **65**, 210-215.

FACTORS AFFECTING THE DISPERSAL OF *AMBLYSEIUS FALLACIS* IN AN APPLE TREE ECOSYSTEM

Donn T. Johnson and B. A. Croft

Pesticide Research Center and
Department of Entomology
Michigan State University
East Lansing, Michigan

INTRODUCTION

A number of factors affect the seasonal dispersal of the predaceous mite, *Amblyseius fallacis* (Garman) into the apple tree and, subsequently, biological control of the European red mite, *Panonychus ulmi* (Koch), and the apple rust mite, *Aculus schlechtendali* Nalepa. Those discussed in this paper include the density of *A. fallacis* in the groundcover, the density of *P. ulmi* or *A. schlechtendali* in the tree and the cumulative degree-days (DD_{54}) in the orchard.

METHODS

In 1978, 16 small nursery trees (prince red delicious, 5 ft tall) were sprayed with Sevin® to eliminate all insect and mite forms, planted beneath orchard trees, and infested with *Tetranychus urticae* Koch. At weekly intervals, all the leaves of each tree were scanned for the presence of *A. fallacis* which had dispersed into the tree. There were 4 treatments: 1) allowed unlimited mite movement (air or ambulatory via the trunk) into the canopy; 2) mite proof screen cages allowed only trunk dispersal; 3) the trunks were covered with Stickem® which allowed for only air dispersal; 4) combined treatments 2 and 3 to most greatly restrict movement into the canopy. These data were transformed by a log (x + 1), and the treatment means were differentiated by a Student-Newman-Kuel's multiple range test.

From 1975 to 1978, data were collected from commercial apple orchards to determine the effects of certain factors on dispersal of *A. fallacis.* Weekly

Vol. I: ISBN 0-12-592201-9

estimates of the ground cover density of *A. fallacis* were taken by scanning 10 apple sucker leaves/tree (1975) or by scanning broad leaf plants for 6 minutes/tree (1976-77) as described by McGroarty (1975). In 1978, broad leaf plants were scanned for 30 minutes/tree which increased the accuracy to 10 percent standard error of the mean density of 0.1 mites/one minute count. Relative densities on apple leaves were taken by sampling 50 leaves/tree in 1975-77 or 100 leaves/tree in 1978 and counting the number of *A. fallacis* and prey with a dissecting microscope.

Seasonal dispersal of *A. fallacis* was measured using 10 × 10 cm plates covered with silicone grease. Ten grease plates were placed horizontally (sticky side up) under each tree and sampled weekly. Dispersal distance by *A. fallacis* was measured by placing 10 grease plates vertically (sticky side toward the orchard) at 6 distances away from the orchard (e.g. under the trees and at 4, 8, 19, 42, and 72 meters).

Degree day (base 54°F) accumulations were calculated for each orchard using the Baskerville and Emin (1969) sine curve method. A base of 54°F was used since Lee (1972) reported this to be the developmental threshold for *A. fallacis*. A multiple regression of the independent variables, e.g. DD_{54}, *A. fallacis* in the ground cover, and *P. ulmi* or *A. schlechtendali* in the tree was derived to determine which factors explained the variation in the densities of *A. fallacis* observed in the tree.

RESULTS

Early Season Dispersal Studies

In the 4 treatments it was found that the major mode of dispersal into the nursery trees was via the tree trunk. This is apparent in Table I since treatment 2 is equivalent to treatment 1 which allowed for unlimited mite movement. Total exclusion of mites was not obtained even in treatment 4. These findings support our present ground cover management practices of leaving strips of vegetation in contact or close to the tree trunk (Croft and McGroarty, 1977). Allowing the weeds or grass to grow into the lower branches of the trees in June would readily allow *A. fallacis* to move into the tree via the trunk or the ground cover contacts with the lower branches.

Data taken from a three year study (1975-77) is given in Fig. 1 which contains three linear regressions and the 95% confidence belts for the relationship of cumulative DD_{54} to the log of the cumulative number of *A. fallacis* per apple leaf. Fig. 1A is the linear regression of the relationship of these factors in an orchard without a dormant spray application. The prey density varied from 40 to 130 red mite equivalents/apple leaf at the time *A. fallacis* colonized the apple tree. A red mite equivalent is defined as the number of red mites plus the number of rust mites/15; it has been reported by Croft (unpublished data) that

TABLE I.
Summary of the Exclusion Experiment Giving the Absolute Density of *A. fallacis* Found on Each Treatment. Data Expressed as Log (Mites/Apple Leaf + 1); Data Followed by the Same Letter(s) Were Not Significantly Different at the .05 level.

Treatment	7/4/78	7/11/78
1 (all)	.52a	.72ab
2 (trunk)	.55a	.79a
3 (air)	.21b	.53ab
4 (none)	.15b	.42 b

one red mite equals 15 rust mites in the diet of *A. fallacis*. This unusually high prey density allowed for colonization by the *A. fallacis* entering the trees around 660 ± 80 DD_{54}. The data points in Fig. 1B came from orchards where a dormant oil was applied. The prey density varied from 2 to 60 red mites equivalents/apple leaf at the time *A. fallacis* colonized the tree. It is apparent that colonization under these conditions occurred 140 DD_{54} later at 800 ± 80 DD_{54} (circled). The data represented by the black squares in Fig. 1B shows an earlier colonization of *A. fallacis* after 550 ± 60 DD_{54}. This earlier colonization is believed to be due to an increased efficiency of detecting *A. fallacis* by col-

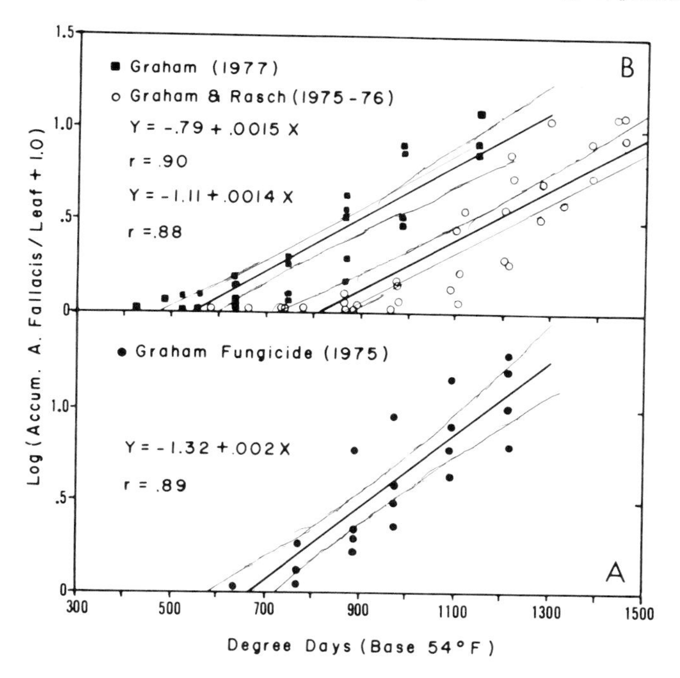

Fig. 1. Summary of 3 years data showing the linear relationship between degree days (base 54°F) and the log transformation of the accumulation of *A. fallacis* per apple leaf up to its peak population. In A, the orchard had no applications of a dormant oil whereas in B, the orchards were sprayed with a dormant oil.

lecting leaves only from the tree interior where *A. fallacis* appear first.

The next step was to use weekly data of DD_{54}, *A. fallacis* density in the ground cover, and *P. ulmi* equivalents/apple leaf to derive a multiple regression to predict the density of *A. fallacis*/apple leaf. It was found that the ground cover density of *A. fallacis* gave the best prediction of the density of *A. fallacis*/apple leaf when prey in the tree were not limiting colonization.

Mid Season Dispersal Studies

During the within-tree interaction between *A. fallacis* and its prey, the stage distribution of *A. fallacis* dispersing from the tree via the air changes as its food becomes limiting. Table II (top portion) describes the percent stage distribution of *A. fallacis* in commercial delicious apple trees throughout the season. The bottom line gives the corresponding density of apple rust mites/apple leaf. Table III shows the percent stage distribution of *A. fallacis* in the air. These percentages refer to the 7 days following the last apple tree leaf counts. On July 26, one week after the peak of the rust mite density, *A. fallacis* was still effectively reproducing on the prey available. Of the predatory mites present in the tree, 14% were larvae, 30% were nymphs, 10% were adult males, and 15% were adult females. Active dispersal at this time was restricted mostly to adult females (75%) and males (19%). It was assumed that the nymphs and larvae (collectively 6%) were passivly being blown off leaves. On August 20, the prey density of 7 mites/apple leaf was limiting reproduction of *A. fallacis* and starvation was occurring locally within the tree. This factor resulted in a change in the stage distribution of *A. fallacis* dispersing. Although the adult female still was the dominant active dispersing stage (50%), a significant percentage of the other stages were assumed to be actively dispersing (e.g. males (19%), nymphs (20%), and larvae (6%)). Table IV il-

TABLE II.
Stage Distribution of *A. fallacis* and Rust Mite Density in Commercial Apple Trees (1976).

Stage[a]	Date (Graham 28)				
	7/12	7/26	8/9	8/20	8/30
E	43.5	37.1	43.8	6.5	0.0
L	13.0	14.3	7.2	0.0	0.0
P-D	13.0	22.9	19.6	23.9	8.0
♂	0.0	10.5	12.9	34.8	4.0
♀	30.4	15.2	16.5	34.8	88.0
Tot. *A. f.*[b]	0.23	2.10	3.88	0.92	0.50
Rust Mite[b]	269.7	198.0	31.9	7.0	0.0

[a] Weekly percentage of total *A. fallacis*/stages/apple leaf; E = eggs, L = larvae, P-D = proto/deutonymphs

[b] Mites/apple leaf/week

TABLE III.
Stage Distribution of *A. fallacis* Dispersing Out of the Apple Tree in the Air (1976).

Stage[a]	Date (Graham 28) 7/19	8/2	8/16	8/30	9/7
L	0.0	2.8	16.7	6.1	0.0
P-D	0.0	3.7	13.9	19.5	11.4
♂	0.0	18.5	21.9	19.3	11.4
♀	100.0	75.0	47.5	55.1	77.1
Tot. *A. f.*[b]	0.45	5.40	39.5	50.92	8.75

[a] Percent of total *A. fallacis* caught that week; L = larvae, P-D = proto/deutonymphs.
[b] Total number dispersing/grease plate/week.

lustrates the seasonal percent of the total number of *A. fallacis* life stages captured on the grease plates. These proportions were larvae (7%), nymphs (22%), adult males (18%), or females (53%).

In order to maximize the colonizing of the dispersing mite landing in a new environment, females are mated immediately after molting and prior to dispersing. Under these conditions, adult females, and to a much lesser extent the males, have a tendency to disperse when food is prevalent. This has both genetic and adaptive advantages for the population. Even when food is prevalent, the mites can spread their individual genetic makeup to other locations. When food becomes limiting, all stages begin dispersing which increases the likelihood that an individual reaches a new site from which the population may survive and reproduce. The results of this study support the statements made by Mitchell (1970) concerning the behavioral adaptations of spider mites.

Dispersal Distance Studies

The last study deals with how far *A. fallacis* disperses from its source population during the season. Table V gives a summary of the relative density

TABLE IV.
Seasonal Percent of Total Population of *A. fallacis* That are Dispersing in the Air as Larvae (L), Proto- or Deutonymphs (P-D), Males (♂), or Females (♀) During 1976 in Two Commercial Apple Orchards in Michigan.

Stage	*Rasch Trees* 112	304	*Graham Trees* 25	28	$\bar{x}$ and S. D.
L	6.4	8.5	5.0	8.8	7.2 ± 1.8
P-D	29.0	32.0	11.6	15.1	21.9 ± 10.1
♂	20.0	15.5	14.6	21.9	18.1 ± 11.9
♀	44.4	43	68.8	54.2	52.6 ± 11.9

of *A. fallacis* dispersing in the air at various distances from the orchard border. Grease plates were set up and monitored weekly. The weeks of July 13th and 20th were periods where food was available for *A. fallacis* in the tree. Normally the density of *A. fallacis* is negligible in the grassy field where the grease plates were positioned (e.g. below 3 mites/10 × 10 cm plate). Data showed a logarithmic decline in the number of mites impacting on the plates relative to the orchard source. Food was limiting in the tree from July 27th on, and it was assumed that there existed a significant density of *A. fallacis* in the grassy field which may also be dispersing onto the grease plates. It was for these reasons that the number of *A. fallacis* captured at the successively greater distances showed some decline during the third week, but were essentially equal during the last sampling period.

SUMMARY

In summary, *A. fallacis* disperse into the tree via the trunk or low branches in contact with the ground cover. Colonization of *A. fallacis* within the tree was found to be affected by cumulative DD_{54}, ground cover density of *A. fallacis,* and the prey density within the apple tree. If food was available, *A. fallacis* colonized the tree after an accumulation of 600 ± 100 DD_{54}. A multiple regression was presented which contained the major factors affecting the early tree densities of *A. fallacis.* Given that food in the tree was not limiting, the major factor appeared to be *A. fallacis* density in the ground cover. During the predator-prey interaction in the tree, the females and to a much lesser extent the males were the active dispersing stages when food was available, but when food became limiting the other stages also actively dispersed. Finally, the density of *A. fallacis* dispersing in the air declines logarithmically at successively greater distances from the orchard source of *A. fallacis.* This dispersal capability enhances its ability to successfully colonize new sites some distance from the orchard.

TABLE V.
Summary of the Density of *A. fallacis* That Dispersed Via the Air to Various Distances from the Orchard Border Given in Mean Number of Mites/10 × 10 cm Grease Plate (Graham Orchard, 1977).

Date	Distance From Orchard in Meters					
	Origin	3.8	8.4	19.1	41.9	72.4
7/13-7/20	24.0a	19.1b	8.4c	5.9c,d	3.3c	3.0d
7/20-7/27	25.8a	19.6b	8.9c	7.6c	4.4d	1.9e
7/27-8/03	27.1b	41.6a	28.1b	23.5b,c	18.5c	9.3d
8/03-8/10	4.9a,b,c	3.5b,c	5.7a,b	5.8a,b	6.7a	3.3c

a-e Represent significant differences at .05 level, with *a* being the largest and *e* being the smallest value.

REFERENCES

Baskerville, G. L., and Emin, P. (1969). *Ecol.* **50**, 514-17.
Croft, B. A., and McGroarty, D. L. (1977). *M. S. U. Agr. Expt. Sta. Res. Rep.* 333. 22 pp.
Lee, T. D. (1972). Ph.D. thesis, Rutgers University, 41 pp.
McGroarty, D. L. and Croft, B. A. (1975). *Proc. N. C. B.-Entomol. Soc. Amer.* **30**, 49-52.
Mitchell, R. (1970). *The Amer. Nat.* **104**(939), 425-431.

LABORATORY STUDIES ON THE FOOD HABITS OF THE PREDACEOUS MITE *TYPHLODROMUS EXHILARATUS*

S. Ragusa

Istituto di Entomologia Agraria
Universita di Palermo
Palermo, Italia

INTRODUCTION

During a survey on phytoseiid mites in Sicily, a new species, *Typhlodromus exhilaratus* (Ragusa, 1977), was found inhabiting various cultivated and wild plants. It was also observed on citrus trees associated with *Panonychus citri* (McGregor) which since 1975 had appeared in our citrus orchards, causing serious damage. The study conducted during 1977/78 at the Istituto di Entomologia Agraria di Palermo, was concerned with the development of *T. exhilaratus* and its oviposition rate on mites and various kinds of pollen.

MATERIALS AND METHODS

Predators were collected from a tangerine plantation infested by *P. citri* and transferred into the laboratory, where breeding was initiated with *Carpobrotus* pollen on black, plastic cages as described by Swirski *et al.* (1970). Data on female longevity and fecundity were obtained by isolating young adults (1-2 days old), counting the number of laid eggs, and replacing dead males. In trials on postembryonic survival every egg (1 day old) was isolated in a 2 cm dia. cage and observed daily. Fresh and abundant diet was added daily. Trials were conducted in a conditioned room at 25-26°C and above 60% R. H.

RESULTS

The effect of various diets on postembryonic survival is given in Fig. 1A. *P. citri* and *Tetranychus urticae* Koch had a positive influence, while *Tydeus* was

Vol. I: ISBN 0-12-592201-9

found less positive. A high percentage of young stages reached maturity on *Carpobrotus, Rosmarinus, Oxalis, Papaver, Eriobotrya, Rosa,* and *Duranta.* Predators were in good condition, with coloured idiosomas and usually living among pollens and prey animals; a lower percentage matured on *Borrago, Bougainvillea,* and *Jasminum. Chorisia* was far from satisfactory and, finally, a negative influence was noted for the remaining tested pollens, on which young attained the protonymphal stage. Predators were thin, pale, nervous, either running around the cage or staying still on the border near the cotton providing water.

The effect of various diets on oviposition rate is given in Fig. 1B. *P. citri* appeared to be the favorite food of adult predators; their daily oviposition rate was the highest, nearly 1 egg/♀/day. *T. urticae* gave lower results (0.48), while *Tydeus* had quite a negative influence.

Among pollens, egg laying was medium to relatively high on *Carpobrotus, Borrago, Rosmarinus, Oxalis, Papaver,* and *Rosa;* it was low on *Jasminum* and *Duranta* and nil or almost nil on the remaining pollens. Maximum oviposition in pollens was on *Rosmarinus* (0.96). All the pollens but *Bougainvillea, Jasminum, Eriobotrya,* and *Duranta,* and the prey mite *P. citri,* showed a positive correlation between postembryonic development and egg production.

Behaviour of predators was the same as in the survival tests, but it should be mentioned that mites were in good condition even on those pollens which had a positive influence on postembryonic survival, but a negative one on the oviposition rate. On pollens which had a positive influence on oviposition rate during the 10 day experiment, it was decided to check female survival and oviposition through the whole lifespan. Trends of curves in Fig. 2A, 2B, and 3 were similar, with no marked differences. The mean number of eggs/female was as follows: 17.4 on *Carpobrotus;* 26.2 on *Borrago* (showing a slightly bet-

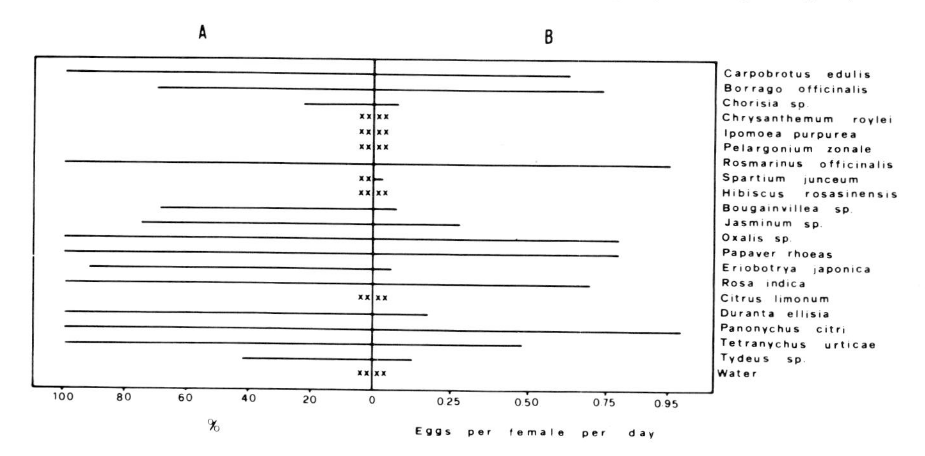

Fig. 1. Effect of various diets on postembryonic survival (A) and oviposition rate (B) of *Typhlodromus exhilaratus.*

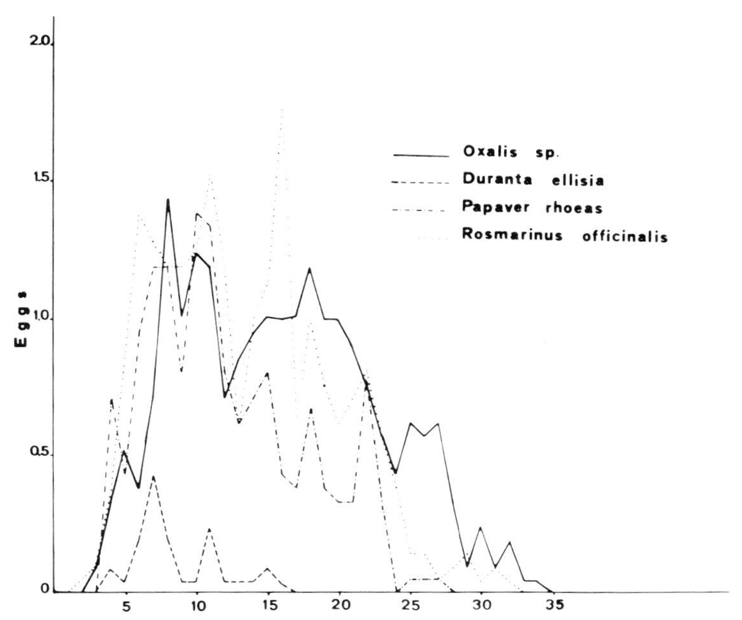

Fig. 2A. Effect of various pollens on the ecological oviposition rate of *Typhlodromus exhilaratus.*

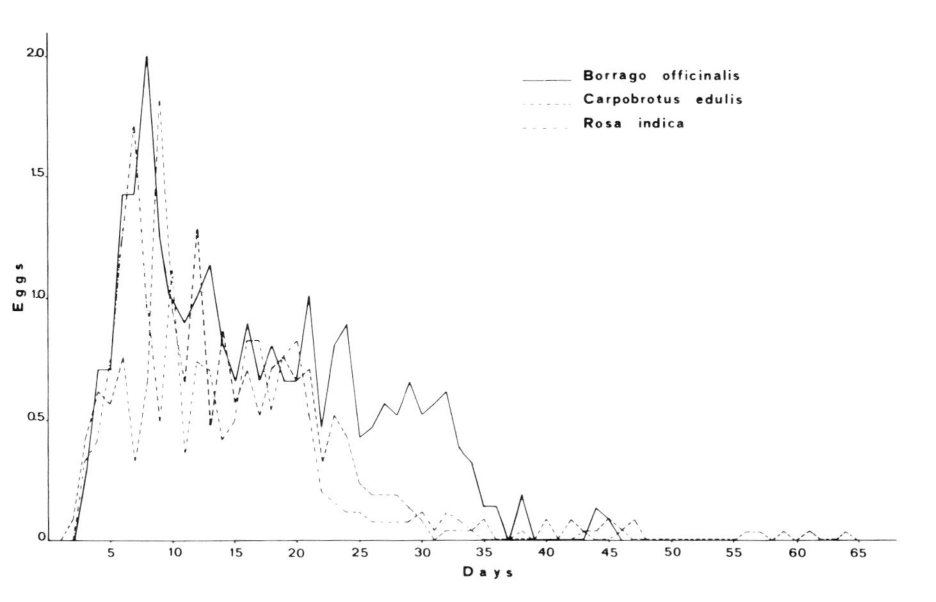

Fig. 2B. Effect of various pollens on the ecological oviposition rate of *Typhlodromus exhilaratus.*

ter effect); 21.24 on *Rosmarinus;* 21.8 on *Oxalis;* 14.67 on *Papaver;* 15.56 on *Rosa;* 1.5 on *Duranta.* Fifty percent of the egg complement was laid in an average of 13 days on *Carpobrotus* and *Papaver,* 15 days on *Borrago* and *Oxalis,* 7 days on *Duranta,* and 12 days on *Rosmarinus* and *Rosa.* Females usually stopped laying after 33-36 days; the maximum of 64 days was observed on *Carpobrotus,* but with quite a low number of eggs after the 30th day. Oviposition of females fed on *Duranta* was low in contrast. It was noted that 50% of survival was between 35 and 53 days (Fig. 3), while on *Ipomea* it was only 7 days, showing in contrast a very short survival on this pollen.

DISCUSSION AND CONCLUSIONS

Several research studies have been carried out on the influence of pollens as alternative food to prey animals (Chant, 1959; McMurtry and Scriven, 1964a; Swirski, Amitai and Dorzia, 1970; Swirski and Dorzia, 1968; Sciarappa and Swift, 1977). Results were usually satisfactory, with the exception of *Phytoseiulus persimilis* Athias-Henriot (Laing, 1968).

Lower oviposition rate of *T. exhilaratus* on *T. urticae* than on *P. citri* may possibly be explained by the abundant webbing produced by *T. urticae.* Predators, impeded in their movements by it, were unable to reach the prey

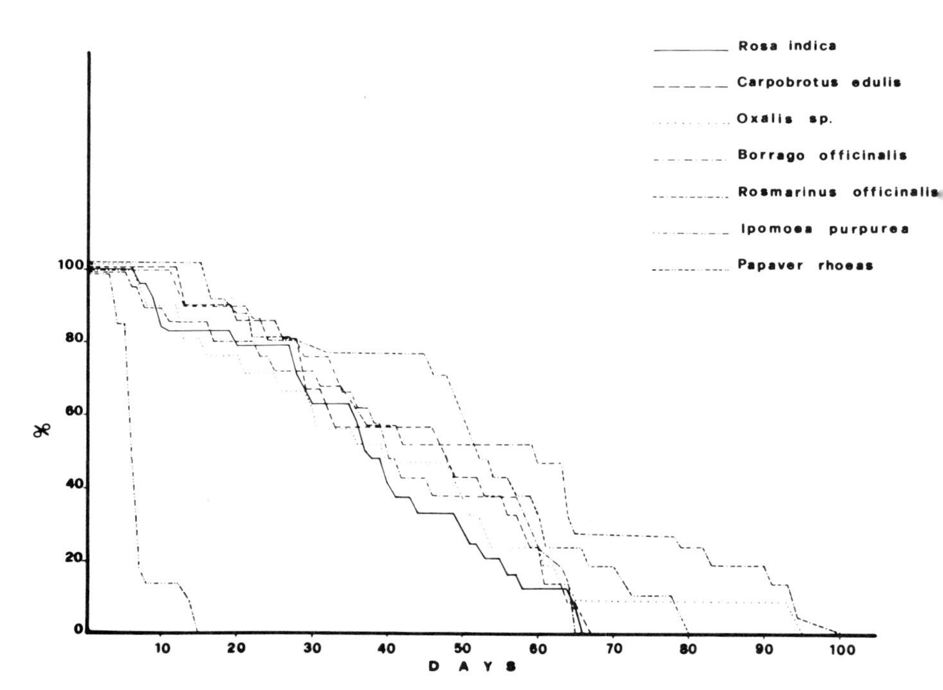

Fig. 3. Survival of females of *Typhlodromus exhilaratus* fed on different kinds of pollen.

and often died. Larvae of *T. exhilaratus* were observed feeding on no other substances but water; 26 out of 35 and 24 out of 25 molted to the protonymph when only water or different stages of *P. citri* were supplied (in the last case, in fact, the idiosoma remained uncoloured). Therefore, considering this parameter, we have the following groups: 1) larvae which do not feed as those of *Typhlodromus rickeri* Chant (McMurtry, and Scriven, 1964a), *Metaseiulus validus* (Chant) and *Typhloseiopsis pini* (Chant) (Charlet and McMurtry, 1977); 2) larvae which can either feed or not feed as *Typhlodromips sessor* De Leon (Sciarappa and Swift, 1977); 3) larvae which need prey in order to molt into protonymphs as *Amblyseius rademacheri* Dosse (Dosse, 1958). Larvae of *T. exhilaratus* are to be included in the first group.

Citrus pollen gave clearly negative results in these tests. Similar results were obtained with *Amblyseius hibisci* (Chant) (McMurtry and Scriven, 1964b) and *Amblyseius limonicus* Garman and McGregor (Swirski and Dorzia, 1968). A low reproductivity and survival was also found for *Amblyseius rubini* Swirski and Amitai and *Typhlodromus athiasae* Porath and Swirski (Swirski, Amitai and Dorzia, 1967a, b), while for *Amblyseius swirskii* Athias-Henriot citrus pollen had a satisfactory nutritional value for postembryonic survival, but not for oviposition rate. *Amblyseius largoensis* Muma gave same results as *A. swirskii* (Kamburov, 1971). *T. exhilaratus* living on citrus trees, in absence of prey, has to look for other kinds of food to survive. Its good survival on other pollens is, therefore, an important factor in that such pollens that can be classified as optimal food (Muma, 1971) are without doubt an important alternative food in absence or low presence of prey. Therefore, if prey increased, *T. exhilaratus* might have a quick response. It has to be pointed out that this predator was found in the field associated with and preying on red spider mites, and that it reproduced and developed on this diet in the laboratory. However, from the biocontrol standpoint of this interesting Sicilian species, which is also found in other parts of Italy and Crete, additional laboratory tests and especially field tests have still to be carried out.

SUMMARY

The author observed the effect of various kinds of pollen and spider mite on the postembryonic survival and oviposition rate of *T. exhilaratus*. Among mites *P. citri* gave best results with the highest percentage of laid eggs (0.99/♀/day); among pollens the maximum was attained on *Rosmarinus* (0.96/♀/day). On citrus pollen *T. exhilaratus* neither developed nor oviposited. Larvae of *T. exhilaratus* do not feed.

ACKNOWLEDGEMENTS

The author greatly acknowledges Prof. E. Swirski (Volcani Center, A. R. O., Bet-Dagan,

Israel) for his invaluable help and advice; Dr. G. Lucido for preparing some of the drawings and Sig. V. Ciulla for his assistance during laboratory tests.

REFERENCES

Chant, D. A. (1959). *Can. Entomol. Suppl.* **12**, 1-166.
Charlet, L. D. and McMurtry, J. A. (1977). *Hilgardia* **45**, 173-236.
Dosse, G. (1958). *Pflanzenschutzberichte* **21**, 44-61.
Kamburov, S. S. (1971). *J. Econ. Entomol.* **64**, 643-648.
Laing, J. E. (1968). *Acarlogia* **10**, 578-588.
McMurtry, J. A. and Scriven, G. T. (1964a). *Ann. Entomol. Soc. Amer.* **57**, 362-367.
McMurtry, J. A. and Scriven, G. T. (1964b). *Ann. Entomol. Soc. Amer.* **57**, 648-655.
Muma, M. H. (1971). *Fla. Entomol.* **54** (I), 21-34.
Ragusa, S. (1977). *Acarologia* **18**, 379-392.
Sciarappa, W. J. and Swift, F. C. (1977). *Ann. Entomol. Soc. Amer.* **70**, 285-288.
Swirski, E. and Dorzia, N. (1968). *Israel J. Agric. Res.* **18**, 71-75.
Swirski, E., Amitai, S., and Dorzia, N. (1967a). *Israel J. Agric. Res.* **17**, 101-119.
Swirski, E., Amitai, S., and Dorzia, N. (1967b). *Israel J. Agric. Res.* **17**, 213-218.
Swirski, E., Amitai, S., and Dorzia, N. (1970). *Entomophaga* **15**, 93-106.

EVALUATION OF PREDATORY MITES FOR CONTROL OF SPIDER MITES

Mous W. Sabelis

Agricultural University
Department of Theoretical Production Ecology
Wageningen, The Netherlands

Marinus van de Vrie

Research Station for Floriculture
Aalsmeer, The Netherlands

INTRODUCTION

Release experiments with predatory mites, aiming to control spider mites, have provided successful cases of biological control. Observations in the field and laboratory experiments (life tables, functional and numerical response) with many predatory species have formed a basis for the understanding of their regulative capacity. Although the importance of the laboratory observations with respect to the interactions between predator and prey is often suggested, it is scarcely proved by lack of verification. This problem may be solved by a system analysis, offering a quantitative verifiable framework. The ultimate aim of this analysis is to give an explanation of the population fluctuations in the field and to make an interspecific comparison of predatory mites, especially with regard to possibilities for long term or short term control of spider mites.

Because of the non-linear relations underlying the age and state dependent interactions between predator and prey, numerical simulation models should be used as an analytic instrument. The computer programs used are of the state-variable type; state and rate variables are distinguished and mathematical expressions are given to calculate the value of each rate variable from the state of the system. The new state variables are calculated by numerical integration over a short time interval. Examples of state variables are the number of eggs, the gut content, and the webbed area.

Mathematical formulation of the interactive growth processes forces the researcher to measure the biological components in such a way that they fit in

Vol. I: ISBN 0-12-592201-9

the context of the model. Then the importance of the input data, relative to each other, may be determined by studying the overall effect of a standard perturbation of a parameter value in the model. This so called sensitivity analysis gives at the same time an indication of the accuracy with which the input data must be measured, and thus an indication of the number of replicates needed. Finally, the output of the model has to be compared with field experiments to determine the explanatory value of the input data and the structure of the model.

DESCRIPTION OF THE SYSTEM: A CASE STUDY

A greenhouse culture of ornamental roses has been selected for a case study. This substrate for predator-prey interaction consists of leaflets, composed leaves, shoots, bushes and hedges. The rose shoots, rising above the hedges, are cut off at a beginning stage of flowering and it is this part of the crop that must be protected against spider mite infestation. On the other hand, the hedge beneath the economically important part is allowed to have some low degree of infestation. The adult females of the spider mite, *Tetranychus urticae,* are the founders of colonies on the underside of the leaf, near the edge or a rib. The juvenile and adult stages find shelter in a self-produced web, the substrate of the colony. Some of the preovipositional females disperse after mating to other leaves, which obviously results in a compact focus of infestation. Upward dispersal may cause damage to the upper shoots and should be controlled by the action of predatory mites. Those predatory mites species which prefer to stay in the webbed areas disperse from colony to colony, thereby possibly moving from leaflet to leaflet, shoot to shoot, bush to bush, or hedge to hedge. Predatory mite species without such a preference may disperse to parts of the plant other than the webbed areas, e.g. the area near the ribs.

HIERARCHICAL MODELING

This structure of the system is suitable for a hierarchical approach. Firstly, attention should be focused on the predator-prey interaction in the mite colony, and secondly on dispersal. The structure may be further subdivided as follows:

1) rate of encounter of the predator with its prey in the mite colony
2) predation and reproduction rate of an individual female predator at a certain density of a prey stage (egg, larva, proto- and deutonymph, adult male and female, and the different moulting stages)
3) interactions between the population growth of the predator and that of the prey in the mite colony

4) dispersal of the prey and that of the predator, both in relation to predator and prey density and the morphological heterogeneity of the crop.

To keep the final model of the complex interactive growth processes within limits, each hierarchical level is modeled separately; the results of a hierarchical lower model are used in a more descriptive form in a higher order model (constant, one-, two- or three-dimensional functions, etc.). The same hierarchical approach applies to the dispersal level with regard to the degree of heterogeneity included in the model; is it necessary to include dispersal of the predator between the colonies *or* is it sufficient to consider this dispersal between bigger entities, like infestation focuses?

In the following text the essence of models will be discussed, thus offering a framework for the evaluation of predatory mite species.

RATE OF ENCOUNTER

The rate of encounter E of a predator with its prey is determined by the diameter of both predator and prey and their walking velocity. Assuming that the walking directions of the mites are mutually independent, Skellam (1958) showed that:

$$E = (\phi_{prey} + \phi_{predator}) . v . D_{prey}$$

D is the density (number/cm^2) and ϕ is the diameter (cm); v is the resultant velocity of the velocities of predator and prey (cm/min) and is given by:

$$v^2 = v^2_{prey} + v^2_{predator} - 2.v_{prey}.v_{predator}.\cos\Theta$$

where Θ is the angle between the momentary directions. Taking expectations we have:

$$\Sigma v^2 = \Sigma v^2_{prey} + \Sigma v^2_{predator}, \text{ since } \int_0^{2\pi} \cos\Theta \, d\Theta = 0$$

However complications arise when the predator has a tortuous walk, as with most predatory mites in webbing, and every prey contacted is removed (eaten). Then it is felt intuitively that the rate of encounter is depressed. To test this hypothesis, a numerical simulation model has been made which simulates the walking behaviour of predator and prey on basis of the following assumption: that the direction of each step deviates from the direction of the previous step with an angle chosen at random from an experimentally defined frequency distribution, autocorrelations being accounted for. The animals are represented as a circle and the contacts between the circles are registered. The effect of a tortuous walk on the rate of encounter is studied in situations where every prey contacted is removed.

A basic assumption in these calculations is the absence of any form of remote sensing, as suggested to be the case by Mori and Chant (1966). However it may be that the web threads interfere with searching by acting as a warning signal for the prey or by drawing the attention of the predator to active prey. The absence of these effects has been shown by comparison of an encounter experiment and calculation with Skellam's formula. This result means that predatory preference for a prey stage or escaping capacity of the prey may, for computational purposes, be considered to arise at the moment of contact.

The activity of predator and prey and the coincidence in space are also of importance with respect to the rate of encounter. The activity is defined as the percentage of the observation-time the animal is walking. The coincidence in space accounts for the phenomenon that predator and prey in the web may pass over and under each other. Multiplication of the coincidence in space with the sum of the rates of encounter in the three activity combinations of predator and prey resp. (walk-walk, walk-rest and rest-walk) gives the average rate of encounter.

All possible relations with the satiation deficit of the predator, webbing density, and contact frequency with predator or prey are studied and supplied to the predation model in the form of constants or functions.

PREDATION AND REPRODUCTION RATE OF A FEMALE PREDATOR IN RELATION TO PREY DENSITY

Only some of the encounters result in predation, depending on the gut content of the predator. The gut content changes continuously as a consequence of food intake and digestion. Holling's disc equation does not suffice because it accounts only for the effect of the feeding time on the time available for catching more prey. Therefore Fransz (1974) constructed a numerical simulation model, in which the handling of the prey and the food intake are treated separately from searching for prey. (See also Curry and DeMichele (1977) for an application of queuing theory to the problem of the functional response of a predator to prey density.) The number of prey encountered by the predator during a time period t follows a Poisson distribution, if the prey is distributed at random over the surface. The probabilty of zero encounters is then equal to:

$$P(E_t = 0) = e^{-E \cdot t}$$

Within a short time interval the probability of an encounter during the search is equal to $1 - P(E_t = 0)$. Replacing E by the number of successes per time unit (S = success ratio × E), the probability of a success during a short time interval is determined. The success ratio may now be related to the gut content, which is simulated by integration of the ingestion and digestion rate.

The digestion rate is equal to the relative digestion rate, multiplied by the actual gut content. The ingestion rate is proportional with the satiation deficit of the gut or the amount of food in the prey. The relative digestion rate is measured as the relative breakdown of β-carotenoids in the gut, which process is visible as a colour decrease through the transparent body wall (Rabbinge, 1976). The relative ingestion rate is obtained from the uptake of P^{32}-labeled prey content by the predator. The maximum amount of food in the various prey stages and the maximum gut content of the predator are determined by weighing of hungry predators before and after feeding, on an electrobalance.

The relation of the success ratio, activity and feeding time with gut content is obtained from continuous observations of a predator, of which the gut content is estimated by computation of digestion and observation of ingestion.

Simulation with the predation model showed that the effect of temperature on the activity of the prey and the digestion rate of the predator explains ±90% of the temperature effect on predation in the range of 15-30°C. This result allows considerable simplification in experimentation. Separate validation justifies the use of the model output in the model for interactive population growth. The output consists of the relative predation rate (= predation rate divided by prey density) in relation to prey stage, prey density, gut content and temperature, and the prey utilization in relation to the gut content of the predator.

The relation between reproduction rate of the predator and food intake may be simulated by assuming, for reasons of simplicity, that the digested food passing the gut wall is utilized for the maintenance of its weight and for egg production. Weight loss and the predatory egg are again determined by weighing with the electrobalance. The simulated relation between reproduction, food intake and temperature has been verified by comparison with experimental results on numerical response of a female predator.

All measurements for the predation and reproduction model are done with female predators of 3-10 days old. The predation rate at other ages is assumed to be proportional to the age dependent reproduction rate, which has been measured, and thus also proportional to the relative digestion rate which appeared to be the key factor in the reproduction process, the maximum gut content remaining almost constant with age. The predation rate of the juvenile and male stages is estimated on the basis of small scale experiments, because the contribution to total predation is low.

POPULATION GROWTH OF PREDATOR AND PREY

To account for the dependence of reproduction, mortality, sex ratio, etc. on the developmental stage, the number of animals per stage must be monitored at any moment. This is done by means of a boxcar train technique, where the stream of individuals developing from egg to maximum age is subdivided in development classes (boxcars). In principle the content of each class is pushed into the next class after the lapse of appropriate residence time. Dispersion in developmental time is simulated according to a method of Goudriaan (1973), which allows the mimicking of the observed standard deviations of development. Hence, the number of mites per class is monitored by accounting for the inflow from the preceding class and the outflow to the following class. In addition, the mortality is accounted for by means of a relative mortality rate (= $(\ln N_o - \ln N_t)/t$ which can be related to relevant factors. In the case of prey, developmental and relative mortality rates depend on temperature, assuming that relative humidity has a minor influence in the range of 50-90%. Predation rate is added to the mortality rate. In case of population growth of the predator, developmental rate and relative mortality rate depend on the gut content and temperature (R.H.). The relative predation rate depends on the developmental class, prey stage, prey density, gut content of the predator, and temperature (R.H.). The relative reproduction rate depends on the developmental class, temperature (R.H.) and, in the predator, on its weight, intake and relative digestion rate. Newly laid eggs are transferred to the first class of the boxcar train, which closes the developmental cycle.

The average gut content of the predators is again simulated by integration of the ingestion and digestion rates. The ingestion rate however is now equal to the sum of food intake over all prey stages. Food intake per prey stage is equal to the relative predation rate, multiplied by number of prey and predators and by prey utilization.

This model is used without simplification in the dispersal model as a representation of the predatory-prey interactions within a colony. On the very same colony level, verification experiments have been performed.

DISPERSAL OF PREDATOR AND PREY IN RELATION TO PLANT HETEROGENEITY

According to Johnson and Croft (1977) predatory dispersal by wind is improbable below the wind velocity of 45cm/sec, which is much higher than the average wind velocity in greenhouses (= 5-15 cm/sec). Therefore dispersal by walking and, in the case of spider mites, dispersal with the aid of web threads also may be most important.

Preovipositional dispersing spider mites may move on the shoot and start an upward or downward dispersal. The probability of invading a new leaf is equal to the ratio of stalk attachment and the circumference of the shoot (= 50%). If the food conditions are satisfactory, colonization takes place. Measurement of the dispersing tendency from a leaf in relation to the percentage of uncolonized leaf surface is the fundamental input to the model of prey dispersal. This model is validated in a population experiment, where shoots with a fixed number of leaves are taken as a unit to study leaf-to-leaf dispersal. Similar experiments have been conducted to study shoot-to-shoot dispersal. These experiments proved that the dispersal by walking was predominant, so that the number of contacts between shoots (plant distance) is an important factor. This fact, together with the 50% probability of invading a leaf, gives an explanation for the compact character of the infestation focus.

Predatory mites disperse throughout the adult female stage. Those with a preference for the webbed areas leave a colony only at lower prey densities.

After departure, their activity increases and they move to the upperside or to a rib on the underside of a leaf, where their walking velocity becomes much higher. At the shoot, the probability of invading another leaf will not differ from that described for the preovipositional spider mites. Walking along the edge or on a rib, they may find a spider mite colony on the underside of a leaf and in contact with the edge or a rib. The detection of the web threads and/or the presence of kairomones seem to cause the entrance of the predator into the colony. This behaviour causes the predator to spend little time outside the webbed areas if they do not leave the infestation focus, and to stay longer at localities of high prey density. This description of predatory dispersal is true for *Typhlodromus occidentalis* and *Phytoseiulus persimilis*. However *Amblyseius potentillae* differs to a large extent; this predator prefers the neighborhood of the thickest parts of the ribs as a resting place. This is the most important reason for the ineffectiveness of this predator which is known to suppress the non-web producer, *Panonychus ulmi* (Rabbinge, 1976; McMurtry and v. d. Vrie, 1973). Another factor influencing distribution of the predators over the colonies is mutual interference (Kuchlein, 1966). This factor may cause an even exploitation of the unevenly distributed food source.

In the numerical dispersal model the relative departure rate is connected to the prey and predator density. Leaf-to-leaf and shoot-to-shoot dispersal is accounted for in the transition probabilities from each colony to all the others. Because of the exchange of predators between colonies, possibly differing in prey density, gut content classes are introduced in the model besides age classes. After each time interval of integration the changes in age and gut content are calculated per colony and next reclassed to the appropriate classes. Verification of this model is done by comparison of its output with a population experiment in glasshouse culture of ornamental roses.

CONCLUSION

The approach described here is developed as a quantitative framework for the evaluation of four predatory mite species (*Amblyseius potentiallae, Amblyseius bibens, Typhlodromus occidentalis* and *Phytoseiulus persimilis*). Because of the central role of the food supply to the predator in these models, an extension in the direction of the food quality possibly should be a future consideration. The work of Kozai, Goudriaan and Kimura (1978) offers the possibility of connecting the interactive growth model to a model that simulates the leaf temperature distribution in a greenhouse crop.

REFERENCES

Curry, G. L. and DeMichele, D. W. (1977). *Can. Ent.* **109**, 1167-1174.
Fransz, G. H. (1974). *Simulation Monographs.* Pudoc, Wageningen,. 1974.
Goudriaan J. (1973). *Neth. J. agric. Sci.* **21**, 269-281.
Johnson, D. W. and Croft, B. A. (1977). *Ann. Entomol. Soc. Amer.* **69**, 1019-1023.
Kuchlein, J. H. (1966). *Med. rijksfac. Landbouwwet. Gent,* XXXI nr. 3.
Kozai, R., Goudriaan, J. and Kimura, M. (1978). *Simulation Monographs,* Pudoc, Wageningen.
Mori, H. and Chant, D. A. (1966). *Can. J. Zool.* **44**, 483-491.
McMurtry, J. A. and van de Vrie, M. (1973). *Hilgardia IV* **42**(2), 17-34.
Rabbinge, R. (1976). *Simulation Monographs,* Pudoc, Wageningen.
Skellam, J. G. (1958). *Biometrics* **14**, 385-400.

EFFECTIVENESS OF ARTHROPODS IN THE CONTROL OF THE CARMINE MITE

S. Kantaratanakul, K. Sombatsiri and P. Tauthong

Department of Entomology
Kasetsart University
Bangkok, Thailand

INTRODUCTION

The possible value of predators for control of phytophagous mites has been a subject of considerable interest and controversy for many years. The predatory mites of the family Phytoseiidae have received the most recent and widespread attention as controlling agents for Tetranychidae on agricultural crops (Oatman *et al.*, 1968; McMurtry *et al.*, 1970; Oatman, 1970). Certain groups of insect predators have also been reported (Robinson, 1951; Collyer, 1964; Putman, 1965; McMurtry *et al.*, 1970; Lord, 1971).

Although many species of mites and insect predators are recognized as natural enemies of tetranychids, only three native predators of the carmine mite, *Tetranychus cinnabarinus* (Boisduval), are commonly found in Thailand. They are the predaceous thrips, *Scolothrips indicus* Priesner; the phytoseiid, *Amblyseius (Amblyseius) longispinosus* (Evans); and the coleopterous predator, *Stethorus vagans* Blackburn. The laboratory studies reported here were undertaken to gain a better understanding of various interesting factors which determine the effectiveness of these three predator species.

MATERIALS AND METHODS

All experiments were performed in the laboratory at 25 ± 2°C and 55%-65% RH. Large numbers of the mite prey *T. cinnabarinus* at different stages were required daily for predator rearing, as well as for experimental purposes. The prey cultures were therefore established in the insectary on the yardlong bean, *Vigna sesquipedalis* grown in 20 cm dia pots. The heavily infested beans were then used for mass culture of the three predator species.

Gravid *T. cinnabarinus* were placed on a square-inch section of leaf and

Vol. I: ISBN 0-12-592201-9

allowed to oviposit for one day. The females were then removed. The numbers of eggs were counted and the required numbers of active *T. cinnabarinus* were placed directly on the test leaf.

In all tests, except a test for competition among three predator species, a square-inch leaf with a definite prey density was arranged with the upper surface down on wet cotton in a 5 × 7 × 2.5 cm plastic box. The edge of the leaf was coated with a thin film of Tanglefoot® which served as a barrier to prevent escape of the mites and predators from the leaf. Only one predator was randomly selected from the cultures and released on the leaf. Counts were then made at a given period to determine the number of prey consumed. A new leaf with a predetermined prey density was substituted as needed.

TABLE I.
Developmental Stages of Three Predator Species; *Amblyseius, Scolothrips* and *Stethorus.*

Stages	Development (days)		
	Amblyseius	*Scolothrips*	*Stethorus*
Egg	2	5.0	2.9
Larvae			
First instar	1	2.0	1.9
Second instar or N_1	1	2.0	1.3
Third instar or N_2	1	—	1.0
Fourth instar	—	—	1.0
Prepupa	—	1.0	1.0
Pupa	—	2.0	3.0
Adult	39.5	14.40	53.2
Total life cycle	44.5	26.40	65.3

RESULTS AND DISCUSSION

Development and Reproduction

The results in Table I show the developmental stages of the three predator species. The life stages of the phytoseiid mite, *A. longispinosus,* consisted of the egg, six-legged larva, protonymph, deutonymph, and adult. In the laboratory, duration of the egg stage was about 2 days. The larval stage and the two nymphal stages each lasted about 1 day. Mating occurred soon after the adult females emerged, and a preoviposition period of 2-3 days followed. The time required for development from egg to adult was only 5 days.

There were 2 larval instars in the predator thrips, *S. indicus,* each lasting about 2 days. It took 1 and 2 days for prepupal and pupal stages, respectively. The adult lived for an average of 14 days. It took 12 days to develop from egg to adult.

The coleopterous predator, *S. vagans,* developed from egg to adult within 12 days with 4 larval instars. The first stadium lasted about 2 days while the second, third and fourth stadia each lasted 1 day. The prepupal and pupal stages took 1 and 3 days, respectively. Adult longevity was about 53.2 days.

The *Stethorus* sp. had a relatively long oviposition period with a potentially high daily oviposition rate (Table II). *Scolothrips* appeared to have a slightly higher oviposition rate, but a shorter oviposition period than that of the coleopterous predator. The oviposition rate of *Amblyseius* was less than those of the others.

Mating was necessary for oviposition in *Amblyseius* and *Stethorus,* but not in *Scolothrips.* However, fertile female *Scolothrips* produced more eggs than nonfertile females (unpublished data).

TABLE II.
Oviposition Rate of Three Predator Species; *Amblyseius, Scolothrips* and *Stethorus.*

Species of Predator	Eggs/Female	Oviposition Period (Days)	Eggs/Female/Day
Amblyseius	53.7	18.9	2.9
Scolothrips	88.5	11.1	8.1
Stethorus	401.9	48.9	7.5

Feeding Ability

The average prey consumption for *Stethorus* (Fig. 1) showed a linear response to the prey density up to a maximum of 100 eggs per leaf. Maximum egg consumption rate was 26.57 eggs per hour. The average feeding rate of *Scolothrips* and *Amblyseius* was not nearly as great as that of *Stethorus.* The maximum number of eggs consumed was 1.93 and 2.65 eggs per hour for *Scolothrips* and *Amblyseius* respectively. The results also indicated that the consumption rate of *Scolothrips* was not as dependent on prey population density as were those of *Amblyseius* and *Stethorus,* even though it showed a density response when eggs were abundant. The rate of prey consumption decreased at densities of over 60 per leaf. Approximately the same number of eggs were consumed at prey densities of 60, 80 and 100 eggs per leaf. The average numbers of prey consumed would be a factor determining the ability of a predator to obtain its daily food requirements. These studies lead to the possibility that *Stethorus* may be of major consequence as a control agent for *T. cinnabarinus.* It is a voracious egg predator, and has a long life span (Table I).

Feeding Responses to Various Stages of Prey

Results indicate that all stages of the mite prey were readily attacked and consumed. Data in Table III show that all stages of phytoseiids preferred feeding mainly on eggs rather than on the other stages.

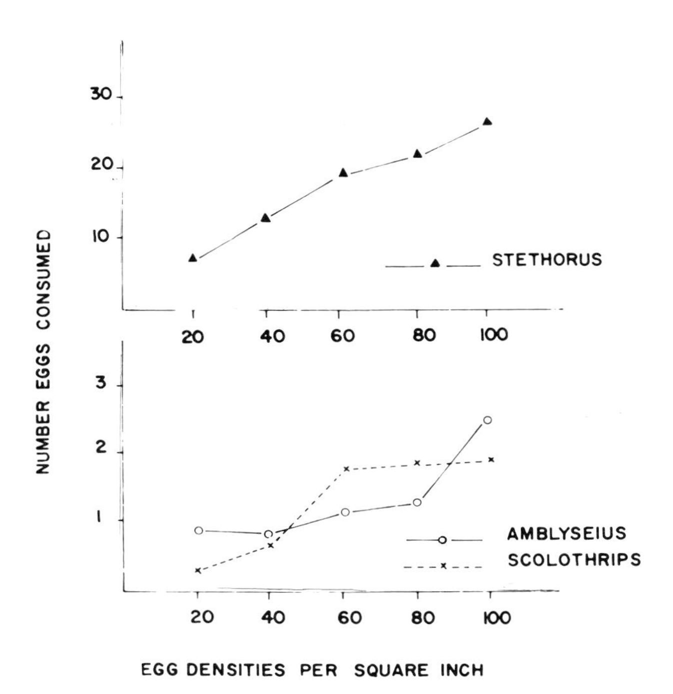

Fig. 1. Consumption rate/adult/hour of three predator species: *Amblyseius, Scolothrips* and *Stethorus* at various egg densities of *T. cinnabarinus.*

In the feeding test for *Scolothrips,* it was observed that, although the larvae did not feed extensively on the mite prey, the adults consumed fairly large quantities. The adult males consumed less prey than the adult females, but rather more than the larval stage. The results in Table III indicate that the adults of predator thrips displayed no difference in feeding response to various stages of the prey. They consumed eggs and nymphs as well as adult males. However, the larvae were found to consume mainly nymphs and eggs of the prey.

A comparison of the results in these experiments revealed that all stages of *Stethorus* could consume a far greater quantity of the mite prey than could the previously discussed species. (Fig. 1). However, it did not feed extensively on mite eggs during the first two instars, but fed on fairly large quantities in the third and fourth instars. Egg consumption was 101.6 and 187.2 eggs per day, respectively. *Stethorus* was also found to feed extensively on eggs rather than nymph or adult prey.

Competition Among the Three Predator Species

When several species of predators are searching for a common prey in a given area, there must be a competition between them for prey. The purpose of

TABLE III.
Average Number of Various Stages of *Tetranychus cinnabarinus* Consumed in Any One Day by Each Stage of Three Predator Species (10 Replicates Per Stage).

Species	Stages	No. Prey Consumed[a]		
		Eggs	Nymphs	Adult Males
Amblyseius	Larva	2.6a	1.7a	0 b
	Protonymph	3.7a	1.7b	1.5b
	Deutonymph	6.5a	3.5a	2.9a
	Adult female	39.6a	4.0b	2.4b
	Adult male	15.2a	5.2b	2.0b
	Total	67.6	16.1	8.8
Scolothrips	First instar	3.9b	5.5a	0.4c
	Second instar	4.0b	5.4a	1.8c
	Adult female	20.8a	20.8a	21.6a
	Adult male	7.2a	7.2a	6.8a
	Total	35.9	38.9	30.6
Stethorus	First instar	35.2a	20.8a	29.6ab
	Second instar	59.2a	46.4ab	28.8b
	Third instar	101.6a	53.6b	35.2b
	Fourth instar	187.2a	76.0b	56.0b
	Adult female	76.8a	42.4b	36.8b
	Adult male	46.4a	56.0a	22.4b
	Total	506.4	294.4	208.8

[a]Numbers in the same rows followed by the same letter are not significantly different at 5% level (Duncan's new multiple range test).

this experiment was to determine which of the predator species, if any, survived most successfully and maintained its own density when competing with other predators at a given uniform prey density.

Five individuals of each predator species were selected at random from stocks of newly hatched larvae and then released simultaneously to a square-inch leaf surface containing 100 eggs of *T. cinnabarinus*. A fresh square-inch leaf with 100 eggs was presented to the predators each day. Surviving predators were counted at daily intervals for five days.

The results in Fig. 2 show that the most efficient predator species from a practical point of view was *Stethorus*. After five days, this species could survive in the greatest number. Even though its consumption capacity was about ten times greater than the other two species (previous experiments), it was rarely found to starve at this restricted prey population density. This indicated that *Stethorus* had the highest degree of competition among predators. It was also found to prey on *Amblyseius* and *Scolothrips*, with more frequency on the latter species. Thus, the average surviving efficiency of *Stethorus* was not decreased, perhaps due to its consumption of the other competitive predators. This permitted the density of *Stethorus* to be maintained even at low prey density.

The predaceous thrip was the least competitive species. Because of its suc-

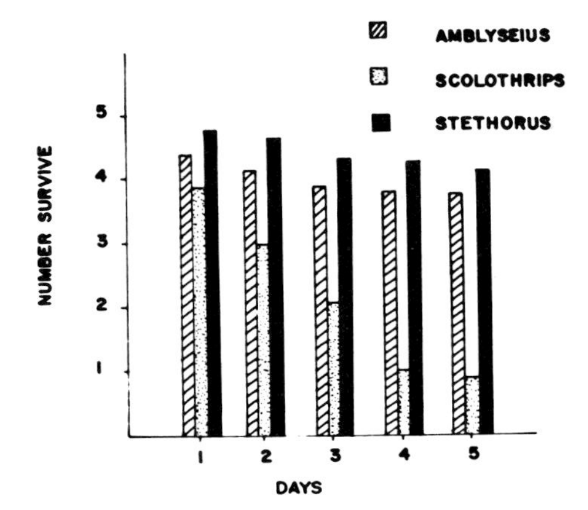

Fig. 2. Average survival rate of three predator species: *Amblyseius, Scolothrips,* and *Stethorus* after preying on 100 eggs of *T. cinnabarinus*/ square inch/day.

culent body and slow movement, it was sensitive to attack by *Stethorus.* Consequently, after a five day experimental period, this species had rarely survived.

The results also show that *Amblyseius* had a high rate of survival even though it was found to be attacked by *Stethorus.* It could develop to adult stage and oviposit several eggs. The advantage of this species may relate to its relatively small size (slightly larger than its prey), rapid movement, and low food demand (2.65 eggs/hr).

SUMMARY

The effectiveness of the three predator species, *Amblyseius longispinosus, Scolothrips indicus,* and *Stethorus vagans* was studied under laboratory conditions by releasing each predator to prey on a specific number of various stages of *Tetranychus cinnabarinus* on a one square-inch leaf surface. *Stethorus* was far more efficient as a predator of *T. cinnabarinus* than were *Amblyseius* and *Scolothrips.* It fed extensively on large numbers of prey because it is voracious with a long life span. The adult fed on an average of 26.57 eggs/hr. The maximum consumption occurred in the third and fourth larval instars. Moreover, this predator species had a relatively long oviposition period, oviposited large numbers of eggs (401.9 eggs/female) and had the highest degree of competition among species. It could also prey on the other two predator species.

Of the three predator species, the development period in *Amblyseius* was the shortest. It took only five days to develop from egg to adult. Its consumption rate was only one tenth that of *Stethorus* (2.65 eggs/hr). It also illustrated a high degree of competition, and could escape and develop to a healthy adult even after attack by *Stethorus.*

Scolothrips seemed to be the least effective predator for control of the carmine mite if *Stethorus* were present. The larvae were found to be consumed by this coleopterous species. Its feeding capacity was less than one tenth that of *Stethorus* (1.93 eggs/hr).

REFERENCES

Collyer, E. (1964). *New Zealand J. Agr. Res.* **7**, 551-568.
Lord, F. T. (1971). *Can. Entomol.* **103**, 1663-1664.
McMurtry, J. A., Huffaker, C. B., and van de Vrie, M. (1970). *Hilgardia* **40**, 331-390.
Oatman, E. R. (1970). *J. Econ. Entomol.* **63**, 1177-1180.
Oatman, E. R., McMurtry, J. A., and Voth, V. (1968). *J. Econ. Entomol.* **61**, 1517-1521.
Putman, W. L. (1965). *Can. Entomol.* **97**, 1208-1221.
Robinson, A. G. (1951). *Ann. Rept. Entomol. Soc. Ontario* **82**, 33-37.

POPULATION DENSITY EFFECTS ON BIOLOGY OF *TETRANYCHUS ARABICUS*, THE COMMON SPIDER MITE IN EGYPT

M. A. Zaher, K. K. Shehata, and H. El-Khatib

Faculty of Agriculture
Cairo University
Giza, Egypt

INTRODUCTION

Population density produces marked and peculiar effects on morphology, biology and physiology of several arthropods. Such effects are of greater significance than generally realized. Uvarov (1921) was the first to state the phase variation in locusts, brought about by crowding. However, according to the literature, crowding produces variable effects on various insect species. On the other hand, relatively little is known concerning mites. Attiah and Boudreaux (1964) found that crowding reduced longevity and oviposition of *Tetranychus urticae* Koch and *T. cinnabarinus* (Boisd.). Similar results were obtained by Kuchlein (1968) on the effect of increasing densities on fecundity of females of *Typhlodromus longipilus* Nesbitt. On the other hand, Rasmy (1972) found that crowding of different densities had no effect on oviposition rate of *T. cinnabarinus*.

Therefore, the objective of the present study was to determine the effect of population density of the common spider mite in Egypt, *T. arabicus* Attiah, on some biological criteria.

MATERIALS AND METHODS

Seedlings of common bean *(Phaseolus vulgaris)* planted in small plastic pots were used as host plants. Newly hatched larvae of the same age were confined to plant leaves in circular arenas 2 cm in dia. and demarcated by barriers composed of a mixture of Canada balsam and citronella and clove oils, at densities of 1, 5, and 15 larvae per circle, and left to develop. Treatments were replicated 8, 10, and 15 times for each of the above mentioned densities. When

Vol. I: ISBN 0-12-592201-9

reaching adulthood, 1, 3, and 5 males were added to each replicate of each density level. Males were allowed 3 days to insure copulation, and then removed. Surplus food was guaranteed by transferring individuals to fresh leaves when needed. Examinations were made twice daily to obtain the duration of immature stages, the oviposition period, and number of eggs. Sex ratio estimate was based on 500 eggs at each density level. Experiments were carried out at a temperature of 24 ± 2°C and a relative humidity of 65 ± 5%.

RESULTS AND DISCUSSION

For studying the influence of population density on mite biology, states of solitude, crowding, and overcrowding were attained by the confinement of 1, 5, or 15 individuals in a fixed area of plant leaf surface, 2 cm in dia. The three biological phenomena observed, i. e. duration of immature stages, oviposition, and sex ratio, were considered as being significant regulators of population trends in nature.

Results (Table I) showed that crowding prolonged the duration of the immature stages (7 days) one day more than solitude (6 days), while overcrowding shortened this period to 5 days. This prolongation of total period of immature stages under crowded conditions might be due to the crowding level being insufficient to give optimum conditions, whereas overcrowding appeared to irritate immatures and to accelerate development so that they reached maturity faster and produced new generations earlier. However, crowding affects the biology of various arthropods in varied ways. For example, total duration of immature stages was shorter under crowded conditions of *Brathra brassicae* (Hirata, 1956), *Plusia gamma* and *Pieris brassicae* (Zaher and Long, 1959), and *Spodoptera littoralis (=Prodenia litura)* (Zaher and Moussa, 1961). On the contrary, this period was longer for *Agrotis ypsilon* (Zaher and Moussa, 1962).

A negative relationship occurred between density levels and the reproduction potential of *T. arabicus* (Table I). Females oviposited an average of 87.4, 50.3, and 15.0 eggs during a period of 11.5, 8.2, and 4.8 days for solitary, crowded, and overcrowded conditions, respectively. This phenomenon was supported by the finding of Attiah and Boudreaux (1964) on *T. urticae* and *T. cinnabarinus* and also of Kuchlein (1968) on the phytoseiid mite, *T. longipilus*. Nevertheless, by crowding females of *T. cinnabarinus* at different densities, Rasmy (1972) found no significant differences in egg oviposition. However, such results were obtained by rearing adults only, whereas in the present investigation population density levels were measured during both immature and adult stages. This difference might indicate that crowding of immatures rather than of adults was the main factor affecting egg oviposition later in the adult stage. However, this point needs more clarification.

Crowding also showed a significant effect on sex ratio as male percentage of offspring increased with increased density. This percentage reached 11.0%,

13.6%, and 47.0% for solitary, crowded, and overcrowded conditions, respectively. The high percentage of males in overcrowded conditions might have been due to irritation caused by close contact which in turn reduced mating opportunities so that greater numbers of parthenogenetic males were produced. Thus, it could be concluded that population density had a negative relationship with oviposition and a positive relationship with male reproduction. This finding might explain the sudden decrease in mite populations in nature following high peaks. Thus, it could be stated that *T. arabicus* population under natural conditions tends to regulate its density.

TABLE I.
Effect of Population Density on Various Life Phenomena
of the Spider Mite, *T. arabicus,* at 24 ± 2°C and 65 ± 5% RH.

	Number of Mites/Fixed Leaf Area		
	1	5	15
Duration of immatures (days)	6.0 ± 0.15	7.0 ± 0.03	5.0 ± 0.29
Oviposition period (days)	11.5 ± 1.26	8.2 ± 0.60	4.8 ± 1.54
No. of eggs/female	87.4 ± 3.43	50.3 ± 12.86	15.0 ± 5.95
Percentage of males	11.0	13.6	47.0

SUMMARY

Three density levels of *T. arabicus;* namely, solitary, crowding, and overcrowding were employed by confining 1, 5, or 15 larvae or adult females to common bean leaves in a circular arena of tanglefoot 2 cm in diameter. Crowding prolonged the duration of immature development compared to that for solitary mites, but overcrowding shortened this period. A negative relationship occurred between density level and oviposition. Oviposition averaged 87.4, 50.3, and 15.0 eggs/female during a period of 11.5, 8.2, and 4.8 days for solitary, crowded and overcrowded conditions, respectively. However, male offspring increased with the increase in density, reaching 11.0%, 13.6%, and 47.0% respectively for the three levels tested.

REFERENCES

Attiah, H. H. and Boudreaux, H. B. (1964). *J. Econ. Entomol.* **57**, 53-57.
Hirata, S. (1956). *Researches on Population Ecology 3. Entomol. Lab.*, Kyoto University, Japan.
Kuchlein, J. H. (1968). *Rev. Appl. Entomol.* **56**
Rasmy, A. H. (1972). *Appl. Ent. Zool.* **7**, 238.
Uvarov, B. P. (1921). *Bull. Ent. Res.* **12**, 135-163.
Zaher, M. A. and Long, D. B. (1959). *Proc. R. Ent. Soc.*, (A), **34**, 97-109.
Zaher, M. A. and Moussa, M. A. (1961). *Ann. Entomol. Soc. Amer.* **2**, 145-149.
Zaher, M. A. and Moussa, M. A. (1962). *Bull. Soc. Ent. Egypte,* **46**, 365-372.

THE INFLUENCES OF AGE OF FEMALE *TETRANYCHUS KANZAWAI* ON SEX RATIO AND LIFE CYCLE OF ITS PROGENY

Chaining Thomas Shih

Department of Entomology
National Chung Hsing University, Taichung
Taiwan, Republic of China

INTRODUCTION

The influences of extrinsic factors on population increase and developmental rates of spider mites have been well documented. The important factors are temperature, humidity, photoperiod and light intensity (Boudreaux, 1958; Hobza and Jeppson, 1974; McEnroe, 1961, 1963). The combined effect of humidity and temperature and the effect of optimal temperature on spider mite population densities have been summarized by van de Vrie *et al.* (1972). A photoperiod with short days was found to be crucial to the development of spider mite (Bondarenko, 1950).

At an optimal temperature for development, although life cycles of some species of spider mites are completed in the shortest time, longevity and population increase rate are not their greatest. Instead, these are greatest at somewhat lower temperatures (Boudreaux, 1963). Furthermore, host plant condition and physiology may also induce changes in population density in mites (Henneberry, 1962, 1963; Poe, 1971).

On the other hand, the effects of intrinsic factors (e.g. physiological condition, genetic make-up and age of female parent) on developmental rate, change in population density, and fertilization of spider mites are not well-known.

Females of *Tetranychus kanzawai* Kishida tend to oviposit in a localized area; most of the eggs are produced during a peak period of a few days after a preoviposition period. Mated females produce about 20% males and males are the first to emerge as adults (Shih *et al.,* 1978). However, the proportion of male progeny as influenced by ages of mothers has not been documented for spider mites. Population senescence, which may be expressed as retardation rate of population increase or developmental rate of individuals, is evaluated

Vol. I: ISBN 0-12-592201-9

through complex experiment. However, in practice, an estimation of population senescence by some index, such as sex ratio, age structure or age distribution, may be sufficient for prediction of future population densities.

The present study focuses on the effect of age of female *T. kanzawai* on developmental rate, duration of preoviposition period, life cycle, and sex ratio of progeny, and on the effect of sex ratio on life cycle.

MATERIALS AND METHODS

Spider mites, *T. kanzawai,* from field-infested mulberry at Chung Hsing University, Taichung, Taiwan, were maintained as a stock colony on a 15 cm plant of potted pole bean *(Phaseolus vulgaris).* The mite colonies for observation were reared on excised bean leaves. Primary bean leaves with petioles removed were excised from plants 3-5 days after expansion. Each leaf was placed, top surface down, on an absorbent 80 mm dia cheese cloth cushioned with a pad of cotton, of same diameter and 5 mm thick, in a 90 mm petri dish. The cotton and cheese cloth were wetted daily with water to maintain the leaves' vitality.

Each mite colony consisted of 5-10 newly molted females and 2-3 males (F_1) that were progenies of 3-day old females (P) from the bean plant. Each colony was reared in a separate petri dish. All the mites were reared in a constant temperature and humidity chamber maintained at 25 ± 2°C, 65 ± 5% RH, and a photoperiod of 14 hr light. Five eggs (F_2) were collected from each of 6 colonies each day and these were reared through a complete life cycle. Any excess eggs from the colonies were moved back to the stock colony. Records of duration of each developmental stage, duration of life cycle, and sex ratio of adults were kept for each colony. Observations and assessments were made at 12 hr intervals under 20x magnification. Adult mites (F_1) of each colony were transferred to a new leaf daily or whenever the leaf showed signs of deterioration, until all the females had died. Whenever all the adult males (F_1) in a colony died before all the females, additional males were added from the stock colony on the bean plant. Whenever the mites had molted twice between successive observations, the duration of the unobserved stages were recorded as 0.25 day (= 6 hrs).

The relationship between percentage of males and duration of immature stages or time required to complete the life cycle were estimated by fitting mathematical models to observed data using regression analysis with least square technique.

The development of the models was based on the assumption that the percentage of males in the population correlates positively with durations of immature stages and time required to complete the life cycle. A simplified hypothetical related network is depicted as follows:

Durations of immature stages
or
Time required to complete the life cycle] Percentage of males in the population

These hypothetical relationships may be represented by equations:

$$Y_1 = \beta_0 + \beta_1 (X_1)_i + u_i$$
$$Y_2 = \beta_0 + \beta_1 (X_2)_i + u_i$$

Where

Y_1: Duration of immature stages.
Y_2: Time required to complete the life cycle.
X's: Percentage of male (Male/(Male + Female))
β_0: Estimated constant, i.e. calculated statistic constant.
β_1: The effect or average influence on X's of a unit charge in X.
u: Disturbance term or random error; assumed to be normally distributed with mean equal to 0 and variance equal to δ^2.

RESULTS AND DISCUSSION

The percentage of male progeny from 1-day old mothers was 23.8%. With increasing age of mothers, the male percentage first declined to zero percent, i.e. only female progeny were reproduced when mothers were 4 days old. The male percentage then increased, with heavy fluctuations of up to 30% within 3 to 4-day cycles, until it reached 100%, i.e. no female was reproduced, when mothers were 14 days old (Fig. 1). Most females of *T. kanzawai* copulate only once or twice immediately after emerging as adults and no copulation was observed 24 hr later after emerging.

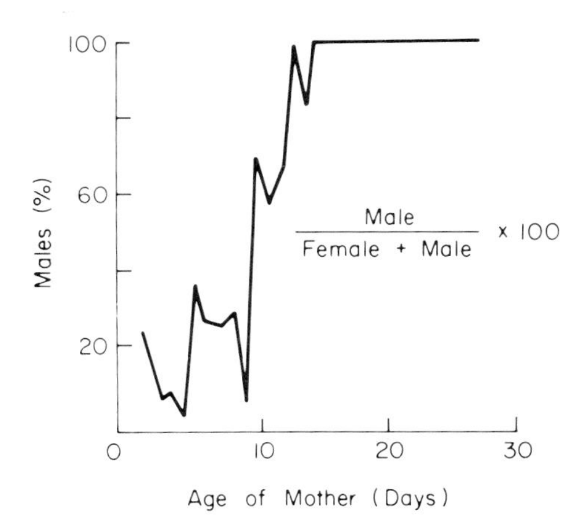

Fig. 1. The influence of age of female *Tetranychus kanzawai* Kishida on percentage of male progeny.

Boudreaux (1963) suggested that the sex ratio of progeny depends on the quantity of sperm the female received at mating. The fertilized eggs of spider mites develop into females, while unfertilized eggs develop into males. Therefore, fertilized eggs, and thus percentage of female progeny, would increase immediately after mating and would become highest within a few days. In addition, the number of fertilized eggs would decrease as the mother aged. When the sperm in the spermatheca were depleted only male progeny would be produced. In the present study, the sperm were depleted when female *T. kanzawai* were 15 days old (Fig. 1).

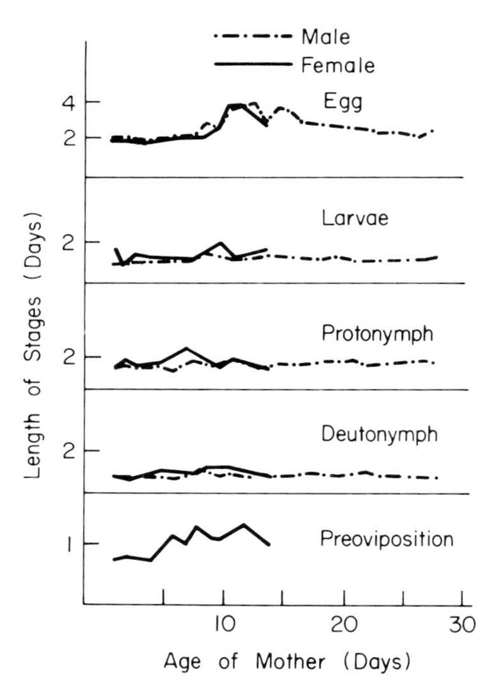

Fig. 2. The influence of female *Tetranychus kanzawai* Kishida on duration of developmental stages and adult preoviposition period in its progeny.

The lengths of time required for the development of immature stages in progeny were affected by the age of the female parents (Fig. 2, 3). Durations of egg and adult preoviposition stages lengthened with increase in age of female parents and these stages were longest when the progeny was from mothers 12 days old (Fig. 2). The durations of larval, protonymphal, and deutonymphal stages of the progeny apparently were not influenced by the age of mothers (Fig. 2). The durations of all immature stages together and the life cycle of progeny increased with the age of mothers and were longest when they were from 11 days old female parents (Figs. 3, 4).

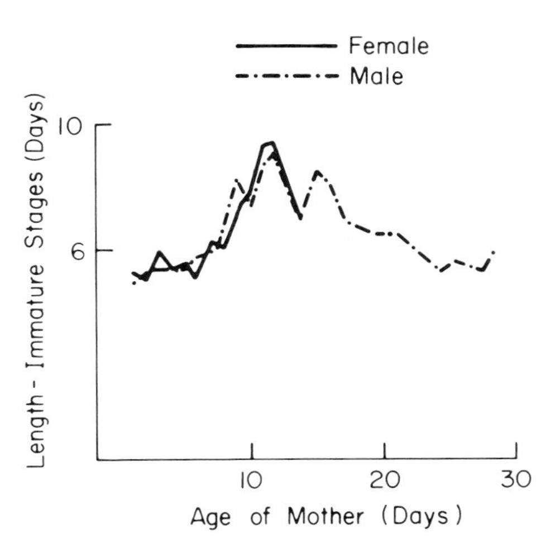

Fig. 3. The influence of age of female *Tetranychus kanzawai* Kishida on duration of immature stages of its progeny.

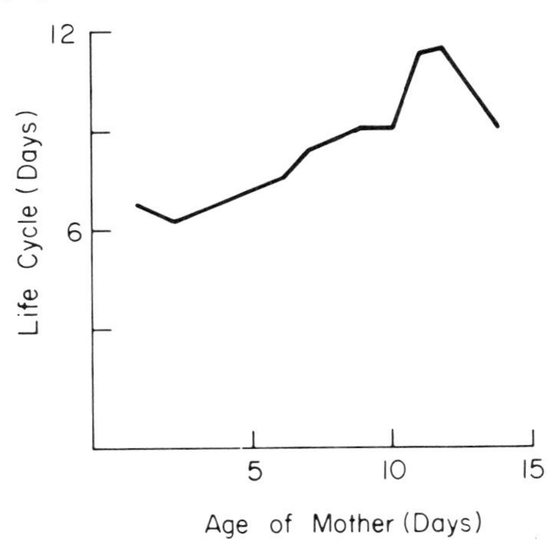

Fig. 4. The influence of age of female *Tetranychus kanzawai* Kishida on the life cycle of its progeny.

Linear regression analysis using least squares produced 2 regression lines that depict some relationships of the duration of immature stages and the lengths of time required to complete a life cycle with percentage of males in the population (Figs. 5, 6). These two linear regression lines are:

Model 1. The duration of immature stages.

$Y_1 = 4.67 + 0.048\ X$

$R^2 = 0.97$

$P = 0.001$

Where Y_1: Duration of immature stages.
X: Percentage of males in the population.

Model 2. The lengths of time required to complete a life cycle.

$$Y_2 = 6.25 + 0.052\,X$$
$$R^2 = 0.96$$
$$P = 0.001$$

Where Y_2: The lengths of time required to complete a life cycle.
X: Percentage of males in the population.

The lengths of time required to complete a life cycle and duration of im-

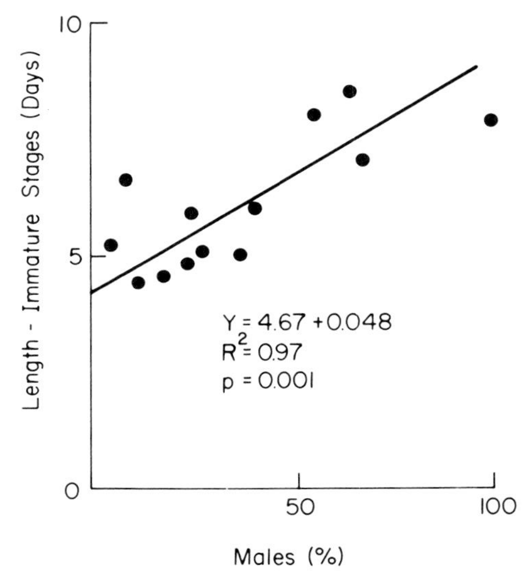

Fig. 5. Relationship between percentage of males and duration of immature stages in populations of *Tetranychus kanzawai* Kishida.

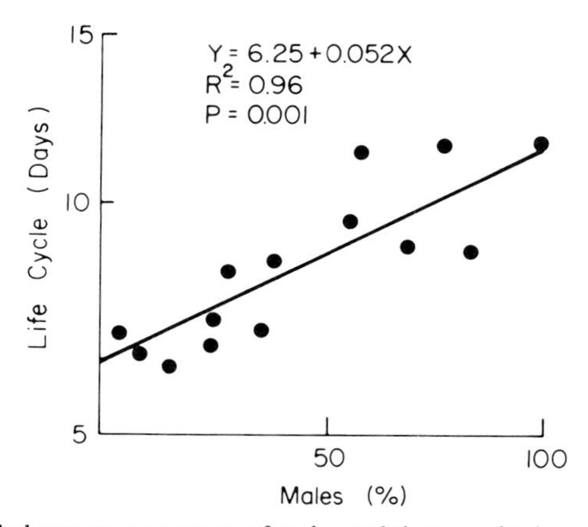

Fig. 6. Relationship between percentage of males and time required to complete a life cycle in populations of *Tetranychus kanzawai* Kishida.

mature stages are positively correlated with percentage of males in the population. Population increase rate decreased as the developmental time of immature stages and time for completing a life cycle increased. In conclusion, percentage of male or sex ratio of a population may be used as an indicator for predicting population developmental rate and population senescence of the spider mites.

SUMMARY AND CONCLUSIONS

The influence of age of female *Tetranychus kanzawai* on life cycle, percentage of male, and developmental rate of its progeny were studied in order to understand the effect of female age on development of population.

1) The duration of immature stages and life cycle of progeny were positively correlated with age of mothers.
2) Only females were produced in progeny of four day old mothers.
3) Percentage of males increased as age of mothers increased, and reached 100% when the mothers were 15 days old, i.e. no female progeny were produced by mothers 15 days old or older.
4) The relationships of percentage of males (PM) in a population with duration of immature stage (DIS) and life cycle (LC) were explained by two regression lines: DIS (days) = $4.67 + 0.048 \times (PM)$ ($R^2 = 0.97$, $P = 0.001$), and LC (days) = $6.25 + 0.052 \times (PM)$ ($R^2 = 0.96$, $P = 0.001$).
5) It was concluded that sex ratio of a population may be used as an indicator for predicting population developmental rate and population senescence of spider mites.

REFERENCES

Boudreaux, H. B. (1958). *J. Insect Physiol.* **2**, 65-72.

Boudreaux, H. B (1963). *Ann. Rev. Entomol.* **8**, 137-154.

Bondarenko, N. V. (1950). *Dokl. Akad. Nauk. SSSR.* **70**, 1077-1080.

Henneberry, T. J. (1962). *J. Econ. Entomol.* **56**, 503-505.

Hobza, R. F. and Jeppson, L. R. (1974). *Environ. Entomol.* **3**, 813-822.

McEnroe, W. D. (1961). *Ann. Entomol. Soc. Amer.* **54**, 883-887.

McEnroe, W. D. (1963). "Advances in Acarology." (J. A. Naegele ed.) Comstock Publ. Assoc. Ithaca, New York.

Poe, S. L. (1971). *Fla. Entomol.* **54**, 183-186.

Shih, C. I., Huang, S. M., and Shieh, J. N. (1978). *Plant Protection Bull.* **20** (In press).

van de Vrie, M., McMurtry, J. A., and Huffaker, C. B. (1972). *Hilgardia.* **41**, 343-432.

BIOECOLOGY AND BEHAVIOUR OF YELLOW JUTE MITE

Abdul K. M. F. Kabir

Division of Entomology
Bangladesh Jute Research Institute
Dacca, Bangladesh

INTRODUCTION

The jute mite *Polyphagotarsonemus latus* (Banks) also occurs on tea in Ceylon and is commonly known as yellow tea mite. In recent years it has become a very destructive pest of jute in Bangladesh. It attacks both the cultivated species of jute, *Corchorus capsularis* L. and *Corchorus olitorius* L. The infestation normally appears on the apical leaves and as a result the plants become stunted in growth. The yield is reduced to the extent of about two maunds of dry fibre per ecre.

P. latus was first recognized as a pest of cotton in Congo in 1936 where the symptons were known as 'acariose' (Ingram, 1960). In 1945, the mite caused damage to cotton in northern Uganda. It was widespread and common, but its distribution was sporadic even within plots and its seasonal incidence was variable (Devies, 1963).

MATERIALS AND METHODS

Jute plants were grown in earthen pots and the seedling plants were brought to the laboratory. All the leaves were removed from the plant except the first or second leaf. In some cases the apical half of the selected leaf was cut off with scissors. Each leaf was examined carefully under the binocular microscope and the mite-free plants were selected for the rearing of the jute mite. Single plants were held in glass tubes containing tap water. Each plant was inoculated with an adult male and female *P. latus* collected from the field. A fine camel hair brush was used for transferring the mites on the leaf. Ten plants were inoculated at a time. The plants thus inoculated were examined at one hour intervals. As soon as eggs were laid on a leaf, the adult mites were

Vol. I: ISBN 0-12-592201-9

transferred to another plant. This process was repeated as long as the female mite lived. The number of eggs laid, the time of oviposition, the incubation period and the duration of life stages were recorded. Total number of eggs laid by a female in her lifetime, and the longevity of male and female mites, also were determined.

For studying the sex ratio, a healthy plant was inoculated with a female nymph and one adult male. When the leaf was found to carry some eggs, the male and female mites were transferred to a fresh plant. This procedure was followed daily as long as the female mite lived. All the eggs laid by a female were kept under constant observation. The sex ratio was determined by counting the number of males and females developed from 50 females.

Populations of yellow jute mite were studied in both *C. capsularis* and *C. olitorius* fields from the beginning of June to harvesting time. Three localities and four plots were selected in each field and five 2′ × 2′ samples were taken from each plot. The total number of plants per sample was counted and the number of mite-infested plants per sample was determined. The number of leaves infested per plant was counted to ascertain the degree of infestation. Weekly observations were made from the selected fields. The mite-infested plants were placed under the following categories:

Very low—from one to a few mites per leaf.
Low —where several mites were present on a leaf.
Medium —where mites were found on a number of leaves of a plant occupying a greater surface area on the leaf.
Heavy —where mites were present in large numbers and the entire leaf was attacked.

Four cultivated varieties of jute (*C. capsularis*—D154 and C6; *C. olitorius*—C. G. and 04) were studied to find out the varietal susceptibility to jute mite. Seeds of these varieties were sown in randomized plots (5 ft x 5 ft) with a spacing of 2 ft between the plots. Each treatment was replicated 6 times, as was the control. The plants were checked weekly and varietal susceptibility was determined according to the degree of infestation.

Life History

The adult male and female mites differ in size and shape. The male is smaller than the female. The body of the male is broadest near the third pair of legs, and gradually tapers towards the posterior end. It normally holds the posterior end of the body above the body surface. Many hair-like projections arise from the tip of the fourth pair of legs. The adult male measures on an average 120μm in length and 70 μm in breadth. The colour of the newly emerged male is whitish or pinkish with a white stripe on the dorsal surface. When mature it turns yellow.

The female is oval in shape. The dorsal surface is curved and the ventral surface is flat. There is a distinct dusky white band corresponding to the mid-dorsal line of the body. Newly emerged female is translucent, and pinkish in colour. The full grown female is brown and measures 180 μm in length and 90μm in breadth. The longevity of the male varies from 5 to 9 days while that of the female varies from 8 to 13 days.

Eggs are laid singly in shallow depression on the ventral surface of young jute leaf. The egg is white, translucent and oval in shape measuring 80μm long and 60μm wide at the broadest region. The dorsal surface is sculptured with round, white tubercles arranged in 5 to 6 rows. There are 7 to 8 tubercles in the middle row and the total number varies from 29 to 37 (Kabir, 1975). The ventral surface of the egg is flat and remains firmly adhered to the leaf. The minimum and maximum number of eggs laid by a single female in her lifetime was recorded to be 30 and 76 respectively. The egg hatches into a larva in about 30 to 36 hrs. The larva has three pair of legs and a distinct white stripe on the dorsal surface of the body. It is sluggish and begins to feed on the plant sap immediately after hatching. The larval stage lasts from 13 to 17 hrs after which it enters into a quiescent "pupal" stage (Gadd, 1946). At this stage the body becomes spindle-shaped. These quiescent forms settle along the midrib, lateral and sub-lateral veins of the leaf and remain inactive until the time of moulting. Female pupae often are carried by the precocious adult males. Young females are round and measure an average 150μm in length and 90 μm in breadth. They have four pair of legs and a dusky white band on the dorsal surface. When they come in contact with males they stop moving and remain stationary for copulation. Mated females produce both males and females while unmated females produce only male offspring. Progeny obtained from laboratory-reared females showed a sex ratio of 1:4.

The minimum number of eggs laid by a female mite was found to be 30 while the maximum number was 76 during the oviposition period of 7 and 11 days respectively. Gadd (1946) reported that the largest number of eggs laid by a female mite was 56 in 12 days. The incubation period was found to be 30 to 36 hours and the mites became adults in 60 to 74 hrs. Kundu *et al.,* (1959) found that the incubation period varied from 27 to 32 hrs. According to Gadd (op. cit.) the eggs of yellow tea mite hatched in about 2 to 3 days and the mites became adults in 2 to 3 days after hatching.

The number of generations of yellow jute mite was studied under laboratory conditions and as many as 29 generations were found throughout the jute season. There is no previous report regarding its number of generations.

Nature of Damage

All stages of the mite are found on the young leaves of jute, mainly on the lower surface. Mites attack the young apical leaves extending downwards to the tenth leaf. Following the infestation the leaves curl down, their colour changes to coppery and purplish, and finally dry up and fall. In case of severe infestation, the apical leaves display a sort of burning effect from a distance. Ultimately the internodes of the affected shoot become shortened and side branches develop, leading to loss in yield and deterioration of the fibre quality. Moutia (1958) reported that the presence of *P. latus* was revealed by the sudden curling and crinkling of leaves followed by blister patches. Kundu *et al.*, (1959) gave an account of the nature of damage caused by *P. latus*. They stated that due to its infestation, the leaves cannot grow to their proper size and turn coppery brown and drop prematurely.

Late in the season, the mite attacks the flower buds and young seed pods of the seed crop. The infested flower buds cannot bloom properly and the colour changes from greenish to blackish. Young seed pods fail to develop to their normal size. Badly infested plants do not bear fruit and give very poor yield.

Seasonal History

The initial infestation of yellow jute mite occurs in mid-May and the population reaches peak periods at the end of June and again in late July. However, there is considerable variation from year to year. Heavy rainfall reduces its population considerably. The years of most serious mite infestation of jute are those of dry periods prevailing in June and July (Kabir, 1975). There is a relationship between the population of yellow jute mite and the season. It has been found that high temperature and semi-humid conditions play an important role in the development and population build-up of this species.

P. latus also attacks a number of food plants, such as potato (*Solanum tuberosum*), tomato (*Lycopersicum esculentum*), egg plant (*Solanum melongena*), chili (*Capsicum frutescens*), mango (*Mangifera indica*), papaya (*Carica papaya*) and tea (*Thea sinensis*). Although Bangladesh produces large quantities of tea, the mite has not yet been found on tea in that country. The population of *P. latus* becomes quite heavy on garden plants during the winter season. During the off jute season, the mite is found abundantly on garden plants, such as dahlia (*Dahlia jaurezee*) and *Jasminum duplex*.

Varietal Susceptibility

The susceptibility of *capsularis* varieties (D154 and C6) was compared with that of *olitorius* varieties (C. G. and 04) over a 3 month period (June-September) in 1976. It was found that the *capsularis* varieties were more susceptible than the *olitorius* varieties. When the susceptibility of D154 was compared with that of C6, 04 and C. G., it was found that there was no significant difference in susceptibility between D154 and C6. However, the differences are significant between D154 and 04 at 5% level. A highly significant value (3.540**) as obtained when the susceptibilty of D154 was compared with C. G. The average (percent) mite infestation was: D154, 63.1; C6, 49.5; 04, 41.0 and C. G., 29.3.

SUMMARY

The yellow mite of jute, *Polyphagotarsonemus latus* (Banks), is a very destructive pest of jute in Bangladesh. It attacks both the cultivated species of jute, *Corchorus capsularis* L. and *Corchorus olitorius* L., and all the stages of the mite are found on the young leaves of the plants. As a result of infestation the leaves curl down, the colour changes to coppery or purplish and finally the leaves dry up and fall off. Late in the season the mite attacks the flower buds of the seed crop, and normal development and blooming of flowers are hampered. Ultimately the seed yield is reduced. The initial infestation of jute occurs in mid-May. The population reaches maximum at the end of June and again during the last week of July. Dry periods are most suitable for rapid multiplication of the mite, while damp weather and heavy rainfall are unfavourable. *Capsularis* varieties are more susceptible to *P. latus* than are *olitorius* varieties.

REFERENCES

Davies, J. C. (1963). *Bull. Ent. Res.* **54**, 425-431.
Gadd, C. H. (1946). *Bull. Ent. Res.* **37**, 157-162.
Ingram, W. R. (1960). *Bull. Ent. Res.* **51**, 577-582.
Kundu, B. C., Basak, K. C., and Sarcar, P. B. (1959). "Jute in India." Indian Central Jute Committee, Calcutta. pp. 169-170.
Kabir, A. K. M. F. (1975). *Jute Pests of Bangladesh.* Bangladesh Govt. Printing Press, Dacca. 59 pp.
Moutia, L. A. (1958). *Bull. Ent. Res.* **49**, 59-75.

NAIADACARUS ARBORICOLA, A "*K*-SELECTED" ACARID MITE

Norman J. Fashing

Department of Biology
College of William and Mary
Williamsburg, Virginia

INTRODUCTION

For the past several years, the subject of *r*- and *K*-selection (MacArthur and Wilson, 1967) has received much attention. In general, organisms which are *r* strategists have high reproductive potential, occupy unpredictable environments, and are poor competitors; whereas *K* strategists demonstrate low reproductive potential, occupy more stable environments, and are good competitors. As Pianka (1970) pointed out, these are extremes and a species would generally not be completely *r*- or *K*-selected, but rather would occupy a place somewhere on an *r*-*K* continuum. Most free-living acarid mites that have been extensively studied appear to be on the *r*-selected end of the continuum; however, *Naiadacarus arboricola* Fashing, an acarid mite that occupies water-filled treeholes in the eastern half of North America (Fashing 1975a), is predominantly *K*-selected. I do not intend to discuss the pros and cons of the theory of *r*- and *K*-selection, but rather the objective of this study is to identify those characteristics of *N. arboricola* that make it a *K*-selected species. As pointed out by Force (1974), the attributes of a species or population can be labeled those of an *r* or a *K* strategist only when compared with another species or population. Whenever possible, *N. arboricola* will be compared with other acarid mites that are *r*-selected.

COMPARISONS WITH OTHER SPECIES

The most obvious results of *r*- or *K*-selection are seen in the demographic characteristics of a species. An *r* strategist generally demonstrates the Type III survivorship curve of Slobodkin (1961), such as Boczek and Stepien (1970) found for *Rhizoglyphus echinopus* (Fumouze and Robin). In contrast, *K*-selected species tend to be characterized by a Type I or II survivorship curve.

Vol. I: ISBN 0-12-592201-9

In general, *N. arboricola* approximates a Type I survivorship curve, i.e., most individuals live to be a specific age determined by environmental conditions, and die more or less simultaneously (Fashing, 1975b).

Short life, rapid immature development, early reproduction, and a high fecundity over a short time period are characteristics of *r*-selected species. In contrast, *K* strategists demonstrate long life, slow immature development, delayed reproduction, and a low fecundity over a long time period. Table I compares these demographic statistics for *N. arboricola* with three *r*-selected acarid species. Longevity is much greater in *N. arboricola,* as is developmental time from larva to adult. The period from ecdysis to adult until first reproduction (prereproductive period) is considerably longer, indicating a delayed reproduction. At least part of the delay in reproduction is due to the fact that *N. arboricola* is larviparous and embryonic development takes place within the female (Fashing, 1975a). Most other acarids are oviparous and development takes place outside the parent. Laboratory studies have demonstrated a high degree of egg mortality in oviparous species (Table I), which is virtually eliminated in *N. arboricola* since the egg is protected within the mother until hatching. A characteristic of *K*-selected species is parental care, and larviparity could be placed in this category.

In *N. arboricola,* the reproductive period is quite long and fecundity quite low when compared with the other species in Table I. In fact, the largest number of larvae produced by *N. arboricola* during a 24 hour period was 9, whereas *Caloglyphus anomalus* Nesbitt has been observed to produce 133 eggs in one day and *Caloglyphus berlesei* (Michael) 145. The average number of larvae per day of reproduction for *N. arboricola* under ideal conditions was 2.5, whereas *C. anomalus* produced an average of 35.4 eggs per day and *C. berlesei* 47.15.

TABLE I.
Comparative Life History Data for Four Species of Acarid Mites.
1) Rivard, 1961a, 1961b; 2) Pillai and Winston, 1969;
3) Rodriguez and Stepien, 1970; 4) Fashing, 1975a, 1975b.

		Mean ($\bar{x}$) Days				Eggs	
Species	°C	Longevity	Larva to Adult	Preovi-position Period	Reproductive Period	Mean per ♀	%Mortality
Tyrophagus (1)	20	44.2	9.7	3.1	29.1	358	26.0
putrescentiae	25	27.8	7.3	2.1	23.2	322	44.5
	30	19.0	6.0	1.9	10.8	101	49.5
Caloglyphus (2)	15	—	8.05	3.88	—	—	48.0
anomalus	20	31-45[a]	6.25	2.13	14-28[a]	738	27.8
	25	—	5.42	—	—	—	21.8
	30	—	3.60	1.25	—	—	50.0
Caloglyphus (3) *berlesei*	25	23.6	5.65	1.5	12.5	588	25.0
Naiadacarus (4)	20	214.6	30.23	11.38	55.52	35	0.0
arboricola	25	101.7	17.21	7.33	57.61	132	0.0

[a] range

All of the above demographic characteristics can be summarized in a single statistic, the intrinsic rate of increase (*r*) (Birch, 1948). Under ideal conditions *r*, based on weekly intervals, was found to be 0.813 for *N. arboricola*. Comparative values for other species are 2.030 for *Carpoglyphus lactus* (L.) (Rodriguez and Stepien 1973), 2.295 for *Tyrophagus putrescentiae* (Schrank) (Barker, 1967), and 2.629 for *C. berlesei* (Rodriguez and Stepien, 1973). To view it from another perspective, a population of *N. arboricola* would multiply itself by only 2.25 times in one week, whereas a population of *C. lactus* would multiply 7.614 times, *T. putrescentiae* 9.922 times, and *C. berlesei* 13.86 times. When compared with these species, *N. arboricola* is certainly at the *K* end of the *r*-*K* continuum in regard to reproductive capacity.

Habitat characteristics also differ between *r* and *K* strategists. The habitats of most acarids appear to be transient and unpredictable, such as decaying substrates that can dry out or be used up rather quickly and are subject to wide temperature fluctuations—a characteristic indicative of *r*-selected species. Since their habitat is transient, *r*-selected species tend to be opportunistic and feed on a number of substrates under a wide range of environmental conditions. *T. putrescentiae* provides a good example, since it might be found almost anywhere the humidity is high and mold is growing. *K*-selected species tend to occupy fairly constant and predictable habitats. The water-filled treehole habitat of *N. arboricola* fits this criterion. A treehole, though only rarely formed, usually remains a part of the forest ecosystem for many years. The water temperature seldom rises about 24°C and changes in temperature are gradual. Living in a constant and predictable habitat, *K*-selected species evolve toward specialization to more efficiently utilize that habitat. *N. arboricola* is certainly a specialist since it is found only in water-filled treeholes (Fashing, 1975a).

Since their habitat is unpredictable, species that are *r*-selected tend to have a high tolerance to harsh environmental conditions. This is reflected in *T. putrescentiae* by the wide range of temperatures under which development can take place. It has been reared at temperatures as low as 11.2°C and as high as 33.9°C (Barker, 1967). *K*-selected species generally have evolved a restricted tolerance to harsh environmental conditions since such conditions would seldom occur in a predictable habitat. *N. arboricola* has been successfully reared only at temperatures of 20 and 25°C. At temperatures of 30°C molting is inhibited and death occurs, and a temperature of 15°C allows development only to the tritonymphal stage (Fashing, 1975b).

Another characteristic of *K*-selected species is the ability to exist with limited food resources. The main source of energy flow into the treehole is in the form of leaves from the autumn leaf fall, and this food source is often exhausted by the various treehole organisms by middle to late summer. *N. arboricola* is adapted to withstand these long periods with limited food. Under laboratory conditions, the number of days at 25°C without food required for 50% mortality of tritonymphs is 84, females 49, and males 35 (Fashing, 1976a).

A characteristic of *r*-selected species is their high mobility or ability for dispersal which is an adaptation for utilizing transient and unpredictable habitats. In most acarids the facultative hypopial instar is a dispersal stage and forms only in response to adverse or declining environmental conditions, allowing for dispersal to more favorable areas. Hypopial formation is, therefore, a function of environmental conditions and will be induced anytime the environment declines. Living in a constant and predictable habitat, *N. arboricola* does not form hypopodes in response to adverse conditions, but rather at a time of year (spring) which is optimal for the success of the dispersers. It does not have to chase a "fleeting environment" as do most other acarids, and hypopodes serve purely as dispersal agents for colonizing new treeholes and outcrossing among populations (Fashing, 1976b).

K-selected species are also good competitors. Although direct studies on competition have not been carried out with *N. arboricola,* it appears to be well adapated for competition with other treehole organisms. For example, during a dearth of food which sometimes occurs by late summer and also during low temperatures (winter), immatures develop only to the tritonymphal instar. When adequate environmental conditions return, the tritonymphs transform to adults almost simultaneously. This results in a population of predominantly young adults which, being at its highest reproductive potential, can best compete with other saprophagous organisms for the available food supply (Fashing, 1976a).

SUMMARY

Most free-living species of Acaridae and related families appear to be *r*-selected. *Naiadacarus arboricola,* an acarid mite restricted to water-filled treeholes, is, however, a *K*-selected species. Characteristics which indicate it as a *K*-strategists include a Type I survivorship curve, long life, late onset of reproduction, larviparity, low fecundity over a long time period, slow rate of post-embryonic development, low capacity for population increase (low *r*), restricted and relatively stable habitat, low availability of resources, and specialized dispersal behavior.

REFERENCES

Baker, P. S. (1967). *Can. J. Zool.* **45**, 91-96.
Birch, L. C. (1948). *J. Anim. Ecol.* **17**, 15-26.
Boczek, J. and Stepien, Z. (1970). *Biul. Inst. Ochr. Rosl.* **47**, 267-275.
Fashing, N. J. (1975a). *J. Nat. Hist.* **9**, 413-424.
Fashing, N. J. (1975b). *Acarologia* **17**, 138-152.
Fashing, N. J. (1976a). *Acarologia* **18**, 704-714.
Fashing, N. J. (1976b). *Amer. Midland Nat.* **95**, 337-346.
Force, D. C. (1975). **In** "Evolutionary Strategies of Parasitic Insects and Mites." P. W. Price, ed. pp. 112-129. Plenum Press, New York.

MacArthur, R. H. and Wilson, E. O. (1967). "The Theory of Island Biogeography." Princeton University Press, New Jersey.
Pianka, E. R. (1970). *Amer. Natur.* **104**, 592-597.
Pillai, P. R. P. and Winston, P. W. (1969). *Acarologia* **11**, 295-303.
Rivard, I. (1961a). *Can. J. Zool.* **39**, 419-426.
Rivard, I. (1961b). *Can. J. Zool.* **39**, 869-876.
Rodriguez, J. G. and Stepien, Z. A. (1973). *J. Kansas Ent. Soc.* **46**, 176-183.
Slobodkin, L. B. (1961). "Growth and Regulation of Animal Populations." Holt, Rinehart and Winston, New York.

LIFE CYCLE OF *APLODONTOPUS MICRONYX* (SARCOPTIFORMES: CHORTOGLYPHIDAE) FROM *SPERMOPHILUS TRIDECEMLINEATUS* (SCIURIDAE) IN NORTH AMERICA

Edwin J. Spicka

Department of Biology
State University of New York
Geneseo, New York

INTRODUCTION

Mites of the family Chortoglyphidae are inhabitants of grain, flour, and other plant products in stores and granaries, and in house dust (Zakhvatkin, 1941; Hughes, 1976). Recently, studies by Tadkowski and Hyland (1974) and Fain and Spicka (1977) have shown the close relationship of certain endofollicular deutonymphs (hypopodes) with members of the family Chortoglyphidae.

The purpose of this study is to describe the developmental stages of an endofollicular hypopus, *Aplodontopus micronyx* Fain and Spicka, 1977, with emphasis on previously undescribed stages.

MATERIALS AND METHODS

Hypopodial mites, *Aplodontopus micronyx,* were collected from the tail-hair follicles of a male Thirteen-lined Ground Squirrel, *Spermophilus tridecemlineatus,* from Indiana (Vigo County, Terre Haute, Rea Park) which was snap-trapped on 4 May 1975. The ground squirrel was refrigerated at 4°C for 2 days when the hypopodes were squeezed from the follicles and placed in groups of 10 to 40 in rearing chambers and then cultured, preserved, mounted and described as in a recently completed study (Spicka and OConnor). However, scabs from the tails were added to the yeast in the chambers and the chambers were examined every day.

Vol. I: ISBN 0-12-592201-9

RESULTS

Both active and inert mites were observed in the rearing chambers one day after being set-up. On the 3rd day, most hypopodes were inert. The first tritonymph was seen on the 4th day. By day 7, tritonymphs started to become inactive, and on day 9, the first adult (female) was observed. On day 11, a copulating pair was seen, along with the first egg. The first larva hatched on day 16, and 2 protonymphs were seen on day 23; when the culture was preserved on day 28, eggs, larvae, protonymphs, deutonymphs (from the first day), males, and females were present.

All stages except the hypopus ate both the granulated yeast and the scabs from the tail with alternate bites of their chelicerae.

When copulating, the male faced a direction opposite the female and assumed a dorso-posterior position. His legs III and IV clasped the female dorso-laterally. Legs I and II weakly clasped her opisthosoma postero-ventrally or legs I dangled freely, occasionally touching the substrate. The female was active, and fed and climbed about the chamber; the male was quiescent. Total mating time is unknown; however, a pair of mites generally remained *in copula* during the examination period (5 to 10 minutes) unless disturbed.

Five laid eggs measured 164 μm (155-175) long, 110 μm (100-113) in dia., were translucent, and oval. Only one preserved female was gravid and contained one egg 155 μm long and 104 in dia.

Family Chortoglyphidae Berlese, 1897
(= Aplodontopinae Fain, 1969)
Genus *Aplodontopus* Fain, 1967
Aplodontopus micronyx Fain and Spicka, 1977

Female (Table I): Idiosomal length averages 336 μm (range 300-390), width 245 μm (230-265). Leg chaetotaxy: tarsi 12-11-9-9, tibiae 2-2-1-1, genua 2-2-1-0, femora 1-1-0-0, trochanters 1-1-1-0. Leg solenidiotaxy: tarsi 2-1-0-0, tibiae 1-1-1-1, genua 1-1-0-0.

Male (Table I; Figs. 1-18): Idiosomal length averages 288 μm (range 280-300), width 217 μm (210-225). Penis 63 μm (60-68) long, with a trifurcate base, pointed and recurved posteriorly at its distal extremity and emerging from 2 lateral and 1 smaller posterior membranous flaps. Copulatory sucker on either side of anal slit, with sclerotized base on which is inserted a rounded membranous funnel 18 μm (17-19) in dia. Tarsus IV with copulatory suckers. Leg chaetotaxy like female except tarsi 12-11-9-7. Leg solenidiotaxy like female.

Tritonymph (Table I): Idiosomal length averages 336 μm (range 300-390), width 245 μm (230-265). Genital slit 13 μm (12-15) long, located between coxae IV, and with 2 small, bullet-shaped genital acetabula on each side. Leg chaetotaxy and solenidiotaxy like female.

Deutonymph (Table I; Figs. 7-12): Idiosomal length averages 295 μm (range 280-310), width 175 μm (145-210). Two large bullet-shaped genital acetabula emerge from beneath genital plate through terminal incision. Leg chaetotaxy:

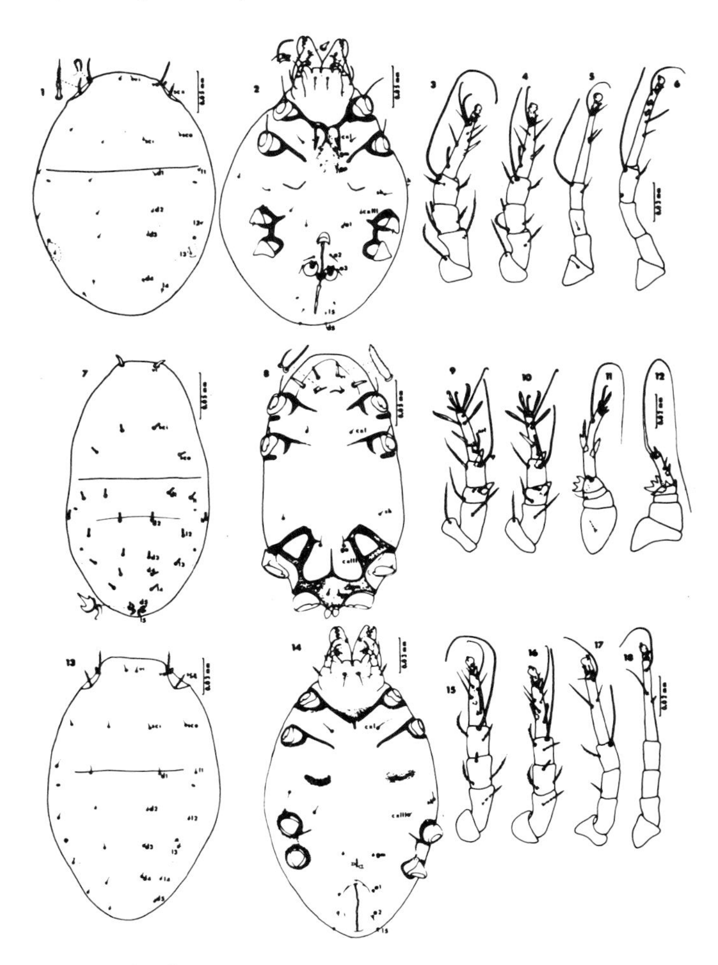

Fig. 1. Dorsum of male.

Fig. 2. Venter of male.

Figs. 3-6. Legs I-IV of male (left to right).

Fig. 7. Dorsum of deutonymph.

Fig. 8. Venter of deutonymph.

Figs. 9-12. Legs I-IV of deutonymph (left to right).

Fig. 13. Dorsum of protonymph.

Fig. 14. Venter of protonymph.

Figs. 15-18. Legs I-IV of protonymph (left to right).

TABLE I.
Aplodontopus micronyx Fain and Spicka, 1977: Length of Idiosomal Setae of Females (7 Specimens), Males (6), Tritonymphs (9), Deutonymphs (10), Protonymphs (2), and Larvae (10). Average Length and Range in μm.

	Female	Male	TrN	DN	PrN	Lrv
vi	9(7-10)	8(7-8)	7(5-7)	16(15-18)	6-7	6(5-7)
ve	36(32-41)	30(27-33)	23(19-32)	15(14-15)	18-20	16(13-20)
sci	10(9-12)	8(7-9)	9(8-10)	8(8-9)	7-7	6(5-7)
sce	10(8-11)	9(8-9)	9(8-10)	8(8-9)	6-7	7(6-7)
scx	20(19-22)	20(19-22)	13(9-15)	—	12-12	10(8-11)
d1	9(8-9)	8(7-8)	9(7-10)	9(8-9)	7-7	7(6-7)
d2	10(9-10)	7(6-9)	8(7-8)	9(8-9)	7-7	6(6-7)
d3	9(8-9)	8(7-8)	8(6-9)	9(8-9)	7-7	6(6-7)
d4	8(8-9)	7(6-8)	8(7-9)	9(8-10)	7-7	7(6-7)
d5	8(7-8)	8(7-8)	7(6-8)	9(8-10)	7-7	—
11	9(8-10)	8(7-9)	8(6-10)	9(8-10)	7-7	6(6-7)
12	8(7-8)	8(7-8)	7(6-8)	9(8-9)	7-7	6(6-7)
13	7(6-8)	7(7-8)	7(7-9)	9(8-10)	7-7	7(6-7)
14	8(7-9)	8(7-9)	7(6-7)	9(8-10)	7-7	7(6-7)
15	8(7-8)	8(8-10)	6(6-7)	6(5-7)	8-8	—
h	9(8-10)	7(7-8)	8(7-9)	9(8-10)	7-8	7(6-7)
sh	9(8-10)	8(7-10)	7(6-8)	12(10-13)	8-8	7(7-8)
cxI	30(28-35)	25(22-28)	18(15-21)	12(11-14)	15-17	14(11-16)
cxIII	17(15-19)	18(15-22)	12(11-13)	7(6-7)	12-15	9(8-11)
ga	14(12-16)	—	—	11(9-14)	—	—
gm	11(10-12)	14(12-16)	8(5-10)	6(6-7)	8-10	—
gp	16(14-18)	14(11-17)	8(6-9)	14(13-15)	—	—
a1	8(6-9)	14(13-16)	5(4-6)	—	5-6	—
a2	8(7-9)	16(15-17)	5(4-6)	—	5-6	—
a3	7(5-7)	14(13-16)	—	—	—	—
a4	6(4-7)	—	—	—	—	—
a5	6(5-7)	—	—	—	—	—

Like female except tarsi 8-7-7-5. Leg solenidiotaxy like female.

Protonymph (Table I; Figs. 13-18): Idiosomal length 210-225 μm, width 140-150 μm. Genital slit 8 μm long, located between coxae IV and with 1 small bullet-shaped genital acetabulum on each side. Leg chaetotaxy: like female except tarsi 12-11-9-7, tibiae 2-2-1-0, trochanters devoid of setae. Leg solenidiotaxy like female except tibiae 1-1-1-0.

Larva (Table I): Idiosomal length averages 155 μm (range 140-165), width 113 μm (108-125). Genital slit, suckers, legs IV, and organs of Claparede absent. Leg chaetotaxy: tarsi 11-11-9, tibiae 2-2-1, genua 2-2-1, femora 1-1-0. Leg solenidiotaxy: tarsi 1-1-0, tibiae 1-1-1, genua 1-1-0.

DISCUSSION

Aplodontopus micronyx has been removed from the family Glycyphagidae and placed in the family Chortoglyphidae, with the subfamily Aplodontopinae being relegated to synonymy with Chortoglyphidae (Fain and Spicka, 1977). However, *Chortoglyphus arcuatus* (Troupeau) (= *Chortoglyphus nudus* Berlese) is not known to have a hypopodial stage.

Aplodontopus micronyx developed from deutonymphs through adults to protonymphs in 23 days. Tadkowski and Hyland (1974) reared hypopial *A. sciuricola* from the Eastern Chipmunk, *Tamias striatus,* to tritonymphs in 45 hours but molting times from tritonymphs to adults were inconsistent.

Larval *Aplodontopus micronyx,* as compared to the female, lack a genital opening, legs IV, and the following setae: d5, 15, ga, gm, gp, anals 1-4, and 1 tarsal I. Larvae also lack the omega 3-I solenidion. Protonymphs lack ga, gp, anals 3-5, 2 tarsal IV setae, and 1 tibial seta. Solenidion phi IV also is absent. The heteromorphic, endofollicular deutonymph lacks mouthparts (replaced with a palposoma). The scx, anals 1-5, 4 tarsal I, 4 tarsal II, 2 tarsal III, and 4 tarsal IV setae are missing. The tritonymphs lack the ga and anal 3-5 setae. Males differ from females in genitalia and size (males are smaller). Also, they lack the ga, anal 4-5, and 2 tarsal IV setae. However, 2 copulatory suckers are present on tarsus IV and most likely are modified setae as seen in the family Pyroglyphidae (Wharton, 1976).

Since only the deutonymphs are found in the tail-hair follicles of the Thirteen-lined Ground Squirrel, it is assumed that the remaining stages occur freely in the nest, where detritus and fungi probably are eaten. When deutonymphs develop, they probably attach to the host, climb down the hairs, and force themselves into the follicles. Parasitism of tissue fluid is probable, as other endofollicular hypopodes have been found to enlarge after entering follicles (Lukoschus *et al.,* 1972). If the host abandons a nest, the mites can be perpetuated by the subsequent development of reproductive stages from the hypopodes in the new nest.

Hypopodial mites have adults which usually are free-living forms in rodent and insectivore nests (Fain, 1969). In one group, setae became modified for hair clasping to aid in attachment to mammal hairs (Fain, 1967). However, Fain (1969) believes some hair-clasping hypopodes moved down the hairs, eventually forcing themselves into the follicles with the subsequent loss of hair-claspers and development of short, heavy, barbed spines for anchorage. *Aplodontopus micronyx* fits well into Fain's hypothesis as an advanced species modified for follicle life.

DIAGNOSIS

Male *Aplodontopus micronyx* are distinguished from male *A. sciuricola* by the form of the copulatory suckers of tarsus IV. In *A. micronyx* they are well developed and funnel-shaped, while in *A. sciuricola* they are short and thin setae. Additional diagnostic characters for the female and hypopus are given by Fain and Spicka (1977).

SUMMARY

In vitro culturing of deutonymphs of *Aplodontopus micronyx*, squeezed from tail-hair follicles of the Thirteen-lined Ground Squirrel, *Spermophilus tridecemlineatus*, produced tritonymphs after four days. Tritonymphs cultured on active dry yeast and tail-scabs produced adults by the ninth day. The first egg was laid on day 11, larvae hatched by day 16 and protonymphs developed by day 23. Adults are similar to those of *A. sciuricola* and *Chortoglyphus arcuatus*.

REFERENCES

Fain, A. (1967). *Ann. Mus. R. Afr. Cent.* no. 156, 1-89.
Fain, A. (1969). *Bull. Inst. Roy. Sci. Nat. Belg.* **45**(33), 1-262.
Fain, A. and Spicka, E. J. (1977). *J. Parasitol.* **63**, 137-140.
Hughes, A. M. (1976). Ministry of Ag., Fish. and Food. *London Technical Bull.* 9, 400 pp.
Lukoschus, F. S., Fain, A., and Driessen, F. M. (1972). *Tijdschr. Entomol.* **115**, 326-339.
Spicka, E. J. and OConnor, B. M., (In press). *Acarologia.*
Tadkowski, T. and Hyland, K. E. (1974) *Proc. IV Intern. Cong. Acarol. 12-19 August 1974.*
Wharton, G. W. (1976). *J. Med. Entomol.* **12**, 577-621.
Zakhvatkin, A. A. (1941). *Zool. Inst. Acad. Sci. USSR new ser. 28* (English translation, 1959).

THE ECOLOGY OF RHAGIDIIDAE

Miloslav Zacharda

Department of Systematic Zoology
Charles University
Prague, Czechoslovakia

INTRODUCTION

Mesoedaphic mites belonging to the family Rhagidiidae are considered predators of other edaphic invertebrates (Evans *et al.*, 1961; Krantz, 1970). As to their distribution the family seems to be cosmopolitan. Only the fauna from the tropics has not been reported up to the present but it probably exists. North America and Central Europe have many common species, e. g. *Rhagidia weyerenensis* (Packard), *R. pratensis* (C. L. Koch), *R. saxonica* Willmann, *R. mucronata* Willmann and *Coccorhagidia pittardi* Strandtmann. It seems that there are no principal differences in the framework of Palearctic fauna, but faunas of Subantarctic and Antarctic regions are unique and different. No austral species have been found in the Palearctic region up to now.

Rhagidiids usually are found in the uppermost layers of the soil profile and most species are considered hemiedaphic. Some species, however, are strikingly adapted morphologically to life within the soil and therefore they can be considered euedaphic. Many rhagidiids are troglophilous, preferring moist, dark and cool habitats, but genuine troglobitic species are relatively scarce. Rhagidiids are known to occur even under extreme conditions of an alpine zone in high mountains and in Arctic and Antarctic tundra (Strandtmann, 1967, 1971). They may be found also in littoral on the sea shore (Schuster, 1958). Rhagidiids are always scattered in their habitats, rather than clustered (Strandtmann, 1971). No detailed data have been gathered regarding their ecology except for brief comments on their autecology in taxonomic papers. Our objective was to study seasonal occurrences of particular species and their developmental stages.

The life cycle of Rhagidiidae comprises an egg, a calyptostatic prelarva, an elattostatic larva, a proto-, deuto-, and tritonymph and an adult (Ehrnsberger, 1974). However only nymphs and adults are found in practice because they are movable so that they can be extracted by means of a Tullgren apparatus. The

Vol. I: ISBN 0-12-592201-9

larva is movable only for a few days after hatching, then it spins a cobwebby moulting nest in the soil (Ehrnsberger, *ibid.*). Therefore the larvae are found very rarely.

I investigated the occurrence of the developmental stages in rhagidiid communities on Oblik-Hill in the volcanic territory Ceske stredohori-Mountains: Czechoslovakia, Bohemia bor., Oblik-Hill, 7 km N. to Louny-Town, 509 m, inclinations of slopes 30-50°; nephelinitic basanite and its moulderings create a basis of substratum. Mean annual temperature 8.5°C, mean annual total of precipitation 500-550 mm. Phytogeography—district of Central European thermophilous vegetation (*Pannonicum*) (Dostal, 1960); here it is a grassy xerotherm steppe. The material of rhagidiids was collected on six sites in the season from April 13, 1974 to May 7, 1975. The following are characteristics of the particular sites of collection:

PL_1—A summit plateau of the hill: a moist habitat with a mean annual soil moisture of 50% (annually fluctuating in the range of 33-62%), mull-ranker soil type, plant community—*Carici humilis—Festucetum sulcatae,* cover 100%.

Note: The data on soil moisture were taken from Rydlo (1973) which were obtained from measurements taken in 1971-1972. The data apply to a dry and a moist period on Oblik-Hill, respectively. The dry period was found to be from 7/7 to 30/10/1971, the moist period from 22/3 to 19/5/1972. The data on plant communities and soil types were taken from Studnickova and Studnicka (1975).

V_1—Eastern slope, about 10 m below the summit with a mean annual soil moisture of 36% (17-39%), moder-ranker soil type, plant community as in PL_1, cover 90%.

V_9—Eastern slope, the foot of the hill, about 90 m below the summit with a mean annual soil moisture of 35% (19-41%), ranker soil type, plant community as in PL_1, V_1, cover 100%.

J_1—Southern slope, about 10 m below the summit: a mean annual soil moisture of 35% (16-37%), moder-pararendzina soil type, plant community—*Erysimo crepidifolii-Festucetum valesiacae,* cover 80%.

Z_1—Western slope, about 10 m below the summit with a mean annual soil moisture of 25% (13-29%), soil type and plant community as in J_1, cover 80%.

N—South-eastern slope, about 40 m below the summit; mean annual soil moisture is 28% (13-30%), moder-ranker soil type, plant community as in J_1, Z_1, cover 70%.

METHODS AND MATERIALS

The application of pitfall traps proved to be the best method for collection of a sufficient number of rhagidiids on observation sites. Almost no rhagidiids were found when I used a usual soil sampling cylindrical tool and desiccated soil samples in the Tullgren funnels.

The pitfall traps were small jars approximately 7 cm in diameter and containing 3% formaldehyde solution. The jars were sunk to a ground level of the soil. Ten jars were always placed randomly (Lewis and Taylor, 1967) in a 5 x 5 m square on each site; the catch was collected after 14-60 days.

The evaluation of species dominancy was based on work at Krogerus (1932). A standard unit C. E. U. (catch per effort unit) expressing a number of individuals caught in only one pitfall trap per day was introduced. As the absolute value of the C. E. U. was always very low, it had to be multiplied by a co-efficient K = 1000.

RESULTS

The species communities of rhagidiids are almost the same on all the sites. I have found *Rhagidia diversicolor* (C. L. Koch), *R. ruseki* sp. n., *R. pratensis, R. mucronata, R. osloensis* Thor, *Robustocheles robusta* gen. n., sp. n., *Evadorhagidia oblikensis* gen. n., sp. n., *Latoempodia macroempodiata* gen. n., sp. n., and *Shibaia vulgata* gen. n., sp. n. On site J_1, however, only one species *Thoria brevisensilla* gen. n., sp. n., has been found. (New taxa not described here.)

The dominant species are *R. diversicolor, R. osloensis* and partly also *E. oblikensis.* The others are influent or recedent. The sex ratio of all the species is strikingly disproportionate:

Species	Total no. of individuals	Females	Males
R. osloensis	44	43	1
R. diversicolor	499	499	—
E. oblikensis	103	103	—
R. ruseki	25	22	3

(No males have been found in the other species.)

There is a striking phenomenon common for the occurrence of the developmental stages of the majority of particular species on all the sites: the nymphs occur only in autumn and then they are replaced by the adults later in winter and in the beginning of the next spring. The adults successively disappear so that in summer they cannot be found at all. The life cycle of *Thoria brevisensilla* is shifted slightly to spring when the nymphs occur as late as from January to the end of March and the adults are abundant in April, whilst the adults of the other species have almost disappeared by that time. I have found in the dominant species *R. diversicolor, R. osloensis* and *E. oblikensis* that the time succession of the particular developmental stages is accompanied by quantitative changes of their occurrence in the population (Fig. 1).

However, not all of the rhagidiid species found on Oblik-Hill have their adults occurring only in winter or in the beginning of spring. For instance, scarce adults of *R. mucronata* were found unusually early (at the beginning of autumn) and the adults of *R. pratensis* and *S. vulgata* in August and June, respectively. Upon listing and comparing dates of collections of these and many other species originating from various localities in Czechoslovakia, all developmental stages in many species, e. g. *R. pratensis, S. vulgata* or *Coccorhagidia clavifrons* (Canestrini) were found to occur throughout the year. On the other hand, the adults of *Rhagidia gigas* (Canestrini) occur at the end of summer, in autumn and in winter, while the adults of *R. mucronata* can be encountered only from summer till autumn.

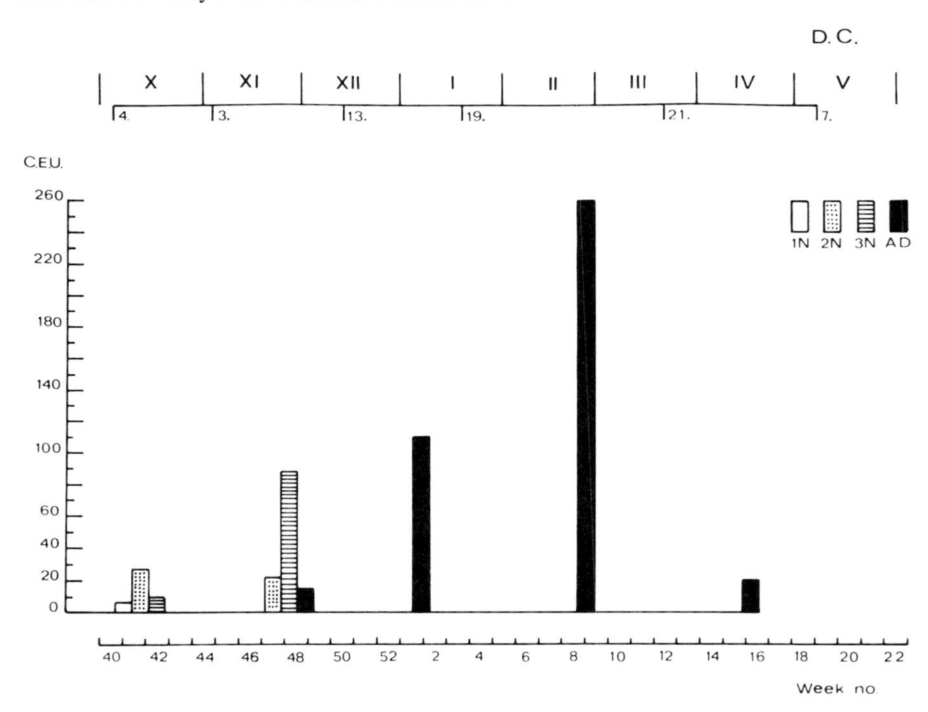

Fig. 1. Seasonal occurrence changes of the developmental stages in *Rhagidia diversicolor* on the site V_9. 1N—protonymph, 2N—deutonymph, 3N—tritonymph, AD—adult, C. E. U.—catch per effort unit, D. C.—date of collection.

DISCUSSION

The life cycles of the dominant rhagidiid species inhabiting Oblik-Hill are season-dependent. These species have not been found at all in summer when the soil moisture is low. I must stress that these species have been found only exceptionally, never dominantly, in different moister habitats in Czechoslovakia where they could eventually find more favourable conditions all through the year. Ehrnsberger (1977) described a similar season-dependent development in *Rhagidia danica* Ehrnsberger. In this species the parthenogenetic females occurred only in autumn and in spring whilst none of them could be found in summer. The breeding of this species (Ehrnsberger, ibid.) has revealed that there is a genuine egg-diapause which lasts 4 months at 18°C. The phenomenon of season-dependent occurrence of particular stages of most species on Oblik-Hill can be similarly explained. The populations of these species survive the hot and dry summer time either in the egg-diapause (most probably) or as small euedaphic larvae or nymphs within the soil.

The environmental factors affecting the life cycles in Rhagidiidae are probably various. Ehrnsberger (1977) found season-independent development in the parthenogenetic species *R. pratensis* and *R. reflexa* (C. L. Koch) which commonly occur in the forest litter and mosses in moist habitats. He suggested that these species were affected by temperature, fluctuations of which were only facultative in those habitats, so that a continuous development was possible. However, the explanation may be more complicated. For instance, I have observed *R. gigas* with its season-dependent life cycle occurring together with *R. pratensis* and *R. reflexa,* (i. e., the species with the season-independent life cycles) in the same moist forest habitats and at the same time for a few seasons.

Tritonymphs and adults of *R. diversicolor, R. osloensis* and *E. oblikensis* were collected on the very dry site Z_1, whilst younger nymphs of the same species have been found on the other sites at the same time. The nearest such site is PL_1. I have supposed that the tritonymphs and adults found on site Z_1 were immigrants originating from PL_1.

The disproportional occurrence of the sexes, when the males are scarce or missing, indicates parthenogenesis. Ehrnsberger (1977) found thelytoky and coincidental absence of males in a few rhagidiids. In the bisexual species, however, the sex ratio was always 1:1 in the natural populations.

SUMMARY

Some rhagidiids are polyvoltine and the occurrence of their developmental stages is season-independent. On the other hand, however, some species are monovoltine with strictly season-dependent life cycles. Especially the adults of species inhabiting dry habitats, e. g. xerotherm grassy steppes and rocky dry steppes, occur predominantly in winter whilst they disappear in spring and summer. Parthenogenesis is quite common in the natural populations of many species. Pitfall traps are very suitable for rhagidiid collection.

REFERENCES

Dostal, J. (1950). *Sborn. Cs. spol. zemep.* **65**, 193-202.

Ehrnsberger, R. (1974). *Osnabrucker Naturw. Mitt.* **3**, 85-117.

Ehrnsberger, R. (1977). *Acarologia* **19**, 67-73.

Evans, G. O., Sheals, J. G., MacFarlane, D. (1961). "The Terrestrial Acari of the British Isles." Vol. I. 219 pp. London.

Krantz, G. W. (1970). "A Manual of Acarology." 335 pp. O. S. U. Book Stores, Inc., Corvallis, Oregon.

Krogerus, R. (1932). *Acta zool. fenica* **12**, 1-308.

Lewis, T. and Taylor, L. R. (1967). "Introduction to Experimental Ecology." 401 pp. Academic Press, London and New York.

Rydlo, J. (1973). (In Czech, unpubl., a thesis of Dept. of Botany, Charles University, Prague).

Schuster, R. (1958). *Vie et Milieu* **9**, 88-109.

Strandtmann, R. W. (1967). *Antarct. Res. Ser.* **10**, 51-80.

Strandtmann, R. W. (1971). *Pacif. Insects* **13**, 75-118.

Studnickova, J. and Studnicka, M. (1975). *Sborn. Severoces. Mus., Ser. Natur., Liberec* **7**, 3-27.

EFFECT OF BURNING CAST CROPS ON SOIL MICROFAUNA WITH AN EVALUATION OF THE FOLLOWING CROP

Mohsen Shoukry Tadros

Plant Protection Department, Faculty of Agriculture
Tanta University
Kafre El-Sheikh, Egypt

INTRODUCTION

Effects of wild fires on fauna are more severe and longer lasting than mild ones. Their effect depends upon the intensity of the fire and susceptibility of the site to erosion. Most investigators indicated that these wild fires often destroy much of the forest floor. Moreover, a reduction occurs in meso and microfauna due to heat and suffocation, or destruction of their foodstuff and soil porosity.

Some of the many investigators who have discussed the relationships between fauna and fire are Heyward and Tissot (1936), Buffington (1957) and Lussenhop (1976) from the U.S.A.; Hill *et al.*, (1975) from Canada; Jahn *et al.*, (1950, 1951) from Austria; Huta *et al.*, (1969) and Karppinen (1957) from Finland; Forssland (1951) from Sweden; and others.

In Egypt, farmers always burn crop residues on the spot after harvesting. This process is done so as to lessen transport expenses; moreover, they believe that fire kills the majority of injurious insect stages occurring either in the remaining plant stems or in soil, and that the ashes left after burning act as a fertilizer to the following planted crop.

The objects of the present investigation were:

1) To record the effect of mild fire in two cast crops (i.e. *Orisa sativa* and *Triticum sp.* (Indian), plus *Imprata cylindrica,* a wild grass grown on field margins and drainages) on soil fauna, and to measure the effect of mild fire on fauna in the disturbed and undisturbed soil layers.
2) To find out the effect of fire on soil structure and its components.
3) To measure the time needed for fauna to recover to its normal level.
4) To compare the effect of two main fertilizers with that of the ashes left after the burning process, on the cultivated crop.

Vol. I: ISBN 0-12-592201-9

MATERIALS AND METHODS

To evaluate the distribution of soil fauna and the effect of fire on it, two plots of about 1/8 feddan (1 Egyptian feddan = 1.038 A) were chosen from two plantations; one represented the winter season (i.e. *Triticum sp.*) and the other the summer one (i.e. *Orisa sativa*). These two plots were selected at random from large plantations of about 80 feddans in the circle of the Faculty of Agriculture at Kafre El Sheikh, Egypt.

It was also planned to consider fauna existing under wild blady grass, *Imprata cylindrica* that grew either on field margins or drainage banks in about 3 feddans, and to determine effect of fire upon it in the second season. In the first season the plot planted with rice was sampled up to 50 cm deep, with no attention given to particular levels. However, in the winter season, the plot implanted with wheat was sampled as two layers: the upper one (0-25 cm from soil surface) representing the disturbed soil strata, and the lower one (25-50 cm from soil surface) representing the undisturbed strata.

It was arranged to divide a selected spot recultivated with rice into four plots for examination of 3 fertilizers. In the first plot, balady manure (organic matter partly degraded) was added at a rate of 20 m^3/feddan before cultivation; in the second one, ash residues of a mild fire were raked and harrowed in soil. Superphosphate at a rate of 100 kg/feddan was added to the third plot after one month of cultivation date, while the fourth plot received nothing and was left for control. It was arranged also to do 6 replicates in plots 1 and 3.

Temperature was measured before fire at two strata and was remeasured directly after the fire was extinguished at the same spot. The experiment lasted for about one yr. beginning in October and ending by April of the following year, but the process of burning and examining fertilizers took about 5 months. Random samples were taken at weekly intervals at a rate of 4/replicate. The sampling and extraction procedure have been described by Tadros (1965).

Wheat stems left after harvesting the crop were about 35 cm high and measured 21.90% moisture on the burning date. Burning took place about 2 months after harvesting. On the other hand, rice stems after harvesting were about 15-20 cm in length and measured 83.22% moisture at burning time, which also took place 2 months after harvest. Blady grass was about 90 cm high and measured 16.23% moisture on the burning date.

The mild fire height ranged from 20 cm in rice stems to 40 cm in wheat residues, while it reached 80 cm in blady grass. Fire remained flaming about 15, 20 and 30 minutes in rice, wheat and blady grass residues respectively.

RESULTS AND DISCUSSION

Effect of Mild Fire on Fauna and Soil

The soil arthropod fauna tended to decline owing to the rise in temperature caused by fire in the three tested plantations. From the data in Table I, it appears that the decrease in fauna caused by fire varied from one plantation to another, but the decrease in fauna under *Imprata* (46.2%) was higher than under the 2 tested residue crops (22.7% to 16.6% in wheat and rice respectively). On taking depth as a factor, it appears from the same Table that a decrease in fauna occurred in the upper strata, while there was an increase in the lower strata in all the 3 tested plantations.

Finding the °C reduction either in the 3 tested plantations, or in the 2 levels, it appeared that the decrease in fauna occurring in the upper strata under blady grass equaled nearly 2 times that under the other 2 tested materials, whereas an increase occurred in the lower soil layer. This increase was about the same in both wheat and rice. The author attributes this result to 2 factors; first, to the variation in the moisture percentage in the 3 tested plantations (83.22%, 21.90%, 16.23% in rice, wheat and blady grass respectively), and second, to the time and height of fire that was effected by the variation in moisture.

Although fire decreased total fauna at different percentages as shown in Table II, it was clear that the resistance of different groups was not the same. Some groups were not much affected in the 3 tested plantations, i.e. the decrease in Oribatei was minimal in the 3 tested plots and ranged between 28-30%. On the other hand soft mites were much affected by fire and suffocation and suffered decreases ranging from 75-97%. This result may be due to the structure of the mite and its ability to resist high temperatures long enough to emigrate to lower soil strata. The decrease in insects varied from one plantation to another, but it was considerable in *Orisa* and *Imprata* (73.53% and 69.74% respectively) and little affected under *Triticum* (22.33%). This result

TABLE I.
Reduction in Fauna Means Due to Mild Fire in Three Plantations.

	Fauna			Temperature °C		
Strata	Before Burning	After Burning	% Reduction	Before Burning	After Burning	Increase or Decrease in °C
			Triticum sp.			
Upper	136	50	63.2	46.4	60.4	2.51[b]
Lower	207	215	37.2	42.4	45.0	3.17[a]
Total	343	265	22.7	—	—	—
			Orisa sativa			
Upper	39	20	48.7	28.7	37.9	2.56[b]
Lower	18	34	47.1	25.3	29.6	3.72[a]
Total	57	54	16.6	—	—	—
			Imprata cylindrica			
Upper	303	89	70.6	24.4	57.4	6.48[b]
Lower	121	139	12.9	21.8	39.2	1.03[a]
Total	424	228	46.2	—	—	—

[a] Increase in fauna [b] Decrease in fauna

TABLE II.
Effect of Mild Fire on Six Groups of Fauna and Three Plantations

Faunal Groups	*Orisa sativa* Before Burning	*Orisa sativa* After Burning	*Orisa sativa* % Reduction	*Triticum sp.* Before Burning	*Triticum sp.* After Burning	*Triticum sp.* % Reduction	*Imprata cylindrica* Before Burning	*Imprata cylindrica* After Burning	*Imprata cylindrica* % Reduction
Oribatei	104	72	30.76	42	30	28.57	968	940	28.93
Soft mites	408	16	96.08	154	139	9.74	1312	92	92.98
Total Acari	512	88	82.81	196	169	13.75	2280	1032	54.74
Insecta	272	72	73.53	103	80	22.33	476	144	69.74
Myriapoda	00	00	00	33	9	72.73	104	56	46.15
Other organisms	00	00	00	10	7	30.00	48	36	25.00
Total Fauna	784	160	79.59	342	265	22.51	2908	1268	56.40

TABLE III.
Effect of Three Fertilizers on Fauna With an Evaluation of the Crop, *Tripolium alexandrinum*.

Faunal Groups	Balady Manure Mean	Balady Manure %	Ashes Mean	Ashes %	Super Phosphate Mean	Super Phosphate %	Control Mean
Acarina							
Oribatei	12.1	44.5[b]	14.2	52.2[b]	36.6	13.5[b]	27.2
Soft mites	33.2	6.9[a]	56.5	56.6[a]	78.8	40.6[a]	32.0
Total Acari	52.2	8.8[b]	70.7	11.9[a]	115.4	19.5[a]	59.2
Insecta	41.8	16.1[a]	42.7	16.4[a]	87.7	33.7[a]	26.0
Myriapoda	2.8	46.6[b]	1.4	23.3[b]	4.7	78.3[b]	6.0
Total organisms	89.8	9.8[b]	114.8	12.6[a]	207.8	22.8[a]	91.2
pH value	8.4		8.2		6.8		7.7
Organic Matter %	6.4		1.6		1.5		1.4
Increase in crop weight, %	13.5		12.4		8.1		

[a] increase in fauna [b] decrease in fauna

may be due to differences in insect orders that occur in each crop situation. Some are prevalent under one crop and may be absent altogether under another one. Some individuals may be resistant (Coleoptera) while others may be strongly affected (Collembola). This appeared to be the case, since most insects under wheat were Collembola. Myriapoda existed under wheat and blady grass; their reduction was high under wheat (72.73%), and low under blady grass (46.15%). The absence of this group under the rice crop and its variation from one field to another may be due to the practices of cultivation and irrigation which may prevent their establishment in this habitat.

Other organisms were affected nearly at the same percentage (30.0% and 25.0%) under wheat and blady grass respectively. This group contains highly mobile individuals which may more easily escape unsuitable conditions.

It appeared that the upper strata (0-25 cm from soil surface) was affected by fire and its temperature raised at a range of 33-39°C from one plantation to another. This range was due to both the percentage of R. H. found in either plant stems or soil, and also to the period in which fire was flaming. On the other hand, the lower surface was less affected and the temperature rose only about 3°C in soil under rice to 10°C under wheat. However, its maximum reached 33°C under blady grass. This variation may be due to the fact that rice fields receive irrigation for a long period and soil becomes wet and saturated even after harvesting the crop, hence the range in temperature rise is low. On the other hand, wheat does not receive irrigation before harvesting for a long period, which leaves the soil nearly dry and enhances temperature transfer to deep layers. Blady grass is grown on field margins, a fairly dry type of soil which receives a low percentage of irrigation water or gets its needs from the water table. This makes RH in the soil quite similar in the two tested strata and permits the penetration of heat.

We notice that, while there was a decrease of fauna in the disturbed layer caused by fire, there was an increase of soil microfauna in the lower surface, giving a chance for organisms to emigrate to subsurface. This fact was clear with Oribatei, soft mites and other organisms (Coleoptera . . .).

Soil Structure and Components

The pH value in the soil before burning was 7.70, but after the mild fire broke down, the pH value raised and reached 8.20. This result indicates that soil becomes somewhat alkaline following burning. From the analysis of soil strata before and after burning to find the percentage of important cations and anions, it appeared from the figures that the ash residue makes the soil somewhat alkaline. This last characteristic makes soil hold a high percentage of moisture and micropores will be prevalent. The soil also becomes consolidated and very firm. On the other hand organic matter degrades in soil and produces sodium humate. These new features in soil lessen soil capacity for changing soil air and non-aerobic circumstances become prevalent. These are unsuitable for fauna to exist.

Time Factor

From the figures collected in this investigation, fauna began to recover to its normal level after 1 to 3 months depending on group. This result varied from one group to another owing both to irrigation and life cycle of organisms. Acarina and insects were recovered in higher numbers than were Myriapoda.

The recovery of animals varied from one plantation to another, but the data showed that fauna under blady grass recovered sooner than the fauna under wheat, while recovery under rice was intermediate. This result may be due to factors related to soil nature itself.

Fertilizers and Fauna

As shown in Table III, balady manure (organic) decreased fauna either as a whole or as groups, with the exception of certain Insecta. Ashes and superphosphate, on the other hand, tended to decrease total fauna slightly, but their effect varied from one group to another. This result may have been due to the alkaline pH that transforms organic matter into sodium humate and reduces fungal growth, so that fauna does not find sufficient food stuff to survive.

Considering each faunal group, data indicated that Myriapoda and Oribatei were sensitive groups and that they decreased with the 3 selected fertilizers, while Insecta increased at slightly different rates. On the other hand, soft mites increased with the 3 tested materials but their response was higher with ashes.

Taking crop weight into consideration, data indicated that the 3 tested fertilizers increased the planted crop (*Tripolium alexandrinum*) at different rates. This increase was 8.1% in superphosphate, 12.4% in each plot and 13.5% in organic matter.

SUMMARY

Mild fire increased fauna, but this increase varied from one group to another. Fire affected the fauna in the disturbed layer (25 cm from soil surface) more than in the undisturbed strata (up to 50 cm deep). The ashes resulting from fire altered the pH value of soil and made it somewhat alkaline, which in turn altered soil structure to unsuitable conditions for fauna to flourish. The time factor for fauna recovery varied from one group to another.

Fertilizers altered soil acidity and its effect varied from one faunal group to another; the resultant crop varied in accordance with the fertilizer used.

REFERENCES

Buffington, J. D. (1957). *Ann. Entomol. Soc. Am.* **60**, 530-553.
Forsslund, K. H. (1951). *Entomol. Medd.* **26**, 144-147.
Heyward, F. and Tissot, A. N. (1936). *Ecology* **17**, 659-666.
Huta, Narminen, V. M., and Valpas, A. (1969). *Ann. Zool. Fenn.* **6**, 327-334.
Hill, S. B., Metz, L. J., Farrier, M. H. (1975). *Proc. 4th North Amer. Forest Soil Conf.* Laval Univ. Quebec 1973, 119-135.
Jahn, E. (1950). *Z. Angew. Entomol.* **32**, 208-274.
Jahn, E. and Schimitschek, G. (1950). *Forstwes.* **91**, 214-224.
Jahn, E., and Schimitschek, G. (1951). *Forstwes.* **92**, 36-46.
Karppinen, E. (1957). *Ann. Entomol. Fenn.* **23**, 181-203.
Lussenhop, J. (1976). *Ecol.* **57**, 88-98.
Tadros, M. S., Wafa, A. K. and El-Kifl, A. H. (1965). *Bull. Soc. Ent. Egypte* **49**, 1-37.

NOTES ON THE ECOLOGY OF CORTICOLOUS EPIPHYTE DWELLERS 1. THE MITE FAUNA OF FRUTICOSE LICHENS.

H. Andre

Laboratoire d'Ecologie Animale
Universite Catholique de Louvain
Louvain-la-Neuve, Belgium

INTRODUCTION

The importance of the habitat has recently been emphasized by Southwood (1977). The habitat approach remains a basic step in the concrete study of any ecosystem. However, the description of an ecosystem—as of any system—depends on the chosen space-time resolution level. As noted by Burges (1960), the problem is essentially one of scale, and of the size of the units used in the examination of the communities. In the context of the corticolous habitat and the fauna it shelters, the centimeter or millimeter scale has been selected; i.e., a study of small microhabitats has been fixed on. On such a choice will depend every property of the corticolous medium and fauna. This explains why Gilbert (1971) considers the trunk bark as being homogenous whereas Andre (1976) judges the same environment as heterogenous (see also Lebrun, 1976) and reveals the importance of epiphytes as a heterogeneity factor. However, since no systematic study of the arthropod communities in connection with the epiphytes has been carried out, it was felt that such a survey should be undertaken.

MATERIALS AND METHODS

This survey has been undertaken in Lorraine in southern Belgium. The area is well known for its air salubrity and corticolous lichen richness (Lambinon, 1969). Three media received special attention. *Evernia prunastri* (L.) Ach. was sampled on *Fraxinus excelsior* L. in a site described previously (Andre, 1975). Moreover, the same lichen and *Ramalina farinacea* (L.) Ach. were sampled on *Populus* sp., 5 km further to the west, on the same cuesta and at the same

Vol. I: ISBN 0-12-592201-9

elevation. Twenty-five samples from each medium were taken quarterly during 1974. Furthermore, samples were taken in June: 15 from *Ramalina fastigata* (Liljeb) Ach. and 15 from *Ramalina fraxinea* (L.) Ach. on *Fraxinus* and, lastly, 4 from *Anaptychia ciliaris* (L.) Koerb., a very rare lichen, on *Populus*.

The lichens were weighed before and after a Berlese funnel extraction of one week. They also were examined, crumpled and smoked out before extraction. The samples were sorted under a dissecting microscope and the specimens were mounted for species and stase identification.

RESULTS

Mite Populations

In the fruticose lichens, mites represent only a small part of the microarthropod community, i.e. about 22%. Some Collembola such as *Entomobrya nivalis* (L.), some Psocoptera such as *Hyperetes guestfalicus* Kolbe, and Oniscoidea (Isopodes), are particularly abundant. Mites comprise no fewer than 55 species representing 4 orders. These are listed in Table I. This table gives the abundance, i.e. the number of individuals/100 gm of dried lichen following funnel extraction, and the frequency, i.e. the number of samples inhabited/100 samples.

Gamasida. Four species of *Typhlodromus* were found, and represented about 91% of the gamasid total. Unfortunately, 2 of these species (*T. pyri* and *T.* cf. *tubifer*) were confused during the early phases of this investigation, with the result that I am unable to define their respective distributions. All that may be said is that the ratio *T. pyri*/*T.* cf. *tubifer* is *ca* 12/65 and that both are found only on *Fraxinus*. *T. rhenanus* was found in all 3 media while *T. richteri* was obviously more abundant on *Populus*. Two other phytoseiid mites were observed (*Amblyseius cf. cucumeris* and *Phytoseius macropilis*), along with some other species.

Oribatida. Two species, *Carabodes labyrinthicus* and *Eremaeus cf. oblongus,* were abundant but both were found only on *Fraxinus*. On the other hand, two species, *Phauloppia lucorum* and *Trichoribates trimaculatus,* were also abundant but were found almost exclusively on *Populus*. These 4 species represented some 86% of the Oribatida. Lastly, *Oribatella quadricornuta* seemed more abundant on *Fraxinus* than *Populus*. Eight other species were found but were too rare to deserve further comment.

Actinedida. Some 27 different species belonging to 13 families have been identified. Tydeidae were the most important family, both in the number of individuals (near 62% of the Actinedida) and the number of species (several species are new, and descriptions are in preparation). However, only *Metatriophtlydeus lebruni* (Fig. 1) seemed to show a clear difference of repar-

TABLE I.
Mite Populations in the 3 Studied Communities. Ab. = Abundance/100 gm of Dried Lichen. F = Frequency (in %).

	Taxa	*Evernia prunastri*				*Ramalina farinacea*	
		Populus		*Fraxinus*		*Fraxinus*	
		Ab.	F.	Ab.	F.	Ab.	F.
GAMASIDA:	*Typhlodromus pyri* Scheuten *Typhlodromus* cf. *tubifer* Wainstein	—		26.5	22	43.6	24
	Typhlodromus rhenanus (Oudemans)	11.5	10	19.6	14	21.2	14
	Typhlodromus richteri Karg	30.3	20	2.3	2	5.6	4
	Amblyseius cf. *cucumeris* (Oudemans)	—		—		1.3	1
	Phytoseius macropilis (Banks)	—		1.2	1	—	
	Arctoseius sp.	0.7	1	—		—	
	Parasitinae (immatures)	0.7	1	4.6	2	3.4	5
	Unidentified nymphs	2.1	3	4.6	3	2.2	2
	Uropodinae	2.2	2	—		1.1	1
	Subtotal	47.6	29	58.7	37	78.2	40
ORIBATIDA:	*Carabodes labyrinthicus* (Michael)	—		78.3	17	93.8	22
	Eremaeus cf. *oblongus* Koch	—		18.4	12	55.8	23
	Phauloppia lucorum (Koch)	18.7	16	—		1.1	1
	Trichoribates trimaculatus (Koch)	26.7	19	—		—	
	Oribatella quadricornuta (Michael)	0.7	1	4.6	2	8.9	3
	Dometorina plantivaga (Berlese)	3.6	5	2.3	2	6.7	6
	Tectocepheus sarekensis Tragardh	0.7	1	6.9	5	—	
	Camisia segnis (Hermann)	—		—		7.8	3
	Ceratoppia bipilis (Hermann)	2.2	1	—		—	
	Cymbaeremaeus cymba (Nicolet)	—		—		1.3	1
	Oppia sp.	0.7	1	—		—	
	Subtotal	53.3	36	110.6	34	175.3	47
ACTINEDIDA:	*Lorryia bedfordiensis* Evans	26.7	16	21.9	15	14.5	9
	Lorryia n. sp. 1	4.3	6	18.4	12	12.3	9
	Lorryia n. sp. 2	—		3.5	3	10.1	9
	Lorryia n. sp. 3	—		2.3	2	—	
	Lorryia armaghensis Baker	—		1.2	1	—	
	Tydeus n. sp.	6.5	8	6.9	3	3.4	3
	Metatriophtydeus lebruni Andre	13.7	14	49.5	28	36.9	24
	Tydaeolus sp.	—		1.2	1	—	
	Microtydeus sp.	0.7	1	—		—	
	Coccotydaeolus sp.	—		—		1.1	1
	Eupodes sp. 1	17.3	14	1.2	1	2.2	2
	Eupodes sp. 2	0.7	1	—		—	
	Cunaxa sp.	6.5	4	11.6	9	11.2	7
	Cunaxoides sp.	2.9	4	—		—	
	Bryobia kissophila van Eyndhoven	0.7	1	17.3	10	17.9	12
	Anystis sp.	—		2.3	2	3.4	3
	Bdellodes sp.	6.5	9	—		—	
	Mediolata mariaefrancae Andre	1.4	2	—		—	
	Stigmaeus sp.	—		1.2	1	—	
	Unidentified Stigmaeidae	0.7	1	—		—	
	Rhagidia sp.	—		—		1.1	1
	Ricardoella sp.	0.7	1	4.6	2	—	
	Nanorchestidae	—		—		2.2	2
	Pachygnathidae	0.7	1	—		—	
	Tarsonemidae	0.7	1	—		—	
	Pyemotidae (sp. 100 + sp. 656)	—		1.2	1	1.1	1
	Unrecovered specimens	—		3.5	2	4.5	4
	Subtotal	90.8	55	145.2	64	121.7	56
ACARIDIDA:	Saproglyphidae, nymphs sp. 1152	—		—		485.8	2
	Others (6 species)	5.7	5	—		1.1	1
	Subtotal	5.7	5	—		486.9	3
TOTAL		197.5	77	314.5	82	862.1	75

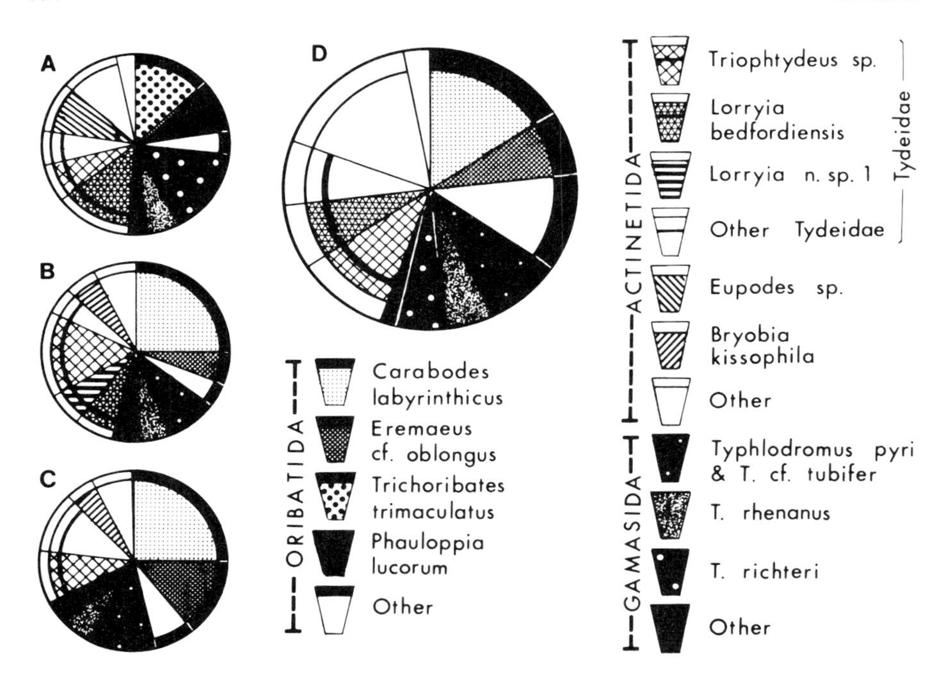

Fig. 1. Diagrams expressing the relative abundance of mites in *E. prunastri* on *Populus* (A), in *E. prunastri* and *R. farinacea* on *Fraxinus* (B, C) and in fruticose lichens as a whole (D). The white sectors correspond to Acaridida.

tition and was more abundant on *Fraxinus*. Two other families were important and represented nearly 10% of the Actinedida. The Eupodidae were obviously more abundant on *Populus* whereas Cunaxidae seemed more important on *Fraxinus*. Another species, *Bryobia kissophila,* also was abundant although no ivy was observed on our trees; however they correspond to the description of Mathys (1957). The other families generally included one species (except Stigmaeidae with 3 species and Pyemotidae) and were too rare to lead to any conclusion.

Acaridida. Obviously this taxon was poorly represented. However, in a single sample were found 432 Saproglyphidae nymphs. This high value may be considered as an oddity and, consequently, these nymphs will not be taken into account later.

Mite Communities

Figure 1 illustrates the relative abundance for each community (diagrams A, B and C) and fruticose lichens as a whole (diagram D). This latter shows that Oribatida represent nearly 35% of the mites, Gamasida, 21 and Actinedida, 42. The most important families are Tydeidae (25%), Phytoseiidae (19%), Carabodidae (16%) and Eremaeidae (7%). On *Populus* (with *E.*

prunastri as an epiphyte), the characteristic species are *Typhlodromus richteri* (16%), *Trichoribates trimaculatus* (13.5%), *Lorryia bedfordiensis* (13.5%) and *Phauloppia lucorum* (9.5%). The diagrams B and C, both illustrating the community composition on *Fraxinus,* are very similar. The main species are *Carabodes labryinthicus* (25%), *Eremaeus* cf. *oblongus* (6 to 15%, *Metatriophtydeus lebruni* (10 to 16%) and *Typhlodromus pyri* and cf. *tubifer* together (8 to 12%). On *Populus,* the second main actinedid family is Eupodidae whereas on *Fraxinus,* it is Tetranychidae.

Table II illustrates the diversity of the 3 communities and gives the number of species, Shannon-Weaver's index of heterogeneity (type I index *sensu* Peet (1974), i.e., most sensitive to changes in the rarest species), Pielou's index of evenness and Simpson's index of heterogeneity (type II index, i.e., most sensitive to changes in the most abundant species). The 2 communities studied on *Fraxinus* have the same heterogeneity, very similar evenness and number of species. On the other hand, the community on *Populus* is quite distinct and is more diverse owing to a higher richness than in the 2 other communities.

Table III expresses the similarity between the 3 communities based on 2 indices. The first one, the function T of Kullback, is derived from the Shannon-Weaver's index while the other is based on Simpson's index. As expected, the 2 communities from *Fraxinus* are very similar whereas the 2 communities from different trees and lichens are the most dissimilar. Comparison of the 2 indices shows the importance of the abundant species as a factor in the similarity of

TABLE II.

Number of Species (*S*), Shannon & Weaver's Index of Heterogeneity (*H*), Pielou's Evenness (*J*) and Simpson's Index of Hererogeneity (ϕ) for the 3 Studied Communities.

	Evernia prunastri Populus	*Evernia prunastri Fraxinus*	*Ramalina farinacea Fraxinus*
S	34	26	28
H	3.97	3.71	3.72
J	.77	.78	.76
ϕ	.91	.88	.88

TABLE III.

Similarity Between the 3 Communities (Upper Triangular Matrix: Index Derived from the Simpson's Index; Lower Matrix: Index Derived from the Shannon-Weaver's Index).

	Evernia prunastri Populus	*Evernia prunastri Fraxinus*	*Ramalina farinacea Fraxinus*
Evernia prunastri Populus	X	.29	.20
Evernia prunastri Fraxinus	.44	X	.93
Ramalina farinacea Fraxinus	.40	.91	X

the communities on *Fraxinus*. On the other hand, the similarity between the community on *Populus* and the other two is based more on relatively rare species.

CONCLUSIONS

Previously, I speculated on the identity of the key factors determining microarthropod distribution, i.e., species, major taxon, and structure of the morphological type of epiphyte (Andre, 1976). Comparison of the above results with the fauna sheltered in crustose lichens (Andre, 1975 and unpublished data) reveals that the differences are great even at the ordinal level. However, when the fruticose lichen type is considered, results suggest that the differences are minor and that the tree species—and particularly the bark properties—should be a determining factor in the distribution of several mite species (among Oribatida as well as Gamasida and Actinedida), and on the community structure. This would seem to emphasize the importance of the lichen morphological type.

These results are only a part of a wide project and should be more understandable in the light of future investigations. This is only a step in the inquiry on feeding and sheltering specificity between epiphytes and microarthropods, one of the most important areas in which information is in demand, as noted by Gerson and Seaward (1977).

SUMMARY

The mite fauna of three corticolous communities (*Evernia prunastri* on *Populus; E. prunastri* and *Ramalina farinacea* on *Fraxinus*) were studied. The results suggest the importance of the three communities on mite distribution and community structure. Moreover, these communities seem quite different from those observed on crutose lichens.

ACKNOWLEDGEMENTS

I am grateful to Drs. G. W. Krantz and Ph. Lebrun for reviewing the manuscript. For the loan of paratypes, I thank Drs. E. W. Baker (U. S. D. A., Beltsville, Md.) and K. H. Hyatt (British Museum, London). I also thank Dr. W. Karg (Institut fur Pflanzenschutzforschung, Kleinmachnow, DDR) for help in the determination of *Typhlodromus* and Mr. G. Wauthy (Universite Catholique de Lounvain, Belgium) for advice on Oribatida identification. This work was initiated at the Fondation Universitaire Luxembourgeoise, Arlon and was continued at the Universite Catholique de Louvain, Laboratorie d'Ecologie Animale, Belgium and at the Department of Entomology, Oregon State University, Corvallis, U. S. A.

REFERENCES

Andre, H. (1975). *Bull. Ecol.* **7**, 431-444.
Andre, H. (1976). *Notes de recherche de la F. U. L.* **4**, 1-31.
Burges, A. (1960). *J. Ecology* **48**, 273-285.
Gerson, U. and Seaward, M. R. D. (1977). *In* "Lichen Ecology," (Seaward, M. R. D., Ed.), Academic Press. London. 69-119.
Gilbert, O. L. (1971). *J. appl. Ecol.* **8**, 77-84.
Lambinon, J. (1969). "Les Lichens." *Les Naturalistes belges.* Bruxelles. 196 pp.
Lebrun, Ph. (1976). *Bull. Ecol.* **7**, 417-430.
Mathys, G. (1957). *Bull. Soc. Entom. Suisse* **30**, 189-204.
Peet, R. K. (1974). *Ann. Rev. Ecol. System.* **5**, 285-307.
Southwood, T. R. E. (1977). *J. Anim. Ecol.* **46**, 337-365.

THE ACARINE COMMUNITY OF NESTS OF BIRDS OF PREY

James R. Philips and Daniel L. Dindal

Department of Environmental and Forest Biology
SUNY College of Environmental Science and Forestry
Syracuse, New York

INTRODUCTION

In the nests of birds and mammals the acarine community includes a variety of parasites, predators, and stored product, soil and litter mites. However, few extensive studies of birds' nest fauna have been made. Nests of birds of prey (Falconiformes and Strigiformes) are of special interest since they may contain prey parasites, such as fleas from mammals, in addition to avain nidicolous parasites.

The only major study of mites found in raptor nests was done by Nordberg (1936). There are apparently only four species of mesostigmatic mites, 11 species of prostigmatic mites, 10 species of astigmatic mites and 19 species of oribatid mites known from nests of birds of prey. North American records are limited to two species of prostigmatic mites, *Acaropsella schmidtmanni* Price and *Tarsonemus fusarii* Coor.

Acarina are known to cause raptor nestling mortality. Details of this and other arthropod parasitism in raptor nests have been reviewed by Philips and Dindal (1977). The objective of the present research was to investigate the microarthropod community of raptor nests.

METHODS

From the central New York area, we collected nests of three species of birds of prey—the great horned owl (*Bubo virginianus*), the screech owl (*Otus asio*) and the American kestrel (*Falco sparverius*). Nest material was extracted in Tullgren funnels for 2 wks.

Vol. I: ISBN 0-12-592201-9

RESULTS

The great horned owl nest was collected (16 April 1976) from a ledge (23 m high) on a rock crusher at a quarry. Nestlings present were 2-3 wks old, and the nest debris consisted mainly of regurgitated pellets. The water content of the nest material was 6.75% (dry wt); 116 g (dry wt) of this material yielded 1 specimen of *Androlaelaps fahrenholzi* (Berlese).

The second nest consisted mainly of leaf litter from a tree-hole used by screech owls. The nest also contained pellet fragments, feathers, eggshells, twigs and wood chips. The hole was 9 m high in a white oak tree and was used annually by the owls. The sample was collected just before egg-laying (12 March 1976). Extraction of 175 g of nest matter yielded 22,725 mites, a density of 130 mites/g of nest material. Mites from the following groups represented 98.9% of the total nest microarthropod fauna; Oribatei 12,214 individuals (4 species), Astigmata 5,720 (12 species), Prostigmata 4,168 (11 species), Mesostigmata 622 (17 species) and Ixodoidea one.

The American kestrel nest was collected (24 June 1976) from a nestbox 5 m high in a dead tree. Sticks and leaves comprised most of the nest material along with a few fragments of pellets, chitin and excreta; the nestlings fledged the day before the nest was collected. Based on an analysis of about 50% of the extracted samples, the nest contained slightly over 20,000 mites at a density of 30 mites per gram dry weight of material. Mites represented 89.3% of the invertebrates extracted as follows; Astigmata 8,775 (13 species), Prostigmata 1,318 (7 species), Mesostigmata 104 (14 species) and Oribatei 12 (7 species).

DISCUSSION

The presence of one *A. fahrenholzi* in the great horned owl nest was attributed to nest prey remains since a fresh rat (*Rattus norvegicus*) carcass was observed nearby. The extremely dry conditions of this nest were very unfavorable for invertebrates in general, and the site was free of nest parasites.

A total of 8 species were common to both screech owl and kestrel nests—*A. fahrenholzi, Hyperlaelaps microti* (Ewing), *Laelaps alaskensis* Grant, *Dendrolaelaps* sp. nr. *presepum* (Berl.), *Orycteroxenus soricis soricis* (Ouds.), *Tydeus* sp. and 2 species of uropodines. Four of these species are parasitic or phoretic on rodents which are major prey items of these birds. Other mesostigmatic mites found included ascids, ameroseiids, parasitids and macrochelids. Also the screech owl nest contained an undescribed species of *Kleemannia.*

The dominant species in the screech owl nest was *Oppia clavipectinata.* This species represented over half the nest mite fauna (12,056 specimens). The highly decomposed leaf litter in the tree-hole environment apparently provided near optimum conditions for this species. The oribatid mite density in the screech owl nest appears to be unmatched in any reported nest data. Gembestky and Andrechikova (1969) examined 32 nests of 7 passerine species and collected only 98 oribatid specimens. Bukva *et al.* (1976) found 979 oribatids in 278 small mammal nests, and Kramarova and Mrciak (1971) collected 13,364 from 303 small nests.

A new species of *Lardoglyphus* (Philips and Norton, 1978) was the dominant species in the kestrel nest; it represented over 80% of the nest mites (8,407 specimens). Hypopi of *Tytodectes cerchneis,* a hypoderid, were also found in the kestrel nest. This genus was not previously known from the USA and the species was known from only one specimen associated with the European kestrel (*Falco tinnunculus*) in Rwanda, Africa (Fain, 1967). Other astigmatic mites from the kestrel included glycyphagids, acarids, saproglyphids, anoetids, feather mites and one pyroglyphid specimen (*Dermatophagoides* sp).

Pyroglyphids were not found in the screech owl nest, and their scarcity in these raptor nests is surprising considering their abundance in many other types of bird nests, including swallow nest-boxes near the kestrel nest. It is possible that the carrion and pellets in the raptor nest or the moisture in raptor nest holes, created an unfavorable environment for pyroglyphids.

The screech owl nest contained many new taxa of astigmatic mites which have been described elsewhere (Fain and Philips, 1977a; Fain and Philips, 1977b; Fain and Philips, 1978a). Mammalian associates included pilicolous hypopi (*Neoxenoryctes* n.g.), endofollicular hypopi (*Echimyopus* n. sp.) and nidicoles (*Fusacarus* n. sp.). Taxa with entomophilic hypopi included Euglycyphaginae n. subfam., *Sapracarus* n.g., *Acotyledon paradoxa* Ouds. (Fain and Philips, 1978b), *Histiogaster* and anoetids.

Although the rodent associates *Echimyopus* and *Neoxenoryctes* may have been brought to the nest via the screech owl's prey, we believe these species, as well as *Fusacarus,* were associated with grey squirrels (*Sciurus carolinensis*) that were very abundant in the area. These squirrels occupied the nest hole in 1978 on a semi-permanent basis and probably frequented it before that time. Both the squirrel tick and squirrel flea were found in the owl nest material, as well as squirrel chiggers.

Tydeids, cheyletids and tarsonemids were found in both the kestrel nest and screech owl nest, while the latter also contained a variety of pygmephoroids and a new ganus of tydeid.

The kestrel nest, in addition, contained a female *Demodex* of an undescribed species. Since there were no carrion remains in the nest, and there was little opportunity for a mammal to visit the nest after the birds left, we believe that these mites may survive for days in the nest environment without a host. If so, this provides some evidence that nest transmission of demodicids could occur. Perhaps this mite dropped off its host during avian feeding, or it may have survived ingestion and been regurgitated alive within a pellet by the bird. We have observed, for example, that some other parasitic mites of prey may have been egested alive in pellets on rare occasions.

In general, these raptor nests contained an acarine faunal composition widely different from most avian and ground mammal nests reported in the literature. Large faunal differences were observed between nests, but each contained mammal-associated mites. Very few avian ectoparasites were found. Further studies should elucidate the ecology of the raptor nest microcommunity and are likely to find new mites associated with both birds and mammals.

SUMMARY

Three nests of birds of prey from Central New York were examined for invertebrates. A great horned owl nest that was very dry yielded only one mite, *Androlaelaps fahrenholzi* (B.). Forty-five acarine species, totaling 22,725 individuals represented 98.9% of the microarthropod fauna of a screech owl nest. Forty-one species, totaling about 20,000 mites (89% of total nest fauna) were collected from an American kestrel nest. In the screech owl nest, oribatid individuals were numerically dominant (53%), followed by astigmatic mites (25%); whereas, in the kestrel nest, astigmatic mites dominated (77%) followed by prostigmatic mites (12%). The greatest number of species in each nest, however, was represented by the mesostigmatic mites. Among the mites found, there were new genera, genera not previously known from the USA, hypopi and adults not previously associated, new host records, and many mammal associates.

ACKNOWLEDGEMENTS

We are very grateful to Drs. A. Fain, R. Norton, J. Gaud, E. Lindquist, N. Wilson and F. Athias-Binche for their invaluable taxonomic assistance.

REFERENCES

Bukva, V., Daniel, M., and Mrciak, M. (1976). *Vest. Cesk. Spol. Zool.* **40**, 241-254.
Fain, A. (1967). *Bull. Inst. Roy. Sci. Nat. Belg.* **43**, 1-139.
Fain, A. and Philips, J. R. (1977a). *Intl. J. Acarol.* **3**, 105-114.
Fain, A. and Philips, J. R. (1977b). *Acta Zool. Path. Ant.* **69**, 155-162.
Fain, A., and Philips, J. R. (1978a). *Acta Zool. Path. Ant.* **70**, 227-231.
Fain, A., and Philips, J. R. (1978b). *Zool. Med.* **53**, 29-39.
Gembestky, A. S. and Andrechikova, E. E. (1969). *Prob. Parasitology* **2**, 87-88.
Kramarova, L., and Mrciak, M. (1971). *Proc. 3rd Int. Cong. Acarol. Prague, (1971)* 427-433.
Nordberg, S. (1936). *Acta Zool. Fenn.* **21**, 1-168.
Philips, J. R. and Dindal, D. L. (1977). *Raptor Res.* (in press).
Philips, J. R. and Norton, R. A. (1978). *Acarologia* (in press).

THE DEVELOPMENTAL CYCLE OF SPONGE-ASSOCIATED WATER MITES

Robert M. Crowell

Department of Biology
Saint Lawrence University
Canton, New York

Cornelis Davids

Zoologisch Laboratorium
Universiteit van Amsterdam
The Netherlands

INTRODUCTION

Some of the mites within the genus *Unionicola* have been known since 1839 to have associations with freshwater sponges (Arndt and Viets, 1938). Many of the records are such that species confirmations are impossible to make, but presently at least six species of *Unionicola,* all occurring in the subgenus *Unionicola, s.s.,* are known to be associated with freshwater sponges. The nature of that relationship is not entirely clear, but it does seem to be obligatory.

Jones (1965) reported the first records of *Unionicola* larvae outside a molluscan or sponge host when he found two species parasitizing chironomid midges. One of those he identified as *U. aculeata,* a mite which deposits its eggs in the tissues of freshwater mussels. Bottger (1972) found larvae which he identified as *U.crassipes,* a sponge mite, on midges in the vicinity of Kiel, Germany, and summarized the life cycle accordingly: ova deposited in sponge tissue hatch and emerge from the sponge, taking up a phoretic association with chironomid pupae in their cases. When the adult midge emerges the mite larvae are or become attached. After a period of parasitism on the midge the engorged larva detaches and re-enters the chambers of a sponge where it spends the protonymphal stage. The deutonymph which then emerges leaves the sponge, presumably for a predatory feeding period, after which it re-enters

Vol. I: ISBN 0-12-592201-9

the sponge and undergoes the tritonymphal transformation to the imaginal stage. The adult then leaves the sponge and after another feeding period returns to the sponge for a third time. At this time ova are deposited in the sponge tissue.

There are few published records of *Unionicola* larvae on midges. Hevers (1975) shows larvae of *U. aculeata* on abdominal sternites of *Chironomus thummi* and Bottger (1972) shows *U. crassipes* attached to the femora of the second and third legs of the same host species. Smith and Oliver (1976) report *Unionicola* larvae from femora and from the abdomen of midges. Ferrier (1978: unpublished student project report) found the same occurrence of *Unionicola* larvae on unidentified midges.

The principal European sponge mites are *Unionicola crassipes* (Muller, 1776) and *U. minor* (Soar, 1900). Identification of the related North American species is treated in another paper presently in preparation in which they are considered separate and distinct species (Crowell and Davids, 1979).

METHODS AND RESULTS

In an attempt to elucidate the ecological cycle of the sponge mites we have made collections of the non-parasitic stages of these mites in The Netherlands (Het Hol, Kortenhof) approximately 16 km southeast of Amsterdam, in an Adirondack Mountain lake in northern New York (Joe Indian Pond, Parishville), and in two lakes in eastern Ontario (Opinicon, Leeds Co. and Pine Lake, Frontenac Co.). The Netherlands study will be reported in detail in a paper now in preparation. The North American study is continuing.

Data from The Netherlands were collected during the period from mid-summer 1976 to November, 1977. They are based on mites collected by netting, by emergence from sponges held in aquaria, by dissection of sponges, and by submerged light trap collections.

The North American data were collected during the autumn of 1977 through July, 1978. Those data are based on mites collected by netting and by submerged light traps.

The data obtained in Holland for both *U. crassipes* and *U. minor* show a peak of egg production in late July and August, followed by increased numbers of larvae in early September.

From late September to mid-October, our data show deutonymphs to be the most prevalent post-larval stage (2.8:1 on 23 September; 2.6:1 on 14 October).

From mid-October to mid-November, nymphs and adults are present in approximately equal numbers, but in late November and December collections only adults are taken. This is interpreted to mean that the transformation from deutonymph to adult has taken place during this period of time. Another possible explanation, considered less likely, is that nymphal stages are already beginning to hibernate in the mud. However, from a simulated "over-wintering" experiment only adults emerged in February.

Net collections in March show ovigerous females of both species, with *U. crassipes* preceding *U. minor* slightly. These are followed by larval and nymphal peaks and a lesser peak of ovigerous adults in June and July.

For the related North American sponge mites our data show that by mid-September (1977) the sponge mite population in Joe Indian Pond is 99% nymphs and 1% adult. Net collections in late April (1978) have yielded no *Unionicola* and in mid-May only nymphs and adults of the *minor*-like mite were collected. This was also the case in early June in Joe Indian Pond and Lake Opinicon.

Collections at Pine Lake continuing through July (1978) have shown an increase in the proportion of adults in the population from as low as 40% (6 July 78) to as high as 95% (25 July 78). The *minor*-like species has continued to constitute approximately 85% (82.3 to 85.7%) of the population of these species with the larger *crassipes*-like species representing 15% (14.2 to 17.6%).

DISCUSSION AND CONCLUSIONS

The rather complete data that we have for *U. crassipes* and *U. minor* in Holland have led us to conclude that there are two generations per year, suggesting that the fertilized females hibernate over winter and that the males generally do not survive the winter.

Our data for the related North American sponge mites have led to a tentative conclusion that those mites exhibit a single generation per year in our latitude, and that it is the deutonymphal instar which hibernates over the winter.

The habitat in Holland remains open nearly all of the year with at most a few centimeters of ice cover for a few days in December or January. The lakes we have collected from in New York and Ontario are covered from December to late April with ice which may reach a meter or more in thickness. Variations in the mite developmental cycle may also be related to events in the ecological cycle of the sponges with which the mites are associated. In Holland it appears that the sponges may retain functional tissues through most if not all of the winter whereas, at comparable latitudes in Vermont, Simpson and Gilbert (1973) have shown that the freshwater sponges do not retain functional tissues through the winter.

It seems most likely that these differences in the life cycles of the sponge mites may be accounted for primarily by climatic considerations.

SUMMARY

Collections of sponge associated water mites have been made to elucidate the ecological cycle of their development in The Netherlands and in North America. The data show that the European species *Unionicola minor* (Soar, 1900) and *U. crassipes* (Muller, 1776) have two generations per year. Fertilized females over-winter, producing a population of offspring which mature by mid-summer. A second generation matures by mid-October.

Collections of two related North American species indicate a single generation per year with the deutonymph as the over-wintering stage, maturing in early summer and producing offspring which reach the next deutonymphal stage by mid-September. It is concluded that these differences in the cycle are related to climatic differences in the two regions.

REFERENCES

Arndt, W. and Viets, K. (1938). *Z. Parasitenk,* Berlin, **10**, 67-93.
Bottger, Klaus. (1972). *Int. Revue ges. Hydrobiol.* **57**, 263-319.
Crowell, R. M. and Davids, C. (1979). *Ohio J. Sci.* In Press.
Ferrier, K. A. (1978). Unpublished student project report. IFWAI-II.
Hevers, Jurgen. (1975). *Diss. Christian-Albrechts-Universitat zu Kiel.* 1-354.
Jones, R. K. H. (1965). *Nature.* **207**, 317-318.
Simpson, T. L. and Gilbert, J. J. (1973). *Trans. Amer. Micros. Soc.* **92**, 422-433.
Smith, I. M. and Oliver, D. R. (1976). *Can. Entomol.* **108**, 1427-1442.

EFFECTS OF SOME SOIL FEATURES ON A UROPODINE MITE COMMUNITY IN THE MASSANE FOREST (PYRENEES-ORIENTALES, FRANCE)

Francoise Athias-Binche

Laboratoire Arago
Banyuls Sur Mer, France

INTRODUCTION

This paper presents the first results of a two year sampling (June 1975-May 1977) of the soil mesofauna in the Massane beech-wood, a Natural Reserve in the Alberes mountains (Eastern Pyrenees). The sampling plot is a long and narrow "catena" (110 m long, 6 m wide) situated along the left slope of the Massane valley, at an altitude of about 650 m. The present work comprises the description of a community of uropodine mites and its relation to the topography, the soil, and the litter of the plot.

SAMPLING PLOT CHARACTERISTICS

From the top of the slope to the bottom, on the bank of the Massane river, 22 levels are labelled at 5 m intervals. The slope varies (Fig. 1): level 1 is horizontal, then the slope steepens strongly down to level 7. The angle reaches only 10° from level 8 to level 14. Several steps occur between levels 15 and 18; then the slope increases strongly to an angle of about 20° to level 21. Level 22 is the flat bank of the river.

The soil is rather acid (Fig. 6); except for the sandy bank of the stream, it is a brown soil overlying schist. The blackish soil of level 1 is silty (Fig. 4); the soil becomes a rocky ranker with a mull to level 7, then an earth mull to level 14, and a coarse mull down to level 18. From level 19 to 21, the soil is a sandy brown eroded ranker (see silt and clay values versus coarse sand in Fig. 4). All these brown soils are gravelly, and the soil thickness rarely exceeds 30 cm. Level 22 is an alluvial sandy soil with pebbles. In relation to the topography, erosion, and the occurrence of litter, organic matter ratio and nitrogen content

Vol. I: ISBN 0-12-592201-9

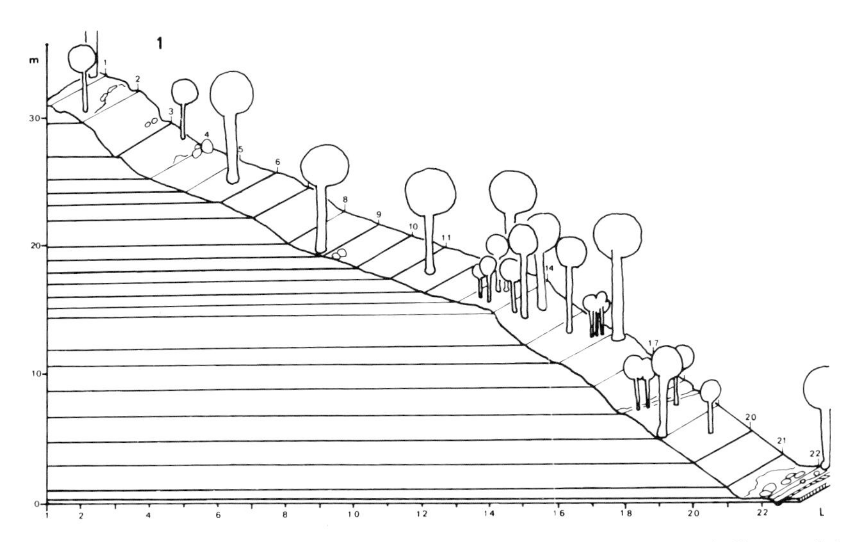

Fig. 1. Diagram of the sampling plot (110 m × 6 m); the numbers 1 to 22 indicate each level (L) and the 22 transversal sampling lines. The ordinal scale (m) for the slope is twice the horizontal scale in the abscissa.

are highest between levels 1 and 15 (Figs. 3 and 5).

The vegetation mainly consists of beech (*Fagus sylvatica*). The density of trees is rather low except for levels 14 and 16 (Fig. 1). The herbaceous layer is very thin, except at level 22. The occurrence of beech at this altitude in a Mediterranean area may appear surprising. Probably the particularly fresh climate of the Massane valley is responsible for the natural conservation of beech, which is also considered as a witness of some quaternary climatic phases. The average annual rainfall is about 1200 mm and the average temperature is 10.2°C (1.5°C winter minimum in February, and 18°C summer maximum in August). The number of frosty days is about 45 per year. This climate contrasts with that of the surrounding Roussillon plain where the annual rainfall reaches about 600 mm, and the mean temperature lies around 14-16°C.

The wind is an important ecological factor. In the Massane valley, it can carry the fallen leaves from the bottom of the valley to upper levels. The steps at levels 18 to 14 constitute as many obstacles to the wind current; by losing its power, the current deposits the leaves carried up. The highest litter weights are consequently observed between levels 8 and 14 (Fig. 7). Obviously, when the slope decreases, the litter is more constantly retained on the soil. The ratio of samples containing litter (Fig. 2) shows the occurrence of litter during the year. A 100% ratio means that litter is present all of the time (and consequently permits a good biological activity for microorganisms and fungi); on the contrary, a low percentage indicates that the fallen leaves are carried away by wind, especially in winter.

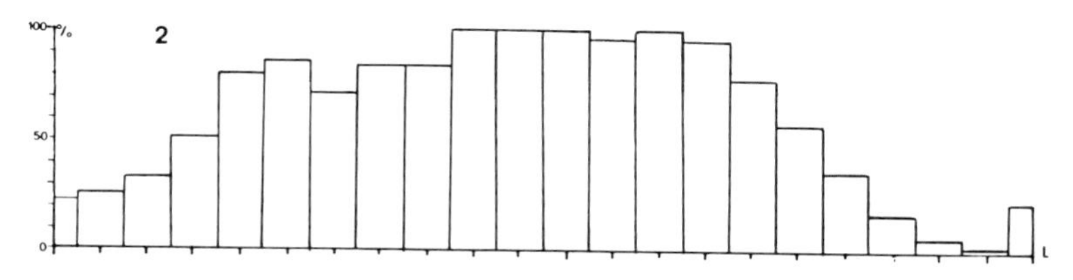

Fig. 2. Percentage of samples containing litter (%) per level (L); number of samples with litter/number of total samples at each level; average values of the 2 year sampling.

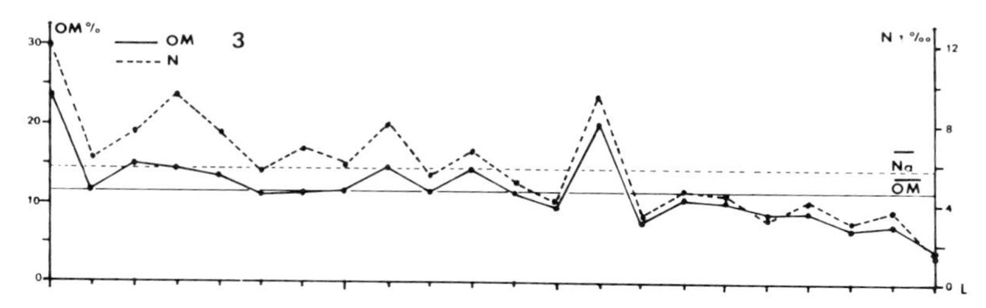

Fig. 3. Organic matter ($\overline{OM}$%) and nitrogen ($\overline{N}$%) contents for each level (L); average values of organic matter ($\overline{OM}$) and nitrogen ($\overline{N}$) for the entire plot.

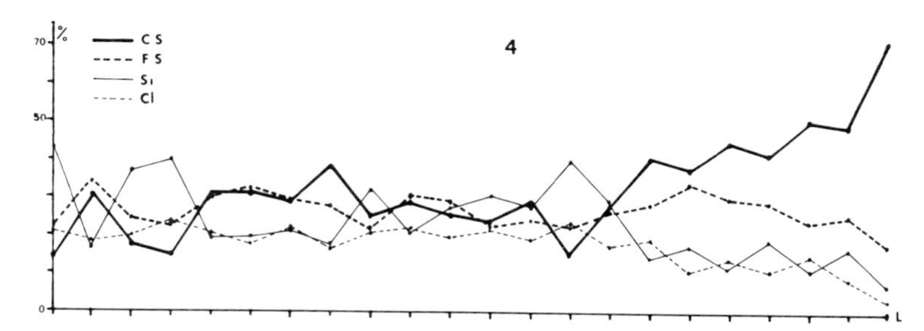

Fig. 4. Granulometry (%) of each level (L): coarse sand 200-2000 μm (CS); fine sand 50-200 μm (FS); silt 2-50 μm (Si) and clay less than 2 μm (Cl).

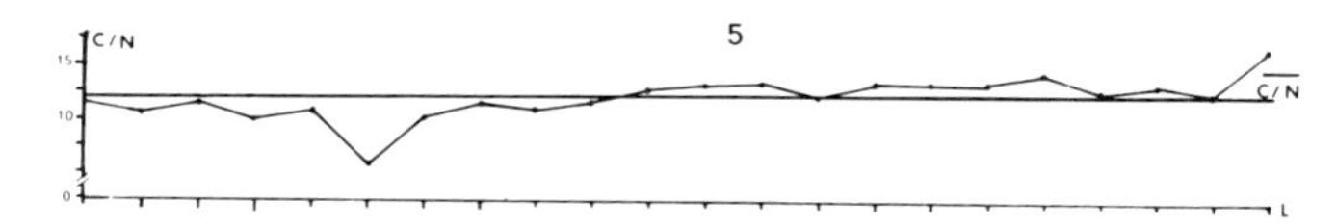

Fig. 5 C/N ratio of each level (L) and average value for the entire plot.

METHODS AND MATERIALS

Six sampling points were labelled at meter intervals on the 22 transverse sampling lines (Fig. 1). Thus the plot contained 132 sampling points. Every week, 7 points were sampled at random; 233 samples were taken during the first year of the study. Soil samples were collected with a core borer (5 cm diameter, 2 cm height), and litter was sampled, when it existed, on a 80 cm^2 surface. The mesofauna was extracted by means of a Tullgren funnel ap-

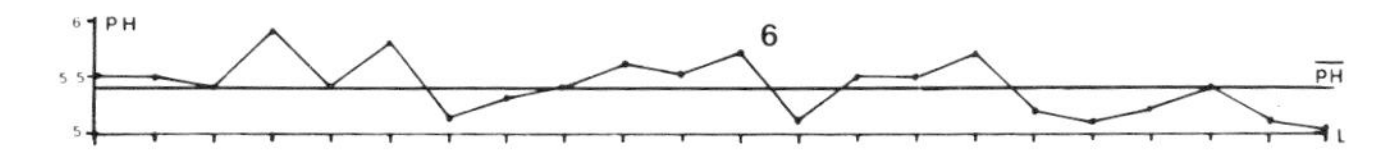

Fig. 6. pH value of each level (L) and average value for the entire plot.

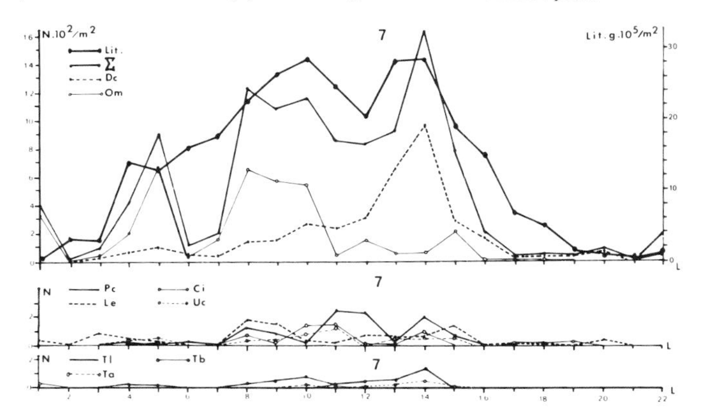

Fig. 7. Litter weight (g/m²) for each level (L, mean of 2 years); density of the total uropodine community (Σ, in number of individuals per m²); abundance of each species (Dc: *Dinychura sp;* Om: *O. minimus;* Pc: *P. cf. higginsi;* Le: *Leiodinychus* (?) sp; Ci: *C. cf sellnicki;* Uc: *U. carinatus;* T1: *T. lamda;* Ta: *T. aegrota;* Tb: *T. cf baloghi*).

paratus. After the extraction, the litter samples were dried (70°C, 24 h) and weighed. The microarthropods were sorted out and and counted in 75% ethanol under a dissecting microscope. The Uropodina were studied under the microscope using the open slide technique of Grandjean (1949). All the species and stases (larvae, PN, DN, adults) were identified.

RESULTS

Ten uropodine species occurred in the plot; 2 of them were dominant, these were *Olodiscus minimus* (Kramer) which represented 35.6% of the total uropodine fauna, and *Dinychura* sp. which comprised 31.7% of the community. The 8 other species were:*Polyaspinus* cf *higginsi* (9.4%), *Leiodinychus* sp. (8.2%), *Cilliba cf sellnicki* (5.1%), *Urodinychus carinatus* (Berl.) (4.2%), *Trachytes lamda* (Berl.) (3.7%), *T. aegrota* (Koch), *T. cf. baloghi* (0.3%) and *Oodinychus cf. ovalis*[a] (0.07%).

The average uropodine density in the plot during the first year of sampling was only 507 ind./m², but the respective numbers depended on the sampling line. Figure 7 shows that the number of mites may have been related to the litter quantities—except for level 1, the soil of which is particular (see Figs. 3 and 4). The abundance was generally higher between levels 8 and 14, i.e. in the part of the plot where the litter weight was highest. The maximum uropodine density occurred at level 14 with 1600 ind./m². Even this value is particularly low

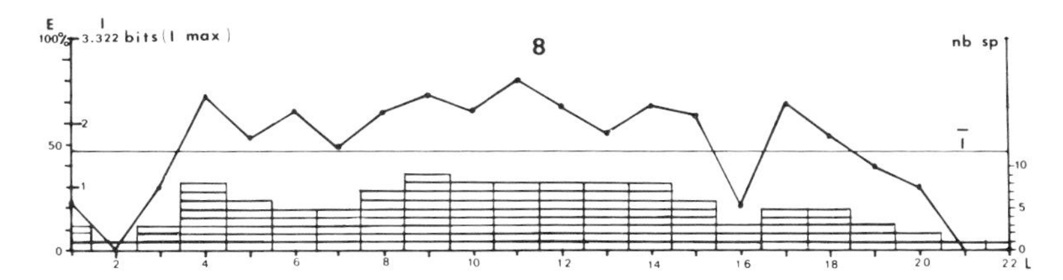

Fig. 8. Shannon diversity index ($\bar{I}$ bits), "equitability" (E%), number of species of Uropodina (sp) observed on each level (L) and average value of $\bar{I}$ for the entire plot ($\bar{I}$).

compared with data of a temperate forest ecosystem, the average numbers of which usually lie between 3500 and 9000/m^2 (Athias-Binche, 1978). The density was extremely low at the lower levels (17-22).

The highest numbers of uropodine species were encountered between levels 8 and 14, with the maximum of 9 species found at level 9 (Fig. 8). From level 4 to 7, and between levels 15 and 18, the number of species was about 5. The other parts of the plot were very poor, with only 1 to 3 species. These results show that the number of ecological niches offered to the community increases with the presence of litter. The Shannon diversity index ($I_{bits} = -\Sigma(ni \div N) \times (\log_2 ni \div N)$) is at its maximum when the abundance for each species of a community is equal. Here, with 10 species, one obtains $I_{max} = \log_2 10 = 3.322$ (Fig. 8). The diversity index and the "equitability" ($E\% = 100\ I/I_{max}$) are highest between levels 4 and 15. The maximum value occurs at level 14 (I = 2.63, E = 79.1%), due to the occurrence of 4 species with similar abundance (*Dinychura, Leiodinychus, Polyaspinus* and *U. carinatus,* see Fig. 7). It can be noted also that the density of *O. minimus* decreases clearly from level 11 to the bottom of the plot (except for level 22). Concomitantly, *Dinychura* becomes the main species.

The diversity index does not take into account the absolute density but only the relative frequency. Thus, part of the information content of the data is neglected. The correlation analysis is used in order to include the absolute density in the analysis.

The interclass correlation coefficient was used for the study of similarity between levels (= classes). Each class was represented by 10 variables (= abundance of each species). The data were normalized by using logarithmic transformation ($X = \log(x + 1)$). The Bravais-Pearson correlation coefficients were calculated between the levels with a computer; results are given in a 22 × 22 matrix. The abundance of the dominant species influenced the correlation coefficients: one recognizes a first group (levels 1 to 10), which is clearly correlated with level 22. In all these levels, *O. minimus* is the main species. On the other hand, level 11, where *Dinychura* becomes the main species and *Polyaspinus* is the second one, is not correlated with the upper levels. These features allow one to distinguish a second correlation group (11-18). The correlation values are not as high as in the first group. This may be due to the variability of the abundance of *Dinychura* from level 12 to 16. The lower levels

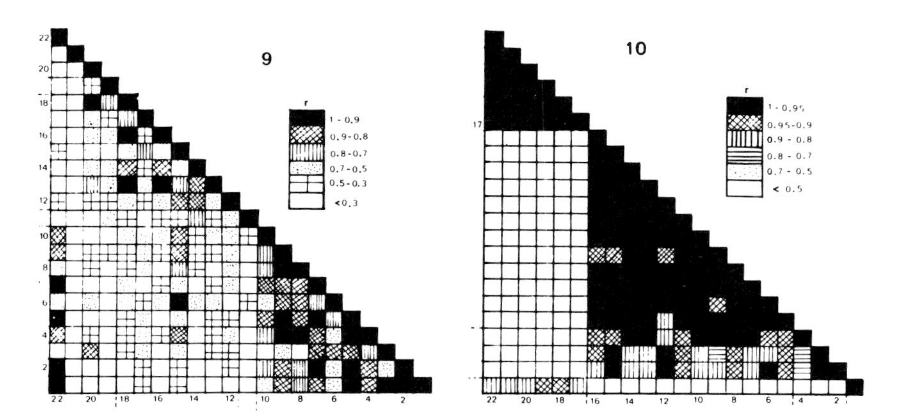

Fig. 9. Matrix of intraclass correlation coefficients between the levels for the species density.

Fig. 10. Matrix of Spearman rank correlation coefficients between each level for the litter weights.

are not correlated with the 2 upper groups (except for level 22), and the levels with a poor fauna (19-21) are not correlated with one another.

The Spearman rank correlation coefficient method was used in the case of litter weights because the average values differ between first and second year. The ranks of litter quantities were used for each level (i.e. rank for level 14, rank 2 for level 10, etc. . . .). The correlation coefficients were computed between ranks for the first year, ranks for second year and average values for the 2 years together. The resulting correlation coefficients are very high (Fig. 10). Four groups may be distinguished: 1) level 1 is correlated only with the last group (17-22), where litter is very poor (see Fig. 2 and 7); 2) group 2-3, which presents rather low correlation with 3) the following group (4-16). The levels 4 to 16 are highly correlated between one another (only 2 coefficients lower than 0.9); this group is completely different from 4) the lower group 17-22. The lower levels again are highly correlated between one another (no coefficient less than 0.95). The matrix of rank correlation for litter shows 4 main soil biotopes: 1) horizontal blackish silty soil with little litter; 2) the heterogenous ranker (group 2-3, and possibly level 4); 3) earth mull with permanent litter and feeble slope; 4) sandy eroded soil, with almost no litter (cf Fig 1 to 7).

DISCUSSION

The abundance of uropodine mites appears to be correlated with litter quantities (most of them inhabit the litter layer (Athias-Binche, 1978)) as the number of species increases in relation of litter mass and persistence. The presence and the importance of the litter depends on the topography of the slope; they are also related, perhaps to a lesser degree, with the effects of wind and rain. At first sight, the plot seems to be homogenous between levels 7 and 14 because of the litter and soil features. However, the correlation analysis of the faunistic composition of the community shows 2 types of fauna, the limits of which lie at level 11. The first fauna is dominated by *O. minimus,* a widely distributed European species, and the second is dominated by *Dinychura* sp. (probably a Mediterranean species).

These results show that the faunal pattern is a very sensitive ecological index that may be used as a precise ecological indicator in soil studies. Further research is necessary in order to find the ecological causes of the faunistic features of this uropodine community, such as temperature, soil water, or quality of litter. Different statistical analyses, especially cluster analysis, should be used in search of possible relationships between faunal and ecological factors.

SUMMARY

A community of uropodine mites was studied in the soil of a valley slope in the Massane beech-wood.

Several soil features (such as organic matter, C/N ratio and litter mass) depended on the steepness of the slope. The density and the diversity of the community was influenced by these soil features. The highest number of mites and the highest diversity index was observed in places where litter was the most concentrated, i.e. where the steepness of the slope decreased. Correlation analysis showed two different types of fauna in the plot.

REFERENCES

Athias-Binche, F. (1978). *Rev. Ecol. Biol. Sol.* **15**, 67-88.
Grandjean, F. (1949). *Bull. mus. Hist. Nat. Paris, 2e ser.*, **21**, 155-169.

MITES ASSOCIATED WITH SWEAT BEES (HALICTIDAE)

George C. Eickwort

Department of Entomology
Cornell University
Ithaca, New York

INTRODUCTION

The aculeate Hymenoptera (wasps, bees, and ants) provide stable habitats within their nests for diverse taxa of Acari, many of which have adjusted their life cycles to those of the Hymenoptera and are phoretic upon nest-founding host females. Provisions intended for host larvae, feces and debris, nest material itself, immature and adult host instars, and other inquilines all are potential food sources for mites. The mites associated with ants, higher social bees, and cavity-dwelling bees and wasps are better known than those associated with the more primitive soil-nesting wasps and bees. This paper summarizes life history information on mites associated with soil-nesting sweat bees (Halictinae). Except for some excellent papers by E. A. Cross and his associates, most previous studies of mites associated with Halictidae (Table I) have been taxonomic descriptions of the stages phoretic on adult hosts. Taxonomic descriptions and detailed biological accounts of mites from my studies will be published elsewhere.

MATERIAL AND METHODS

Cell contents were excavated from host nests and brought live to the laboratory, each cell in a separate vial or depression in a wax-lined petri dish. The mites' developmental stages were determined, their positions on the cell contents noted under a dissecting microscope, and a sample was preserved. Mites were reared along with post-feeding larval and pupal bees in wax-lined petri dishes, and with various components of the nests as food in plaster of paris-charcoal-lined petri dishes.

Vol. I: ISBN 0-12-592201-9

RESULTS

The Hosts

Halictidae is a large, cosmopolitan family of rather primitive bees. Of the 3 subfamilies, 2 have been examined for associated Acari: the Nomiinae, which includes the alkali bee, *Nomia melanderi* Cockerell, and the Halictinae, which includes the sweat bees that I studied. Halictinae may nest solitarily or in primitive societies (see Michener, 1974). Most nest in the ground although a few nest in dead wood. A female sweat bee digs a main burrow into the soil or wood, from which she digs lateral burrows, each leading to a cell that will contain just one offspring. The cell is an ovoid excavation in the substrate lined with a wax-like, waterproof glandular secretion. The bee makes a provision mass of pollen and dilute honey and lays an arched egg on top. The lateral burrow is then filled with soil; it is through this closure that many mites can enter and leave the cell. The bee larva hatches and consumes the provision mass, resting on top of it until nearly finished. It then rests on its dorsum, head towards the cell mouth, and begins defecating, plastering the feces on the upper and posterior cell walls. The post-defecating larva (called the prepupa) then pupates, the pupa resting in the same position. The emerging adult bee digs out through the lateral burrow into the main nest burrows. The duration of developmental stages of *Dialictus umbripennis* (Ellis) is: egg, 2 days; feeding larva, 4-5 days; prepupa, 4 days; pupa, 11 days (Wille and Orozco, 1970).

Mites are generally phoretic upon both sexes of their hosts, although sometimes less frequently so on males (Cross and Bohart, 1969). Male bees usually do not return to nests after emergence and mating occurs away from the nests. Unless phoretic mites transfer between hosts during copulation, only mites phoretic on female hosts succeed in reproducing. Phoretic mites disembark when their female hosts provision or oviposit in new cells.

The Mites

Anoetidae. Anoetid mites are the most frequent and highly co-evolved acarine inquilines in sweat bee nests, and indeed the type-species of *Anoetus* was named for a hypopus (deutonymph) collected on a halictine. Because most anoetids have been described only from hypopodes, the supra-specific classification is in need of revision. Mites from my studies belong to *Histiostoma* and *Anoetus* based on hypopodes and have been described by Woodring (1973) and by Delfinado, Baker, and Eickwort (unpubl.). The account below is based largely on *Histiostoma halictonida* Woodring, associated with *Halictus rubicundus* (Christ), and an undescribed *Anoetus* associated with *Dialictus lineatulus* (Crawford).

Hypopodes attach to the host's wings. On female bees they also often attach in shingle-like rows to gastral (= metasomal) tergum I or II, while on males they may occur on the venter of the head, thorax, and sometimes the

TABLE I.
Acari Associated with Halictid Bees

Family	Genus	Host[a]	Taxonomic references
Laelapidae	*Laelaspoides*	H	Eickwort, 1966
Anoetidae	*Anoetus, Histiostoma*	H	Mahunka, 1969b, 1974b; Woodring, 1973; Delfinado, Baker & Eickwort, unpubl.
	Glyphanoetus	N	Cross, 1968
Acaridae	*Sancassania*	H, N	Cross, 1968; Delfinado, Baker, & Eickwort, unpubl.
	Rhizoglyphus	H	Eickwort, unpubl.
	Halictacarus	H	Mahunka, 1975
	Schulzea	H	Delfinado & Baker, 1976a
?Saproglyphidae	*"Nanacarus"*	N	Woodring, 1966
Pygmephoridae	*Parapygmephorus*	H, N	Cross, 1965; Mahunka, 1974a; Rack & Eickwort, unpubl.; Rack & Delfinado, unpubl.
	Trochometridium	H, N	Cross, 1965
	Siteroptes	H	Mohamed & Soliman, 1974
Scutacaridae	*Scutacarus*	H	Delfinado & Baker, 1976b
	Imparipes	H, N	Mahunka, 1969a, 1974a; Cross & Bohart, 1969; Delfinado & Baker, 1976b
	Nasutiscutacarus	N	Beer & Cross, 1960

[a]H = Halictinae; N = Nomiinae

gaster. Hypopodes move from their phoretic host onto the provision mass in a new cell, and transform into tritonymphs that soon moult into adults. The adult females stay on the provision mass and swell greatly. The males remain very small and crawl upon the dorsum of the female when the host larva is about half grown. The females begin oviposition when the host larva finishes feeding, laying eggs on the cell walls and on the bee larva itself, where the mite larvae also occur. Protonymphs begin to appear when the bee larva pupates and crawl on the pupa, especially ventrally. Moulting to hypopodes begins about half way through the pupal stadium. The hypopodes preferentially cluster on the pupa's dorsal propodeal surface and about the wing bases. They transfer to the adult bee when it emerges.

The tritonymphs and adult females imbibe surface liquids of the provision mass, and the larvae and protonymphs imbibe surface liquids of the developing bees, although hypopodes can be reared from eggs in the absence of immature bees. The mites probably feed on microorganisms and do not harm the bees, a conclusion reached by Cross (1968) for *Glyphanoetus nomiensis* Cross on the alkali bee. Indeed, the relationship could be mutualistic rather than just commensal. Female *Lasioglossum coriaceum* (Smith) have a specialized glabrous area fringed by long hairs on gastral tergum I that functions as an acarinarium; 85% of museum specimens bear anoetid hypopodes.

Acaridae. I have reared *Sancassania* spp. from cells of *Lasioglossum leucozonium* (Schrank) and *Agapostemon radiatus* (Say) (Delfinado, Baker, and Eickwort, unpubl.) and *Rhizoglyphus* spp. from several species of *Dialictus*. Acarids are scavengers on dead bees, fecal remains, and especially on moldy provision masses (on which the bee eggs have died), which support large colonies of all stages. *Sancassania* literally tears apart the moldy provision mass and appears to ingest the pollen. Cross and Bohart (1969) report similar habits for *Sancassania boharti* (Cross) in alkali bee nests and hypothesize that hypopodes move through the soil to locate healthy hosts in the nest burrows. Hypopal *S. boharti* attach preferentially to gastral intersegmental spaces of the alkali bee.

Pygmephoridae. *Parapygmephorus* (subgenera *Parapygmephorus* and *Sicilipes*) are frequently found on halictine and nomiine bees (Cross, 1965; Mahunka, 1974a; Rack and Delfinado, unpubl.). The following account is of a new species of *P. (Sicilipes)* associated with *Agapostemon nasutus* Smith in Costa Rica (Rack and Eickwort, in preparation).

Phoretic adult females of *Parapygmephorus* attach preferentially to gastral tergum I and the propodeum of female *A. nasutus*. Female mites could not be recovered from cells with feeding bee larvae and may move through the soil at this time. When the bee larva defecates, the female mites move onto the feces and oviposit. The mite larvae feed on the feces or some contaminant thereof and swell greatly. Inactive, pharate adult mites occur about ⅓ through the bee's pupal stadium on pupa, feces, and cell walls. Adult male mites emerge first and frequently carry pharate adult females about with their enlarged hind legs. Adult female mites emerge about ⅔ through the pupal stadium and move freely about the cell. Just before host ecdysis, adult females concentrate on the pupa and attach to the emerging adult bee.

Parapygmephorus do not feed on developing bees nor cause them any harm. Adult females may feed only after they leave their phoretic hosts. The larva is an important feeding stage. Fungus may form the food for larvae (fungal contamination of larval bee feces always occurs) and adult females, although I did not demonstrate this.

Trochometridium tribulatum Cross and "*Siteroptes cerealium* Kirchner" differ from the above in that the mites are usually in cells with dead bee offspring (Cross, 1965; Cross and Bohart, 1969; Cross and Moser, 1971; Mohamed and Soliman, 1974). The female mites are reported to kill the immature host and feed upon it, and also upon the fungus that contaminates the provision mass. The female mite swells greatly and larvae are quiescent *(Trochometridium)* or contained within the egg choria *(Siteroptes)*. New adult females move through the soil to locate other cells. *Trochometridium* adult females (but apparently not *Siteroptes*) are phoretic on adult hosts.

Scutacaridae. *Imparipes* spp. are frequently associated with halictid bees, perhaps second in abundance to Anoetidae. Mahunka (1969a) and Delfinado and Baker (1976b) have described the scutacarids from my studies. This account is based largely on *Imparipes apicola* (Banks), an associate of diverse halictid and andrenid bees that I have reared with *Evylaeus quebecensis*

(Crawford). The life cycle is very similar to that of *Parapygmephorus*.

Adult female *Imparipes* clasp the hairs of their phoretic halictine hosts, typically ventrally on the thorax and anterior gaster and on the lower surfaces of the propodeum. As the bee larva feeds, the female mites may occasionally be found on the bee or the cell walls, but they move readily between cells in the laboratory. When the bee larva begins to defecate, the slightly swollen female mites move to the feces and begin oviposition. Larvae feed on the fecal surface (probably on fungal contaminants) while the bee is a prepua and early pupa, then become inactive pharate adults, attached to the feces and pupa. Adult males emerge about half way through the pupal stadium and carry pharate adult females with their enlarged hind legs. Adult females emerge before the end of the pupal stadium and attach to the emerging adult bee.

When I removed bee eggs from cells, female *I. apicola* were often on top of them, apparently probing the surface with their gnathosomas. In the laboratory, these bee eggs died and the female mites became quite swollen. The provision masses quickly became moldy and the mites commenced feeding on the fungus and began oviposition. Multiple generations and very large colonies resulted, the mites feeding on the fungus (Mucorales) in the absence of pollen and immature bees. I found similarly heavily infested moldy provision masses in field nests. While the mites may have killed the eggs, it is also possible that they functioned as saprophytes on already dead hosts (as might *Trochometridium* and *Siteroptes*). Many host eggs perished without mites in the moist nest site and halictine eggs are very difficult to maintain in the laboratory.

As fungivores, scutacarids could be mutualistic in bee nests. *Imparipes americanus* (Banks) is associated with healthy bee brood and 87% of adult female alkali bees carry these mites (Cross and Bohart, 1969). Similarly, 90% of *Lasioglossum titusi* (Crawford) carry *Imparipes vulgaris* Delfinado and Baker (Linsley and MacSwain, 1959, mites referred to as acarid hypopodes).

Laelapidae. In halictine nest cells, I occasionally encounter gamasine mites that are probably predators—*Cosmolaelaps vacuus* (Michael) and *Hypoaspis queenslandicus* (Womersley) (Laelapidae), *Proctolaelaps* spp. (Ascidae), and Digamasellidae. However, these are not phoretic on adult bees and I view them as soil or wood mites that incidentally enter halictid nests. The only known mesostigmatid that is an inquiline with Halictidae is *Laelaspoides ordwayae* Eickwort, a hypoaspidine mite related to the ant-associated genus *Laelaspis* (Ordway, 1964; Eickwort, 1966). Adult female *L. ordwayae* are phoretic on the thoraces and legs of female bees (*Augochlorella* spp.) at the beginning of the flight season. Ordway found all stages of the mites except larvae in cells containing bee larvae and pupae, but not in cells with dead hosts. Within single nests (although not necessarily single cells; Ordway did not preserve contents of each cell separately), all stages are found synchronously. The mites eat pollen and do not harm the developing bees.

DISCUSSION

The four groups of Acari that have evolved to become inquilines in halictid bee nests are basically commensal or even mutualistic in their relationships with their hosts. The only possible exception is the association of some Scutacaridae and Pygmephoridae with dead bee eggs and larvae, although the principal food source is the fungus that contaminates the bees' provision masses. All groups have life cycles adjusted so that food is available within the nest cells when the appropriate stage in the mite's life cycle is reached and so that the appropriate phoretic instar appears when the adult bee emerges.

The acarine associates of the soil-nesting bees in the Andrenidae, Colletinae, and Anthophorinae appear to be generally similar to those of the Halictidae. In contrast, the predominantly cavity- and vegetation-nesting Megachilidae and Xylocopinae and the social Apidae have different arrays of acarine associates, especially the Saproglyphidae, Chaetodactylidae, and diverse Mesostigmata. This suggests independent invasions by Acari into different types of bee nests.

SUMMARY

Fifteen genera in six families of mites have been associated with halictid bees. *Imparipes* (Scutacaridae) and probably *Parapygmephorus* (Pygmephoridae) feed on fungi, *Anoetus* and *Histiostoma* (Anoetidae) feed on surface liquids of bee provisions and immature bees, *Laelaspoides* (Laelapidae) feed on pollen, and *Sancassania* (Acaridae) are saprophytic scavengers. The life cycles of the mites are correlated with the development of their hosts and mutualism may occur.

ACKNOWLEDGEMENTS

I thank Drs. J. P. Woodring of Louisiana State University, Mercedes Delfinado of the New York State Museum and Science Service, Edward Baker of the Systematic Entomology Laboratory, USDA, Sandor Mahunka of the Hungarian Natural History Museum, and Gisela Rack of the Universitat Hamburg for describing mites collected from my halictine nests. I also thank Mr. Barry OConnor of Cornell University for comments on the manuscript and assistance in identifications. This research was supported by N. S. F. grants GB-35954, BMS-72-02386, and DEB-78-03151.

REFERENCES

Beer, R. E., and Cross, E. A. (1960). *J. Kans. Entomol. Soc.* **33**, 49-57.
Cross, E. A. (1965). *Univ. Kans. Sci. Bull.* **45**, 29-275.
Cross, E. A. (1968). *Southwest. Nat.* **13**, 325-334.
Cross, E. A., and Bohart, G. E. (1969). *J. Kans. Entomol. Soc.* **42**, 195-219.
Cross, E. A., and Moser, J. C. (1971). *Acarologia* (Paris) **13**, 47-64.
Delfinado, M. D., and Baker, E. W. (1976a). *J. N. Y. Entomol. Soc.* **84**, 76-90.
Delfinado, M. D., and Baker, E. W. (1976b). *Acarologia* (Paris) **18**, 264-301.
Eickwort, G. C. (1966). *J. Kans. Entomol. Soc.* **39**, 410-429.
Linsley, E. G., and MacSwain, J. W. (1959). *Univ. Calif. Publ. Entomol.* **16**, 1-46.
Mahunka, S. (1969a). *Parasitol. Hung.* **2**, 153-157.
Mahunka, S. (1969b). *Reichenbachia* **12**, 179-186.
Mahunka, S. (1974a). *Ann. Hist.-Nat. Mus. Natl. Hung.* **66**, 389-394.
Mahunka, S. (1974b). *Folia Entomol. Hung.* **27**, 99-108.
Mahunka, S. (1975). *Acta Zool. Acad. Sci. Hung.* **21**, 39-72.
Michener, C. D. (1974). "The Social Behavior of the Bees." Harvard Univ. Press, Cambridge, Mass., 404 p.
Mohamed, M. I., and Soliman, Z. R. (1974). *Ann. Agric. Sci.* (Moshtohor) **2**, 155-162.
Ordway, E. (1964). *J. Kans. Entomol. Soc.* 37, 139-152.
Wille, A., and Orozco, E. (1970). *Rev. Biol. Trop.* **17**, 199-245.
Woodring, J. P. (1966). *Proc. La. Acad. Sci.* **29**, 76-84.
Woodring, J. P. (1973). *J. Kans. Entomol. Soc.* **46**, 310-327.

6. Recent Advances in Soil Mite Biology

Section Editor

R. A. Norton

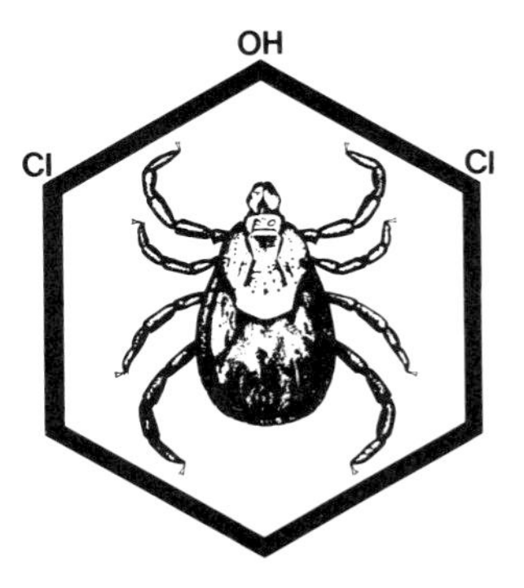

EFFECTS OF PHYSICAL PARAMETERS AND FOOD RESOURCES ON ORIBATID MITES IN FOREST SOILS

Myron J. Mitchell

State University of New York College of Environmental Science and Forestry Syracuse, New York

INTRODUCTION

The decomposition of the organic matter of forest soils and the recycling of nutrients is accomplished by a vast assemblage of organisms including bacteria, fungi, protozoa, nematodes, annelids and arthropods (Burges and Raw, 1967; Dickinson and Pugh, 1974; Anderson and Macfadyen, 1976). Among the soil arthropods, mites represent the most diverse and numerous taxon with oribatids (Cryptostigmata) forming the majority of the Acari. Because of their numerical importance, an understanding of their role within forest soils is necessary in order to ascertain the overall biotic processes within this portion of the forest ecosystem. This paper will focus on the relationships of oribatids to their physical environment and food resources within specific soils. The role of vegetation, climate and soil type in affecting the global distribution and abundance of oribatids has been considered elsewhere (Wallwork, 1976).

The forest floor is heterogeneous with a complex mosaic of biotic and abiotic components. This heterogeneity is an extension of the structural diversity of the overlying canopy and herbaceous layer which alters both the quality and quantity of litter, precipitation and incident radiation inputs. This heterogeneity, in conjunction with the inherent variation of the mineral soil, produces a distinct array of abiotic and biotic parameters which vary both temporally and spatially. The interaction between these parameters and oribatids within forest soils is discussed below.

Vol. I: ISBN 0-12-592201-9

ABIOTIC COMPONENTS

Temperature

For all organisms their thermal environment is critical with regard to basic physiological functions. Oribatids, as most invertebrates, are probably ectothermic and would have a body temperature close to that of their environment. The majority of oribatids live cryptic lives in which they are shielded from direct solar radiation although some tree and litter dwelling individuals may be directly exposed to sunlight.

Within soils there are both seasonal and diurnal temperature fluctuations, the amplitudes of which decrease with depth. Thus, those oribatids near the surface will be more subject to temperature extremes. These fluctuations have been hypothesized to be partially responsible for the vertical migration of certain oribatids (Wallwork, 1959; Mitchell, 1978). Most oribatids, however, live a relatively immobile life in which movement is over a short distance and infrequent (Pande and Berthet, 1973).

The temperature of forest soil varies not only vertically but also horizontally owing to the overlying vegetation. For example, in those soils covered with snow, the regions adjacent to trees lose snow more rapidly on account of microclimatic conditions (Mitchell, 1978). In such areas, any temperature dependent process would be accelerated.

Temperature has a direct effect on the physiological activity of oribatids as reflected in respiration, ingestion, egestion, growth (secondary production) and survival. The response often follows a power or exponential function. Respiration rates generally show Q_{10}'s from 2.8 to 4.0 (Wood and Lawton, 1973; Luxton, 1975; Mitchell, in press). Ingestion rates may have a Q_{10} of 2.2 (Kowal and Crossley, 1971), although this value may vary drastically over a given temperature range (Mitchell and Parkinson, 1976).

The effect of temperature on growth and thus secondary production has been demonstrated indirectly by laboratory studies on development. For example, *Ceratozetes gracilis* (Michael) has a maturation time of 36 weeks at 15°C (Mitchell, 1977) and 19 weeks at 20°C (Hartenstein, 1962). The growth of *Damaeus onustus* (Koch) and *D. clavipes* (Herm.) with respect to temperature approximates sigmoid curves which show Q_{10}'s of 4.0 and 3.6 respectively below 20°C (Lebrun, 1974).

The survival of oribatids is also highly temperature dependent (Madge, 1966). Although the actual lethal temperature varies between species, a maximum temperature of about 45°C has been found.

Since many oribatids are dependent on the microflora, especially fungi, for food (Luxton, 1972; Mitchell and Parkinson, 1976) and the growth of microflora is temperature dependent, this parameter may indirectly affect food availability. In addition, the seasonal changes of the soil community which include organic input from higher plants are highly correlated with temperature changes.

Moisture

The effect of temperature is inextricably linked with that of moisture since these two parameters are major determinants in the potential transpiration of an organism. The generally high relative humidity of the soil has enabled numerous invertebrates, including oribatids, to utilize the interstitial habitat. Those oribatid species able to withstand desiccation such as *Carabodes labyrinthicus* (Michael), *Steganacarus magnus* (Nicolet), and *Platynothrus peltifer* (Koch) are more common in the upper layers which are subject to moisture depletion (Madge, 1964). Moisture sensitivity also varies with the life stage of an oribatid (Baumler, 1970a, b; Mitchell, 1978).

The range in moisture tolerances is probably a major factor in the differential distribution of oribatids. This moisture effect has also been shown in studies on horizontal distributions where those species most sensitive to desiccation were found in areas of higher moisture (Popp, 1970; Mitchell, 1978).

Moisture may also have an indirect effect in altering food availability by enhancing the palatability of wood and litter to macrophytophages such as phthiracarids (Hayes, 1963). Also, it would alter the food availability for microphytophages since numerous studies have shown that the dynamics of microbial populations is highly moisture dependent. The effects of sensitivity of life stages and food availability have been shown to be reflected in the life histories of some oribatids (Bäumler, 1970b; Mitchell, 1977).

Depth

Concomitant with changes in depth are alterations in moisture, temperature, organic matter quality and quantity, pore space and soil atmosphere. The amplitude or range of these components decreases with depth as the organic matter becomes more homogeneous and the physical environment more stable. This complex of factors leads to a distinct profile within a forest soil. This can be seen clearly in a mor soil in which the upper litter consists of an intact layer of organic materials derived from the above ground system. As this material is fenestrated and comminuted by various soil invertebrates, including oribatids, and attacked by the microflora, it forms a fermentation layer. This material is further altered by catabolism and repolymerization to form a stabilized substance defined as humus. Each of these various layers, as well as subdivisions within these layers, produces specific microhabitats which support different species of oribatids (Pande and Berthet, 1973, 1975; Mitchell, 1978).

With depth there is a tendency for smaller sized oribatids to become more predominant. For example, within an aspen woodland soil, *Ceratozetes gracilis* (550 μm length) had a more upward distribution than its smaller congeneric *C. kananaskis* (Mitchell) (403 μm length) (Mitchell, 1978). Space limitations and potential transpiration may be important factors in these differential distributions. The latter factor would be a function of body size since larger forms would have a lower surface area to body weight ratio and thus a potentially lower transpiration loss per unit of body weight.

In addition, intraspecific differences in the depth distributions of oribatids have been found. The immatures of some taxa may be more concentrated than the adult forms in the fermentation horizon of mor soils (Anderson, 1975; Mitchell, 1978).

Inorganic Nutrients

In forest soils inorganic nutrients have a heteroegeneous vertical and horizontal distribution and may thus indirectly affect the distribution of oribatids. Usher (1976) found that the distibution of another major forest arthropod taxon, Collembola, was correlated to certain nutrients (N, P, K, Ca). Also for both mites and Collembola, he found that fixed nitrogen was an important factor in predicting abundance since it had a large loading in a principal component analysis in which a variety of factors were analyzed.

The oribatids may have an indirect effect on the distribution of nutrients by their overall impact on decomposition (Edwards *et al.*, 1970; Dickinson and Pugh, 1974). Also they, along with members of the other biota, may serve as foci for nutrient cycling (Carter and Cragg, 1977; Crossley, 1977).

BIOTIC COMPONENTS

Predators and Parasites

The effect of parasites and predators on oribatids is poorly documented. However, their importance has been indicated by a study on the population dynamics of *Hermannia gibba* (Koch) in which a gregarine may have been a major mortality agent (Bäumler, 1970b).

Food

There have been numerous studies on oribatid feeding preferences with recent reviews by Luxton (1972) and Harding and Stuttard (1974). Some oribatids such as *Achipteria, Steganacarus,* and *Phthiracarus* may feed directly on trachaeophyte tissue.

However, most forest oribatids cannot utilize litter unless some microbial attack has occurred. Many oribatids are known to feed predominantly on fungal hyphae which are abundant in forest soils. The distribution of fungi has been related to the vertical distribution of oribatids based on preference studies (Luxton, 1972). However, feeding specificity among oribatids may depend on the mite species, soil type and season (Anderson, 1975). Although feeding on specific fungal species cannot be ascertained from field studies, there is evidence that different morphological types of fungi are consumed by specific oribatids. This is reflected in the vertical distribution of both the fungi and oribatids (Mitchell and Parkinson, 1976).

There is a complex set of interactions between the oribatids and their fungal food source in which the mites may affect the secondary production and community structure of the fungi while the fungi may affect the population dynamics and community structure of the oribatids (Mignolet, 1971; Mitchell and Parkinson, 1976).

DISTRIBUTION

As discussed previously both the abiotic and biotic components of forest soils are distributed unevenly. This heterogeneity has been shown in the horizontal and vertical distribution of the organic layers of an aspen woodland soil (Fig. 1). The oribatids generally show a contagious horizontal distribution as indicated by the density of adult *Ceratozetes kananaskis* in the same aspen woodland (Fig. 2) (Mitchell, 1978). This clumping has often been expressed as a negative binomial distribution. The cause of these aggregations may be due to not only interactions among oribatids and their abiotic and biotic environment but also to intraspecific interactions (Usher, 1975; Mitchell, 1978).

The use of soil sectioning techniques has shown that oribatids are restricted to specific microhabitats which vary among species (Anderson, 1975; Pande and Berthet, 1975). The formation and subsequent arrangement of these microhabitats would be important factors in affecting distributional patterns, both vertically and horizontally.

POPULATION BIOLOGY AND COMMUNITY DIVERSITY

In comparison with other faunal components such as protozoa, nematodes, and even Collembola, the life cycles of the oribatids are more protracted and they show a more stable population structure (Lebrun, 1970; Mitchell, 1977). This protraction is also reflected in their low weight-specific metabolic rates (Wood and Lawton, 1973; Mitchell, in press) and may enhance their role in stabilizing nutrient fluxes.

A forest soil clearly shows marked differences both in space and time. The different soil microhabitats allow the oribatids to utilize a variety of resources and potentially partition space and food. Although there may be a wide

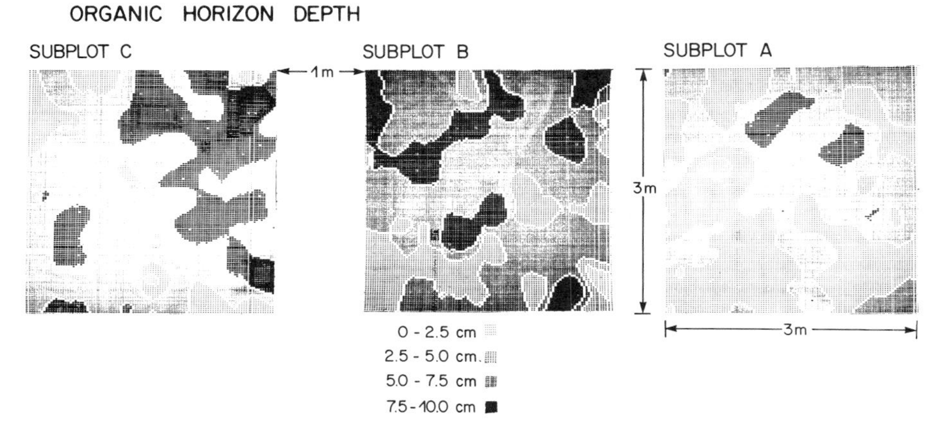

Fig. 1. Horizontal distribution of total organic horizons (litter, fermentation and humus) in an aspen woodland soil. Adapted from Mitchell (1978). Copyright (1978) by the Ecological Society of America.

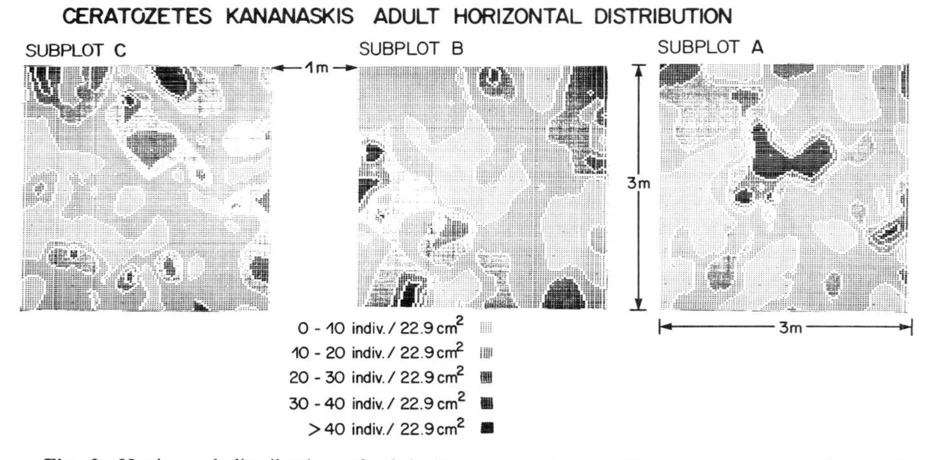

Fig. 2. Horizontal distribution of adult *Ceratozetes kananaskis* in an aspen woodland soil. Adapted from Mitchell (1978). Copyright (1978) by the Ecological Society of America.

overlap in the food utilized, if the organisms are restricted to specific microhabitats, the competition for this resource would be decreased. It has been shown that within the different layers of a forest soil the oribatids are partitioned vertically and their feeding and population parameters may be synchronized to the availability of fungal food resources (Mitchell and Parkinson, 1976; Mitchell, 1977; Anderson, 1978).

The high species richness of oribatid mites in forest soils may be due to variation in their population parameters, soil complexity, high habitat productivity and stability (Mitchell, 1977; Anderson, 1978). To better understand the basic biological attributes of oribatids more emphasis should be placed on the determination of the physiological and population parameters of specific species. This type of knowledge is necessary before we can more fully understand the biological variation inherent within the oribatids. This information can be integrated into formulating a more complete picture of the functioning of the oribatid community and its interactions with the other members of the soil biota.

SUMMARY

A forest soil is a mosaic of biotic and abiotic components. These components are arranged differentially with respect to horizontal and vertical distribution and temporal patterns.

Of the abiotic parameters temperature and moisture seem especially critical in affecting the physiological activity of oribatids and their distribution. Depth is a complex variable which is linked with a number of components, all of which may affect both the inter- and intra-specific distribution of oribatids. Inorganic nutrient cycling may also be linked with the oribatids.

With regard to the biotic components of a forest soil, food is probably the most important in affecting the biology of oribatids. The distribution and population dynamics of microphytophages may be directly related to the availability of microbial food resources.

The high species richness of oribatids in forest soil may be explained by their stable population structure, separation in microhabitats, and partitioning of food as well as the stability and high productivity of their habitat.

REFERENCES

Anderson, J. M. (1975). *J. Anim. Ecol.* **44**, 475-495.

Anderson, J. M. (1978). *Oecologia* **32**, 341-348.

Anderson, J. M. and Macfadyen, A., eds. (1976). "The role of terrestrial and aquatic organisms in decomposition processes." Blackwell Sci. Publ., Oxford.

Baumler, W. (1970a). *Z. Angew. Entomol.* **66**, 257-277.

Baumler, W. (1970b). *Z. Angew. Entomol.* **66**, 337-362.

Burges, A. and Raw, F., eds. (1967). "Soil Biology." Academic Press, London.

Carter, A. and Cragg, J. B. (1977). *Pedobiologia* **17**, 169-174.

Crossley, D. A., Jr. (1977). *In* "Biology of Oribatid Mites" (D. L. Dindal, ed.) pp. 71-85. Publications Office, CESF, SUNY, Syracuse, NY.

Dickinson, C. H. and Pugh, G. J. F., eds. (1974). "Biology of plant litter decomposition." Academic Press, London.

Edwards, C. A., Reichle, D. E. and Crossley, D. A., Jr. (1970). *In* "Analysis of temperature forest ecosystems" (D. E. Reichle, ed.), pp. 147-172. Springer-Verlag, NY.

Harding, D. J. L. and Stuttard, R. A. (1974). *In* "Biology of plant litter decomposition" (C. H. Dickinson and G. J. F. Pugh, eds.) pp. 480-532. Academic Press, London.

Hartenstein, R. (1962). *Ann. Entomol. Soc. Am.* **55**, 583-586.

Hayes, A. J. (1963). *Entomol. Exp. Appl.* **6**, 241-256.
Kowal, N. E. and Crossley, D. A., Jr. (1971). *Ecology* **52**, 444-452.
Lebrun, P. (1970). *Oikos* **20**, 34-40.
Lebrun, P. (1974). *Acarologia* **16**, 343-357.
Luxton, M. (1972). *Pedobiologia* **12**, 434-463.
Luxton, M. (1975). *Pedobiologia* **15**, 161-200.
Madge, D. S. (1964). *Acarologia* **6**, 566-591.
Madge, D. S. (1966). *Acarologia* **8**, 155-160.
Mignolet, R. (1971). *In* IV. Colloquium Pedobiologiae. C. R. 4éme Coll. Int. Zool., pp. 155-162. Paris, I. N. R. A.
Mitchell, M. J. (1977). *Pedobiologia* **17**, 305-319.
Mitchell, M. J. (1978). *Ecology* **59**, 516-525.
Mitchell, M. J. *Pedobiologia,* in press.
Mitchell, M. J. and Parkinson, D. (1976). *Ecology* **57**, 302-312.
Pande, Y. D. and Berthet, P. (1973). *Oecologia* **12**, 413-426.
Pande, Y. D. and Berthet, P. (1975). *Trans. R. Entomol. Lond.* **127**, 259-275.
Popp, E. (1970). *Oikos* **21**, 236-240.
Usher, M. B. (1975). *Pedobiologia* **15**, 355-363.
Usher, M. B. (1976). *In* "The role of terrestrial and aquatic organisms in decomposition processes" (J. M. Anderson and A. Macfadyen, eds.), pp. 61-94, Blackwell Sci. Publ., Oxford.
Wallwork, J. A. (1959). *Ecology* **40**, 557-563.
Wallwork, J. A. (1976). "The distribution and diversity of soil fauna." Academic Press, London.
Wood, T. G. and Lawton, J. H. (1973). *Oecologia* **12**, 169-191.

SOIL MITES IN THE MARINE ENVIRONMENT

Reinhart Schuster

Zoological Institute
University of Graz
Austria

INTRODUCTION

Soil mites are a characteristic component of the littoral fauna (Schuster, 1962, 1965). They are to be found in all climatic zones in the supralittoral as well as in the intertidal area. There exist especially several families and genera which have littoral representatives in all or at least in several continents. This fact could be interpretable as an indication of the existence of specific adaptations in these taxa for the life in the marine environment. Examples of such a wide distibution are the family Selenoribatidae (Schuster, 1977), the genus *Dendrouropoda* (Schuster, unpubl.) and even species such as *Nanorchestes amphibius, Rhodacaropsis inexpectatus* (Schuster, 1965) and some *Ameronothrus* species (Schulte, 1975).

Our knowledge of the biology of littoral soil mites has been greatly expanded by investigations during the last 15 year period. In this paper a survey of what has been accomplished so far is given—mainly in the field of overflow tolerance, feeding biology and reproduction—based on literature and our own research program, including hitherto unpublished results.

OVERFLOW TOLERANCE

The dominating factor for soil arthropods in the littoral zones is the rhythmical, sometimes arhythmical, overflow with seawater. Littoral mites have a high tolerance as flood experiments demonstrate.

In a study conducted to demonstrate tolerance of mites to seawater, several groups (A-D, Table I) of mites that originated in the Mediterranean, North Sea, Baltic Sea and inland regions were subjected to total submersion in seawater/freshwater. The mites were immersed in water, 18 cm^3/mite, 18°C ± 0.5°C, in an environmental chamber under faint light. The results indicate that they can withstand submersion for several months; the absolute maximum was

Vol. I: ISBN 0-12-592201-9

TABLE I.
Overflow Tolerance of Seawater at 32 Parts per Thousand (SW) and Freshwater (FW) among Littoral and Terrestrial Soil Mites from the Mediterranean Sea (MS), North Sea (NS), Baltic Sea (BS), and Inland (IN).

				Tolerance in Days			
		Number of Individuals		50% Mortality		Maximum Mortality	
Species	Origin	SW	FW	SW	FW	SW	FW
Group A							
Hydrogamasus salinus	MS	20	20	136	2	272	6
Hydrogamasus giardi	MS	20	20	125	1	238	4
Ameronothrus lineatus	NS	20	20	74	48	249	132
Ameronothrus marinus	NS	20	20	68	60	143	109
Urosternella neptuni	MS	10	10	61	12	136	45
Haloribatula tenareae	MS	8	6	–	–	108	72
Macrocheles superbus	NS	10	10	–	–	84	27
Group B							
Ameronothrus maculatus	BS	20	20	43	84	160	177
Hermannia subglabra	NS	12	12	–	84	160	177
Group C							
Platynothrus peltifer	NS	20	12	18	68	41	162
Platynothrus peltifer	IN	20	20	17	74	45	197
Group D							
Steganacarus magnus	IN	20	20	7	97	23	226
Damaeus onustus	IN	20	20	–	–	36	191
Xenillus tegeocranus	IN	20	20	3	62	18	108
Euzetes globulus	IN	14	8	–	–	19	64
Oribatula sp.	IN	10	10	–	–	12	88
Cilliba sp.	IN	10	10	2	26	12	88
Nothrus silvestris anauniensis	IN	20	20	2	25	6	74
Pergamasus runciger	IN	20	20	–	–	1	18

ca nine months, reached by one specimen of the thalassobiont, *Hydrogamasus salinus* (Table I).

As can be seen in Table I (including unpublished data obtained in investigations made after earlier laboratory experiments—Schuster, 1962, 1965, 1966), there exist considerable differences between littoral and terrestrial (inland) species in tolerance of seawater and freshwater. All species which exist only in the marine littoral zones (Group A = thalassobiont) are able to survive a very long overflow with seawater. But most of them also withstand freshwater for a long period; they are water-tolerant in general. Exceptional is the quick lethal effect of freshwater in *Hydrogamasus salinus* and *H. giardi* (this fact will be discussed later). The tolerance of terrestrial species (Group D), including the indifferent *Platynothrus peltifer* (Group C, wide distribution in terrestrial biotopes but also abundant in salt marsh), is contrary: overflow with freshwater is much better tolerated than with seawater. The thalassophile species (Group B), *Ameronothrus maculatus* and *Hermannia subglabra* have generally a high water tolerance similar to that of the thalassobiont species, but in contrast, freshwater is better tolerated. Concerning overflow tolerance, both thalassophile species rank between thalassobiont and terrestrial species. This result demonstrates a notable correlation with the ecological requirements of these compared groups of soil mites. Among investigated Ameronothridae, *A. maculatus* has the highest freshwater tolerance. There is also a remarkable correlation with the results obtained in field investigations, because this species is the only one distributed from the sea coast far inland along the shore lines of rivers (Schulte *et al.*, 1975).

Overflow experiments with Selenoribatidae are not finished. Interim results, omitted from Table I, also indicate a high seawater tolerance (Schuster, unpubl.).

Investigations on *Hyadesia fusca* by Ganning (1970), *Hermannia subglabra* by Weigmann (1973) and *Ameronothrus* by Schulte (1978) demonstrate additionally the influence of salinity, temperature and flooding rhythm on overflow tolerance. It is of special interest that rhythmic overflows reduce tolerance. In his detailed studies of Ameronothridae, Schulte (1978) elicited differences between males, females, nymphs and larvae with regard to seawater tolerance. The lower tolerance of juvenile stages could be an explanation for the occurrence of nymphs and larvae in upper zones of the littoral, above the habitats of adults. Generally, overflow is one of the most important factors regulating the vertical zonation of littoral soil mites, as Schulte (1978) demonstrated.

Another decisive factor for soil mites in the littoral zones is the influence of salinity, especially effective during the period of submersion. In the last few years attempts to obtain measurements of concentration of the haemolymph by microcryoscope were successful for *Hermannia subglabra* (Weigmann, 1973) and some species of Ameronothridae (Schulte, 1978). The results can be summarized as follows: littoral soil mites possess the ability of hypertonic regulation to water of lower osmotic values than their haemolymph; however to water of higher values they are poikilosmotic. But the efficiency of osmoregulation seems to be unequal among littoral species, as seen in the aberrant behavior of *Hydrogamasus salinus* and *H. giardi* in resistance experiments—several months survival time in seawater but only a few days in freshwater (see Table I). That leads to the assumption that in these mites the osmoregulation is insufficient if the salinity of water reaches a low degree. It is probably an indication that both species have a particular physiological adaptation for living in the marine environment. A comparable example is known among littoral Collembola, proved by Weigmann (1973) in experiments with *Archisotoma pulchella*.

The mechanism of osmoregulation in soil mites is quite unknown. The results of histological investigations (Woodring, 1973) and detailed electronmicroscopical investigations (Alberti and Storch, 1977) indicate that coxal glands are probably organs with such a function.

The ability of littoral soil mites to survive a long period of permanent flood presupposes that these air-breathing animals are capable of removing oxygen from the surrounding water. Covered with seawater the mites become immobile, and in this physical state oxygen consumption seems to be very low. In marine insects plastron respiration is very common (Hinton, 1976) but in littoral soil mites morphological features interpretable as adaptations for underwater respiration are hitherto rare. In the thalassobiont oribatid genus *Fortuynia,* v.d. Hammen (1963) described a system of canals near the base of the legs, connected with spiracula. He interpreted this extraordinary structure—named "van der Hammen's Organ" by Luxton (1967)—as an air reservoir when the mite is submerged during high tide. Last year I found similar structures in other *Fortuynia* species. A detailed morphological and physiological analysis is presently going on. Another example is given by Krantz (1974); the intertidal uropodid species *Phaulodinychus mitis* is characterized by an abnormal morphology of their peritrema which suggests a plastron respiration. The microstructure of the notogaster of *Ameronothrus* (Schubart, 1975) is probably also an indication of the existence of a plastron. Schulte (1978) recently discussed this problem by adding scanning photographs. Meanwhile, I found some further indications for plastron respiration in littoral soil mites e.g. in Selenoribatidae and a special investigation was started.

FEEDING BEHAVIOR

For a long period very little was known about feeding habits in littoral soil mites; detailed studies (e.g. King, 1914) were rare. Our recent knowledge in this field of littoral research is essentially based on a few investigations in the last 25 years.

Mesostigmata—Gamasida

Based on published results, primarily obtained by Glynne-Williams and Hobart (1952), Luxton (1966), Schuster (1962) and unpublished data about feeding habits of *Macrocheles superbus, Cyrthydrolaelaps hirtus, Arctoseius* sp. and other undetermined species, it can be stated that the distinct majority of littoral Gamasida is carnivorous (Table II). Some species seem to be specialized predators, others have a wide range of food sources. Among the littoral gamasid fauna *Thinoseius fucicola* is an exception, because this species feeds on decayed marine phanerogames like *Zostera* and algae, but *Fucus platycarpus* wrack is the best breeding substrate, as Remmert (1956) proved in laboratory experiments.

TABLE II.
Feeding Habits among Littoral Soil Mites. Legend: Nemertini (1), Nematoda (2), Oligochaeta (3), Polychaeta (4), Mollusca (5), Tardigrada (6), Crustacea (7), Collembola (8), Pterygota Adults (9), Pterygota Larvae (10), Acarina (11) and on Plant Material Including Blue-Green Algae (A), Green Algae (B), Fungi (C), Lichens (D) and Decayed Wrack (E). Active Feeding = +, Feeding Only on Damaged Animals = (+); Questionable Results = ?.

	Animal Foods										
Acarine Group	1	2	3	4	5	6	7	8	9	10	11
Gamasina	+		+	+	(+)		(+)	+		+	+
Uropodina		+	(+)	(+)	(+)		(+)			+	
Prostigmata								+	+	+	+
Endeostigmata						+	+				
Oribatei						+	(+)	(+)		(+)	

	Plant Foods				
Acarine Group	A	B	C	D	E
Gamasina					+
Uropodina					?
Prostigmata		+			
Endeostigmata		+			
Oribatei	+	+	+	+	?
Astigmata	?	+			

Mesostigmata—Uropodina

Very little is known about feeding habits of Uropodina especially in littoral species. Experiments suggested that *Dendrouropoda vallei* and *Urosternella neptuni* are carnivorous (Schuster, 1962, unpubl.). They did not attack living prey but they fed on morbid or crushed animals. An undescribed species from the Bermuda Islands was observed catching nematodes and chronomid larvae in the natural substrate while in experiments it also fed on fragments of crushed polychaetes, isopods and copepods (Schuster, unpubl.).

Prostigmata

Since terrestrial species of *Halotydeus* are phytophagous (e.g. *H. destructor* is a pest of crops in several continents), we could find out that adults and also juvenile stages of *H. albolineatus* from the rocky coast of Bretagne feed on green algae (Schuster and Schuster, unpubl.). In contrast, *H. hydrodromus,* a closely related species, should be predaceous, but the statement by Glynne-Williams and Hobart (1952) that "Halbert (1920) observed them feeding on the small mite *Nanorchestes amphibius* . . ." is probably an error, since such information is not contained in Halbert's paper.

All littoral species of Rhagidiidae, Bdellidae and Erythraeidae are carnivorous. In breeding experiments littoral and also terrestrial species of *Rhagidia* were kept alive for a long time with Collembola and juvenile stages of mites (Ehrnsberger, 1973). On the Breton rocky coast, several times an intertidal *Rhagidia* sp. was observed attacking isotomid Collembola such as *Axelsonia littoralis* (Schuster, unpubl.), an indication that Collembola are the main food under natural conditions. Collembola and soft-skinned mites are also the preferred prey of Bdellidae in general (Alberti, 1973). *Neomolgus littoralis,* among mites the biggest predator in the littoral zones, can also be seen hunting large pterygote insects, especially flies. The prey spectrum of Erythraeidae is, however, distinctly different from that of Bdellidae. Field observations and laboratory experiments also indicated differences in feeding habits between the two littoral species of *Abrolophus.* Mites (*Ameronothrus, Hyadesia, Nanorchestes, Halacaridae*) are the prey of *A. rubipes* while *A. passerinii* seems to be specialized, feeding without exception on larvae and pupae of chironomids. Occasionally cannibalism and egg feeding occur in both species. An unpublished summary is given by Witte (1972). In additional experiments *A. rubipes* was observed sucking on damaged, nearly immobile *Rhagidia* and *Neomolgus* (Schuster, unpubl.).

Endeostigmata

The published data about feeding habits of Nanorchestidae are contradictory. In the course of our comparative studies we could prove that littoral as well as terrestrial representatives of the genus *Nanorchestes* are algivorous (Schuster and Schuster, 1977). Quite another feeding habit was observed in the genus *Pachygnathus.* Laboratory experiments were not successful (Schuster, 1962) but subsequent examinations of the gut content of *P. marinus* from the West Mediterranean coast indicated that this species is carnivorous (Schuster, unpubl.). Fragments of arthropods, especially copepods, and very often of tardigrades, could be identified. It is suggested that the slow tardigrades were easily overcome by this particular rapid predator.

Oribatei

Concerning feeding biology, the oribatids are the best known group among all littoral soil mites. Schulte (1976a) recently published a detailed study about Ameronothridae. All species feed on microphyta but with a variable degree of specialization. Green algae are the food of species living on intertidal rocks while salt marsh species feed primarily on fungi. For *Ameronothrus maculatus* and the terrestrial *A. lapponicus* the main food consists of lichens. It is remarkable that feeding and defecation in the intertidal *A. marinus* follow an endogeneous rhythm which is correlated with the tide rhythm. In tideless littoral, such as in the Baltic Sea, the feeding activity is synchronized with diurnal cycles (Schulte, 1976b). The feeding behavior is also connected with distinct vertical migrations (Schulte, 1973).

In contrast to Ameronothridae, the small body size of Selenoribatidae makes feeding observations and choice experiments very difficult. First examinations of the gut content did not bring concrete results (Schuster, 1962; "*Thalassozetes* sp." is *Schusteria littorea* Grdj.). In subsequent studies, fragments of tardigrades were found in the gut (Schuster, 1977). This unexpected result leads to the assumption that Selenoribatidae are carnivorous. But recent investigations in selenoribatids of the Bermuda Islands call this statement into question; several specimens of a still undescribed species were observed feeding on blue-green algae under natural conditions (Schuster, unpub.). Blue-green algae seem to be the main food of *Fortuynia.* This is the first report of feeding experiments and examinations of the gut content which were started last year.

Haloribatula tenareae, a littoral species of the Mediterranean rocky shores, is probably carnivorous. The gut contents sometimes include, among other particles, fragments of Collembola (Schuster 1962). In laboratory experiments feeding on damaged larvae of Diptera (Dolichopodidae) and marine isopods was observed (Schuster, unpubl.). Among the Ameronothridae, as it was discussed before, there exist other littoral oribatids which feed on fungi. In choice experiments Luxton (1966) reported a fungal diet for *Punctoribates quadrivertex* and *Hermannia pulchella.* It is interesting that examinations of the gut content in the related *Punctoribates hexagonus* from European inland salt soils brought the same results (Schuster, 1959). *Passalozetes bidactylus* and *P. perforatus* from the sea-shore, as well as the terrestrial *P. intermedius,* are also typical fungi feeders (Schuster, 1960, 1962). Another feeding habit is shown by *Oribatula thalassophila* which lives in the supralittoral zone along the Breton rocky shore; it feeds on lichens (Schuster, unpubl.).

Astigmata

The world-wide distributed family Hyadesiidae consists of several species. *Hyadesia fusca,* recorded from Sweden, feeds on Enteromorpha (Ganning, 1970). Identical results were obtained in the course of investigations on morphology and ecology of this species in Germany, Denmark and Norway (Schuster and Busche, unpub.). Since the first unsuccessful studies on *Hyadesia sellai* (Schuster, 1962) it was possible to investigate a great number of individuals from localities in the Northern Adriatic. Examinations of gut contents and additional feeding experiments have now proven that green algae are the main food, as in *H. fusca*; rarely the gut also contains diatoms and blue-green algae, but only in low concentrations (Schuster, unpubl.).

A comparative survey on feeding habits of littoral soil mites shows that a great deal of the food comes from the sea; marine animals and plants must be considered as an important base for the existence and abundance of soil mites in the marine environment (Table II). On the other hand, mites are also prey for other components of the littoral fauna, such as pseudoscorpions, spiders, beetles, etc. (Schuster, 1962, fig. 11, and 1965, fig 12). These facts indicate in general the prominent role of soil mites in the littoral web of life.

REPRODUCTION

The amount of published data about sexual biology in littoral soil mites is now relatively great. Especially in the last few years many gaps in our knowledge were filled. Now it is possible to compare mating behavior, mode of sperm transfer, and morphology of spermatophores between littoral and related terrestrial mites to find out if there exists any adaptation which can be correlated with life conditions in the littoral area. The well known sexual biology of several typical littoral mites will be discussed now.

In *Hydrogamasus salinus* and *Macrocheles superbus* the mating behavior and sperm transfer were studied in detail (Schuster, unpubl.; Schuster and Rost, unpubl.). We could not find any indication for such a littoral adaptation.

The sperm transfer of Rhagidiidae is quite different from that of other Prostigmata. *Rhagidia* does not produce stalked spermatophores; rather, sperm drops are fixed on silk threads (Ehrnsberger, 1977). There is no differences between terrestrial and littoral species which would indicate a correlation with life conditions in the marine environment. On the contrary, the more complicated structures around the sperm drops exist in inland species.

Terrestrial as well as littoral species of Bdellidae transfer their sperm indirectly by stalked spermatophores. The morphology of spermatophores differs among genera and species as Alberti (1974) stated; special littoral adaptations are not to be found. Additional results obtained on several terrestrial species (Wallace and Mahon, 1976) confirm this statement. Comparative studies made by Witte (1973, 1977) on Erythraeidae also demonstrated the absence of adaptations; stalked spermatophores with complicated structures are to be found in littoral as well as in terrestrial species.

Spermatophores of the littoral *Tydeus (Pertydeus) schusteri* and *Nanorchestes amphibius* are stalked without special structures (Schuster and Schuster, 1970, 1977). The deposition of spermatophores in the form of a tangled thicket, observed in both species, is extraordinary. But it is not possible to interpret this behavior as an adaptation for living under littoral conditions. Comparable investigations in related terrestrial species are lacking for Tydeidae as well as Nanorchestidae.

A detailed investigation on reproduction of littoral oribatids does not exist, but several authors recorded deposition of spermatophores (e.g. Luxton,

1966). Based on our observations, which included *Ameronothrus, Hermannia subglabra* and also inland species of *Hermannia* (Schuster and Schuster, unpubl.), it can be stated that neither the form of spermatophores nor the mode of deposition indicates any adaptations in littoral species.

SUMMARY

In all climatic regions the littoral fauna includes soil mites as a characteristic component. Life conditions in littoral zones are dominated by flood-tide. The overflow tolerance of mites corresponds very well with their ecological requirements. Littoral species withstand a permanent overflow with seawater for several months (at 32° / oo, a maximum of nine months); terrestrial species, however, demonstrate a much lower tolerance. The overflow tolerance of freshwater is high throughout the majority of the littoral as well as terrestrial species. Among littoral species the degree of osmoregulation seems to be unequal. Morphological features interpretable as adaptations for underwater respiration (plastron) are hitherto rare.

Different feeding habits with a smaller or wider range of specialization exist among littoral soil mites. The mite fauna consists chiefly of carnivorous, algivorous and fungivorous species; among carnivores, the zoophages dominate the saprophages in the number of species. The sea also has a remarkable ecological influence on feeding biology, because marine organisms represent a high percentage of food material.

A detailed comparison of mating behavior, sperm transfer and morphology of spermatophores between littoral and terrestrial soil mites does not indicate any adaptation which could be correlated with the extreme conditions in the marine environment.

REFERENCES

Alberti, G. (1973). *Z. Morph. Tiere.* **76**, 285-338.
Alberti, G. (1974). *Z. Morph. Tiere.* **78**, 111-157.
Alberti, G. and Storch, V. (1977). *Zool. Jb. Anat.* **98**, 394-425.
Ehrnsberger, R. (1973). Dissertation, Universität Kiel.
Ehrnsberger, R. (1977). *Acarologia* **19**, 67-73.
Ganning, B. (1970). *Oecologia (Berl.)* **5**, 127-137.
Glynne-Williams, J. and Hobart, J. (1952). *Proc. Zool. Soc. Lond.* **122**, 797-824.
Hammen, L. v. d. (1963). *Acarologia* **5**, 152-167.
Hinton, H. E. (1976). *In* "Marine Insects" (L. Cheng, Ed.), 43-78. North-Holland Publishing Company, Amsterdam-Oxford.
King, L. A. L. (1914). *Proc. Roy. Phys. Soc. Edinb.* **19**, 129-141.
Krantz, G. W. (1974). *Acarologia* **16**, 11-20.
Luxton, M. (1966). *Acarologia* **8**, 163-174.
Luxton, M. (1967). *New Zealand Journal of Marine and Freshwater Research* **1**, 76-87.

Remmert, H. (1956). *Z. Morph. Tiere.* **45**, 146-156.
Schubart, H. (1975). *Zoologica* **123**, 23-91.
Schulte, G. (1973). *Netherlands Journal of Sea Research* **7**, 68-80.
Schulte, G. (1975). *Veroff. Inst. Meeresforsch. Bremerhaven* **15**, 339-357.
Schulte, G. (1976a). *Pedobiologia* **16**, 332-352.
Schulte, G. (1976b). *Marine Biology* **37**, 265-277.
Schulte, G. (1978). *Veröff. Inst. Meeresforsch. Bremerhaven.* (In press).
Schulte, G., Schuster, R. and Schubart, H. (1975). *Veroff. Inst. Meeresforsch. Bremerhaven.* **15**, 359-385.
Schuster, R. (1959). *Sitzungsberichte der Osterreichischen Akademie der Wissenschaften, Mathem.-naturw. Kl., Abt. I* **168**, 27-78.
Schuster, R. (1960). *Mitteilungen des Naturwissenschaftlichen Vereines für Steiermark* **90**, 132-149.
Schuster, R. (1962). *Int. Revue ges. Hydrobiol.* **47**, 359-412.
Schuster, R. (1965). *In* "Verhandlungen der Deutschen Zoologischen Gesellschaft in Kiel 1964," 492-521. Akademische Verlagsgesellschaft, Leipzig.
Schuster, R. (1966). *Veroff. Inst. Meeresforsch. Bremerhaven Sonderband* **2**, 319-327.
Schuster, R. (1977). *Acarologia* **19**, 155-160.
Schuster, I. J. and Schuster, R. (1970). *Naturwissenschaften* **57**, 256.
Schuster, R. and Schuster, I. J. (1977). *Zool. Anz.* **199**, 89-94.
Wallace, M. M. H. and Mahon, J. A. (1976). *Acarologia* **18**, 65-123.
Weigmann, G. (1973). *Z. wiss. Zool.* **186**, 295-391.
Witte, H. (1972). Dissertation, Universitat Kiel.
Witte, H. (1975). *Z. Morphol. Tiere.* **80**, 137-180.
Witte, H. (1977. *Acarologia* **19**, 74-81.
Woodring, J. P. (1973). *J. Morph.* **139**, 407-429.

SOIL MITE COMMUNITY DIVERSITY

Ph. Lebrun

University of Louvain, Animal Ecology
Louvain-la-Neuve, Belgium

INTRODUCTION

The lecture at hand does not intend to give a global view of all the problems encountered in soil acarology. On the contrary, we want to present a deliberate selection of those aspects we consider most important. Every choice depends on the subjectivity of its author. Therefore, without justifying our options, we can at least express the moti :es and ideas which are our basic points as expressed in the summary of this paper. Let us take these three points and develop certain aspects in the light of specific and concrete facts.

THE ROLE OF SOIL MITES IN BIOLOGICAL FERTILITY

Present Status

It is generally accepted today that saprophagous mites are responsible for the indispensable process of antiphotosynthesis, and this by small, multiple and complementary stages that are a direct result of the species diversity of this group (Wallwork, 1976). The role of detritiphagous species in an ecosystem can be illustrated by the classic model of a deciduous forest published by Duvigneaud (1974). From this general model, it is clear that mites play an essential part in the biological fertility of the soil. Their activity contributes greatly to organic decomposition, the synthesis of humus, the restitution of biogenic elements, and the stimulation of fungal and bacterial metabolism. Today these are recognized and experimentally demonstrated facts. When we concern ourselves with the particular functions carried out by mites we can take into account four general principles.

1) The result of mechanical breakdown is to increase the surface area that can be attacked by microorganisms, as well as to facilitate and accelerate the

Vol. I: ISBN 0-12-592201-9

leaching of hydrosoluble elements and the hydration of the organic matter (Van der Drift and Witkamp, 1960; Witkamp and Crossley, 1966; Curry, 1969).

2) The digestive transit that assures chemical and biological breakdown will be more or less intense depending on the species; this is accompanied by a mixing of mineral and organic elements and microorganisms.

3) The accumulation of largely humifiable excreta that can clearly be seen in soil sections (Minderman, 1956; Haarløv, 1960; Anderson and Healey, 1970; Pande and Berthet, 1973) constitutes a highly fertile environment which facilities the growth of roots and the germination of seeds. Furthermore, the consumption of dead roots by the detritiphagous mites considerably increases soil porosity and the development of humus-rich galleries.

4) The close relations between microorganisms and mites deserve to be particularly developed as nearly 56% of the net production of fungi is consumed by mycophagous species (McBrayer *et al.*, 1974). On the other hand, organic decomposition is five times faster when microorganisms and mites work together than by microorganisms alone (Ghilarov, 1963).

Relations between Mites and Microorganisms

These basically trophic relations present fundamentally different consequences which are very important in the functional dynamics of the edaphic ecosystem, as seen in Table I. Bacterial and fungal stimulation (in the sense of Macfadyen, 1964) by mites is often highly specific. Apart from the fact that each species modifies the quality of the substrate capable of being colonized by fungi, the selective grazing accelerates or slows down the liberation of nutrients and prevents the aging of microbial populations.

TABLE I.
Interactions of Soil Fungi and Oribatid Mites. (After Mitchell and Parkinson, 1976).

Interaction	Process	Effect
Mites influencing fungi	1. Quantitative removal of fungi	1. Decomposition rate; growth rate and prevention of fungi senescence.
	2. Qualitative removal of fungi	2. Community structure of fungi; decomposition rate.
	3. Dissemination of spores (mixing)	3. Community structure of fungi; decomposition rate; enrichment of deeper soil layers.
Fungi influencing mites	1. Source of food	1. Population dynamics of mites (fecundity, survival, etc.) Community structure (distribution, species abundance)
	2. Toxin production	2. Community structure; heterogeneity and stability increase

The regulative role of organic decomposition may seem at times anecdotic and difficult to quantify and present in precise terms (Crossley, 1977). However, thanks to highly refined experimental approaches, these functions, considered for a long time as mere possibilities, have become accepted as reality. In this way Ausmus *et al.* (in press) have demonstrated that mites significantly stimulate microbial metabolism, and Ausmus and Witkamp (1973) have clearly shown the stability of the fungal biomass when developed in soil containing mycophagous microarthropods.

As Mitchell and Parkinson (1976) pointed out, laboratory studies on the choice of diet gave few reliable indications as to the nutritional specificity, when compared to analysis of the gut content, the precision of digestive enzymes, or observations "*in situ*." In spite of these multiple experimental possibilities, precise knowledge of the specific diet of soil mites is lacking, even though it is indispensable for the study of specific functions carried out by each species (Wallwork, 1958; Lebrun, 1971, 1977; Butcher *et al.*, 1971).

Recent Perspectives

Biogeochemical Cycles. Confirming the hypothesis of Crossley and Witkamp (1964) and Wallwork (1971, 1975), Cromack *et al.* (1977a, b) brought to light the essential position occupied by mites in the calcium cycle pertaining to the edaphic environment. Oribatids concentrate calcium in their exoskeleton (Gist and Crossley, 1975). According to Wallwork (pers. comm.) highly sclerotised adult oribatids may contain up to 18% calcium. The same author states that 45% of the exoskeleton of phthiracarids is $CaCO_3$. One must point out the importance of this phenomenon in slowing the leaching and the decline of acid soils where calcium is often an element sorely lacking. On the basis of their research, Cromack *et al.* (1977b) propose an interesting diagram to explain the calcium cycle in the soil. This is a partial and provisional cycle but it clearly shows that oribatids and fungi intervene judiciously at the moment where losses, by leaching or percolation, may appear.

Influence of Mites on the Rhisosphere. It is in cultivated soil that this aspect is most interesting to study as Edwards and Lofty (1977) have recently pointed out. These authors have shown that microarthropods (mites and Collembola) have a positive influence on the density of seeds sown in pots as compared to sowing carried out in soil without microfauna. The presence of microarthropods is comparable to ploughing of the soil in preparation for sowing.

On the other hand, it seems that subtle associations exist in the soil betweeen mites and the root system of many plant species. The study of Curry and Ganley (1977) showed that mites and Collembola of pastured meadows show significant variations in abundance and in species diversity following the microhabitat "weeds" or "grasses." The rhisosphere of *Dactylis* and of *Senecio* are, in this way, mostly populated by mites and Collembola represented by a large number of species. There is no doubt that this particular distribution is directly related to the diverse conditions of organic decomposition, elemental cycling and the potential productivity.

Other Perspectives Concerning the Role of Mites in the Soil. The conceptions of Cornaby (1977) suggest new paths in research concerning the qualitative role of edaphic mites.

"Potentially harmful substances are being introduced at increasing rates into ecosystems. The possible role of saprophagous arthropods and other soil organisms in concentrating and deactivating polycyclic organic matter, carcinogens, and other noxious materials waits investigation. For 300 substances reported as being carcinogenic, mutagenic, or teratogenic (McCann *et al.*, 1975), there is considerable potential for the use of saprophagous animals as intervening agents in the flow of pathological substances from sources to man."

PLACE OF ACAROFAUNA IN MAINTENANCE OF SOIL FERTILITY

Recent Observations

When we undertake the problems relative to the maintenance of biological soil fertility it is obvious that these issues cannot be dissociated from the conservation of fauna, the maintenance of its diversity and the numeric relation between trophic levels.

In the context of modern agroecosystems, the aggressions and perturbations are varied and many. As Figure 1 shows, the primary objective of modern agricultural practices, the increase of production, affects every level of the ecosystem. Man's activity ranges from simple physical interventions (such as crushing of the ground by cattle, erosion, microclimatic modifications) to the introduction of xenobiotics or the inoculation (voluntary or not) of foreign organisms, causing toxic effects and altering the complex biological and biogeochemical cycles. Figure 2, modified from Edwards and Lofty (1969) and Wallwork (1976), gives a synthesis of the principal effects of agricultural practices on the soil fauna.

Among the positive effects, one must quote the study of Zyromska-Rudzka (1977), which shows that in an adequately fertilized field, the rate of CO_2 evolution is increased by 20%. However, the decrease in species diversity, which may be easily observed in the case of pollution, means a loss of complementary functions. The consequences may be deduced from what we have mentioned above, perturbation of biogeochemical cycles (notably by slowing CO_2 evolution), instability in the trophic structure, abnormal proliferation of resistant species, etc. In short, the whole architecture of the ecosystem is modified and its harmonious functioning disrupted. This can go to the extent

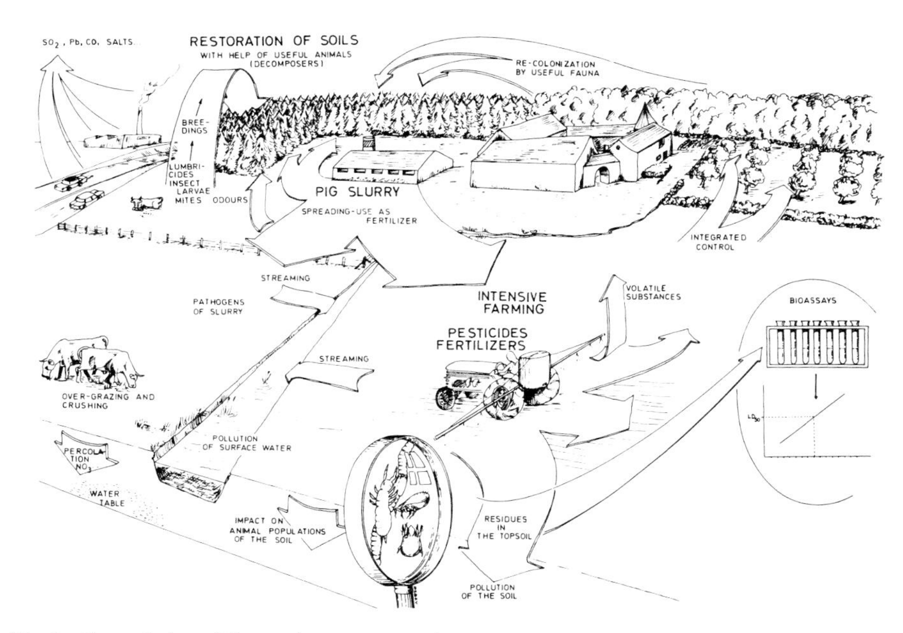

Fig. 1. General view of the modern agroecosystem.

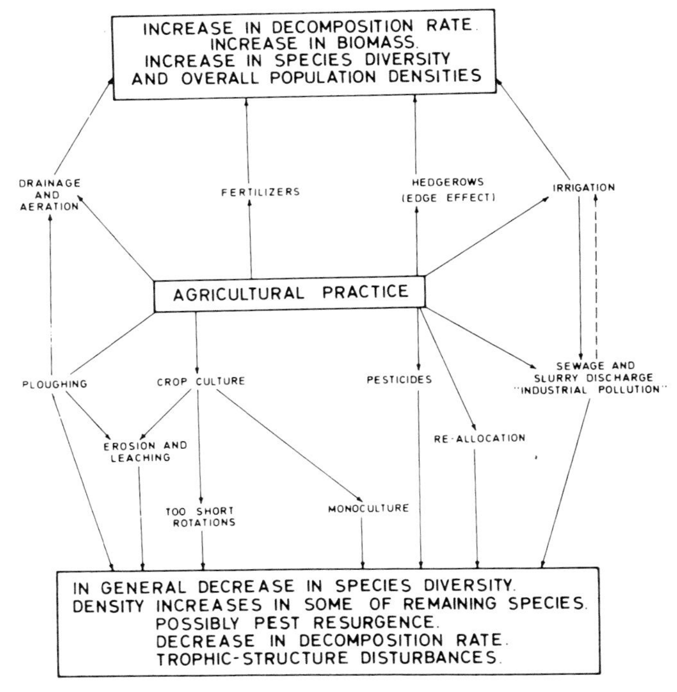

Fig. 2. Ecological effects of agricultural practice on the soil fauna (adapted from Edwards and Lofty, 1969 and Wallwork, 1976).

of significant losses of plant production that may reach 10%, as shown by Perfect *et al.* (1977).

Examples

A recent study, carried out by Weary and Merrian (1978) is a good example of the effect of pesticides on the microfauna of the soil, and is in close relation with the first part of this paper. The experimental use of carbofuran at a normal dose (285 g Ai/ha) reduced the rate of decomposition of a Canadian red maple forest-floor from 1.85 to 1.15 g/m^2/day, a reduction of over 40%.

The eradication of arthropods was demonstrated by the study of Thirumurthi and Lebrun (1977). This was based on a collembolan species, but the phenomenon is identical in the case of mites. Compared to the controls, the overdosing (dose of 10 Kg/ai/ha) caused prolonged reduction; no recovery in population was observed 24 weeks after treatment. Even at a normally applied dose the population only showed a weak tendency to rebuild itself.

Restoration of Biological Fertility

Returning to the diagram previously presented (Fig. 1), we see that there are two principal ways to restoration: the natural way, recolonization based on environments that have not been altered or by homeostasis, and by internal regulations of the ecosystems, and restoration carried out by human intervention.

The precise diagnosis of the impact of certain perturbations permits us to establish the possibilities of natural restoration, especially the mechanisms involved. In this way Dindal and Wray's (1977) study showed that the rehabilitation of abandoned quarries is extremely dependent on the activity of soil animals and that the time involved in restoration can be shortened by planting adequate vegetation or by the use of mulching. These techniques facilitate the activity of edaphic arthropods and accelerate soil formation. With regard to chemical pollution due to pesticides, the tendency is to use molecules that are not persistant and whose action is wiped out in a few weeks. In this respect, modern pesticides are, despite their high toxicity, closer to ecological imperatives (Thompson, 1973).

The studies carried out on this subject make it clear that soil organisms have real possibilities of reconstitution when faced with the new generation of pesticides. Thus, the negative impact of certain substances is only active during one season of vegetation. Furthermore, many authors have already pointed out the immense possibilities of adaptation and resistance of the soil and the animals that live there. Concerning older pesticides, organochlorines for example, we have already pointed out (Lebrun, 1977) the resistance of certain useful organisms as well as their possibilities of biodegradation. However, one must be wary of overexaggerated optimism as the long-term effects of pesticides are the most often ignored. Also, the possibilities of detoxification of the soil by arthropods vary considerably from one taxonomic group to another (even between congeneric species) and also vary with the type of pesticide. This illustrates yet again the large diversity of soil mites. Furthermore, resistance is often accompanied by biological accumulation, therefore running the risk of contamination by means of the trophic chains.

The conclusion is therefore clear. In the agricultural context of today it is not sufficient only to count on the natural possibilities of restoration, even when these are facilitated. It is urgent to have at hand a large scale measure of restoration. In this type of intervention one must have a large dose of imagination. Certain attempts are already well known, such as the introduction of beetles that accelerate the cycle of matter as well as the speed of incorporation of cattle excreta into the soil (McKinney and Morley, 1975). The attempts at introducing, or reintroducing earthworms are also well known and have given very probing results (Van Rhee, 1971, 1977).

These are, however, only partial attempts as they do not take into account the role of microarthropods and thus reconstitute incompletely the diverse and complementary functions carried out by soil animals as a whole. As Cornaby (1977) has pointed out, improved utilization of saprophagous arthropods is essential to the ultimate regeneration of spent land. So, a very interesting study has been carried out by Debry (1978) on the restoration of soils which were becoming exhausted and podzolised by excessive resinous planting. This research showed that a very acid soil (pH = 3.8) planted with larches during sixty years can be restored by pig slurry application. One year after this treatment the density of the most important decomposer mite (*Platynothrus peltifer*) became doubled in the treated plots. This involves a real stimulation of the decomposition rate and has suggested to the author the simultaneous introduction of this mite.

SOIL MITES AND ECOLOGICAL SENSITIVITY

It is in this respect that the diversity of mites is most significant. Their value as bioindicators of specific conditions apart from those aspects we have already discussed, is effective at many different levels. These are the classification of soils, humus and the speed of humification; observation and analysis of fine variations in life-conditions (microclimate, physico-chemical factors); the preventive detection of perturbations (e.g. by the technique of bioassays); and the study of biocenotic laws such as numeric regulation of communities and their spatial distribution and succession in time.

Ecological Sensitivity of Mites in Natural Communities

A recent example of the close relationships between mites and their biotopes is the study carried out by Bonnet *et al.* (1975). This study is interesting as it constitutes a demonstration of parallel variation between a biological system and its environment. Multivariate analysis carried out on 20 microbiotopes occupied by 45 species of oribatids and 26 species of Collembola, showed that the variability may be explained by four principal factors with 33% for the first, 14% for the second, 12.5% for the third and 9% for the fourth. These factors are, in order: (1) the degree of dependence on the edaphic environment; (2) the distance to the least individualized environment, constituted by mosses developing on the soil (Lebrun, 1971); (3) variations in temperature, and (4) variations in humidity. Therefore, 70% of variability in the composition of fauna can be correlated with these four ecological factors.

Value of Oribatids as Bioindicators of the Rate of Humification

The second aspect present is the use of bioindicators of the rate of humification, proposed by Lebrun and Mignolet (1978) in the course of research concerning a comparison of organic decomposition of different substrates and the associated groups of oribatids. The basic idea is that analysis of ecological groups of oribatids (ecological groups defined in natural conditions) and their relative importance parallels the rate of transformation of organic substrates. The relative abundance of humicolous oribatids in a substrate made up of hornbeam leaves shows that humification, and therefore organic decomposition, is much faster in this type of litter. The author have thus been able to compare over 15 litter types and define their specific value as enriching species. It appeared that the rate of humification is extremely dependent on the rate of colonization by mites but is independent of weight losses of the litter at the moment of being placed on the soil.

Use of Mites in "Bioassays"

The technique of bioassays gives a very rapid diagnosis that is often very precise as to the quality of the environment. Today edaphic microarthropods are largely used to detect residual concentrations of pesticides. After having established the tolerance of a sensitive species to one or another residual toxicant, we then put experimental populations in contact with tested pedological samples. Following the mortality in the experimental population, we deduce, by extrapolation, the concentration of a given product. This type of biological trial has already been put into practice with edaphic Collembola (Thompson, 1973; Thirumurthi and Lebrun, 1977). It is a shame that mites have not been used for this type of research as they present an entire range of sensitive species. All groups of edaphic mites present species that are ecologically very dependent on their environment, and consequently could be used as indicators.

CONCLUSIONS

The conclusion of this paper shall be short because we only want to point out the main fields where soil acarology is able to bring a substantial aid to man.

The increase of natural resources is possible only if we care for soil fertility. The soil is, in fact, a real living macroorganism. At a time when the demographic pressure is too high, and when the needs of human populations are intense and immense, it is wise to realize that the soil is a capital for human survival. This soil is a living organism, highly complex and diverse due to the result of animal activity to a very large extent. The diversity of mites is spatial, temporal, microclimatic, biochemical, etc., as illustrated by Wallwork (1976). This is a demonstration, not of an enigma, but of the functional complexity of the edaphic ecosystem. The place of mites in this delicate machine is fundamental, because of their quantitative and qualitative role in energy flow as well as their ecological meaning.

In my view, research must be developed in the following areas:

1. role of mites in mineral cycling and turn-over;
2. role of mites in organic matter decomposition and transformation;
3. use of the ecological sensitivity of mites in detection and quantification of toxicants;
4. use of mites in restoration of spent soils.

SUMMARY

The soil and its microfauna are not purely academic subjects; the soil is the very support of the earth's productivity. In this respect, the contribution of Acarology is immense in the sense that its studies arise from a better understanding of bio-geo-chemical cycles, from the search for qualitative and quantitative factors involved in biological fertility and from the maintenance, conservation and restoration of the latter. This is particularly true in the case where human activity tends to induce irreversible disturbances. To speak of soil mites as an entity is meaningless since, by definition, the edaphic ecosystem is profoundly marked by a great number of dependent interrelations. The acarologist is constantly confronted with a global picture of the functional relationships that mites share with other organisms. The study of soil mites must therefore be fundamentally an ecological one. The large spatial and temporal distribution of soil mites, their great species diversity and their narrow ecological sensitivity makes them a prime candidate for biocenotic studies. As such, the study of soil mite associations and communities may shed light on the degrading of ecosystems over-exploited by man as well as on the consequences of these perturbations.

REFERENCES

Anderson, J. M. and Healey, I. N. (1970) *Pedobiologia* **10**, 108-120.

Ausmus, B. S. and Witkamp, M. (1973). Oak Ridge Nat. Lab., EDFB-IBP, Publ. no. 73-10, 183 pp.

Ausmus, B. S., Ferrigni, R. and McBrayer, J. F. (1978). *Soil Biol. Biochem.* (in press).

Bonnet, L., Cassagnau, P. and Travé, J. (1975). *Oecologia* **21**, 359-373.

Butcher, J. W., Snider, R. and Snider, R. J. (1971). *Ann. Rev. Entomol.* **16**, 249-288.

Cornaby, B. W. (1977). *In* "The Role of Arthropods in Forest Ecosystems" (W. J. Mattson, ed.), pp. 96-100. Springer-Verlag, New York.

Cromack, K., Sollins, P., Todd, R. L., Crossley, D. A., Fender, W. M., Fogel, R. and Todd, A. W. (1977a). *In* "The Role of Arthropods in Forest Ecosystems" (W. J. Mattson, ed.), pp. 79-84. Springer-Verlag, New York.

Cromack, K., Sollins, P., Todd, R. L., Fogel, R., Todd, A. W., Fender, W. M., Crossley, M. E. and Crossley, D. A. (1977b). *Ecol. Bull. (Stockholm)* **25**, 246-252.

Crossley, D. A. and Witkamp, M. (1964). *Acarologia* **6**, 137-145.

Crossley, D. A. (1977). *In* "The Role of Arthropods in Forest Ecosystems" (W. J. Mattson, ed.), pp. 49-56. Springer-Verlag, New York.

Crossley, D. A. and Witkamp, M. (1964). *Acarologia* **6**, 137-145.

Curry, J. P. (1969). *Soil Biol. Biochem.* **1**, 253-258.

Curry, J. P. and Ganley, J. (1977). *Ecol. Bull. (Stockholm)* **25**, 330-339.

Debry, J. M. (1978). *Pedobiologia* (in press).

Dindal, D. L. and Wray, C. C. (1977). *In* "Limestone Quarries: Responses to Land Use Pressures" (E. J. Perry and N. A. Richards, eds.), pp. 72-99. Allied Chem. Corp., Syracuse.

Duvigneaud, P. (1974). "La Synthèse Écologique." Doin, Paris, 296 pp.

Edwards, C. A. and Lofty, J. R. (1969). *In* "The Soil Ecosystem" (J. G. Sheals, ed.), pp. 237-247. Systematics Association, London.

Edwards, C. A. and Lofty, J. R. (1977). *Ecol. Bull. (Stockholm)* **25**, 348-356.

Ghilarov, M. S. (1963). *In* "Soil Organisms" (J. Doeksen and J. van der Drift, eds.), pp. 255-259. North Holland Publ. Co., Amsterdam.

Gist, C. S. and Crossley, D. A. (1975). *Amer. Midl. Nat.* **93**, 107-121.

Haarløv, N. (1960). *Oikos, Suppl. 3,* 1-176.

Lebrun, Ph. (1971). *Mém. Inst. Roy. Sc. Nat. Belg.* **165**, 1-203.

Lebrun, Ph. (1977). *Pédologie* **27**, 67-91.

Lebrun, Ph. and Mignolet, R. (1978). *In* "Proc. 4th Intern. Congr. Acarology, Saalfelden, 1974." (In press).

McBrayer, J. F., Reichle, D. E. and Witkamp, M. (1974). Oak Ridge Nat. Lab., EDFB-IBP, Publ. no. 73-8, 78 pp.

McCann, J., Choi, E., Yamasaki, E. and Ames, B. N. (1975). *Proc. Nat. Acad. Sci.* **72**, 5135-5139.

Macfadyen, A. (1964). *Acarologia* **6**, 147-149.

McKinney, G. T. and Mortley, F. H. W. (1975). *J. Appl. Ecol.* **12**, 831-837.

Minderman, G. (1956). *Plant and Soil.* **8**, 42-48.

Mitchell, M. J. and Parkinson, D. (1976). *Ecology* **57**, 302-312.

Pande, Y. D. and Berthet, P. (1973). *Oecologia* **12**, 413-426.

Perfect, T. J., Cook, A. G., Critchley, B. R., Moore, R. L., Russel-Smith, A., Swift, M. J. and Yeadon, R. (1977). *Ecol. Bull. (Stockholm)* **25**, 565-568.

Thirumurthi, S. and Lebrun, Ph. (1977). *Med. Fac. Landbouww. Rijksuniv., Gent* **42**, 1455-1462.

Thompson, A. R. (1973). *J. Econ. Entomol.* **66**, 855-857.

Van der Drift, J. and Witkamp, M. (1960). *Burm. Arch. Neerl. Zool.* **13**, 486-492.

Van Rhee, J. A. (1971). *In* "Proc. IVth Colloquium Pedobiologiae, Dijon, 1970," pp. 99-107. *Ann. Zool. Ecol. Anim.,* Paris.

Van Rhee, J. A. (1977). *Pedobiologia* **17**, 107-114.

Wallwork, J. A. (1958). *Oikos* **9**, 260-271.
Wallwork, J. A. (1971). *In* "Proc. 3rd Int. Congr. Acarology" (M. Daniel and B. Rosicky, eds.), pp. 129-134. W. Junk Publ. Co., The Hague.
Wallwork, J. A. (1975). *In* "Progress in Soil Zoology" (J. Vanek, ed.), pp. 231-240. Academia, Prague.
Wallwork, J. A. (1976). "The Distribution and Diversity of Soil Fauna." Academic Press, London. 355 pp.
Weary, G. C. and Merriam, H. G. (1978). *Ecology* **59**, 180-184.
Witkamp, M. and Crossley, D. A. (1966). *Pedobiologia* **6**, 293-303.
Zyromska-Rudzka, H. (1977). *Ecol. Bull. (Stockholm)* **25**, 133-137.

ORIBATID MITE COMPLEXES AS BIOINDICATORS OF RADIOACTIVE POLLUTION

D. A. Krivolutsky

A. N. Severtzov Institute of Evolutionary Animal Morphology and Ecology Academy of Science of the USSR

INTRODUCTION

In the Laboratory of Soil Zoology, A. N. Severtzov Institute of Evolutionary Animal Morphology and Ecology, studies have been carried out on the effects of experimental radionuclide pollution of soils on the soil-dwelling invertebrates (Ghilvarov and Krivolutsky, 1972; Krivolutsky and Fedorova, 1973; Krivolutsky and Kozhevnikova, 1972; Krivolutsky *et al.*, 1972). Oribatids provide extremely convenient material for investigating the effect of local experimental pollution of soils with radionuclides. The great number of species in communities, as well as high population density which remains at a relatively stable level, make it convenient to confine experimental areas to 1-2 m^2 plots. In this way the radionuclide hazard for the investigator is avoided.

METHODS

The experiments were made in meadow grey soils two years after treatment with radionuclides on 10 m^2 plots (5 m by 2 m) at a rate of 2-5 mCu/m^2 (Fig. 1, Table I).

RESULTS

The oribatid mite population density amounted to 15-20 thousand adult specimens per square meter in the localities investigated. The oribatid mite density decreased on the polluted plots (Fig. 2) and the structure of the soil invertebrate community tended to become simplified from the effect of pollution. In meadow soils oribatids are rather evenly distributed in the up-

Vol. I: ISBN 0-12-592201-9

TABLE I.
Distribution of Radiation of ^{90}Sr and ^{90}Y Accumulated Doses in Soil Treated with ^{90}Sr (2-3 mCu/m^2).

Biogeocenosis Stratum	Accumulated Dose (r/day)	Dose on the Animals from Incorporated ^{90}Sr (r/day)
Tops of Trees	0.02-0.5	0.001-0.005
Herbage	0.2 -0.5	0.005
Soil (depth in cm)		
0 (soil surface)	0.5 -1.0	
0-1	0.8 -1.2	
1-2	1.4 -2.1	
2-3	2.0 -3.0	
3-4	2.0 -3.0	
4-5	1.3 -1.8	0.2 -0.05
5-6	0.8 -1.2	
6-7	0.5 -0.7	
7-8	0.3 -0.5	
8-9	0.1 -0.15	

permost surface layers. There a negligible decrease in species composition on experimental plots was recorded (9-14 species vs. 15 in the control), as well as a two-fold lowering of the population density of adult mites. An abrupt decrease of population densities in Collembola and Gamasida was observed on the same plots.

No significant differences in the oribatid mite population density have been observed in polluted soils in a birch forest. However, the diversity of mite species complexes on polluted plots markedly decreased, especially with respect to soil surface and litter dwellers (Table II).

In wheat fields on chernozem soils polluted by ^{239}Pu at a rate of 1.78 mCu/m^2 a sharp decrease in abundance and decline of microarthropod

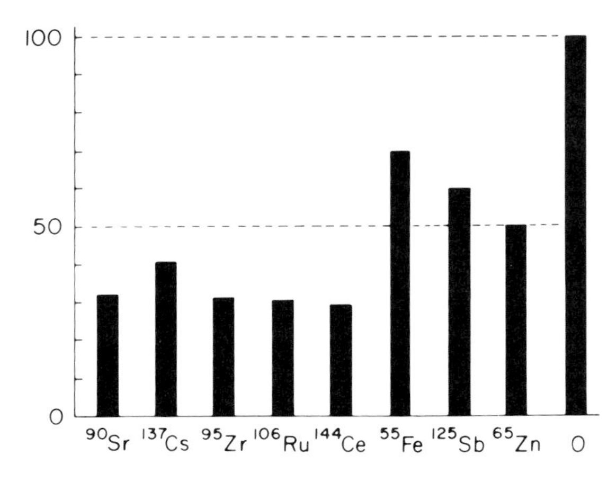

Fig. 1. The vertical distribution of radionuclides in polluted plots of meadow soils.

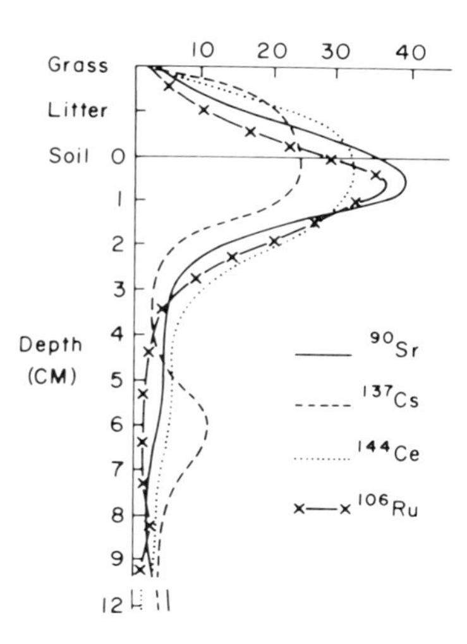

Fig. 2. The number of Oribatei in control and radioactive polluted soils expressed as percentage of control.

TABLE II.
Oribatid Mite Populations on Control and ^{90}Sr-Polluted Plots in a Birch Forest.[a]

Selected Species	^{90}Sr	Control 1	Control 2
Oppia nova Oud.)	+	+	+
Oppia bicarinata (Paoli)	+	+	+
Quadroppia quadricarinata (Michael)	+	+	+
Suctobelbella subcornigera Forsslund)	+	+	+
Liochthonius sellnicki (Thor)	+	+	+
Oribatula tibialis (Nicolet)	+	+	+
Zygoribatula exilis (Nicolet)	+	+	+
Tectocepheus velatus (Michael)	+	+	+
Ceratozetes sellnicki Rajski	+	+	+
Galumna lanceata Oudemans	+	+	+
Steganacarus striculus (C.L. Koch)	+	+	+
Oppia fallax obsoleta (Paoli)	–	+	+
Suctobelba hammeri Krivolutsky	–	+	+
Autogneta traegardhi Forsslund	–	+	+
Scheloribates latipes (C.L. Koch)	–	+	+
Gymnodamaeus bicostatus Kulcz.	–	+	+
Epidamaeus bituberculatus (Kulcz.)	–	+	+
Protoribates capucinus Berlese	–	+	+
Peloptulus phaenotus (C.L. Koch)	–	+	+
Xenillus tegeocranus (Hermann)	–	+	+
Achipteria nitens (Nicolet)	–	+	+
Oribotritia loricata (Rathke)	–	+	+
Total Number of Species	14	30	32
Soil Surface and Litter Dwellers	5	20	19
Mineral Soil Dwellers	9	10	13

[a] Application rate 1.8-3.4 mCu/m^2; + = present; – = absent.

TABLE III.
Microarthropod Populations (Mean No./m^2) on Control and ^{239}Pu-Polluted Plots in a Wheat Field.[a]

Species	Control	^{239}Pu
Total Oribatei	110	70
Liochthonius sellnicki (Thor)	5	-
Brachychthonius berlesei Willmann	-	5
Oppia nova (Oudemans)	40	5
Oppia unicarinata (Paoli)	5	35
Oppia bicarinata (Paoli)	5	-
Suctobelba hammeri Krivolutsky	10	20
Tectocepheus velatus (Michael)	30	5
Scheloribates laevigatus (C.L. Koch)	10	-
Oppia minus (Paoli)	5	-
Gamasida	50	-
Acaridae	630	35
Collembola	150	20
Total microarthropods	930 ± 320	120 ± 30

[a] Application rate 1.78 mCu/m^2.

community diversity were noted (Table III). As a result of pollution, the population density of soil mesofauna (Lumbricidae, Insecta larvae, millipedes, etc.) decreased 2.3 times, with the decrease being especially marked in dwellers of deep soil horizons.

It has previously been shown that the structure and dynamics of oribatid mite communities can be related to soil conditions, and Oribatei can be used as bioindicators of soil type (Krivolutsky, 1975) as well as industrial pollution (Vanek, 1967, 1975). The conclusion from the present study is that the oribatid mites can also be used as bioindicators of radioactive pollution.

REFERENCES

Ghilarov, M. S. and Krivolutsky, D. A. (1972). *Proc. IV Coll. Pedobiol., Dijon.* INRA Publ. 71-7, Paris, pp. 537-544.

Krivolutsky, D. A. (1975). *In* "Progress in soil zoology" (J. Vanek, ed.), pp. 217-221. Academia, Prague.

Krivolutsky, D. A. and Fedorova, M. N. (1973). *Zool. Zh.* **52,** 601-603.

Krivolutsky, D. A. and Kozhevnikova, T. L. (1972). *Ekologiya (Sverdlorsk)* **2,** 69-74.

Krivolutsky, D. A., Tichomirova, T. L. and Turchaninova, V. A. (1972). *Pedobiologia* **12,** 374-380.

Vanek, J. (1967). *In* "Progress in soil biology" (O. Graf and J. E. Satchell, eds.), pp. 331-339. Friedr. Vieweg & Sohn GmbH, Braunschweig.

Vanek, J. (1975). *Quaest. Geobiol.* **14,** 35-116.

INFLUENCE OF HUMAN ACTIVITIES ON COMMUNITY STRUCTURE OF SOIL PROSTIGMATA

Daniel L. Dindal and Roy A. Norton

State University of New York
College of Environmental Science and Forestry
Syracuse, New York

INTRODUCTION

During 1969 though 1972 we initiated several large scale projects with one major objective—to determine the impact of human activities on soil invertebrate communities. As project data were being analysed, a second comparative objective developed, that being to determine if any common predictable stress response patterns existed among the microarthropods. These, therefore, are the objectives of this paper.

One limitation was the taxonomic status of some of the Prostigmata which we collected. Many are new or undetermined species. On the basis of apparent morphological differences we assigned either alphabetical or numerical symbols distinguishing different specific taxa, since their ecological characteristics, the elucidation of which is the main function of this paper, were clear and distinctively specific. Also, we had three projects in progress simultaneously, each with numerous samples and each yielding a high number of microarthropods. Volumes of data have been collected, but for practical purposes of space we present only representative samples here.

METHODS OF COMPARISON

Interpretations of results were made from several general analysis methods. These were 1) comparisons of the total counts, 2) evaluations of species diversity as calculated using the Shannon-Wiener formula ($\overline{H} = -\Sigma p_i \log_{10} p_i$, Shannon and Weaver, 1963) and, the major emphasis of this paper, 3) an evaluation of responses of the Prostigmata to the direct and indirect effects of human activities. Here we considered such activities as selection pressures.

Vol. I: ISBN 0-12-592201-9

These selection pressures were determined by comparing the treated and untreated sites based upon a subjective analysis of dominance to very rare categories. Limits of these categories were selected arbitrarily based upon the percent of total number of individuals from samples of a given period. Categories were as follows: *dominant*—those species comprising 35% of the numbers of individuals present per sampling period; *subdominant*—species comprising 15-34%; *uncommon*—10-14%; *rare*—1-9%, and *very rare*—less than 1%. A specific graphic symbol is associated with each category on the figures.

Directed by the same objectives and using the same methods, the human selection pressure effects on the oribatid mites in the three studies have been published elsewhere (Dindal, 1977).

RESULTS AND DISCUSSION

Effects of DDT

The first study is that of the effects of DDT on soil invertebrates in an old field herbaceous site. A single granular application of technical grade DDT (rate 1.12 kg/ha; 1 lb/A) was made on a 4 hectare (10 A) old field herbaceous vegetative community in west central Ohio. For a two year period after application, 1500 soil cores containing 450,000 invertebrates were extracted and analysed from a treated (T) site and also from an adjacent comparable control (untreated = UT) field. Further site descriptions and initial results on increased microbial respiration and the dynamics of the total microarthropods were presented earlier (Dindal *et al.*, 1975a; Folts, 1972).

General Population Characteristics

Total numbers of Prostigmata on the treated site were not different from those on the untreated area except when the former was greatly increased during October and November of each year. There was no significant difference, in general, in the species diversities (Table I).

Effects of Selection Pressure

Responses from the single application of DDT were definite niche shifts and replacements among the species (Fig. 1) that were formerly dominant (D), subdominant (SD), and uncommon (UC). Those species responding positively to the selection pressure of DDT were tetranychid larvae and *Scutacarus*. Among those species responding negatively to the selection pressure were *Tarsonemus* spp, *Eupodes* A, *Microdispus* and *Microtydeus*. There were several species which were unaffected by the application. When looking at the

TABLE I.
Some Comparative Aspects of Species Diversity of the Prostigmata Communities Subjected to Various Human Activities. Legend: $\overline{H} = -\Sigma p_i \log_{10} p_i$ = species diversity; S = mean no. species/sample = richness; SE = standard error of the mean.

	Untreated (UT)			Treated (T)		
Seasons	S	$\overline{H}$	± SE	S	$\overline{H}$	± SE
	DDT Application (1.12 kg/Ha)					
	Old Field Herbaceous Site (30 yr old)					
1969-1970						
Autumn	25	0.5533	± 0.0430	39	0.5100	± 0.0500
Winter	1	0.1300	± 0.0890	9	0.2367	± 0.720
Spring	7	0.2700	± 0.0400	7	0.3033	± 0.0547
	Urban Street Salting					
	Roadside Turf					
1972-1973						
Winter	17	0.3284	± 0.0602	42	0.3922	± 0.0748
	Wastewater Irrigation (5 cm/wk)					
	Old Field Herbaceous (16 yr old)—White Spruce Site					
1972						
Spring	44	0.7470		47	0.4994	
	Mixed-Oak Hardwood					
	18	0.6297		14	0.2953	

rare (R) and very rare (VR) species, the number of negative respondants was slightly greater than those which responded positively. So, in general, species shifting was the major response.

Equalized shifting supports the fact that we observed no increase in species diversity.

Urban Street Salting

The second selection pressure is one which is unique to regions such as the northwestern United States where winter is characterized by a very cold, snowy season. During these periods salt applications to roadways are commonplace in order to provide clear and safe automobile travel. Such salt applied to icy and snowbound streets ultimately passes through or accumulates in roadside turf-grass and soil.

Samples of sod adjacent to streets were taken in autumn and the following winter in the city of Syracuse, in central New York. During these two seasons, approximately 1211 MT (1335 T) of salt (mostly sodium chloride) were spread over the streets within the study area. Turf samples were taken and examined for comparison from two nearby control sites—grassy areas of a local school

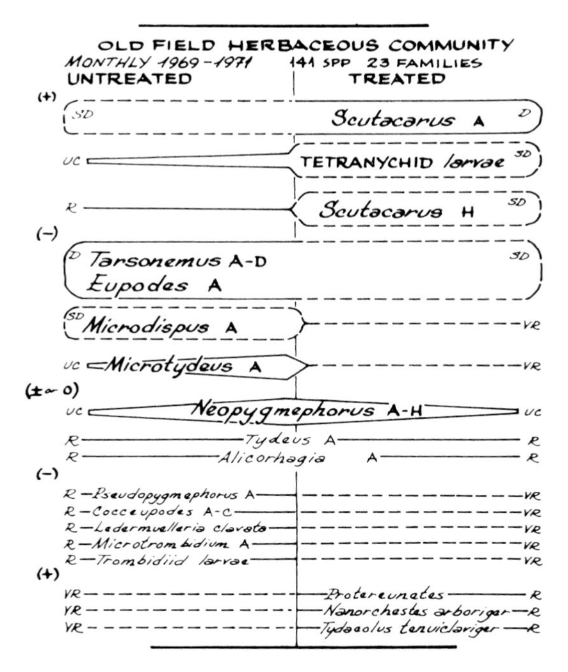

Fig. 1. Responses of soil Prostigmata to selection pressures: direct and indirect effects of single application of DDT (application rate 1 kg/ha).

ground and a playground. These were not subjected to salt or salt spray. A total of 240 turf soil cores yielded about 8,000 microarthropods. Samples taken in autumn were reflective of effects of the previous winter's salting; that salting effect was ameliorated somewhat by the weather conditions of the intervening spring and summer.

General Population Characteristics

In general, the mean numbers of all types of soil microarthropods per sample in autumn were 60 for UT sites, 112 for T turf; in winter there were 35 organisms from UT and 45 for T samples. Figures 2 and 3 more specifically show the overriding trends of the salt treatment that caused definite responses among the Prostigmata. In both the autumn and the winter samples treatment stimulated the Prostigmata the greatest of all of the suborders of mites. Likewise, during both seasons, the species diversity and richness were enhanced by treatment in all cases (Table I).

Effects of Selection Pressure

Figures 2 and 3 show niche shifts and species replacements among the D

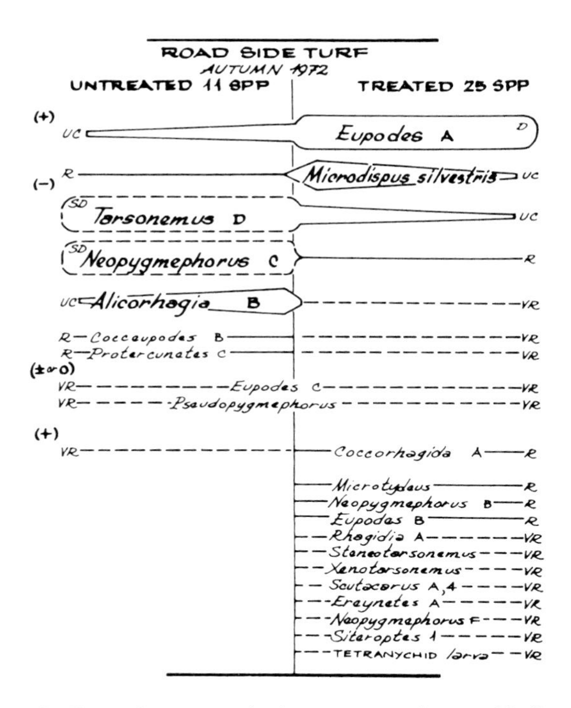

Fig. 2. Responses of soil Prostigmata to selection pressures: direct and indirect effects of urban street salting.

through UC categories. These shifts and replacements are very similar to those seen under DDT conditions. *Eupodes* A responded positively to the selection pressure of salt. *Tarsonemus, Cocceupodes* spp. and *Alicorhagia* all responded negatively to the selection pressure. There were species which were unaffected, such as *Pseudopygmephorus* and several others. The majority of the R-VR forms on both Figures 2 and 3 were found to colonize after the treatment had taken place. This shows one of the major differences between the effects of salt and that of DDT. During the winter 12 R-VR species colonized and in the autumn 11 species appeared after treatment.

The diversity relationships mentioned earlier can be understood more completely based upon selection responses. The D-SD shifts caused very little change among the major species, and therefore, the very positive colonization response among the R-VR organisms completely explains the increase in species diversity caused by treatment.

Effects of Wastewater Irrigation

Since 1962 the municipal wastewater effluent from the town of State College, Pennsylvania has been recycled and renovated by spraying it on various vegetative communities. Numerous positive responses of this activity

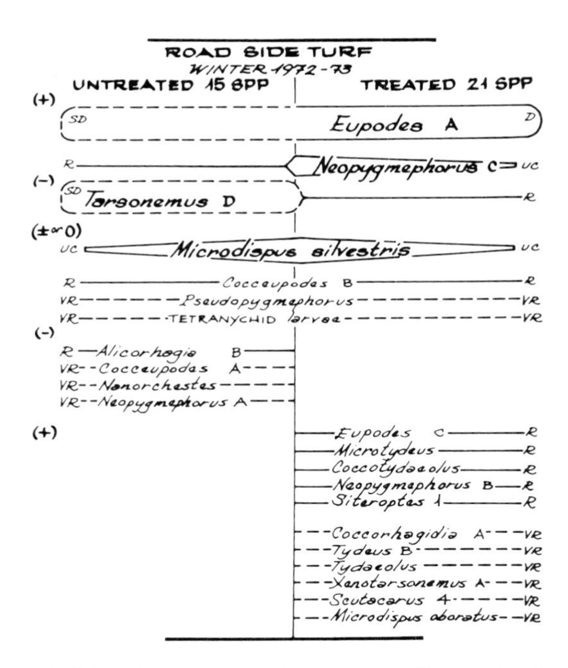

Fig. 3. Responses of soil Prostigmata to selection pressures: direct and indirect effects of urban street salting.

on the physical-chemical factors of the soil, the ground water, and the plant growth have been compiled by Sopper and Kardos, 1973. In 1972, we joined the research team by investigating the impact of spray irrigation on the ecology of soil invertebrates. An overview of our preliminary results concerning all invertebrates is given in Dindal *et al.* (1975b). Also, details on irrigation-earthworm-soil factor interrelationships are presented in Dindal *et al.* (1978).

For the mite studies 20 cores per site per season were collected accumulating 640 samples totally. Each treated site had received effluent applications of 5 cm/wk (2 in/wk) during the frostfree months of the year. Data from two seasons, spring and autumn, from two particular sites, the old field and the mixed oak hardwood community are compared.

Old Field-White Spruce Community

All mite suborders were reduced by the selection pressure of wastewater irrigation. The Prostigmata were very dominant on the UT sites. These were reduced considerably by treatment. Collembola were the only organisms that showed any positive responses although sporadically. Also, the species diversity (Table I) showed a significant decrease among the Prostigmata due to treatment.

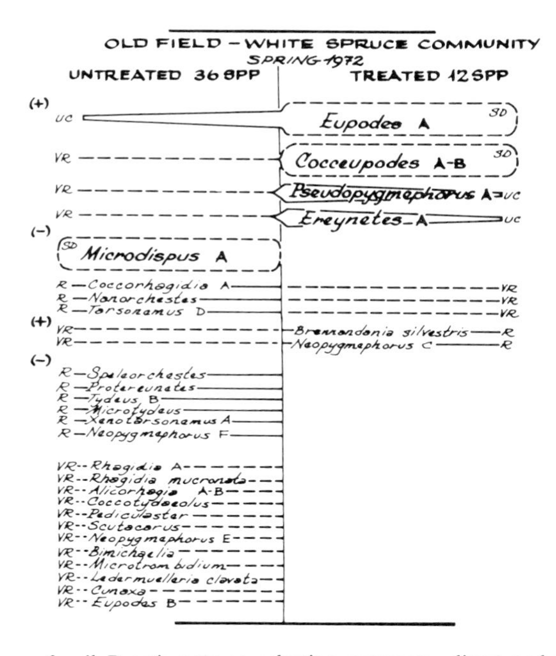

Fig. 4. Responses of soil Prostigmata to selection pressures: direct and indirect effects of municipal wastewater irrigation (application rate 5 cm / wk).

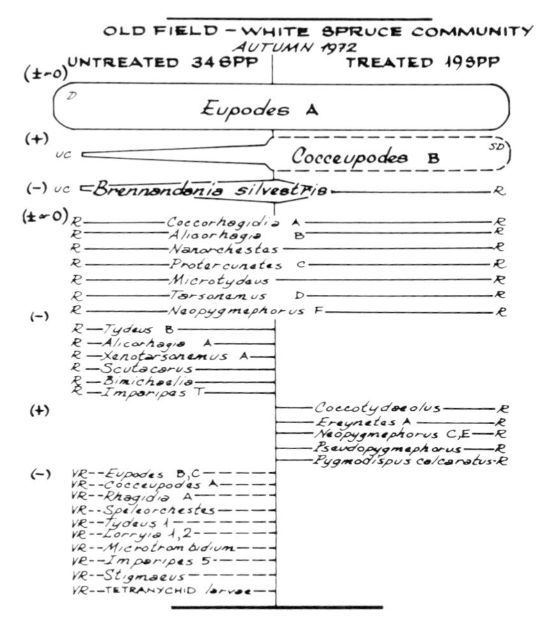

Fig. 5. Responses of soil Prostigmata to selection pressures: direct and indirect effects of municipal wastewater irrigation (application rate 5 cm / wk).

Effects of selection pressures. Regarding the D and SD categories (Fig. 4 and 5), definite maintenance or increase patterns occurred among some of the organisms. *Eupodes* A, *Cocceupodes* and *Pseudopygmephorus* are several of the species which were definitely increased, especially in the spring of the year. Looking at the R-VR species in the spring there was an extensive negative selection pressure with 22 species of R and VR forms completely decimated from the sites. During the autumn, *Eupodes* A was maintained at a dominant status. Seven species were unaffected by the application. Sixteen of the R-VR forms were selected against and five rare genera appeared as colonizers.

The above findings support the observed decrease in species diversity. The increase in dominance always corresponds to an indirect decrease in diversity. This, along with the negative selection pressure response of the rare forms, is the reason for the total drop in diversity.

Mixed-Oak Hardwood Community

As in the old field, all mite suborders in the hardwood site were decreased by the selection pressure of the effluent application. Species diversity again was significantly decreased on this site (Table I).

Effects of selection pressure. As was seen on the old field site the D and SD forms were either maintained or increased by irrigation (Fig. 6 and 7). The species *Cocceupodes* B observed in the spring at a SD level was maintained at that level. *Cocceupodes* A was increased from a R status to a SD status.

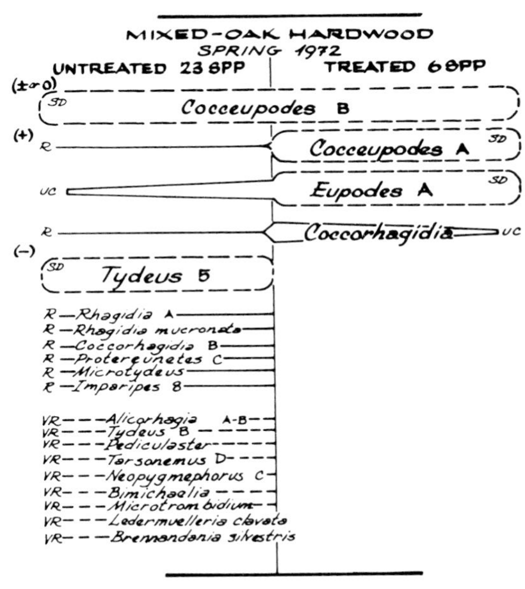

Fig. 6. Responses of soil Prostigmata to selection pressures: direct and indirect effects of municipal wastewater irrigation (application rate 5 cm/wk).

Fig. 7. Responses of soil Prostigmata to selection pressures: direct and indirect effects of municipal wastewater irrigation (application rate 5 cm / wk).

Eupodes A was also greatly increased as well as *Coccorhagidia* thus showing some positive responses to the selection pressure. In the spring *Tydeus* was completely decimated from its SD level; and among the R-VR forms, 15 species responded negatively to the selection pressure in the spring. In the autumn, the overall pattern of responses was similar to that observed in the old field herbaceous sites, *Cocceupodes* B, *Eupodes* A, *Rhagidia* and *Coccorhagidia* all responded positively. Four very rare genera colonized the site after treatment. Six R-VR species remained unaffected by the application, while 18 rare forms responded negatively to the selection pressure.

SUMMARY OF RESPONSE PATTERNS

Looking collectively at all the studies there were a number of positive response trends. *Eupodes* spp., for example, were common to all sites, treated or control, no matter what the selection pressure was. They were very dominant, and therefore we feel that these are good general indicator species. *Cocceupodes* and *Coccorhagidia* were relatively dominant on the wastewater study sites. They should also be considered as good indicator species.

Negative response trends were also noted. Among the D-UC forms

Microdispus always exhibited negative responses. *Microdispus* therefore we feel is a good general indicator species. There are a number of other species which could also be considered as indicator forms but they are found in the R-VR categories and would involve a considerable amount of work to assay their presence. However, they do show a uniform trend of a negative response. These organisms include *Rhagidia, Alicorhagia, Cunaxa, Microtrombidium,* and *Tarsonemus.*

In conclusion, the community structure of the soil litter Prostigmata is modified and altered considerably by man's activities. Some Prostigmata respond positively, others negatively, and still others give no response to various selection pressures caused by human perturbations. Because of these responses there are several species which can be considered as good general indicator species for the purposes of predicting man's impact on his environment.

ACKNOWLEDGEMENTS

The research upon which this report is based was supported by funds as follows: DDT-microarthropod study, provided for by the U. S. AEC (ERDA) Grant Contract No. AT-(11-1)-3474; Sewage effluent disposal-soil invertebrate study, provided for by the Northeastern Forest Experiment Station of the USDA Forest Service, through the Pinchot Institute Consortium for Environmental Studies; Urban street salting-microarthropod study, supported by the New England Interstate Water Pollution Control Commission.

REFERENCES

Dindal, D. L. (1977). *In* "Biology of Oribatid Mites" (D. L. Dindal, ed.) pp. 105-121. SUNY Coll. Environ. Sci. Forestry, Syracuse, NY.

Dindal, D. L., Folts, D. D. and Norton, R. A. (1975a). *In* "Progress in Soil Zoology" (J. Vanek, ed.) pp. 505-513. Dr. W. Junk B. V. Publ. and Academia, The Hague and Prague.

Dindal, D. L., Schwert, D. P. and Norton, R. A. (1975b). *In* "Progress in Soil Zoology" (J. Vanek, ed.) pp. 419-427. Dr. W. Junk B. V. Publ. and Academia, The Hague and Prague.

Dindal, D. L., Moreau, J-P. and Theoret, L. (1978). *In* "Municipal Wastewater and Sludge Recycling on Forest and Disturbed Land." (W. E. Sopper, ed.) The Penn. State Univ. Press, University Park and London. (In press).

Folts, D. D. (1972). "Effects of DDT on the oribatid mite community (Acarina: Oribatei) of an old field herbaceous community." Unpubl. MS thesis, SUNY Coll. Environ. Sci. Forestry, Syracuse, NY.

Shannon, W. E. and Weaver, W. (1963). "The Mathematical Theory of Communication." Univ. Ill. Press, Urbana, Ill.

Sopper, W. E. and Kardos, L. T. (1973). "Recycling Treated Municipal Wastewater and Sludge Through Forest and Cropland." The Penn. State Univ. Press, University Park and London.

INDEX OF CONTRIBUTORS[a]

Volume I

[a]Numbers in bold face type denote senior authorship.

INDEX OF CONTRIBUTORS[a]

Volume II

[a]Numbers in bold face type denote senior authorship.